Probabilistic Graphical Models for Genetics, Genomics, and Postgenomics

Probabilistic Graphical Models for Genetics, Genomics, and Postgenomics

Edited by

CHRISTINE SINOQUET

Editor in chief

and

RAPHAËL MOURAD

Editor

OXFORD
UNIVERSITY PRESS

OXFORD
UNIVERSITY PRESS

Great Clarendon Street, Oxford, OX2 6DP,
United Kingdom

Oxford University Press is a department of the University of Oxford.
It furthers the University's objective of excellence in research, scholarship,
and education by publishing worldwide. Oxford is a registered trade mark of
Oxford University Press in the UK and in certain other countries

Published in the United States of America by Oxford University Press
198 Madison Avenue, New York, NY 10016, United States of America

British Library Cataloguing in Publication Data

Data available

Library of Congress Control Number: 2013953773

ISBN 978–0–19–870902–2

Printed in Great Britain by
Clays Ltd, St Ives plc

A NOTE FROM THE EDITOR

To my loved ones.

The idea of editing a collective book about probabilistic graphical models in genetics arose in the spring of 2011. This project was fortunate to obtain the support of researchers at the forefront of innovation in this domain. From then on, in the back of my mind was always present the concern of honoring the confidence of the invited authors by achieving the project within a decent time frame. May they all be warmly thanked for their trust and their deep investment in this project, as well as for all the intellectually stimulating exchanges we had.

A collective book—not proceedings—is much more than the compendium of the scientific contributions that supports it, however invaluable these contributions are by themselves; and this comes at a cost. The edition and compilation of this book drew on any time reserve that could be ferreted out of a researcher's timetable. Using a metaphor borrowed from carpentry, sanding, smoothing, and polishing again and again the job took quite a while before I was able to apply the undercoat paint layers and the top varnish.

I was therefore converted into a sort of Benedictine monk, of the specific kind that monitors a whole reviewing process, reads two or three successive versions of each chapter, writes a submission package to gain the support of the prestigious publishing group targeted, controls bibliographical references, checks figures, tables, captions, homogenizes the presentation throughout the whole draft, indexes the whole book, and benedictinely runs the LaTeX compiler until it does not scream anymore. As I confess a fierce determination to separate professional and private lives, this book has been elaborated at my office at the university, during innumerable weekends as well as countless late, or even very late, evenings. By the way, this specific time schedule offered me the opportunity to frequently hear the owl living in the little wood in front of the lab, and to catch sight of such furtive animals as badgers and foxes, which one would never think would live in a university campus.

Fortunately, these months of labor have reached their term within the time the tribe I ordinarily belong to was still able to recognize me. May they all be thanked for their patience and their attentive listening and concern about the progress of the project.

I am in special debt to Keith Mansfield from Oxford University Press (OUP), for his support of the project from the very start, and not least for his encouragement and his valuable advice and guidance in the preparation of the proposal dossier for OUP. Complying to high standards is the lot if one wishes to publish with OUP. Driven by the confidence of the invited authors of the project and of my joint editor, I had therefore an obligation: obtain the sesame to be allowed to press ahead.

I also wish to warmly thank Clare Charles from Oxford University Press for her efficient management and attentive monitoring of the production step.

C.S., June, 2014

PREFACE

At the crossroads between statistics and machine learning, probabilistic graphical models provide a powerful formal framework to model complex data. Examples of probabilistic graphical models are Bayesian networks and Markov random fields, which represent two of the most popular classes of such models. With the rapid advancements of high-throughput technologies and the ever decreasing costs of these next-generation technologies, a fast-growing volume of biological data of various types–the so-called omics–is in need of accurate and efficient modeling methods, prior to further downstream analysis. As probabilistic graphical models are able to deal with high-dimensional data and non-linear dependences, it is foreseeable that such models will have a prominent role to play in advances in genome-wide analyses.

Currently, few people are specialists in the design of cutting-edge methods using probabilistic graphical models for genetics, genomics, and postgenomics. This seriously hinders the diffusion of such methods. The prime aim of this book is therefore to bring the concepts underlying these advanced models within understanding of a broader audience of scientists, engineers, and graduate students.

If they are not specialists of probabilistic graphical models, bioinformaticians, statisticians, bio-statisticians, and experts in statistical genetics with an intuition that their solution to a problem should involve such models are compelled to glean incomplete information from publications. We are not even talking of surveys whose consultation will never allow launching out into the design of advanced methods. Some academic courses may well be delivered here and there, that dwell on cutting-edge approaches using probabilistic graphical models for the targeted topics; neither are such courses widely available for the potentially interested audience, nor do they cover a sufficiently illustrative set of models and applications.

The target readers of this book include researchers and engineers as well as graduate students starting a master's or a PhD thesis. Besides, if there is one area where transdisciplinarity is the daily lot, it is the advanced analysis of genome-wide data. Constructive cooperation with a domain specialist requires the ability to hold a productive dialogue, which therefore demands a deep understanding of the models as well as a solid background regarding these models. Often, scientists from different fields such as genetics, statistics, or computer science do not use the same scientific language, and this might lead to confusion and misunderstanding. Bridging the gap between different scientific worlds thus helps scientists to better communicate, and from a higher perspective, contributes to the emergence of new fields of research. Currently, the only solution for such people to gain a deep understanding is finding spare time to gather information to learn from it. The book intends to spare such readers this task.

Hopefully, this book will be of equal interest, if still not higher, for the graduate students supervised by members of the aforementioned audience. Depending on their academic institution, students taught computational methods for genetics, genomics, or postgenomics rarely have access to a course presenting the advanced use of probabilistic graphical models in such fields. One reason for this lies in the fact that these models and their potentialities have only rather recently created renewed interest in genetics in the broad sense. Another reason might be the lack of experts possessing this two-fold skill in these students' institutions. Besides, a few hours taught on the subject are not sufficient to provide both enough material and hindsight on the topic. This book attempts to fill this gap.

This book is also designed to help experts in machine learning grasp the interest in designing advanced methods based on probabilistic graphical models in transdisciplinary collaborations.

This book arises out of a six-year collaboration between its scientific editors. Our various interests in computer science, machine learning, applied mathematics, Bayesian statistics, applications in genetics, genomics, and postgenomics have found in probabilistic graphical models a breeding ground for both our own investigations and the preparation and direction of this book. Besides, coming from different backgrounds, we found a common ground in demanding the highest self-containedness in the contributions of the invited authors. In addition to the intrinsic richness of these contributions, our guiding thread was then providing added value through accessibility for non-specialists of probabilistic graphical models, with no concession on the informativeness of the book's contents.

We have been fortunate to obtain the widest consent regarding invited authors' participation in our project. We subsequently enjoyed a fruitful period of dense exchanges with these authors, who accepted this extra workload.

The book is divided into a general introduction, a tutorial on probabilistic graphical networks, and six main sections devoted to specific application fields in genetics (in the broad sense). The introductory chapter aims at providing a minimal background for readers that are not familiar with biology or need information about the high-throughput biological data addressed by the models described in the book. Moreover, such terms and expressions as genetics, genomics, postgenomics, systems biology, and integrative biology are clarified. Indeed, a leitmotif of the book is the integration of heterogeneous sources of omics data, to boost downstream biological applications. Finally, this introduction provides the motivation for using probabilistic graphical models to handle high-throughput biological data and provides a brief evocation of the use of probabilistic graphical networks in the six applications highlighted by the book: gene network inference, causality discovery, association genetics, epigenetics, detection of copy number variations, and prediction of outcomes from high-dimensional genomic data.

The essentials for understanding probabilistic graphical models are offered in a tutorial at the beginning of the book. This tutorial was carefully designed to be accessible to the largest audience. Since the concepts and techniques presented in this tutorial may require broader and non-trivial knowledge, accessibility and self-containedness were again the targeted objectives.

Together with a thorough review chapter focusing on selected domains in genetics, fourteen chapters illustrate the design of advanced approaches, for the six abovementioned applications. This book offers a lot of new insights that could only be gleaned from the literature available through excruciating labor. The chapters are self-contained, and they can be read independently of each other.

C. S. and R. M.

CONTENTS

ABBREVIATIONS

A	Adenine
aCGH	array comparative genomic hybridization
AIC	Akaike information criterion
AUC	area under the receiver operating characteristic curve
BD	Bayesian Dirichlet
BDe	Bayesian Dirichlet equivalent
BDeu	Bayesian Dirichlet equivalent uniform
BIC	Bayesian information criterion
BN	Bayesian network
BNPP	Bayesian network posterior probability
C	cytosine
cDNA	complementary deoxyribonucleic acid
CGH	comparative genomic hybridization
CNP	copy number polymorphism
CNV	copy number variation
CRF	conditional random field
DAG	directed acyclic graph
DDAG	direct directed acyclic graph
DGM	decomposable graphical model
DNA	deoxyribonucleic acid
D-map	dependence map
EM	expectation-maximization
ER	estrogen receptor
ER^+	estrogen receptor positive
ER^-	estrogen receptor negative
eQTL	expression quantitative trait loci

FDR	false discovery rate
FISH	fluorescence *it situ* hybridization
FLTM	forest of latent tree models
G	Guanine
GGM	Gaussian graphical model
GOGE	genetics of gene expression
GWAS	genome-wide association study
HCGR	homogeneous conditional Gaussian regression
HMM	hidden Markov model
HRMF	Hidden random Markov field
IC	inductive causation
IG	interval graph
I-map	independence map
i.i.d.	identically and independently distributed
KEGG	Kyoto Encyclopedia of Genes and Genomes
LARS	least angle regression
LASSO	least absolute shrinkage and selection operator
LCM	latent class model
LCMS	likelihood-based causality model selection
LD	linkage disequilibrium
LOD	log-odds ratio
LTM	latent tree model
LLS	log-likelihood score
MDL	minimum description length
mRNA	messenger ribonucleic acid
MCMC	Markov chain Monte Carlo
MME	mixed model equations
MRF	Markov random field
PCR	polymerase chain reaction
PDAG	partially directed acyclic graph
PGM	probabilistic graphical model
MAP	maximum *a posteriori*
QTL	quantitative trait loci
RNA	ribonucleic acid
RNA-seq	RNA sequencing
ROC	receiver operating characteristic

SBM	stochastic block model
SCT	stochastic causal tree
SEM	structural equation model
SEM	structural expectation-maximization
SNP	single nucleotide polymorphism
SSR	sum of squares
SSTO	total sum of squares
T	thymine
UG	undirected graph
VLMC	variable length Markov chain

LIST OF CONTRIBUTORS

Abel, Haley J.
Division of Statistical Genomics
Washington University School of Medicine
St. Louis, USA

Antal, Péter
Department of Measurement and Information Systems
Budapest University of Technology and Economics
Budapest, Hungary

Chaibub Neto, Elias
Department of Computational Biology
Sage Bionetworks
Seattle, USA

Charbonnier, Camille
LaMME (Laboratoire de Mathématique et Modélisation d'Evry)
UMR CNRS 8071, USC INRA
Évry, France

Current address:
CNR-MAJ, Rouen, Lille and Paris Salpetriere
University Hospitals
Rouen, France

Chen, Min
Department of Mathematical Sciences
University of Texas at Dallas
Richardson, USA

Chipman, Kyle
Department of Computer Science & Biomolecular Science and Engineering
University of California
Santa Barbara, USA

Chiquet, Julien
LaMME (Laboratoire de Mathématique et Modélisation d'Evry)
UMR CNRS 8071, USC INRA
Évry, France

Cho, Judy
Icahn School of Medicine at Mount Sinai
New York, USA

Deng, Xinwei
Department of Statistics
Virginia Polytechic Institute and State University
Blacksburg, USA

Falus, András
Department of Genetics, Cell and Immunobiology
Semmelweis University
Budapest, Hungary

Gézsi, András
Department of Genetics, Cell and Immunobiology
Semmelweis University
Budapest, Hungary

Guedj, Mickaël
Department of Bioinformatics and Biostatistics
Pharnext
Issy-les-Moulineaux, France

Hajós, Gergely
Department of Measurement and Information Systems
Budapest University of Technology and Economics
Budapest, Hungary

Houseman, Andrés E.
College of Public Health and Human Sciences
Oregon State University
Corvallis, USA

Hullám, Gábor
Department of Measurement and Information Systems
Budapest University of Technology and Economics
Budapest, Hungary

Jeanmougin, Marine
LaMME (Laboratoire de Mathématique et Modélisation d'Evry)
UMR CNRS 8071, USC INRA
Évry, France

Current address:
Department of Immunology, Institut Curie

INSERM U932
Paris, France

Jiang, Xia
Department of Biomedical Informatics
University of Pittsburgh
Pittsburgh, USA

Kiiveri, Harri
CSIRO Computational Informatics
The Leuwin Centre
Floreat, Australia

Li, Jing
Electrical Engineering and Computer Science Department
Case Western Reserve University
Cleveland, USA

Millinghoffer, András
Department of Measurement and Information Systems
Budapest University of Technology and Economics
Budapest, Hungary

Moon, Jee Young
Department of Statistics
University of Wisconsin, Madison
Madison, USA

Current address:
Department of Genetics and Genomic Sciences
Mount Sinai School of Medicine
New York, USA

Mourad, Raphaël
LINA, UMR CNRS 6241
Computer Science Institute of Nantes-Atlantic
Nantes University/Polytechnic Institute
Nantes, France

Current address:
Computational Biology Institute
Mantpellier, France

Neapolitan, Richard E.
Division of Biomedical Informatics
Department of Preventive Medicine
Northwestern University Feinberg School of Medicine
Chicago, USA

Pachter, Lior
Department of Mathematics
University of California Berkeley
Berkeley, USA

Rodriguez Zas, Sandra L.
Department of Animal Sciences
University of Illinois Urbana-Champaign
Urbana, USA

Rosa, Guilherme J. M.
Department of Animal Sciences,
Department of Biostatistics & Medical Informatics,
University of Wisconsin-Madison
Madison, USA

Sárközy, Péter
Department of Measurement and Information Systems
Budapest University of Technology and Economics
Budapest, Hungary

Singer, Meromit
Computer Science Division
University of California Berkeley
Berkeley, USA

Singh, Ambuj
Department of Computer Science
Department of Biomolecular Science and Engineering
University of California Santa Barbara
Santa Barbara, USA

Sinoquet, Christine
LINA, UMR CNRS 6241
Computer Science Institute of Nantes-Atlantic
University of Nantes
Nantes, France

Southey, Bruce R.
Department of Animal Sciences
University of Illinois Urbana-Champaign
Urbana, USA

Szalai, Csaba
Department of Genetics, Cell and Immunobiology
Semmelweis University
Budapest, Hungary

Thomas, Alun
Division of Genetic Epidemiology
University of Utah
Salt Lake City, USA

Valente, Bruno D.
Department of Animal Sciences
University of Wisconsin-Madison
Madison, USA

Visweswaran, Shyam
Department of Biomedical Informatics
University of Pittsburgh
Pittsburgh, USA

Yandell, Brian S.
Department of Statistics and Horticulture
University of Wisconsin–Madison
Madison, USA

Yin, XiaoLin
Electrical Engineering and Computer Science Department
Case Western Reserve University
Cleveland, USA

Zhao, Hongyu
Department of Biostatistics
Yale School of Public Health
New Haven, USA

PART I

Introduction

CHAPTER 1

Probabilistic Graphical Models for Next-generation Genomics and Genetics

CHRISTINE SINOQUET

The explosion in "omics" and other types of biological data has increased the demand for solid, large-scale statistical methods. These data can be discrete or continuous, dependent or independent, and from many individuals or tissue types. There might be millions of correlated observations from a single individual or observations at different scales and levels, in addition to covariates. The study of living systems encompasses a wide range of concerns, from prospective to predictive and causal questions, reflecting the multiple interests in understanding biological mechanisms, disease etiology, predicting outcomes, and deciphering causal relationships in data. Precisely, probabilistic graphical models provide a flexible statistical framework that is suitable to analyze such data. Notably, graphical models are able to handle dependences within data, which is an almost defining feature of cellular and other biological data.

This introductory chapter aims at providing a minimal background for readers that are not familiar with biology or need information about the high-throughput biological data the models described in the book deal with. The chapter also provides the motivation for using probabilistic graphical models to handle high-throughput biological data. The chapter is organized as follows. Section 1.1 describes the fine-grained components studied by molecular biology and provides the definitions of key terms. The biological information allows to conduct studies in the fields of genetics, genomics, and postgenomics. The respective scopes of these three domains are first defined. In these domains, various types of analyses allow inference of knowledge about one or several levels of description of living systems. Section 1.2 then focuses on the multiple levels of biological organization of living systems to which the chapters of this book are connected. This section takes the opportunity to clarify which definition of the expression "systems biology" the book is interested in. The definition of "integrative biology" is also clarified. In the era of modern genomics, the data are provided by high-throughput technologies; Section 1.3 briefly surveys the types of data covered by the book. Finally, this section emphasizes the complexity of the biological data available nowadays, and stresses various issues encountered when handling such data. This emphasis serves as a transition to Section 1.4, which starts advocating the use of probabilistic graphical models in genetics, genomics, and postgenomics: thus can be evidenced and exploited dependences within various biological components, with the aim of explanation and prediction. The chapter ends with a brief evocation of the use of probabilistic graphical networks in the six applications highlighted by the book: gene network inference, causality discovery, association genetics, epigenetics, detection of copy number variations, and prediction of outcomes from high-dimensional genomic data.

1.1 Fine-grained Description of Living Systems

1.1.1 DNA and the Genome

Except in viruses, the cell is the smallest structural unit of all living organisms that is capable of independent functioning, through metabolic activities. The metabolism encompasses all chemical transformations within the cell. In contrast with the procaryotic cell (e.g., bacteria), the eukaryotic cell is typically described as possessing a nucleus isolated by a membrane from the rest of the cell (cytoplasm); the nucleus contains the majority of the hereditary material, called the genome. In prokaryotes, the hereditary material is not bound within a nucleus. All the applications described in this book address the human genome, which explains our focus on eukaryotic cells. Under the influence of environmental factors, the genome plays an important role in the development of the individual's observable features (also called **phenotypes**). For instance, it is known that genes influence race, hair and eye color, gender, height, and weight.

In each eukaryotic cell of a living organism, the same genetic information is encoded in a bio-chemical molecule, the DNA. The DNA molecule is double-stranded, and it is twisted into a helix. Each strand consists of a long polymer of nucleotides (or bases). The genome is encoded through an alphabet of four bases: adenine (A), guanine (G), cytosine (C), and thymine (T). The two strands of the DNA molecule are paired, based on hybridization properties: A and T (respectively C and G), on opposite strands, are physically connected together as *complementary* bases. DNA molecules determine the synthesis of proteins, via intermediary messenger ribonucleic acid (mRNA) molecules: mRNA is produced from DNA through the **transcription** step; proteins are produced from mRNA by the **translation** step. The deshybridization property, which locally frees the two DNA strands, is involved in the replication and transcription processes. The replication step produces a DNA copy from a DNA molecule; the transcription step produces a single-strand RNA molecule from a double-strand DNA molecule. One of the most revolutionary levers in science in the twentieth century is the polymerase chain reaction (PCR), which exploits the hybridization and deshybridization properties; PCR thus allows to obtain several million identical copies from a single DNA fragment.

In eukaryotes, the genome is packaged into chromosomes, each consisting of a specific DNA sequence tightly packed into a complex series of coils thanks to proteins (i.e., histones). The human genome contains approximately 3.4 billion base pairs of DNA packaged into 23 chromosomes. Most cells in the body, except female ova and male sperm, are diploid. Diploidy means that such cells possess two sets of homologous chromosomes. Therefore, each cell contains a total of 6.8 billion base pairs of DNA. If (virtually) laid end to end, the 46 DNA molecules in each human cell would produce a two-meter long sequence.

Between any two humans, genetic variation roughly amounts to 0.1%. Thus, on average, about one base pair out of every 1000 is different between any two individuals. Various types of DNA polymorphism are known: the most common type is the single nucleotide polymorphism (SNP), where genetic variations consist in single base-pair differences; moreover, another characteristic of SNP is that over the four possible nucleotides (A, G, C, and T), only two variants are exhibited over a studied population. Other less frequent types of polymorphisms include insertions, deletions, duplications, and rearrangements of segments of DNA, as well as differences in the numbers of copies of a given segment.

1.1.2 Genes and Proteins

Any DNA region that produces a functional RNA molecule is called a gene. In addition, the most well-known acception of the term "gene" relates to the class of genes that code for proteins. The human genome contains approximately 20 000 such genes. Proteins are large molecules that play most of roles in an organism. In a multicellular organism, proteins are required for the structure, function, and regulation of the organism's tissues and organs. For instance, **enzymes** catalyze an overwhelming part of the thousands of chemical reactions that occur in a cell; such proteins are thus essential to the production of the remaining organic biomolecules necessary for life. For example, the phenylalanine hydroxylase enzyme converts the amino acid phenylalanine into another amino acid, the tyrosine. Another crucial role is that of **transcription factors**. Such proteins bind to specific DNA sequences, alone or in a complex, to promote (activate) or block (repress) the positioning of the **RNA polymerase enzyme** on the DNA molecule. Both previous types of proteins thereby control the flow of genetic information from DNA to mRNA, and thus the formation of new protein molecules. Other proteins form the **structural** components of the cell. On a larger scale, they also allow an organism to move. For instance, actin filaments are structural proteins built up of multiple subunits; they help cells maintain their shape and are also involved in muscle contraction. **Storage** and **transport** are two other crucial functions performed by proteins: the proteins concerned bind to atoms or small molecules; transport throughout an organism is thus made possible. For instance, ferritin, a protein made up of 24 identical subunits, is involved in iron storage. Some proteins are messengers that transmit signals to coordinate biological processes between different cells, tissues, and organs. An example is the growth hormone, which regulates cell growth. We complete the enumeration of the vital functions fulfilled by proteins with the mention of antibodies. An antibody is a protein that binds to a specific foreign particle, such as a virus or a bacterium, to help protect an organism. For instance, immunoglobulin G is a type of antibody present in the blood.

1.1.3 Phenotype and Genotype

The **phenotype** of an organism is defined as the combination of the organism's observable characteristics or traits. In particular, phenotypes can be described at the lowest level of living systems, that is the cellular level: the definition of phenotype can be extended so as to designate characteristics that are only made observable by some technical procedure. Such traits are connected to the various levels of the scale through which a biological system is observed. Higher level traits include biochemical properties, physiological properties, development and morphology, phenology, and behavior. For instance, phenological traits comprise periodic biological phenomena, such as flowering, breeding, and migration, in relation to climatic and habitat conditions. Even this level is subject to the influence of the genetic information carried by an organism. Phenotypes result from the expression of an organism's genes as well as the influence of environmental factors and the interplay between the two.

The **genotype** of an organism is defined as the set of alternate variations of genes expressed in some specific traits. Such traits are often expressed through the synthesis of proteins. In genetics, gene expression is the most fundamental level at which the genotype gives rise to the phenotype. In another common usage, the **genotype** of an organism is more systematically defined as the description of DNA variations, based on a set of genetic markers. **Genetic markers** are well-characterized loci of the genome, which represent many short windows in which to observe DNA polymorphism between individuals. In particular, the genotype of a diploid organism accounts for the DNA variants—or **alleles**—present at opposite loci, on the two homologous chromosomes of a pair of chromosomes. Comparing genotypes among a set of organisms of the same species is the key to deciphering the differences in phenotypes observed.

1.1.4 Molecular Biology, Genetics, Genomics, and Postgenomics

Molecular biology is the branch of biology that describes the molecular characteristics of the genome as DNA, RNA, and proteins. Various definitions can be provided for the terms genetics and genomics. In the scope of the present book, the word **genetics** designates the discipline that studies variations between the genomes of individuals in some population; this analysis of variations may focus on simple units (genetic markers) or on more complex units (genes). The definition of **genomics** generally encompasses the range of biotechnological and computational analyses related to genome sequencing, gene mapping, and genome annotation; **functional genomics** is the appropriate expression for this book. Functional genomics focuses on transcription, translation, and interactions between proteins. Notably, functional genomics includes the study of the transcriptome, through DNA chips, to describe and quantify gene expression: for example, gene expression correlation potentially indicates that genes belong to the same gene interaction network; the identification of differentially expressed genes, for instance between affected and unaffected individuals, allows to identify putative causes for a studied disease.

Beyond functional genomics, **postgenomics** takes a step further to encompass an increasingly large range of topics. All such topics essentially aim at teasing higher functional biological understanding out of raw data. These data allow different viewpoints on living organisms. Such viewpoints may be transcriptomics (analysis of gene expression level through mRNAs), proteomics (analysis of gene expression as proteins), and metabolomics (characterization of the small molecules that are intermediates and products of metabolism), to name but a few.

In genetics, genomics, and postgenomics, various types of analyses enable the inference of knowledge about one or several levels of description of living systems. The next section describes the levels that are addressed in this book.

1.2 Higher Description Levels of Living Systems

The activity and state of a multicellular organism may be described from different viewpoints: cell, organ or tissue, system (e.g., cardiovascular, nervous), and whole organism. In this book, methods based on probabilistic graphical models are described that allow the inference of knowledge about various description levels of living systems. The chapters in this book deal with the following description levels:

- genome,
- transcriptome,
- gene interaction networks,
- phenotype.

Depending on the chapter, knowledge inference addresses a single level or deals with several levels. We introduce this section by giving a flavor of the complexity of the processes and the variety of actors involved in the life of a cell. Then, we focus on the multiple levels of biological organization of living systems to which the chapters of this book are connected.

1.2.1 Complexity in Cells

A eukaryotic cell consists of the nucleus and various organelles—the "organs" of the cell, for short—which are immersed in the cytoplasm (see Fig. 1.1). The cell is surrounded by a semi-permeable membrane. Although no exact number can be provided, the number of cells in an adult human body can be approximated as 10^{14}. A unique cell, the fertilized egg, is the origin of all these cells, through cell division. However, though they bear the same genetic information in their nucleus, the cells of a multicellular organism perform different specific tasks, depending on their location in organs or tissues: red blood cells exchange oxygen, muscle cells expand and contract, and cells in the immune system recognize pathogens. Modifications in gene expression play a key role in guiding and maintaining cell differentiation.

Extrinsic and intrinsic factors regulate gene expression in cells. The first category includes small molecules, secreted proteins, temperature, and oxygen. Within the organism, cells communicate with each other by sending and receiving secreted proteins (e.g., growth factors, morphogens,

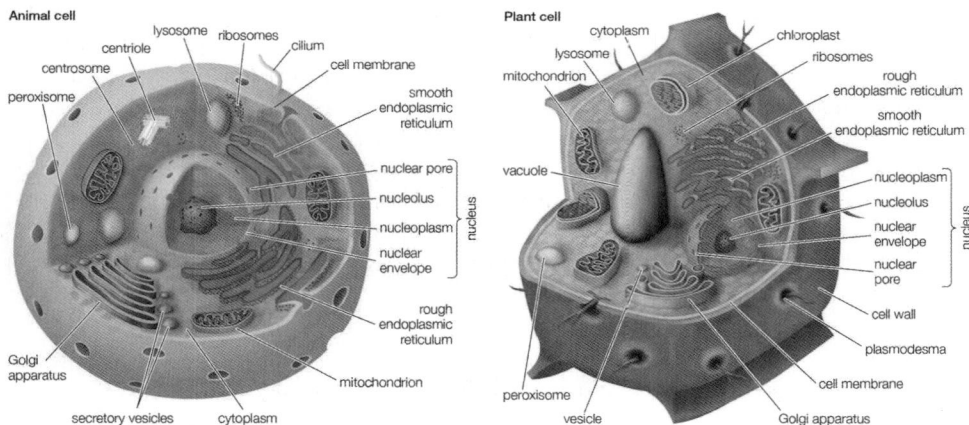

Fig 1.1 Typical animal and plant cells.

2010 Encyclopaedia Britannica, Inc.

cytokines). The receipt of these signaling molecules triggers intercellular signaling cascades that modify the expression of genes. Sequence-specific transcription factors are considered the most important and diverse mechanisms of gene regulation in cells [28]. An example of cell-intrinsic regulation is that of the modification of chromatin (DNA associated with histone proteins) by a cell's own machinery. A possible consequence of chromatin modification is a change in the accessibility of genes to transcription factors; the impact on gene expression may be positive or negative. Two major classes of chromatin modifications include DNA methylation and histone modification.

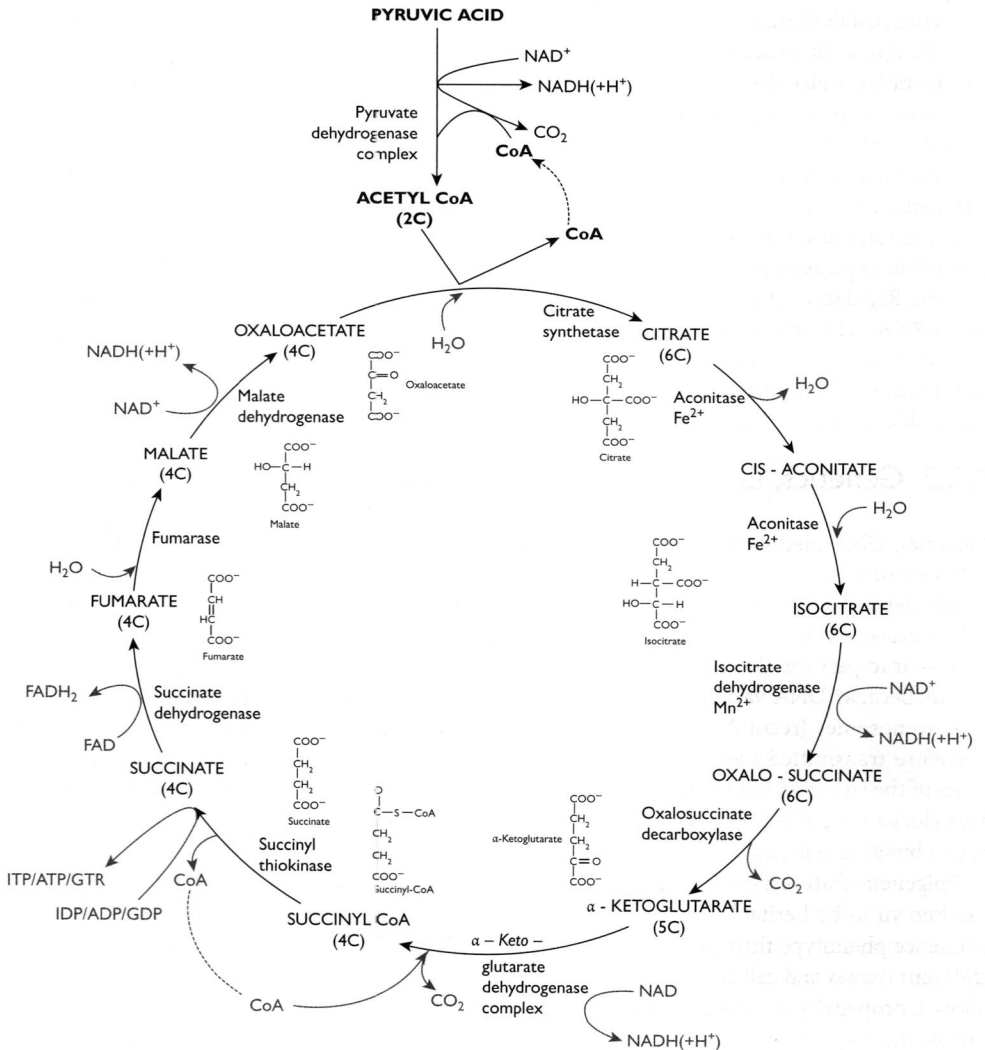

Fig 1.2 Krebs cycle. The first product of Krebs cycle is citric acid (citrate).

Xamplified, Free Online Eucation Resource

In a cell, the metabolism comprises thousands of complex chemical reactions. These reactions are chained in metabolic pathways. A metabolic pathway consists of a series of biochemical reactions, starting from a substrate S to generate a product P. Each intermediate reaction (that is except the first and the last ones) uses at least one product from another reaction in the pathway as a substrate, and generates the substrate of another reaction in the pathway. For instance, the metabolic pathway most widespread in living systems is glycolysis, which breaks down glucose to produce energy and takes place in the cell cytoplasm. Another example of a metabolic pathway is the Krebs cycle (see Fig. 1.2), whose specificity lies in that one of its basic substrates is also the end product of the pathway. Cells produce and transform the organic molecules that supply both the material and the energy requested for life. Metabolism consists of two opposed processes, catabolism, and anabolism. Catabolism extracts energy from complex molecules (e.g., glucids, lipids) by breaking them into smaller molecules. Anabolism requires energy to synthesize complex molecules from simple molecules. Metabolic pathways are controlled by enzymes. Enzymes are proteins that catalyze chemical reactions; namely they accelerate reactions, even in small amounts, and without participating in these reactions. Since many proteins are enzymes, the control of metabolism and the regulation of gene expression are intimately linked. Sometimes, metabolic control aims at homeostasis, that is, maintaining constant levels of some variables or constant rates of some processes; sometimes adaptation demands change. Gene regulation allows the cell to express protein when needed, thus ensuring the versatility and adaptability of an organism. Regulation of gene expression involves a wide range of mechanisms and actors (proteins, microRNAs, chromatin) and complex dynamics (production, storing, degradation). All steps of gene expression can be modulated, encompassing transcriptional initiation, RNA processing, protein synthesis, and post-translational modification of proteins. Fig. 1.3 (page 10) illustrates the hierarchical organization of living systems.

1.2.2 Genetics, Epigenetics, and Copy Number Polymorphism

Genetics, DNA methylation in epigenetics, and Copy Number Polymorphism all deal with the DNA sequence.

The dependences within genetic data (e.g., SNPs) define the linkage disequilibrium (LD). Faithful models of LD are required for the visualization of LD at various scales—including the genome scale—or to perform downstream analyses such as association studies (see Subsection 1.2.6). LD occurs because DNA variants close on the chromosome are scarcely separated by the shuffling of chromosomes (recombination) that takes place during sex cell formation. Such variants are therefore transmitted together (as a haplotype) from parent to child. Such patterns are at the basis of the so-called haplotype block structure [12]: "blocks", where statistical dependences between loci are high, alternate with shorter regions characterized by low statistical dependences, the recombination hotspots.

Epigenetic features, such as DNA methylation and histone modifications, the two most studied, are known to be heritable across cell divisions. It has been shown that epigenetic mechanisms influence phenotype through the regulation of gene expression. Epigenetic features differ across different tissues and cell types. Most of the vertebrate genome is methylated. Unmethylated sites show a propensity to cluster together along the genome; unmethylated clusters are often present in the regulatory regions of many genes. DNA methylation is an important regulator of gene transcription and is tightly linked with cellular differentiation. Besides, in many diseases, abnormal hypermethylation of these clusters results in transcriptional silencing of the nearby genes. Specifically, associations between altered methylation states and various cancers have been

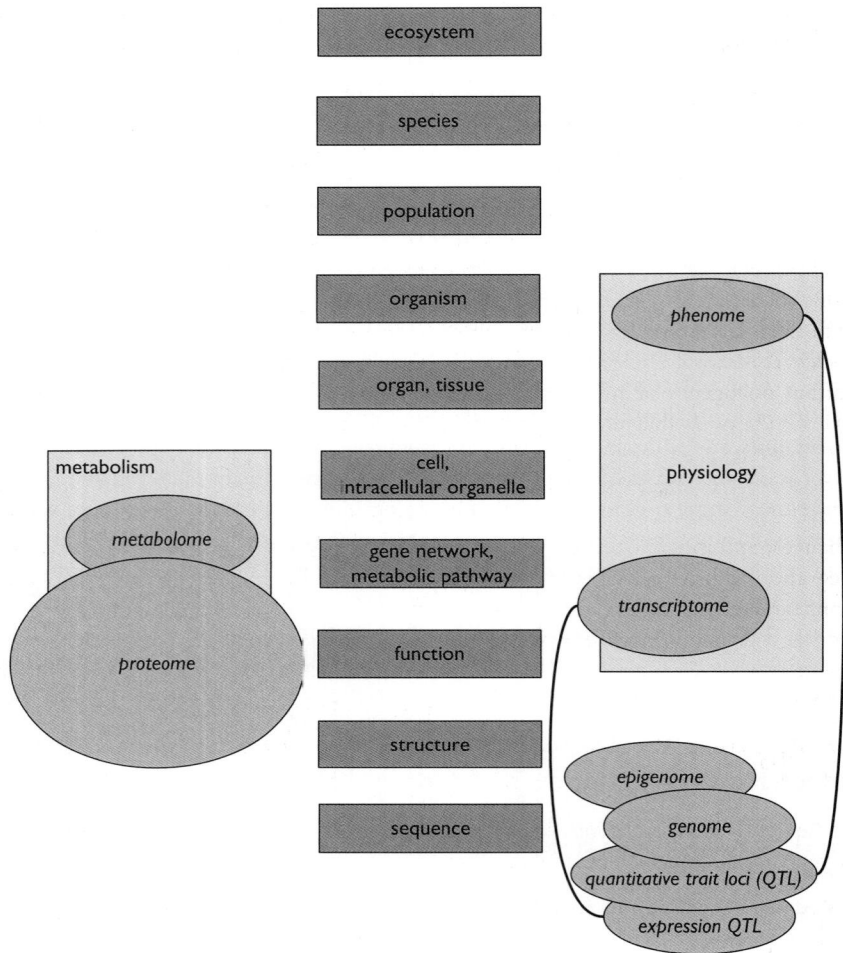

Fig 1.3 Hierarchical organization of living systems.

reported. Moreover, DNA methylation in tumor cells encodes phenotypic information about the tumor or the tumor subtypes. In analogy with the difference between genome sequencing and genotyping, where only a small subset of an individual's nucleotides are assayed, methyltyping suffers from low resolution in comparison with methylome sequencing. Thus, methyltyping poses a challenge in DNA methylation profiling. Modeling DNA methylation to exhibit subtypes in a population is another challenge.

Two chapters are dedicated to the genome-scale modeling of dependences within genetic data:

Chapter 9: Modeling Linkage Disequilibrium and Performing Association Studies through Probabilistic Graphical Models: A Visiting Tour of Recent Advances (C. Sinoquet and R. Mourad), and

Chapter 10: Modeling Linkage Disequilibrium with Decomposable Graphical Models (H. Abel and A. Thomas).

The two chapters above deal with Single Nucleotide Polymorphism. The chapter below addresses another kind of DNA polymorphism, DNA Copy Number Variations:

Chapter 16: Detection of Copy Number Variations from Array Comparative Genomic Hybridization Data using Linear-chain Conditional Random Field Models (X. Yin and J. Li).

In diploid genomes, for each gene, or more generally for each genomic segment, each individual inherits one copy from its father and one copy from its mother. Thus, in principle, the total number of copies is two. However, copy number mutations may occur: the total number of copies may be one (deletion), or three or more (amplifications/insertions).

Copy number alterations have been reported to be associated with numerous diseases. In particular, such chromosal aberrations as amplifications and deletions have led to the discovery of important oncogenes or tumor suppress genes. Array comparative genomic hybridization (aCGH) is a technology that allows the identification of copy number alterations across genomes.

In aCGH, which is an array-based technology, fluorescence is used to measure indirectly the number of copies for each DNA fragment in the array. Analyzing copy number polymorphisms using aCHG data consists of two tasks: detection of the boundaries where the copy number exhibits changes, and inference of the copy number state for any such designated regions. Basic data integration within the genomic level is performed in this case, since it is necessary to align the regions targeted by the array, and thus to refer to the genome sequence.

In another category, but again in the line of methods addressing the lowest level of biological organization—the DNA sequence—two other chapters address DNA methylation profiling. The chapter below analyzes DNA methylation profiles to cluster data:

Chapter 15: Latent Variable Models for Analyzing DNA Methylation (E. Andrés Houseman).

The other chapter will be mentioned in the next section. As it relies on prior genomic knowledge, it is an example of data integration.

1.2.3 Epigenetics with Additional Prior Knowledge on the Genome

In the chapter mentioned below, knowledge integration is performed within the same level of description (DNA):

Chapter 14: Bayesian Networks in the Study of Genome-wide DNA Methylation (M. Singer and L. Pachter).

Therein, more information is incorporated from the genomic data. Basically, the genomic structure is used as a prior on methylation status: in vertebrates, unmethylated sites tend to cluster together. Besides, the so-called CpG sites, which are unmethylated, are more conserved than other sites. Therefore, when experimental annotation of CpG sites is available, the richness of genomic regions in CpG clusters tends to point out unmethylated regions.

1.2.4 Transcriptomics

The sequence of an mRNA mirrors the sequence of the DNA from which it was transcribed. Consequently, by analyzing the entire collection of RNAs (transcripts) in a cell, transcriptomics can determine which gene is turned on or off in the cells and tissues of an organism. Different cells show different patterns of gene expression. Transcriptomics examines gene expression microarrays, in which individuals are observed for a common set of genes.

One aim of transcriptomics is to determine how gene expression changes under the pressure of various factors such as tissue type, stage of development, drugs, or disease status. Differentially expressed genes are genes whose mean expression over a group of individuals sampled under a given condition (treatment, disease status (affected)) is significantly higher or lower than the mean expression over the control group (e.g. unaffected). For rigorous differential expression assessment, there is much more information in a microarray data set than the usual analysis extracts: the correlation structure between genes should be taken into account. The usual simplifying assumption of no correlation is unreasonable as genes are known to be connected in pathways or networks.

Since proteins can be transcription factors for other genes, genes' interplays may be summarized in gene regulatory networks. Genes targeted (in their regulatory regions) by the same transcription factors tend to show similar expression patterns along time. Thus, genes that are simultaneously co-expressed in some experimental or physiological condition (that is, genes that are highly correlated since they have similar expression profiles) are likely to be co-regulated by the same gene or genes.

However, gene network inference is far more complicated than identifying clusters of co-expressed genes. Genes expressed or inhibited in similar conditions or time points are likely to interact together. Yet, gene network reconstruction requires distinguishing between the correlation of two genes due to direct causal relationships and the correlation that originates from intermediate genes. Therefore, it is necessary to evaluate the correlation between genes conditioning on other genes. Through the exhibition of direct causal relationships, gene network inference highlights potential regulations or chains of regulations. For example, it is crucial to identify hubs, those key genes that regulate many other genes. On the other hand, modules—or communities—of genes are main contributors to the robustness and evolvability of biological networks; a module is defined as a set of interacting genes, whose function is separable from the function of other modules. The role of biologists remains to validate the gene network inferred or to clarify which are the exact paths corresponding to regulatory chains.

Observation across various conditions sheds light on the constants or variations of these dependences, that is, on the flexibility of the gene regulation network. In contrast to gene relationships unique to particular conditions or samples, some interactions may be shared across conditions or samples. Potentially complex distributions of gene expression across a wide range of conditions may be described through mixture models.

In some cases, merging different experimental conditions mainly aims at enlarging the number of observations available to infer a gene network. In this case, heterogeneity among microarray experiments represents an issue to cope with. A remedy is to study multiple networks simultaneously with an incentive to share interplays across conditions.

One chapter focuses on the acknowledgment of gene correlation in the assessment of differential expression:

Chapter 3: Graphical Models and Multivariate Analysis of Microarray Data (H. Kiiveri).

Two other chapters address gene network inference:

Chapter 4: Comparison of Mixture Bayesian and Mixture Regression Approaches to Infer Gene Networks (S.L. Rodriguez-Zas and B.R. Southey), and

Chapter 5: Network Inference in Breast Cancer with Gaussian Graphical Models and Extensions (M. Jeanmougin, C. Charbonnier, M. Guedj and J. Chiquet).

1.2.5 Transcriptomics with Prior Biological Knowledge

Chapter 5 is another example of the integrative approach described by this book. Therein, prior knowledge on the latent gene network structure is used. Many sources can be used as a biological prior on the network structure. For instance, prior knowledge may come from the gene level, as information about metabolic pathways or about which genes code for other genes' transcription factors. Metabolic pathways are available from the KEGG (Kyoto Encyclopedia of Genes and Genomes) [20] or BioCarta databases (http://www.biocarta.com/genes/index.asp); the connection between two genes is promoted or penalized depending on whether the genes belong to the same pathway or not. In addition, binding sites of transcription factors point out which genes are potentially regulated by the transcription factors.

All other chapters in the book infer knowledge through the integration of various sources of data.

1.2.6 Integrating Data from Several Levels

This transition provides the opportunity to define the concepts of **integrative biology** and **systems biology**. According to some scientists, integrative biology denotes multidisciplinary research (cross-disciplinary, transdisciplinary) incorporating chemistry, physics, mathematics, and computer science, as appropriate. At the interfaces, significant issues are discussed among scientists bringing together diverse but specific skills. As each chapter in this book takes a machine learning approach to deal with genetics, genomics, or postgenomics, this first definition of integrative biology holds for the book.

To other researchers, integrative biology means using a panel of various techniques and approaches to fulfill their own research programs. The previous definition includes the hierarchical approaches that deal with integration across levels of biological organization. At the extreme, such integrative frameworks describe life from molecules to the biosphere, with diversity across taxa, encompassing viruses, bacteria, plants, and animals. The availability of omics data (genomics, transcriptomics, proteomics, metabolomics, phenomics . . .) allows the implementation of integrative approaches across as many levels of biological organization [19]. In this book, all chapters not already mentioned in the present section fit this specific latter definition of integrative biology.

To some extent, the above definition meets the concept of **systems biology**. Systems biology is an approach in biology and biomedical research meant to understand living systems as wholes, be they an organism, a tissue, or a cell. In the more traditional so-called reductionist biology, a system's pieces are studied separately. In contrast, the purpose of systems biology is to put a system's pieces together, in a holistic perspective. Through this integration, systems biology aims at discovering the emergent properties of cells, tissues, and organisms functioning as systems. Evidencing such emergent properties is ideally addressed by observing multiple components simultaneously and by rigorously integrating data, based on mathematical models. These emergent properties mainly describe the complex interactions within biological systems, as illustrated by gene regulation networks, causal phenotype networks, and associations between genotype and phenotype. All three previous topics are at the core of fourteen of the chapters in this book. The hierarchies of biological levels that are spanned therein may not appear very deep as they connect the genomic and gene levels, relying on genetics, transcriptomics, and phenomics. Nonetheless, the integrative approaches depicted require advanced models.

INTEGRATING GENETICS AND PHENOMICS

Four chapters in this book deal with quantitative genetics, which is the understanding of how genotype contributes to phenotype. In the biomedical research domain, an association study aims at identifying a causal relation between some genomic locus or loci and a disease status (affected/unaffected). Genome-wide association studies (GWASs) tackle the issue of unraveling such genotype-phenotype dependences from massive data. Such data usually describe thousands or ten thousands of subjects with a few hundred thousands to one or two millions of SNPs. The two following chapters address GWAS strategies:

Chapter 9: Modeling Linkage Disequilibrium and Performing Association Studies through Probabilistic Graphical Models: A Visiting Tour of Recent Advances (C. Sinoquet and R. Mourad),

Chapter 11: Scoring, Searching, and Evaluating Bayesian Network Models of Gene-phenotype Association (X. Jiang, S. Visweswaran and R.E. Neapolitan).

In addition, one chapter thoroughly reviews various refined concepts of association, whereas another chapter takes the slightly different viewpoint of *predicting* phenotypes from GWAS data:

Chapter 13: Bayesian, Systems-based, Multilevel Analysis of Associations for Complex Phenotypes: From Interpretation to Decision (P. Antal, A. Millinghoffer, G. Hullám, G. Hajós, P. Sárközy, A. Gézsi, C. Szalai and A. Falus), and

Chapter 17: Prediction of Clinical Outcomes from Genome-wide Data (S. Visweswaran).

INTEGRATING GENETICS, PHENOMICS, AND PRIOR KNOWLEDGE ON BIOLOGICAL PATHWAYS

In the quantitative genetics domain, one chapter of the book illustrates integration across three levels of biological organization:

Chapter 12: Graphical Modeling of Biological Pathways in Genome-wide Association Studies (M. Chen, J. Cho and H. Zhao).

In this chapter, a standard GWAS provides a list of genes associated with a studied disease. On the other hand, some other genes, not surveyed by the GWAS, are known to belong to the same biological pathways as the previous genes. The purpose is to estimate the probability that these other genes may be associated with the disease.

INTEGRATING GENETICS AND TRANSCRIPTOMICS

The phenotypes dealt with in the above cited chapters are discrete variables (affected/unaffected status). In Section 1.1, it was recalled that an organism's phenotype consists in the expression of its genotype, under defined environmental conditions. The expression of observable characteristics includes features that are only observable through the aid of technology. Transcriptomics provides gene expression levels for the genes targeted by a microarray. These expression levels represent as many continuous—or quantitative—phenotypes.

A quantitative phenotype (or trait) is defined as any physical, physiological, or biochemical quantitative feature that may be observed for organisms. The purpose of quantitative trait loci (QTL) mapping is to identify the genomic regions, named QTLs, where genotype variation entails phenotype variation. The definition of QTLs is straightforwardly transposed to expression QTLs (eQTLs) for which the continuous phenotype is a gene expression level.

Dissecting the causal relationships among *expression* traits involved in the same biological pathways—and therefore correlated—is a current research topic. Assumptions about the causal structure of observed variables are often represented in a *directed* acyclic graph. In causality inference, the identification of the eQTLs causal to each phenotype is of prime importance. The genetic architecture (GA) of a given phenotype denotes the locations and effects of its (directly) causal QTLs. The inference of a causal phenotype network (CPN) has to benefit from the knowledge about the genetic architecture: adding causal QTL nodes to a phenotype network allows the inference of causal relationships between phenotypes that could not be distinguishable using phenotype data alone. Conversely, GA inference may be refined based on the information borne by the CPN.

Three chapters in the book are dedicated to the inference of causal phenotype networks. Two of them rely on the mere integration of genetics and transcriptomics:

Chapter 6: Utilizing Genotypic Information as a Prior for Learning Gene Networks (K. Chipman and A. Singh), and

Chapter 8: Structural Equation Models for Studying Causal Phenotype Networks in Quantitative Genetics (G.J.M. Rosa and B.D. Valente).

INTEGRATING GENETICS, TRANSCRIPTOMICS, AND PRIOR BIOLOGICAL KNOWLEDGE

To reconstruct causal phenotype networks, the chapter below implements further data integration:

Chapter 7: Bayesian Causal Phenotype Network Incorporating Genetic Variation and Biological Knowledge (J. Young Moon, E. Chaibub Neto, X. Deng and B.S. Yandell).

Prior biological knowledge is incorporated, which may originate from various sources of biological information. One possible source of information is chromatin immunoprecipitation with microarray experiments (ChIP on chip), which is used to investigate the interaction of proteins and DNA *in vivo*. This technology is employed to generate putative lists of target genes for a given transcription factor; it evidences that a given transcription factor binds to some putative target. Regulation inference from knock-out data and protein–protein interaction can also be used as a prior. Knock-out gene technology allows to inactivate specific genes within an organism. Genes are knocked out by modifying the region of the gene that codes for the protein. Thus can be determined the effect of this gene on the functioning of the organism. Pathway information can also guide to refine the causal phenotype network. Finally, information from the Gene Ontology (GO) [6] may contribute to the biological prior. The GO is a specific vocabulary of terms describing the molecular functions, biological processes, and cellular components of a gene. The GO terms annotate a large fraction of genes. A similarity measure between genes may be defined in this GO framework, that enables the connection of genes in a gene network. This network is subsequently used as a prior for causal phenotype network inference.

Finally, not only does the method described in Chapter 7 perform data integration, it also performs process integration; whereas most approaches conduct GA inference and CNP reconstruction separately, these two processes are intertwined.

1.2.7 Recapitulation

Table 1.1 offers a summarized description of the various data sources involved in the integrative approaches described in this book.

1.3 An Era of High-throughput Genomic Technologies

In the previous section, we emphasized the integrative dimension present in all chapters in this book but one. For readers not familiar with data originating from high-throughput technologies, we now briefly describe the data and the genesis of the data dealt with by the chapters of this book. This section may be skipped by other readers.

Various technologies can be used to generate genome-scale data that provide measurements at various levels of biological organization. These so-called "omics" data offer an unprecedented potential to gain insights on the workings of living systems.

1.3.1 Genotyping

In the broad sense, genotyping is the process of determining the genetic composition of an organism by inspecting its DNA sequence.

Genotyping can be achieved through a variety of methods, depending on the polymorphism of interest (e.g., SNPs, insertions, deletions, duplications, and rearrangements) and the resources available. Copy Number Variations (CNVs), which result from duplications and deletions, will be addressed in Subsection 1.3.2. In the present section, we concentrate on SNPs. SNP-based genotyping focuses on a small subset of nucleotide locations, known to exhibit variety within a population of subjects. The characteristic of SNP lies in that, in each such location, when *refer-ring* to *one* of the two DNA strands, only two variants are observable among the four possible nucleotides. According to the international HapMap project [7], the estimated number of SNPs in the human genome amounts to 10 millions. SNP genotyping is associated with low cost but low resolution techniques. The use of genotyping chips or arrays is an efficient and accurate option for examining many loci simultaneously. In addition, next-generation technologies have reduced the costs of DNA sequencing down to the point that genotyping by sequencing is now feasible.

Subsection 1.3.4 is devoted to the presentation of the principle used in array techniques.

DNA sequencing aims to determine the exact sequence of a given region of DNA. Such regions may cover a short piece, the whole genome, or parts of the genome (e.g., the "exome", which is the 2% or so of the human genome that contains genes). If the targeted DNA stretches encompass SNPs, DNA sequencing may fulfill the purpose of genotyping. The remaining part of this section succinctly explains the technology behind DNA sequencing.

DNA polymerase is the enzyme involved in DNA replication, the biological process that enables the generation of DNA copies from a DNA template molecule. The molecule produced consists

Table 1.1 Levels of biological organization and modalities of integration addressed by the book.

Level	Chapter	Authors	Data	Knowledge Inference/ Biological Application
Genetics	9 10	Sinoquet and Mourad Abel and Thomas	SNPs	Dependences within SNPs, modeling of linkage disequilibrium, at the genome scale
	16	Yin and Li	CNV	Detection of copy number variations from array CGH Data
Epigenetics	15	Andrés Houseman		DNA methylation measurements Finite mixture modeling (discrete) or latent trait modeling (continuous) over a population of individuals
Epigenetics Prior knowledge on genomic structure	14	Singer and Pachter	Methyltype	DNA methylation structure along the genome
Transcriptomics	3	Kiiveri	Gene expression profiles	Identification of differentially expressed genes under the hypothesis of correlation between genes
	4	Rodriguez-Zas and Southey	Gene expression profiles under various conditions	Gene network inference with evidence of a mixture structure
Transcriptomics Phenomics Prior biological knowledge	5	Jeanmougin et al.	Gene expression profiles under various conditions	Gene network inference with **prior** knowledge on structure, gene network inference from various microarrays or time points
Genetics Phenomics	9	Sinoquet and Mourad	SNPs for affected and unaffected subjects	Association study
	11	Jiang et al.		Association study
	13	Antal et al.	SNPs and phenotypes	Multivariate relevance in GWAS
	17	Visweswaran		Prediction of clinical outcomes from genotype
Genetics Phenomics Prior knowledge on biological pathways	12	Chen et al.	SNPs and phenotypes	Integration of **prior knowledge** to enhance GWAS
Genetics Transcriptomics	6 8	Chipman and Singh Rosa and Valente	eQTLs and transcripts QTL and phenotypes	Inference of causal phenotype network
Genetics Transcriptomics Prior biological knowledge	7	Moon et al.	eQTLs and transcripts	Inference of causal phenotype network, enhancement through the integration of **prior biological knowledge**

Note: CGH, comparative genomic hybridization; CNV, copy number variation; eQTL, expression quantitative trait loci; GWAS, genome-wide association study; QTL, quantitative trait loci; SNP, single nucleotide polymorphism.

of one of the strands of the original molecule and one of a novel complementary strand. DNA sequencing reactions are similar to polymerase chain reactions (PCRs) used to replicate DNA at a large scale. The sequencing reaction mix includes the template DNA, free nucleotides, an enzyme, and a primer. The primer is a 20–30 nucleotide long segment of single-stranded DNA meant to hybridize with one segment on a strand of the template DNA. Heating the two strands of the DNA template entails their separation; the primer can then stick to its targeted location on the template strand so that the DNA polymerase may start synthesizing a complementary strand. The elongation consumes the free nucleotides of the reaction mix.

If the elongation were pursued until completion, the process would result in a new strand. DNA sequencing relies on the use of terminator nucleotides. Over millions of starts, a reaction mix with trace amounts of all four A, G, C, and T terminator nucleotides will generate strands stopping at every possible A, G, C, and T, respectively. All terminator nucleotides in the reaction mix bear one of four fluorescent colors meant to distinguish A, G, C, and T. Electrophoresis is then used to separate the resulting fragments by size. Together with the fluorescence in four dyes, this size-order allows to decipher the nucleotide sequence. Fig. 1.4 illustrates this principle.

Automated electrophoresis-based sequencing deals with a limited number of DNA fragments. In contrast, next-generation sequencing first fragments the DNA into a library of small segments. Then, millions of reactions are performed in a massively parallel fashion. Subsequently,

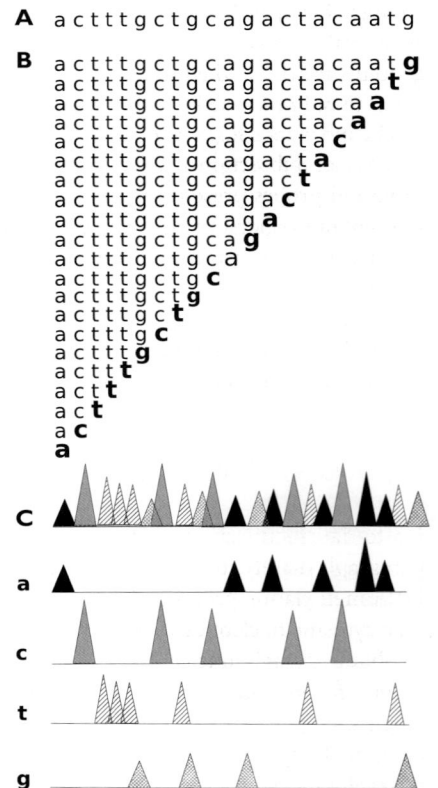

A actttgctgcagactacaatg

B
actttgctgcagactacaat**g**
actttgctgcagactacaa**t**
actttgctgcagactaca**a**
actttgctgcagactac**a**
actttgctgcagacta**c**
actttgctgcagact**a**
actttgctgcagac**t**
actttgctgcaga**c**
actttgctgcag**a**
actttgctgca**g**
actttgctgc**a**
actttgctg**c**
actttgct**g**
actttgc**t**
actttg**c**
acttt**g**
actt**t**
act**t**
ac**t**
a**c**
a

C

a

c

t

g

Fig 1.4 Principle of DNA sequencing. A) The sequence whose nucleotides are unknown. B) The fragments separated by electrophoresis. C) Image compiled from the fluorescence in four dyes and details for each dye.

the newly identified sequences of nucleotides—the reads—are reassembled. If a reference genome is available, it is used as a scaffold to align the reads; otherwise, *de novo* sequencing is performed.

The cost to sequence the first human genome, in 2001, amounted to $100 million. In 2013, this cost has been brought down to around $10,000.

1.3.2 Copy Number Polymorphism

Copy Number Variations (CNVs) account for roughly 12% of the human genome and has been reported to underlie susceptibility to certain diseases [29]. Each variation may range from around 1000 nucleotides to less than 5 megabases.

For a didactical introduction to CNV detection, we mainly focus on fluorescence *in situ* hybridization (FISH) and (array) comparative genomic hybridization (CGH, aCGH). FISH is a molecular biology technique that uses specific probes labeled with a fluorescent marker, at the cell level. Hybridization is thus evidenced in cell slides, by microscopy or molecular imaging. A pitfall of the FISH method lies in that it uses probes and primers that only target specific regions in the genome. In contrast, CGH allows the analysis of CNV at the genome scale and does not resort to cell cultures. The principle is the following: the DNA from the cells of a studied tissue (e.g., tumor) is labeled with a green fluorochrome, whereas reference DNA (i.e., from cells in sound tissue) is labeled with a red fluorochrome. Then, both DNAs are mixed in equal proportions, to compete for hybridization to DNA from sound cells. After hybridization, the green to red fluorescence ratio is measured along the genome. This ratio points out gains or losses across the whole genome, for the studied tissue. In case of loss (gain) in a specific region of the studied tissue, the green to red ratio will be lower (greater) than one.

The more specific form of array CGH was developed, which conjugates DNA microarray techniques (see Subsection 1.3.4) and CGH techniques. Therein, both DNAs are hybridized to several thousand probes present on the array. These probes encompass most of the known genes and non-coding regions of the genome. This conjunction of techniques allows to decrease the resolution of CNV detection down to 100 kilobases. The principle of aCGH is outlined in Fig. 1.5 (page 20).

The aCGH technology is presently the reference method to evidence CNVs. Again, next-generation sequencing offers an alternative: the principle is to compare the number of reads in contiguous windows, for the studied and reference tissues [15].

1.3.3 DNA Methylation Measurements

Up-to-date methylation analysis makes use of microarrays or even sequencing technologies. This paragraph mainly introduces genome bisulfite sequencing, the gold standard for genome-wide DNA methylation profiling. In the DNA sequence, depending on the methylation state of individual cytosine nucleotides, a treatment with sodium bisulfite will entail specific changes. Thus can be obtained single-nucleotide resolution information about the methylation status of a DNA fragment. The previous principle is the basis of bisulfite sequencing, complemented by DNA sequencing techniques, used further to read altered DNA sequences. In contrast to methylome sequencing, methyltyping techniques have been proposed, which emphasize low cost to the detriment of resolution. For instance, a restriction enzyme may be used to fragment the DNA at specific

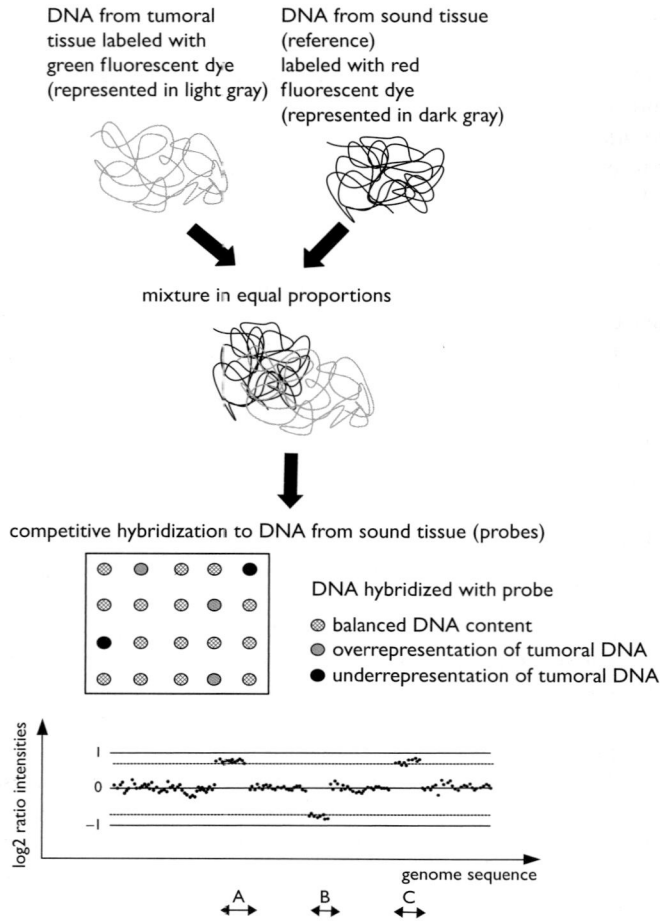

Fig 1.5 Principle of array comparative genomic hybridization. The probes on the DNA microarray (one probe per spot) encompass the whole genome. A and C are regions with a loss, whereas B is a region with a gain.

unmethylated cytosine sites. Due to experimental biases, restitution of the whole methylome is a challenge.

1.3.4 Gene Expression Data

To analyze in parallel the expression of thousands of genes of known and unknown functions, technologies based on DNA microarrays are used. Typically, a microarray experiment aims at the comparison of a sample representing the expression pattern of genes in a specific set of conditions, with a control. In this context, for example, the term "sample" refers to subjects affected by the same disease, or treated with the same drug. The generic designation of "DNA microarray" (or DNA chip, or biochip) covers a wide range of variants. Their common characteristic lies in the orderly arrangement of thousands of known sequences attached to a solid support (e.g., glass, silicon, nylon). Each array spot contains a small amount (picomoles)[1] of the same DNA sequence, such as a fragment of genomic DNA, a PCR product, or a chemically synthesized oligonucleotide. These identical single-stranded sequences are called targets. The robotics used to manufacture arrays either spots prefabricated DNA targets onto the slide or synthesizes oligonucleotides *in situ*.

A typical microarray experiment with two-channel detection starts with the isolation of DNA (or mRNA) from both the sample of interest (e.g., affected cells) and the control (e.g., sound cells). In case of mRNA extraction, this mRNA needs to be converted into complementary DNA (cDNA) using a reverse-transcriptase enzyme. Then, sample and control DNAs are separately amplified (via PCR, see Section 1.1, page 4) and labeled with different fluorescent dyes. Afterward, the mixed labeled cDNAs are subject to competitive hybridization to the DNA targets. Each molecule in the labeled mix will only bind to an appropriate target. After hybridization, unbound cDNAs are washed out of the array. After the array is dried, a laser scanner determines how much labeled DNA is bound to each target spot. This provides a ratio-based indicator to identify up-regulated and down-regulated genes. Usually, the subsequent image acquisition assigns green spots to sample genes up-regulated compared to control; down-regulated sample genes with respect to control are represented as red spots; yellow points out genes of equal abundance. This protocol is illustrated in Fig. 1.6 (page 22).

In contrast, single-channel microarrays provide estimations of absolute levels of gene expression. A consequence is that the comparison of two sets of conditions requires twice as many arrays as for two-channel detection: two separate single-dye hybridizations are performed. An advantage over two-channel detection is that data are more easily compared between different experiments; similarly, the data may be compared between studies conducted months or years apart. Two popular single-channel systems are the Affymetrix "Gene Chip" and the Illumina "Bead Chip". Affymetrix chips contain up to 500 000 targets in a 1.28 cm^2 area. Illumina's bead array technology relies on 3-micron silica beads that self-assemble in the microwells of a substrate (either a fiber-optic bundle or a planar silica slide). Hundreds of thousands of copies of a specific target are attached to each bead. To increase analysis accuracy, any such type of bead is represented many times. The Illumina's latest whole-genome expression array (HumanHT-12 v4 Expression BeadChip) covers more than 47 000 transcripts and known splice variants across the human transcriptome.

Next-generation sequencing has propelled a new alternative to microarrays, which is based on RNA sequencing (RNA-seq) [35]. To quantify gene expression, information about a sample's mRNA content is obtained by first synthesizing cDNA from the mRNA, and then sequencing the cDNA. Subsequently, the short reads of cDNA are aligned against the reference genome. For each gene, the aligned reads are counted. Notably, RNA-seq is able to identify transcripts that have not been previously annotated.

1.3.5 Quantitative Trait Loci

Quantitative trait loci (QTLs) are defined as chromosal regions that are tightly associated with a quantitative character—or trait—(e.g., height, weight, skin color). In particular, such a region encompasses several genes that together influence the phenotypic trait. The number of genes may be

---

[1] 10^{-12}

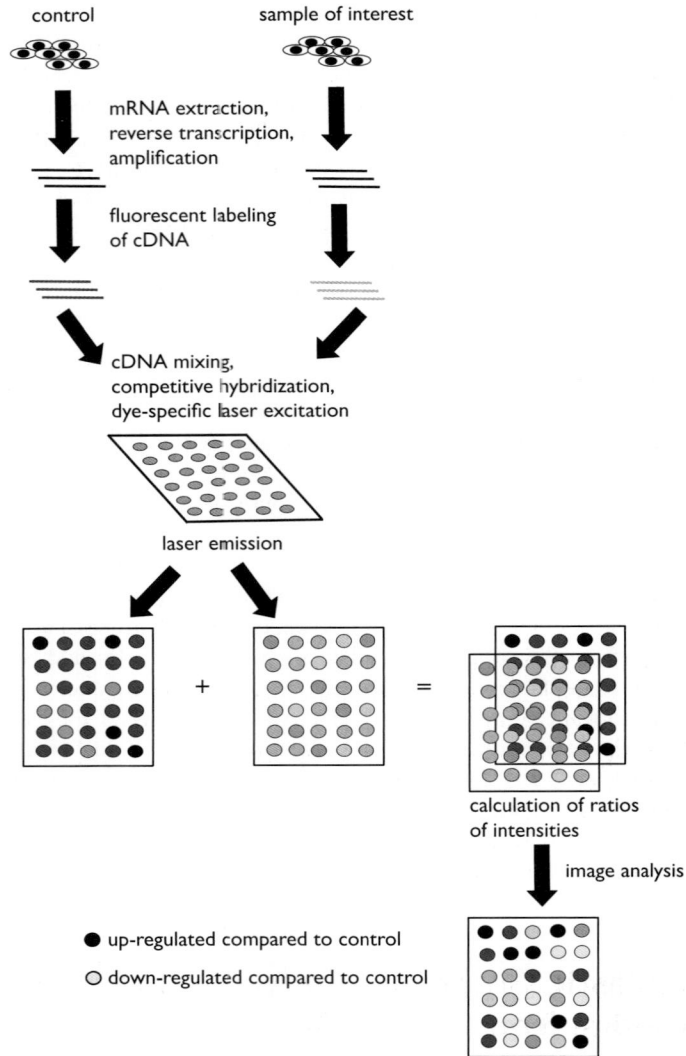

Fig 1.6 Principle of microarrays.

large, with each having a small effect on the phenotype. Together with the environmental influence, the global phenotype results from the many possible allelic combinations. This entails that the distribution for the continuous phenotype shows a bell-shaped curve in the simplest cases, and a superposition of Gaussian curves in the most complex cases. Dissecting the genetic architecture of complex traits is of crucial interest for marker-assisted selection in plant and animal breeding. In these latter domains, experimental line-crosses allow to map the QTLs. All QTL mapping strategies rely on the statistical detection of a dependence between the genotype and the phenotype in large populations of inbred lines (e.g., F2 generation produced by intercrossing inbred strains). Two parents or strains that differ for the trait are first selected. Then, markers that exhibit polymorphism with respect to the two parents are identified. The phenotype for each of the recombinant individuals or lines is assessed, as is the genotype of the markers that vary between the parental strains. One widely used statistical technique to detect QTLs is interval mapping. For each combination of flanking markers, interval mapping estimates the probability that the interval between the two markers is associated with a QTL affecting the trait. With H_0 the hypothesis that no QTL exists in the interval, and H_1 the alternative assumption, a standard statistical test is the log-odds ratio (or LOD score), $LOD = \frac{\mathcal{L}_{H_1}}{\mathcal{L}_{H_0}}$, $\mathcal{L}_{H_i}$ being the likelihood associated with hypothesis H_i. Among likelihood ratios above a given significance threshold, the best estimate of QTL location is provided by the chromosal position with the highest statistics.

1.3.6 The Challenge of Handling Omics Data

When dealing with omics data, researchers must face several issues. First, the genome-wide dimension entails many hurdles: appropriate storage capacities are requested; as handling such data may entail a prohibitive computational burden, grid computing and parallelization may represent an unavoidable step; otherwise, innovative tractable algorithms must be designed to cope with high dimensionality. Second, there has been increasing quality improvement in the provision of data at high throughput; however, missing values, error measurements and noise require such various preprocessing steps as data imputation, curation and correction. Third, omics data are of high dimensionality but of relatively small sample size; thus, meta-analyses may be implemented to enlarge the sample set, and this may require data imputation and correction for biases due to different acquisition conditions; anyway, the small sample size prevents the use of standard statistical methods. Fourth, extracting discernable biological meaning from multiple omics data sources generally demands the design of advanced models and algorithms: the latter must cope with the heterogeneity of the data sources and the data complexity; in addition to the spatial complexity required to store the raw data, the intrinsic complexity of the processed data (e.g., an inferred network of dependences) is highly challenging.

1.4 Probabilistic Graphical Models to Infer Novel Knowledge from Omics Data

At the crossroads between machine learning, statistics, and computer science, probabilistic graphical models (PGMs) provide a formal framework to both represent dependences between variables and model uncertain knowledge about the quantitative dependences between these variables. The graph-based component encodes the dependences over the multidimensional space; typically, it offers a qualitative compact (or factorized) representation of these dependences. In Bayesian networks, one of the most popular categories of PGMs, the quantitative component is a collection of probability distributions that accounts for uncertainty.

PGMs are not new to biology and bioinformatics. However, in the field of research in biology and in medicine, the generation of massive and complex heterogeneous data by high-throughput omics technologies has aroused renewed interest to these models. Their flexibility, scalability, and ability are expected to help infer novel knowledge from heterogeneous sources of data.

In this book, fifteen chapters cover various illustrations of the design and use of PGMs dedicated to research in genetics, genomics, and postgenomics. The book is structured into six parts. Table 1.2 offers a recapitulation of the chapter contributions, in the context of this six-part organization.

We end this introductive chapter with a broad-brush evocation of the use of probabilistic graphical networks in the six applications highlighted by the book.

1.4.1 Gene Network Inference

Inferring—or "reverse-engineering"—gene networks aims at identifying gene interactions from gene expression data through computational analysis [4, 9, 25]. A typical illustration of the inference of dependences within data, gene regulatory network inference has quite naturally resulted in investigations based on Bayesian networks (e.g., [10, 37]). Besides, when it comes to infer the varying structure of a gene network from time series, dynamic Bayesian networks are appropriate tools (e.g., [23, 31]).

1.4.2 Causality Discovery

The difference between a causal Bayesian network and the usual Bayesian network lies in the interpretation of the directed edges. In the usual Bayesian networks, the links between the variables can be explained as correlation or association. In a causal Bayesian network, the links encode the causal influence of parent variables on the values of their child variables. Inferring such causal influence is challenging; observational data alone rarely enable to gain knowledge on causality; the key to causality discovery is to resort to interventional data (in addition to observational data). Adding such interventional data allows to distinguish between directed graphs that otherwise would be equivalent. Equivalence here means that any such graph can account for the observational data. Pioneering works, conducted in the new domain of genetical genomics, led the way to the use of microarray data combined with additional information, to facilitate the identification of the underlying mechanisms of complex traits [17]. In this case, additional information is provided by exploiting the naturally occurring DNA variation observed in segregating populations,[2] meanwhile treating expression profiles as phenotypes governed by eQTLs. In one such technique, promoted by Shadt and coworkers [30, 32], two categories of eQTLs are considered: if the positions of a given gene and of one of its eQTLs overlap, the eQTL is considered to be cis-acting; otherwise the eQTL is considered to be trans-acting. Cis-eQTLs are usually believed to result from a variant in a regulatory region of the gene that influences its mRNA abundance. The identification of cis-acting eQTLs that colocalize with the QTLs of the complex trait of interest is informative, especially in case where the expression and disease traits are correlated. In summary, cis- and trans-acting eQTL data as well as complex-trait QTL data are used to direct edges in the Bayesian network modeling the causal trait network.

The aforementioned work direction has to be distinguished from other integrative approaches. Causality discovery relies on interventional data. Only in plant and animal breeding (e.g., laboratory mice) can line-crosses bring knowledge similar to that gained from interventional data.

---

[2] In plant or animal biology, a segregating population is obtained through a cross. The aim is to study the phenotypic variations in connection with detectable genetic differences.

Table 1.2 Organization of the book.

Part I Gene Network Inference

Chapter 3	Graphical Models and Multivariate Analysis of Microarray Data H. Kiiveri
Chapter 4	Comparison of Mixture Bayesian and Mixture Regression Approaches to Infer Gene Networks S.L. Rodriguez-Zas and B.R. Southey
Chapter 5	Network Inference in Breast Cancer with Gaussian Graphical Models and Extensions M. Jeanmougin, C. Charbonnier, M. Guedj, and J. Chiquet

Part II Causality Discovery

Chapter 6	Utilizing Genotypic Information as a Prior for Learning Gene Networks K. Chipman and A. Singh
Chapter 7	Bayesian Causal Phenotype Network Incorporating Genetic Variation and Biological Knowledge J. Young Moon, E. Chaibub Neto, X. Deng, and B.S. Yandell
Chapter 8	Structural Equation Models for Studying Causal Phenotype Networks in Quantitative Genetics G.J.M. Rosa and B.D. Valente

Part III Association Genetics

Chapter 9	Modeling Linkage Disequilibrium and Performing Association Studies through Probabilistic Graphical Models: A Visiting Tour of Recent Advances C. Sinoquet and R. Mourad
Chapter 10	Modeling Linkage Disequilibrium with Decomposable Graphical Models H. Abel and A. Thomas
Chapter 11	Scoring, Searching, and Evaluating Bayesian Network Models of Gene-phenotype Association X. Jiang, S. Visweswaran, and R.E. Neapolitan
Chapter 12	Graphical Modeling of Biological Pathways in Genome-wide Association Studies M. Chen, J. Cho, and H. Zhao
Chapter 13	Bayesian, Systems-based, Multilevel Analysis of Associations for Complex Phenotypes: From Interpretation to Decision P. Antal, A. Millinghoffer, G. Hullám, G. Hajós, P. Sárközy, A. Gézsi, C. Szalai, and A. Falus

Part IV Epigenetics

Chapter 14	Bayesian Networks in the Study of Genome-wide DNA Methylation M. Singer and L. Pachter

Continued

1.4.3 Association Genetics

In association genetics, intensive investigations around probabilistic graphical models (PGMs) have led to the design of scalable tools. It is now feasible to faithfully model the dependences within genetic data—the linkage disequilibrium—at the genomic scale [1, 26]. Research is intensive in the field of GWASs. As of June 2013, the catalog of published GWASs includes 1640 publications and 10 876 SNPs [16]. PGM-based GWAS strategies have been proposed [5, 36], including approaches dedicated to the association assessment of multi-loci patterns (epistasis) [14, 18]. PGMs are also attractive in that data dimension reduction can be achieved through latent variables [26]. Notably, it has been shown that subsumption through latent variables does not hinder the possibility of detecting associations with a studied disease [34].

1.4.4 Epigenetics

The development of methods explicitly based on PGMs is relatively new in epigenetics. Mapping methylation patterns at the genomic scale [33], detecting specific properties from methylation patterns [12] and identifying DNA methylation subgroups in cancer data [21] are three illustrations of the recent use of PGMs in epigenetics.

1.4.5 Detection of Copy Number Variations

Besides standard approaches [39], genome-wide copy number variation detection has been confined to a specific class of PGMs, that of Hidden Markov Models (HMMs) (i.e., [2, 38]). Only recently, investigations based on conditional random fields have been shown to improve over HMM-based approaches [40].

1.4.6 Prediction of Outcomes from High-dimensional Genomic Data

In the machine learning domain, Bayesian networks have been shown to be valuable tools for making predictions [24]. This competence arises from their flexible structure, which enables them to extract relevant and robust relationships among target variables and related explanatory variables. Not surprisingly, Bayesian networks are often used to make predictions in the health domain [11, 22]. In particular, to predict the future outcome of disease and disease treatment, Bayesian networks have often been used as prognostic models [3, 27]. Notably, Bayesian networks allow predictions in the high-dimensional settings encountered in genomics [13].

· ·

FURTHER READING

[1] H.J. Abel and A. Thomas. Accuracy and computational efficiency of a graphical modeling approach to linkage disequilibrium estimation. *Statistical Applications in Genetics and Molecular Biology*, 10(5), 2011.

[2] K.C. Amarasinghe, J. Li, and S.K. Halgamuge. CoNVEX: copy number variation estimation in exome sequencing data using HMM. *BMC Bioinformatics*, 14(Suppl 2):S2, 2013.

[3] R. Badovinac Ramoni, B.E. Himes, M.M. Sale, K.L. Furie, and M.F. Ramoni. Predictive genomics of cardioembolic stroke. *Stroke*, 40(3 Suppl):S67–70, 2009.

[4] M. Bansal, V. Belcastro, A. Ambesi-Impiombato, and D. di Bernardo. How to infer gene networks from expression profiles. *Molecular Systems Biology*, 3(78), 2007.

[5] B.L. Browning and S.R. Browning. Efficient multilocus association testing for whole genome association studies using localized haplotype clustering. *Genetic Epidemiology*, 31:365–375, 2007.

[6] The Gene Ontology Consortium. Gene ontology: tool for the unification of biology. *Nature Genetics*, 25(1):25–29, 2000.

[7] The International HapMap Consortium. The international HapMap Project. *Nature*, 426:789–796, 2003.

[8] M.J. Daly, J.D. Rioux, S.F. Schaffner, T.J. Hudson, and E.S. Lander. High-resolution haplotype structure in the human genome. *Nature Genetics*, 29(2):229–232, 2001.

[9] R. De Smet and K. Marchal. Advantages and limitations of current network inference methods. *Nature Reviews Microbiology*, 8:717–729, 2010.

[10] A.P. Dempster, N.M. Laird, and D.B. Rubin. Reverse engineering of genetic networks with Bayesian networks. *Biochemical Society Transactions*, 31(6):1516–1518, 2003.

[11] P. Felgaer, P. Britos, and R. Garcia-Martinez. Prediction in health domain using Bayesian network optimization based on induction learning techniques. *International Journal of Modern Physics C*, 17:447–455, 2006.

[12] A.Q. Fu, D.P. Genereux, R. Stöger, A.F. Burden, C.D. Laird, and M. Stephens. Statistical inference of in vivo properties of human DNA methyltransferases from double-stranded methylation patterns. *PLOS ONE*, 7(3):e32225, 2012.

[13] O. Gevaert, F. De Smet, D. Timmerman, Y. Moreau, and B De Moor. Predicting the prognosis of breast cancer by integrating clinical and microarray data with Bayesian networks. *Bioinformatics*, 22(14):e184–e190, 2006.

[14] B. Han, M. Park, and X.-W. Chen. A Markov blanket-based method for detecting causal SNPs in GWAS. *BMC Bioinformatics*, 11(Suppl 3):S5, 2010.

[15] J.L. Hayes, A. Tzika, H. Thygesen, S. Berri, H.M. Wood, S. Hewitt, M. Pendlebury, A. Coates, L. Willoughby, C.M. Watson, P. Rabbitts, P. Roberts, and G.R. Taylor. Diagnosis of copy number variation by Illumina next generation sequencing is comparable in performance to oligonucleotide array comparative genomic hybridisation. *Genomics*, 102:174–181, 2013.

[16] MacArthur J. Morales J. Junkins H.A. Hindorff, L.A., P.N. Hall, A.K. Klemm, and T.A. Manolio. A catalog of published genome-wide association studies. http://www.genome.gov/gwastudies.

[17] R.C. Jansen and J.-P. Nap. Genetical genomics: the added value from segregation. *Trends in Genetics*, 17(7):388–391, 2001.

[18] X. Jiang, M.M. Barmada, G.F. Cooper, and M.J. Becich. A Bayesian method for evaluating and discovering disease loci associations. *PLOS ONE*, 6(8):e22075, 2011.

[19] A.R. Joyce and B.Ø. Palsson. The model organism as a system: integrating "omics" data sets. *Nature Reviews Molecular Cell Biology*, 7:198–210, 2006.

[20] M. Kanehisa and S. Goto. Kegg: Kyoto encyclopedia of genes and genomes. *Nucleic Acids Research*, 28:27–30, 2000.

[21] D.C. Koestler, C.J. Marsit, B.C. Christensen, M.R. Karagas, R. Bueno, D.J. Sugarbaker, K.T. Kelsey, and E.A. Houseman. Semi-supervised recursively partitioned mixture models for identifying cancer subtypes. *Bioinformatics*, 26(20):2578–2585, 2010.

[22] Q.A. Le and J.N. Doctor. Probabilistic mapping of descriptive health status responses onto health state utilities using Bayesian networks: an empirical analysis converting SF-12 into EQ-5D utility index in a national US sample. *Medical Care*, 49(5):451–60, 2011.

[23] S. Lèbre. Inferring dynamic Bayesian network with low order independencies. *Statistical Applications in Genetics and Molecular Biology*, 8(1), 2009. Article 9.

[24] K.C. Lee, H. Cho, and S. Lee. U-BASE: general Bayesian network-driven context prediction for decision support. *Advances in Information Technology, Communications in Computer and Information Science*, 114:63–72, 2010.

[25] D. Marbach, J.C. Costello, R. Küffner, N.M. Vega, R.J. Prill, D.M. Camacho, K.R. Allison, The DREAM5 Consortium, M. Kellis, J.J. Collins, and G. Stolovitzky. Wisdom of crowds for robust gene network inference. *Nature Methods*, 9:796–804, 2012.

[26] R. Mourad, C. Sinoquet, and P. Leray. A hierarchical Bayesian network approach for linkage disequilibrium modeling and data-dimensionality reduction prior to genome-wide association studies. *BMC Bioinformatics*, 12(1):16, 2011.

[27] N. Peek, M. Verduijn, P.M. Rosseel, E. de Jonge, and B.A. de Mol. Bayesian networks for multivariate data analysis and prognostic modelling in cardiac surgery. *Studies in Health Technology and Informatics*, 129(1):596–600, 2007.

[28] B. Pulverer. Sequence-specific DNA-binding transcription factors. *Nature Milestones*, 2005. doi:10.1038/nrm1800.

[29] R. Redon et al. Global variation in copy number in the human genome. *Nature*, 444:444–454, 2006.

[30] E.E. Schadt, J. Lamb, X. Yang, J. Zhu, S. Edwards, D. Guhathakurta, S.K. Sieberts, S. Monks, M. Reitman, C. Zhang, P.Y. Lum, A. Leonardson, R. Thieringer, J.M. Metzger, L. Yang, J. Castle, H. Zhu, S.F. Kash, T.A. Drake, A. Sachs, and A.J. Lusis. An integrative genomics approach to infer causal associations between gene expression and disease. *Nature Genetics*, 37(7):710–717, 2005.

[31] A. Shermin and M.A. Orgun. Using dynamic Bayesian networks to infer gene regulatory networks from expression profiles. In *Proceedings of the 24th Annual ACM Symposium on Applied Computing*, pages 799–803. ACM, 2009.

[32] S.K. Sieberts and E.E. Schadt. Inferring causal associations between genes and disease via the mapping of expression quantitative loci In D. J. Balding, M. Bishop, and C. Cannings, editors, *Handbook of Statistical Genetics*, volume 2, pages 296–326. Wiley Interscience, 2007.

[33] M. Singer, D. Boffelli, J. Dhahbi, A. Schoenhuth, G.P. Schroth, D.I.K. Martin, and L. Pachter. Met-Map enables genome-scale methyltyping for determining methylation states in populations. *PLOS Computational Biology*, 6:e1000888, 2010.

[34] C. Sinoquet, R. Mourad, and P. Leray. Forests of latent tree models to decipher genotype-phenotype associations In J. Gabriel, J. Schier, S. Van Huffel, E. Conchon, C. Correia, A. Fred, and H. Gamboa, editors, *Biomedical Engineering Systems and Technologies*, pages 113–134. Springer, 2013.

[35] A. Sîrbu, G. Kerr, M. Crane, and H.J. Ruskin. RNA-Seq vs dual- and single-channel microarray data: sensitivity analysis for differential expression and clustering. *PLOS ONE*, 7(12):e50986, 2012.

[36] C.J. Verzilli, N. Stallard, and J.C. Whittaker. Bayesian graphical models for genomewide association studies. *The American Journal of Human Genetics*, 79(1):100–112, 2006.

[37] M. Vignes, J. Vandel, D. Allouche, N. Ramadan-Alban, C. Cierco-Ayrolles, T. Schiex, B. Mangin, and S. de Givry. Gene regulatory network reconstruction using Bayesian networks, the Dantzig selector, the Lasso and their meta-analysis. *PLOS ONE*, 6(12):e29165, 2011.

[38] K. Wang, M. Li, D. Hadley, R. Liu, J. Glessner, S.F.A. Grant, H. Hakonarson, and M. Bucan. PennCNV: an integrated hidden Markov model designed for high-resolution copy number variation detection in whole-genome SNP genotyping data. *Genome Research*, 17(11):1665–1674, 2007.

[39] R. Xi, S. Lee, and P.J. Park. A survey of copy-number variation detection tools based on high-throughput sequencing data. *Current Protocols in Human Genetics*, Chapter 7:Unit 7.19, 2012.

[40] Yin X.L. and J. Li. Detecting copy number variations from array CGH data based on a conditional random field model. *Journal of Bioinformatics and Computational Biology*, 8(2):295–314, 2010.

CHAPTER 2

Essentials to Understand Probabilistic Graphical Models: A Tutorial about Inference and Learning

CHRISTINE SINOQUET

The aim of this chapter is to offer an advanced tutorial to scientists with no background or no deep background on probabilistic graphical models. For readers more familiar with these models, this chapter is to be used as a compendium of definitions and general methods, to browse through at will.

Intentionally self-contained, this chapter first begins with reminders of essential definitions. In particular, the distinction between marginal independence and conditional independence, one of the key notions exploited by probabilistic graphical models, is explained in detail.

Then, Section 2.3 briefly surveys the most popular classes of probabilistic graphical models. The references to Markov chains and hidden Markov models gradually prepare the ground for the further presentation of more general models. A unified presentation then connects Markov random fields to Gibbs fields. Two popular variants are illustrated: the hidden Markov random fields and the conditional random fields. The class of Bayesian networks is then presented, with an in-depth illustration of d-separation. Finally, factor graphs and chain graphs are described, as, respectively, unifying models and extended models for both Markov fields and Bayesian networks.

Section 2.4 is devoted to probabilistic inference in Bayesian networks. As an introduction, three main canonical inference queries are mentioned: probability of evidence, most probable explanation, and maximum a posteriori hypothesis. Then this section sketches the principles and techniques underlying a wide range of methods used for inference. Exact inference is illustrated

Probabilistic Graphical Models for Genetics, Genomics, and Postgenomics. First Edition. Christine Sinoquet & Raphaël Mourad (Eds). © Oxford University Press 2014. Published in 2014 by Oxford University Press.

with variable elimination, message-passing algorithms including the sum-product algorithm (or belief propagation), conditioning, and message propagation in junction trees. In particular, this latter topic gives rise to developments around graph moralization, graph triangulation, and junction-tree construction. Approximate inference is depicted through loopy and generalized belief propagation, stochastic sampling, and variational methods. Three standard sampling methods adapted to approximate inference are outlined—Gibbs sampling, the Metropolis-Hastings algorithm, and importance sampling. An emphasis is put on the mechanisms underlying importance sampling and the variational approach.

Section 2.5 addresses parameter and structure learning for the class of Bayesian networks. First, parameter learning from complete data is reviewed through three standard approaches: a statistical method (likelihood maximization) and two Bayesian methods (maximum a posteriori and expectation a posteriori estimations). In the case of incomplete data, a generic framework involving standard expectation-maximization (EM) is depicted. The methods coping with structure learning fall into two categories. The constrained-based approaches rely on dependence analysis. In this category, the IC and PC algorithms are briefly sketched and connected concepts are surveyed (independence maps, dependence maps). In the second category, the corner stone of heuristic strategies is the use of decomposable and Markov equivalent scores. Such scores are described, which either resort to the information theory domain or the Bayesian paradigm. In addition to heuristics searching the space of directed graphs, this section also mentions methods that navigate alternative search spaces; such spaces are the space of orderings of variables (in connection with the standard K2 algorithm) and the space of completed partially directed acyclic graphs navigated by the greedy equivalent search method. To extend the scope of score-based algorithms to incomplete data, structural expectation-maximization provides a generic framework, interpreted as hill climbing through the structure space embedding EM on parameters at each move. Finally, the specific case of latent variables is discussed.

Section 2.6 presents parameter and structure learning for the class of Markov random fields. In the case of complete data and general undirected graphs, both tasks are still more challenging than for Bayesian networks. Iterative approaches such as gradient descent and second-order methods are solutions to tackle the maximization of the likelihood. However, the evaluation of the derivatives requires inference in the Markov network, again a difficult task except for triangulated graphs. On the other hand, the likelihood intractability is efficiently circumvented by using the pseudo-likelihood. In contrast, triangulated graphs are tractable in that they exhibit a specific factorization for the joint probability. In the case of general graphs, structure learning may be handled through using penalized pseudo-likelihood. Beyond score-based heuristics, convex approximations to both a goodness-of-fit metric and a complexity metric ensure the obtention of a global optimum. An L1-regularization technique is illustrated in this context. Finally, the case of triangulated graphs represents a class for which properties coming from the graph theory state how to build the neighbor graphs of the incumbent graph, during a local search process.

In Section 2.7, the focus is set on causal networks, their difference with Bayesian networks, and the strategies used to learn their structure. In particular, two modalities for active learning are presented—batch intervention and sequential intervention.

Finally, a list of references directs the reader to several notable books written about probabilistic graphical models as well as to chapters of books more specifically focused on either inference or learning for such models.

2.1 Introduction

Probabilistic graphical models (PGMs) provide a qualitative framework for representing and reasoning with uncertainty and independences. The qualitative part of a PGM is a graphical representation of dependences between variables, expressed, for example, by a directed acyclic graph (DAG), an undirected graph (UG) or a chain graph, that is, a graph which may have both directed and undirected edges, but without any directed cycles. Uncertain knowledge about the qualitative dependences between the variables is formalized with the aid of probability distributions called the parameters. The parameters represent the quantitative component of the model. In this introductive chapter, we first provide a short presentation of the most popular kinds of PGMs, embracing the Markov random fields (MRFs) and some of their variants, the Bayesian networks (BNs), a unifying model consisting of the class of factor graphs, and the class of chain graphs which represents an extension for both MRFs and BNs. Then we show various mechanisms used to achieve probabilistic inference, mostly in Bayesian networks. Subsequently, we produce an overview of the prominent strategies developed to learn PGMs. In most applications, and especially in the case of high-dimensional data, it is rare that an expert can specify the graphical structure of the graphical model. Therefore, not only is it necessary to learn the model's parameters, learning both the graphical structure and the parameters is required, which represents a daunting challenge in terms of faithfulness to the data as well as tractability. In the course of this tutorial on model learning, we also mention cases when probabilistic graphical models rely on latent - or hidden - variables. Finally, we devote a brief section to the difference between BNs and causal networks. We sketch the principles used to learn causal structure.

The literature on graphical models is vast. The impetus for adding this present contribution arose from the acknowledged need for documentation intermediate between classical survey and extensive compilation. The elliptic style of the former imparts to the reader deeper knowledge. Detailed compilations do not offer a synthetic view. We therefore provide this tutorial.

The references mentioned in this article necessarily represent an arbitrary selection, very much biased by the author's personal perspective and interests.

In the following, unless explicitly stated otherwise, we will constrain the framework to discrete random variables. Capitalized letters will indicate random variables. Non-capitalized letters will be used to denote observations. In addition, to simplify the notation, we will sometimes abbreviate $X = T$ (for true or 1) and $X = F$ (for false or 0) into X and $\bar{X}$, respectively. Moreover, for conciseness, we will occasionally abbreviate $\mathbb{P}(A = a, B = b, C = c)$ into $\mathbb{P}(a, b, c)$.

2.2 Reminders

In this section, we recall essential notions such as joint distribution, marginal distribution, and conditional distribution, together with corner stone elements of the Bayesian framework.

Definition 2.1 (Joint distribution, chain rule (or product rule))
The joint (bivariate) distribution for two binary random variables A and B defines the probabilities for the events in $\{A, \bar{A}\} \times \{B, \bar{B}\}$. This concept generalizes to any number of random variables $X_1, \ldots, X_n$, defining a multivariate distribution.
The chain rule provides the joint distribution:
$\forall x_1, x_2, \ldots, x_{n-1}, x_n \in D_{X_1} \times D_{X_2} \times \cdots \times D_{X_{n-1}} \times D_{X_n}$, where D_{X_i} represents the domain within which X_i takes its values,

$$\mathbb{P}(X_1 = x_1, \ldots, X_n = x_n)$$
$$= \mathbb{P}(X_1 = x_1) \times \mathbb{P}(X_2 = x_2 \mid X_1 = x_1) \times \mathbb{P}(X_3 = x_3 \mid X_1 = x_1, X_2 = x_2) \times \cdots$$
$$\times P(X_n = x_n \mid X_1 = x_1, X_2 = x_2, \ldots, X_{n-1} = x_{n-1}).$$

By abuse of language, we merely write:

$$\mathbb{P}(X_1, \ldots, X_n)$$
$$= \mathbb{P}(X_1) \times \mathbb{P}(X_2 \mid X_1) \times \mathbb{P}(X_3 \mid X_1, X_2) \times \cdots$$
$$\times P(X_n \mid X_1, X_2, \ldots, X_{n-1}).$$

Definition 2.2 (Marginal distribution)
Given $X = \{X_1, \ldots, X_n\}$, some $I \subset \{1, \ldots, n\}$, and X_I the set of random variables obtained when mapping I to X, the joint distribution of X_I is then the marginal distribution of X_I, obtained through summing ("marginalizing") probabilities over the domain of the other variables in $X \setminus X_I$. The dismissed variables are said to be marginalized out.
The marginalization process writes:

$$\mathbb{P}(X_I = x_i)$$
$$= \mathbb{P}(X_{i_1} = x_{i_1}, \ldots, X_{i_r} = x_{i_r})$$
$$= \sum_{x_{i_{r+1}}, x_{i_{r+2}}, \ldots, x_{i_n}} \mathbb{P}(X_{i_1} = x_{i_1}, \ldots, X_{i_r} = x_{i_r}, X_{i_{r+1}} = x_{i_{r+1}}, \ldots, X_{i_n} = x_{i_n}).$$

Inversely, a conditional probability distribution can be calculated by dividing the joint distribution by the marginal distribution of one (or more) of the variables.

Definition 2.3 (Conditional distribution in the bivariate case)
Provided that $\mathbb{P}(X_j) \neq 0$, $\mathbb{P}(X_i \mid X_j) = \frac{\mathbb{P}(X_i, X_j)}{\mathbb{P}(X_j)}$.

Property 2.1 (Relation between marginal and conditional distributions)

$$\mathbb{P}(X_I = x_i)$$
$$= \mathbb{P}(X_{i_1} = x_{i_1}, \ldots, X_{i_r} = x_{i_r})$$
$$= \sum_{x_{i_{r+1}}, x_{i_{r+2}}, \ldots, x_{i_n}} \mathbb{P}(X_{i_1} = x_{i_1}, \ldots, X_{i_r} = x_{i_r} \mid X_{i_{r+1}} = x_{i_{r+1}}, \ldots, X_{i_n} = x_{i_n})$$
$$\times \mathbb{P}(X_{i_{r+1}} = x_{i_{r+1}}, \ldots, X_{i_n} = x_{i_n}).$$

The example shown in Fig. 2.1 and Table 2.1 illustrates these notions and their relations in the trivariate case.

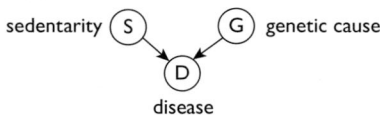

Fig 2.1 The graphical representation of dependences for three boolean random variables whose distributions are described in Table 2.1.

Table 2.1 Joint, marginal, and conditional distributions.

		SG	$S\bar{G}$	$\bar{S}G$	$\bar{S}\bar{G}$
D	T	0.03	0.70	0.03	0.01
	F	0.01	0.05	0.07	0.10

A) Joint distribution

		G	$\bar{G}$
D	T	0.06	0.71
	F	0.08	0.15

B) Marginalization over S

		S	$\bar{S}$
D	T	0.73	0.04
	F	0.06	0.17

C) Marginalization over G

D	T	0.77
	F	0.23

G	T	0.14
	F	0.86

S	T	0.79
	F	0.21

Marginalization
D) over S, G E) over D, S F) over D, G

		SG	$S\bar{G}$	$\bar{S}G$	$\bar{S}\bar{G}$
D	T	0.75	0.933	0.30	0.091
	F	0.25	0.067	0.70	0.909

SG	$S\bar{G}$	$\bar{S}G$	$\bar{S}\bar{G}$
0.04	0.75	0.10	0.11

G) Distribution of D conditional on S, G H) Marginal distribution of S, G

For instance, marginalizing out S and G provides:

$$\mathbb{P}(D = T) = \sum_{S,G} \mathbb{P}(D = T, \ S = . , \ G = .)$$

$$= \mathbb{P}(D = T, \ S = T, \ G = T) + \mathbb{P}(D = T, \ S = T, \ G = F)$$
$$+ \mathbb{P}(D = T, \ S = F, \ G = T) + \mathbb{P}(D = T, \ S = F, \ G = F)$$
$$= 0.03 + 0.70 + 0.03 + 0.01$$
$$= 0.77$$

which may also be computed using the four conditional distributions of D given SG, $S\bar{G}$, $\bar{S}G$ and $\bar{S}\bar{G}$, respectively weighted by prior (i.e., non-conditional) probabilities:

$$\mathbb{P}(D = T) = \sum_{S,G} \mathbb{P}(D = T \mid S = . , \ G = .) \, \mathbb{P}(S = . , \ G = .)$$

$$= \mathbb{P}(D = T \mid S = T, \ G = T) \quad \mathbb{P}(S = T, \ G = T)$$
$$+ \mathbb{P}(D = T \mid S = T, \ G = F) \quad \mathbb{P}(S = T, \ G = F)$$
$$+ \mathbb{P}(D = T \mid S = F, \ G = T) \quad \mathbb{P}(S = F, \ G = T)$$
$$+ \mathbb{P}(D = T \mid S = F, \ G = F) \quad \mathbb{P}(S = F, \ G = F)$$
$$= 0.75 \times 0.04 + 0.933 \times 0.75 + 0.30 \times 0.10 + 0.091 \times 0.11$$
$$= 0.77.$$

We now recall the Bayes theorem, which, in its simplest form, relates two conditional probabilities.

Definition 2.4 (Bayes theorem)
Given two random variables A and B, provided that $\mathbb{P}(B) \neq 0$,

$$\mathbb{P}(A \mid B) = \frac{\mathbb{P}(B \mid A)\,\mathbb{P}(A)}{\mathbb{P}(B)}.$$

The notion of conditional independence is fundamental for probabilistic graphical models. Let us first recall the concept of independence, also called statistical independence or marginal independence.

Definition 2.5 (Independence (or marginal independence))
Given two random variables A and B, independence between A and B ($A \perp\!\!\!\perp B$) is defined as: $\mathbb{P}(A, B) = \mathbb{P}(A)\,\mathbb{P}(B)$. The non-equality entails that both variables are dependent ($A \not\!\perp\!\!\!\perp B$).

The link between the above definition and the intuitive notion of independence can also be expressed through the use of conditional probability. The notion of independence is restated by any of the three equivalent conditions, provided that $\mathbb{P}(B)$ is different from 0 and 1 (which both correspond to trivial cases):

$$\mathbb{P}(A \mid B) = \mathbb{P}(A),$$
$$\mathbb{P}(A \mid \bar{B}) = \mathbb{P}(A),$$
$$\mathbb{P}(A \mid B) = \mathbb{P}(A \mid \bar{B}),$$

which means that A and B are independent if having knowledge about B occurring or B not occurring or having no knowledge about the occurrence of B does not impact the probability of occurrence of A. The interpretation holds when A and B are exchanged.

Together with Table 2.2, and using grid cells, Fig. 2.2 (page 36) illustrates that independence means that the "proportion of A in B" remains the same as the "proportion of A in the universe of possibilities"[1] (namely, $\frac{n_A}{48} = \frac{n_{A,B}}{n_B}$), thus highlighting the absence of influence of B on A. Yet another way to interpret independence is to note that the probability distribution of A is the same

Table 2.2 Marginal independence of two boolean random variables A and B $\mathbb{P}(A \mid B) = \mathbb{P}(A)$ (see Fig. 2.2).

n_A	3		4			6					8						
n_B	16	32	12	24	36	8	16	24	32	40	6	12	18	24	30	36	42
$n_{A,B}$	1	2	1	2	3	1	2	3	4	5	1	2	3	4	5	6	7
$\frac{n_A}{48}$	1/16		1/12			1/8					1/6						
	see Figs 2.2A and 2.2B										see Figs 2.2C and 2.2D for first and last subcases						

[1] In this case, the 48 grid cells.

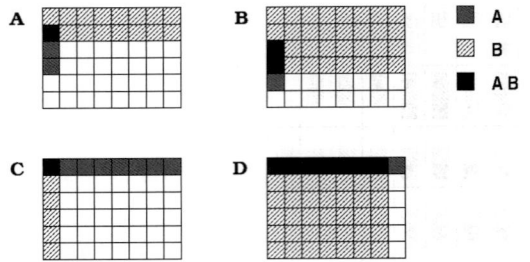

Fig 2.2 Graphical interpretation of marginal independence for two boolean random variables (see Table 2.2).

Table 2.3 Marginal independence of two boolean random variables A and B. The probability distribution of A is the same for all values of B, and symmetrically (see Fig 2.2A).

		B	
		T	F
A	T	1/16	2/32
	F	15/16	30/32

		A	
		T	F
B	T	1/3	15/45
	F	2/3	30/45

for all values of B and the probability distribution of B is the same for all values of A: conditioning on any value of one of the variables does not modify the probability distribution of the other variable. Table 2.3 highlights this last point.

Definition 2.6 (Conditional independence)
Given three variables A, B, and C, conditional independence between A and B given the state of C (A ⊥⊥ B | C) is defined as: $\mathbb{P}(A, B \mid C) = \mathbb{P}(A \mid C)\,\mathbb{P}(B \mid C)$. In the case when $\mathbb{P}(C) > 0$, this is equivalent to $\mathbb{P}(A \mid B, C) = \mathbb{P}(A \mid C)$.

It is important to note, that, for instance, the second proposition in the above definition does mean the following: for all a, b, b', and c with $\mathbb{P}(C = c) > 0$, $\mathbb{P}(A = a \mid B = b, C = c) = \mathbb{P}(A = a \mid B = b', C = c) = \mathbb{P}(A = a \mid C = c)$ (or symmetrically by exchanging A and B); that is, A and B are conditionally independent given C if and only if, given any value of C, the probability distribution of A remains the same for all values of B (which is equivalent to saying that given any value of C, the probability distribution of B remains the same for all values of A). For the marginal independence, it was sufficient to check that $\mathbb{P}(A = T, B = T) = \mathbb{P}(A = T)\,\mathbb{P}(B = T)$. Because the problem is highly constrained, this equality entails that the three other equalities are satisfied. Less formally, A and B are conditionally independent given C if and only if, given knowledge of whether C occurs, knowing whether B occurs provides no information on the probability of occurrence of A (which also means that, symmetrically, knowing whether A occurs provides no information on the probability of the occurrence of B). The non-satisfaction of

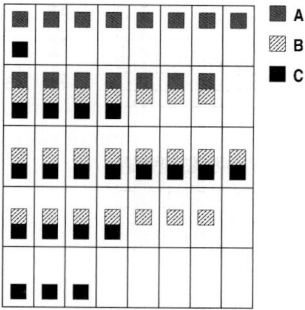

Fig 2.3 Graphical interpretation of conditional independence for two boolean random variables. Conditional independence of A and B given C. In this specific case, we observe that A and B are marginally dependent: $\mathbb{P}(A, B) = \frac{7}{40} \neq \mathbb{P}(A)\,\mathbb{P}(B) = \frac{15}{40} \times \frac{22}{40} = \frac{33}{4} \times \frac{1}{40}$.

Table 2.4 Constraints entailed by conditional independence of two boolean random variables A and B given C. Check for $\mathbb{P}(B = T \mid A = b,\ C = c) = \mathbb{P}(B = T \mid C = c)$. The variable n_X denotes the surface associated with event X. To simplify the notations, $X = T$ and $X = F$ are abbreviated as X and $\bar{X}$, respectively.

$\mathbb{P}(B = T \mid A = T, C = T) = \mathbb{P}(B = T \mid C = T)$		$\mathbb{P}(B = T \mid A = F, C = T) = \mathbb{P}(B = T \mid C = T)$	
$\frac{n_{B,A,C}}{n_{A,C}} = \frac{4}{5}$	$\frac{n_{B,C}}{n_C} = \frac{16}{20}$	$\frac{n_{B,\bar{A},C}}{n_{\bar{A},C}} = \frac{12}{15}$	$\frac{n_{B,C}}{n_C} = \frac{16}{20}$
$\mathbb{P}(B = T \mid A = T, C = F) = \mathbb{P}(B = T \mid C = F)$		$\mathbb{P}(B = T \mid A = F, C = F) = \mathbb{P}(B = T \mid C = F)$	
$\frac{n_{B,A,\bar{C}}}{n_{A,\bar{C}}} = \frac{3}{10}$	$\frac{n_{B,\bar{C}}}{n_{\bar{C}}} = \frac{6}{20}$	$\frac{n_{B,\bar{A},\bar{C}}}{n_{\bar{A},\bar{C}}} = \frac{3}{10}$	$\frac{n_{B,\bar{C}}}{n_{\bar{C}}} = \frac{6}{20}$

the required constraints entails that both variables are conditionally dependent given C, which is denoted $A \not\!\perp\!\!\!\perp B \mid C$.

A graphical illustration is provided in Fig. 2.3. It has been constructed establishing the system of constraints entailed by $\mathbb{P}(C = c) > 0$ for $c \in \{T, F\}$, $\mathbb{P}(B = b) > 0$, for $b \in \{T, F\}$ and for all $b, c \in \{T, F\}$, $\mathbb{P}(A = T \mid B = b, C = c) = \mathbb{P}(A = T \mid C = c)$, then setting the surface of the grid ($n = 40$) and fixing the minimum number of parameters (i.e., set sizes such as $n_{A,C}, n_C \ldots$) to enable the resolution of the constraint system. Table 2.4 checks that the symmetrical system of constraints is verified (for all $a, c \in \{T, F\}$, $\mathbb{P}(B = T \mid A = b, C = c) = \mathbb{P}(B = T \mid C = c)$).

The generalization of conditional independence to a set of random variables S is straight forward:

Definition 2.7 (Conditional independence given a set of variables)
Given a subset of variables $S \subseteq X \setminus \{X_i, X_j\}$, conditional independence between X_i and X_j knowing the state of S ($X_i \perp\!\!\!\perp X_j \mid S$) is defined as: $\mathbb{P}(X_i, X_j \mid S) = \mathbb{P}(X_i \mid S)\,\mathbb{P}(X_j \mid S)$. The non-equality entails that both variables are conditionally dependent knowing (the state of) S, which is denoted $X_i \not\!\perp\!\!\!\perp X_j \mid S$.

2.3 Various Classes of Probabilistic Graphical Models

In this section, we briefly review some popular instances of probabilistic graphical models, namely the undirected Markov network models or Markov random fields, the class of Bayesian networks, as well as a unifying class of models, the factor graphs, and extended models consisting in chain graphs. We recall that we restrain to the case of discrete variables.

2.3.1 Markov Chains and Hidden Markov Models

As Markov random fields are considered to be a generalization of Markov chains, let us first consider Markov chains and related models. Markov chains encode dependences within a temporal or physical sequence of variables X_1, X_2, ..., X_n. A popular example of physical sequences is that of DNA sequences. Using the terminology of the temporal framework, one can informally define Markov chains as models which express that predicting the future does not require extensive knowledge about the past. For instance, the so-called Markov chain of order 1 verifies the weak Markov property:

$$\forall t \geq 1, \mathbb{P}(X_{t+1} = a_2 \mid X_1 = v_1, X_2 = v_2, \ldots, X_t = a_1) = \mathbb{P}(X_{t+1} = a_2 \mid X_t = a_1),$$

meaning that all information required to predict the future state of the process, X_{t+1}, is contained in its present state, X_t.

Moreover, a *homogeneous* Markov chain of order 1, for which the transition probabilities are constant over time, verifies

$$\forall t \geq 1, \mathbb{P}(X_{t+1} = a_2 \mid X_1 = v_1, X_2 = v_2, \ldots, X_t = a_1)$$
$$= \mathbb{P}(X_{t+1} = a_2 \mid X_t = a_1)$$
$$= \mathbb{P}(X_2 = a_2 \mid X_1 = a_1).$$

The generalization to higher orders is straightforward. For an order k, the conditional dependence writes

$$\forall t \geq k, \mathbb{P}(X_{t+1} = a_{k+1} \mid X_1 = v_1, X_2 = v_2, \ldots, X_{t-k} = v_{t-k}, \tag{2.1}$$
$$X_{t-k+1} = a_k, \ldots, X_{t-1} = a_2, X_t = a_1)$$
$$= \mathbb{P}(X_{t-k+1} = a_k, \ldots, X_{t-1} = a_2, X_t = a_1).$$

An illustration of a Markov chain of order 2 is provided in Fig. 2.4A. A generalization of Markov chains leads to hidden Markov models (HMMs). Underlying an HMM are two processes: the non-observed (hidden) process models a series of states Y_1, ..., Y_n along the sequence; this hidden process is a Markov chain of order $k \geq 1$; the observed process generates X_1, ..., X_n, each X_i depending on Y_i. The former process is entirely described by an initial distribution, $\mathbb{P}(Y_1, \ldots, Y_k)$, and transition probabilities (in the line of equation (2.1)). The second process is depicted through emission probabilities $\mathbb{P}(X_i = a \mid Y_i = b)$. An illustration is provided in Fig. 2.4B. HMMs are in essence dynamical versions of *finite mixture models*, in which each observation is generated conditionally on an underlying latent (hidden) state variable. We recall that a finite mixture model is characterized by a probability distribution expressed as $\mathbb{P}(Y = y) = \sum_{k=1}^{K} \pi(k) \, \mathbb{P}_k(Y = y)$, where each $\mathbb{P}_k$ is

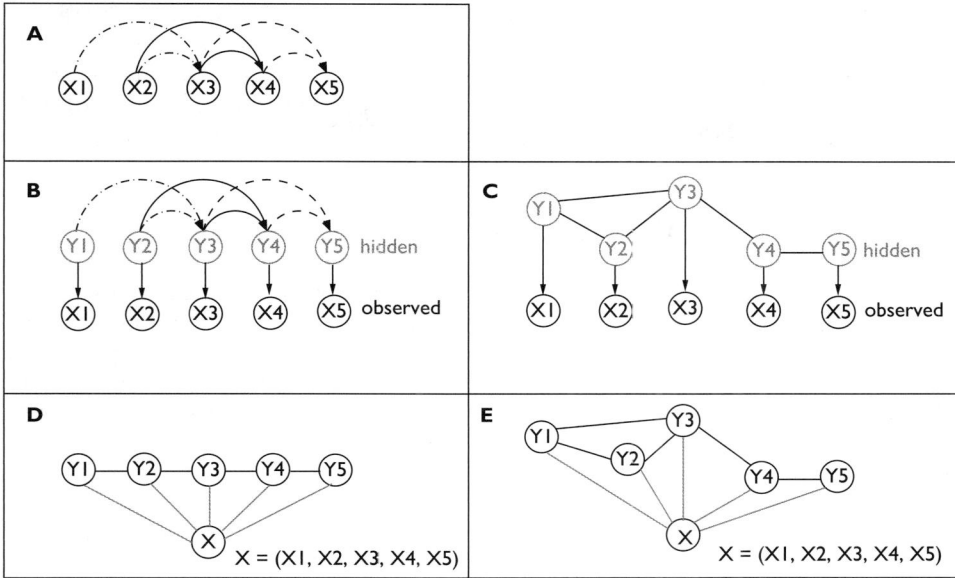

Fig 2.4 Some popular classes of probabilistic graphical models. A) Markov chain of order 2. B) Hidden Markov model of order 2. C) Hidden Markov random field. D) Conditional random field with a Markov assumption of order 1. E) General conditional random field. In cases D and E, each Y_i is globally conditioned by X.

a probability distribution and $\sum_{k=1}^{K} \pi(k) = 1$. HMMs have a widespread use in natural language processing tasks (speech recognition, part-of-speech tagging, and information extraction) and in bioinformatics.

2.3.2 Markov Random Fields

Markov random fields (MRFs) are probabilistic graphical models whose structure is an undirected graph (UG). It has to be noted that, in contrast to Bayesian networks, cyclic dependences are allowed in Markov random fields. On the other hand, Markov random fields cannot represent certain dependences that Bayesian networks can encode, such as induced dependences.

Notation 2.1
If h denotes the set of values each taken by the set of variables X and S denotes a subset of X, then $h_{|S}$ denotes the subset of values in h that correspond to the subset of variables S: $h_{|S}$ is the mapping of h on $S \subseteq X$.

Definition 2.8 (Markov random field (MRF))
A random field X is called a Markov random field with respect to a neighborhood system $\mathcal{N}$ if $\forall X_i, X_i \perp\!\!\!\perp X \setminus \mathcal{N}_{X_i} \cup \{X_i\} \mid \mathcal{N}_{X_i}$, where $\mathcal{N}_{X_i}$ denotes the neighborhood of X_i in the undirected graph.
*The above **local Markov property** states that a variable is conditionally independent of all other variables given its set of neighbors:*

$$\pi_{X_i}(x_i, b) = \mathbb{P}(X_i = x_i \mid X \setminus \{X_i\} = b) = \mathbb{P}(X_i = x_i \mid \mathcal{N}_{X_i} = b_{|\mathcal{N}_{X_i}}).$$

Therefore, in a Markov random field, a variable is shielded by the variables in its neighborhood. In the case of MRFs, the neighborhood coincides with the concept of Markov blanket.

Definition 2.9 (Markov blanket in a Markov random field)

In a Markov random field, the Markov blanket of a variable X_i is the set of its neighbors.

In an MRF, a variable is conditionally independent on all other variables, given its Markov blanket.

We now return to the difference between Bayesian networks and Markov random fields and explain why the latter models cannot account for induced dependences. For example, in a Bayesian network, whose graphical component is a *directed* (acyclic) graph by definition, an induced dependence exists if a child vertex is observed and has several parents (e.g., a converging connection). In this case, information can flow between the parent vertices. In contrast, in a Markov random field, no information can flow across known vertices. Together with the local Markov property, pairwise and global Markov properties are responsible for blocking the information flow in a Markov random field:

Property 2.2 (Pairwise Markov property)

In a Markov random field, any two non-neighboring variables are conditionally independent given all other variables in the network.

Property 2.3 (Global Markov property)

In a Markov random field, any two subsets S_A and S_B of variables are conditionally independent given a separating subset S_C if every path between the two subsets passes through S_C.

A common property of MRFs is their factorization of the joint probability over a set of cliques of the UG, $cl(UG)$, not necessarily maximal neither non-overlapping: $\mathbb{P}(X) \propto \prod_{C \in cl(UG)} \phi_C(X_C)$, where the functions ϕ_C are the so-called potentials and $X_c = \{X_{i1}, \ldots, X_{ic}\} \subset X$.

A family of distributions that play a central role in MRFs arose in the physics literature where Gibbs introduced them. If X is a random field with a Gibbs distribution, then Gibbs random fields are Markov random fields.

Definition 2.10 (Gibbs random field)

The global energy of the system $X = \{X_1, X_2, \ldots, X_n\}$ instantiated with x is defined as $U(x) = \sum_{C \in cl(UG)} \phi_C(x_{|C})$, where the potential ϕ_C of a clique C depends on the values of the variables of the clique (values denoted $x_{|C}$). The energy function U is said to derive from the collection of potentials $\{\phi_C\}_{C \subset X}$ associated with the neighborhood system $\mathcal{N}$ characterizing the UG.

The Gibbs measure characterized with the energy function U is the probability distribution defined as $\mathbb{P}(X = x) = \frac{1}{Z} exp(-U(x))$, where $Z = \sum_x exp(-U(x))$, the so-called Gibbs partition function, is a normalizing constant. (It has to be noted that, in practice, it is generally impossible to compute Z due to the huge number of possible instances x.)

Theorem 2.1 (Theorem of Hammersley-Clifford [5])

For a finite or countable set of variables, a bounded system $\mathcal{N}$ of neighborhoods and a finite state space, X is a Markov random field with respect to $\mathcal{N}$ and the positive distribution $\mathbb{P}$ ($\forall x, \mathbb{P}(x) > 0$) if and only if X is a Gibbs random field associated with $\mathcal{N}$.

Then, X is locally characterized in X_i by:

$$\pi_{X_i}(x_i, b) = \mathbb{P}(X_i = x_i \mid \mathcal{N}_{X_i} = b_{|\mathcal{N}_{X_i}}) = \frac{exp(-\sum_{C \ni X_i} \phi_C(x))}{\sum_{x_i} exp(-\sum_{C \ni X_i} \phi_C(x))}.$$

A subclass of Markov random fields that arises in many contexts is that of *pairwise Markov networks*. Such Markov random fields represent distributions where the potentials are defined either over single variables or pairs of variables. In this case, two sets of potentials are considered: a set of vertex potentials $\{\phi(X_i)\}_{i=1,\dots,n}$ and a set of edge potentials $\{\psi(X_i, X_j)\}_{(X_i, X_j) \in UG}$. The joint distribution is the normalized product of all the potentials (both vertex and edge). Pairwise MRFs are appealing because of their simplicity, and because interactions on edges are an important special case that often arises in practice. In particular, the prototypical Markov random field is the Ising model, a kind of pairwise Markov random field which is popular in image processing and computer vision, and whose graph is a regular lattice (with vertices organized in a planar rectangular grid).

2.3.3 Variants around the Concept of Markov random field

HIDDEN MARKOV RANDOM FIELD

Hidden Markov random fields (HMRFs) are generalizations of hidden Markov models. Both classes of models assume that the probabilistic nature of the observed X_i's is determined by the unobservable Y_i's. In a hidden Markov model, the Y_i's are ruled by a Markov chain whereas in a hidden MRF, the Y_i's define an MRF. Fig. 2.4C on page 39 illustrates the HMRF class.

CONDITIONAL RANDOM FIELD

In an MRF whose nodes describe variables X_i's, the Markov property states that a given X_i is independent of any other X_j conditional on the neighbors of X_i. The concept of conditional random field (CRF) is defined as follows:

Definition 2.11 (Conditional random field (CRF))
Let $G = (V, E)$ be an undirected graph such that Y (unobserved data) is indexed by the vertices of G. (X, Y) is a conditional random field if, when conditioned on X (observed data), the random variables Y_i's satisfy the Markov property $\mathbb{P}(Y_i \mid X, Y_j, j \neq i) = \mathbb{P}(Y_i \mid X, Y_j, j \in \mathcal{N}_{Y_i})$, where $\mathcal{N}_{Y_i}$ denotes the neighborhood of Y_i in G.

Fig. 2.4D shows a CRF which makes a first-order Markov assumption among the Y_i's, resulting in a linear chain. Fig. 2.4E provides an illustration for the general case.

2.3.4 Bayesian networks

One of the most popular kinds of probabilistic graphical models is the Bayesian network (BN). BNs play a central role in a wide range of automated reasoning applications, including diagnosis, sensor validation, probabilistic risk analysis, information fusion, and decoding of error-correcting codes. The graphical component of a BN is a directed acyclic graph (DAG). This DAG determines the conditional decomposition of the joint probability distribution, thus much simplifying the computation of the joint distribution. This property reduces the number of parameters that are required to describe the joint probability distribution.

Definition 2.12 (Local Markov property in a Bayesian network)
We consider the set of n random variables $X = \{X_1, \dots, X_n\}$ represented by n nodes in the DAG, and the set of parameters θ, which describes conditional probability distributions $\theta_i = \mathbb{P}(X_i/Pa_{X_i})$ where Pa_{X_i} denotes node i's parents, and a priori distributions $\theta_i = \mathbb{P}(X_i)$ if node i has no parent. X is a BN with respect to the DAG if it satisfies the local Markov property stating that each variable is conditionally independent of its non-descendants given a known state of its parent variables:

$X_i \perp\!\!\!\perp X \setminus desc(X_i) \mid Pa_{X_i}$ for all $i \in \{1, \ldots, n\}$, where $desc(X_i)$ is the set of descendants of X_i. We recall that in a DAG, the set of descendants of a given vertex are all vertices that can be reached by tracing arrows forward from this vertex.

The local Markov property entails that the joint probability distribution writes as a product of individual distributions, conditional on the parent variables

$$\mathbb{P}(X) = \prod_{i \in \{1, \ldots, n\}} \theta_i = \prod_{i=1}^{n} \prod_{j=1}^{q_i} \prod_{k=1}^{r_i} \theta_{i,j,k},$$

where r_i is the number of possible values of X_i, q_i denotes the number of possible configurations (or instantiations) of Pa_{X_i}, and $\theta_{i,j,k} = \mathbb{P}(X_i = k \mid Pa_{X_i} = j)$. For conciseness, from now on, we will use identifiers of instantiations ($X_i = k, Pa_{X_i} = j$) instead of explicit instantiations (e.g., $X_i = x_k \mid Pa_{X_i} = pa_{ij}$).

It is important to note the difference between the above local Markov property and the conditional independence statement inherent to the definition of the Markov blanket in a BN:

Definition 2.13 (Markov blanket in a Bayesian network)
In a Bayesian network, the Markov blanket of a variable X_i is the set consisting of the parents of X_i, the children of X_i, and the parents of X_i's children. The children of a vertex are the descendants that are directly connected with this vertex.

A variable is conditionally independent on all other variables, given its Markov blanket.

On the one hand, the edges in the graph represent the direct dependences between the variables. On the other hand, the local Markov property exhibited in each vertex of the graph (other than a source) encodes an independence assumption. The joint distribution embodies all conditional independences portrayed by the DAG. Alternatively, the Markov property can be expressed through the concept of d-separation.

Definition 2.14 (d-separation)
In the DAG of a Bayesian network, we define as a path a sequence of distinct vertices such that in each of its vertices there is an arrow to or from the next vertex in the sequence. A path is said to be d-separated by a set of nodes S if and only if one of the two following conditions holds:

- *the path contains a serial connection $A \rightarrow C \rightarrow B$ or a diverging connection $A \leftarrow C \rightarrow B$ such that C belongs to S;*
- *the path contains a converging connection (or collider or v-structure) $A \rightarrow C \leftarrow B$ such that neither C nor its descendants belong to S.*

It is important to note that in this definition, only uncovered connections are addressed, which means that no direct dependence exists between A and B.

In the DAG, if S d-separates A and B, the path is said to be closed or blocked when we condition on S (in other words, when we observe S). If S does not d-separate A and B, then the path is said to be open, and A and B are d-connected.

A set S_C d-separates the sets S_A and S_B if and only if S_C d-separates every path from a variable vertex in S_A to a variable vertex in S_B.

Table 2.5 recapitulates the cases for which vertex C closes or opens the path between vertices A and B, when conditioning on the set S.

Table 2.5 Closing or opening serial, diverging, and converging connections, depending on the membership of vertex C to the conditioning set S. In these simple cases, closing or blocking the connection means a d-separation; opening the connection means a d-connection.

	Serial Connection	Diverging Connection	Converging Connection
	$A \rightarrow C \rightarrow B$	$A \leftarrow C \rightarrow B$	$A \rightarrow C \leftarrow B$
Open	$C \notin S$	$C \notin S$	$C \in S$ or $\exists$ descendant(C) $\in S$
Close	$C \in S$	$C \in S$	$C \notin S$ and $\nexists$ descendant(C) $\in S$

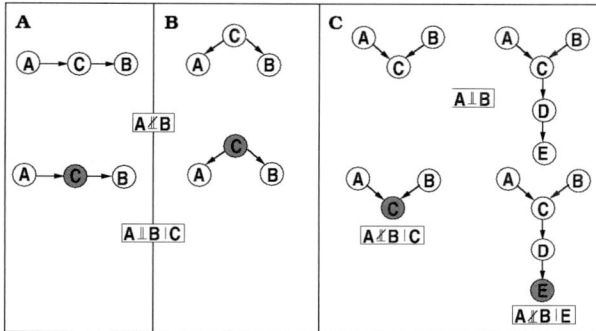

Fig 2.5 Interpretation of d-separation and d-connection. A) Serial connection. B) Diverging connection. C) Converging connection. A) and B) illustrate the concept of d-separation when conditioning (on variable C); C) illustrates the concept of d-connection when conditioning (on variable C).

The concept of d-separation provides a graphical test of independence between two sets of variables in a DAG. S_A is independent of S_B given S_C if and only if S_C d-separates S_A and S_B. The fact that S_A and S_B are d-separated by S_C means that S_C blocks the propagation of information between S_A and S_B in the case when S_C *is the only information known in the graph.*

We explain the link between d-separation and conditional independence through the toy graphs of Fig. 2.5, where we highlight under which schemes the flow of information can be blocked along a path.

In Fig. 2.5A (serial connection), information is transmitted from A to B unless the state of C is known: if the state of C is known, then the state of B depends only on the state of C (B is conditionally independent of A given the state of C). In Fig. 2.5B (diverging connection), information is propagated unless C is instantiated: if C is not instantiated, evidence on A affects the state of B (from the evidence on A, the likely state of C may be inferred, which in turn impacts the belief on the state of B). On the contrary, knowing the state of C is sufficient to predict the state of B, which therefore does not depend on that of A: again, B is conditionally independent of A given C. In the case of the converging connection (Fig. 2.5C), information is transmitted only if C (or one of its descendants in a more general context) is instantiated; otherwise A and B are independent. Table 2.6 interprets the concepts of d-separation and d-connection from the probabilistic viewpoint.

Table 2.6 Interpretation of d-separation and d-connection in terms of probability distributions for the serial, diverging, and converging connections shown in Fig. 2.5.

Connection	Factorized Joint Distribution	No Conditioning (Marginalizing out C)	Conditioning on C	Conclusion
Serial	$\mathbb{P}(a,b,c)$ $= \mathbb{P}(a)\,\mathbb{P}(c\mid a)\,\mathbb{P}(b\mid c)$	$\mathbb{P}(a,b)$ $= \mathbb{P}(a)\sum_{D_C}\mathbb{P}(c\mid a)\,\mathbb{P}(b\mid c)$ $\neq \mathbb{P}(a)\,\mathbb{P}(b)$	$\mathbb{P}(a,b,c) = \mathbb{P}(a,b\mid c)\,\mathbb{P}(c)$ (product rule) $\mathbb{P}(a,b,c) = \mathbb{P}(a)\,\mathbb{P}(c\mid a)\,\mathbb{P}(b\mid c)$ $\implies \mathbb{P}(a,b\mid c) = \frac{\mathbb{P}(a)\,\mathbb{P}(c\mid a)\,\mathbb{P}(b\mid c)}{\mathbb{P}(c)}$ $\implies \mathbb{P}(a,b\mid c) = \mathbb{P}(a\mid c)\,\mathbb{P}(b\mid c)$ (Bayes rule)	$A \not\perp\!\!\!\perp B$ $A \perp\!\!\!\perp B \mid C$
Diverging	$\mathbb{P}(a,b,c)$ $= \mathbb{P}(c)\,\mathbb{P}(a\mid c)\,\mathbb{P}(b\mid c)$	$\mathbb{P}(a,b)$ $= \sum_{D_C}\mathbb{P}(c)\,\mathbb{P}(a\mid c)\,\mathbb{P}(b\mid c)$ $\neq \mathbb{P}(a)\,\mathbb{P}(b)$	$\mathbb{P}(a,b,c) = \mathbb{P}(a,b\mid c)\,\mathbb{P}(c)$ (product rule) $\mathbb{P}(a,b,c) = \mathbb{P}(c)\,\mathbb{P}(a\mid c)\,\mathbb{P}(b\mid c)$ $\implies \mathbb{P}(a,b\mid c) = \mathbb{P}(a\mid c)\,\mathbb{P}(b\mid c)$	$A \not\perp\!\!\!\perp B$ $A \perp\!\!\!\perp B \mid C$
Converging	$\mathbb{P}(a,b,c)$ $= \mathbb{P}(a)\,\mathbb{P}(b)\,\mathbb{P}(c\mid a,b)$	$\mathbb{P}(a,b)$ $= \mathbb{P}(a)\,\mathbb{P}(b)$	$\mathbb{P}(a,b,c) = \mathbb{P}(a,b\mid c)\,\mathbb{P}(c)$ (product rule) $\mathbb{P}(a,b,c) = \mathbb{P}(a)\,\mathbb{P}(b)\,\mathbb{P}(c\mid a,b)$ $\implies \mathbb{P}(a,b\mid c) = \frac{\mathbb{P}(a)\,\mathbb{P}(b)\,\mathbb{P}(c\mid a,b)}{\mathbb{P}(c)}$ $\neq \mathbb{P}(a\mid c)\,\mathbb{P}(b\mid c)$	$A \perp\!\!\!\perp B$ $A \not\perp\!\!\!\perp B \mid C$

2.3.5 Unifying Model and Model Extension

Markov random fields as well as Bayesian networks may be represented by the unifying class of *factor graphs*. On the other hand, the class of *chain graphs* represents an extension for both Markov random fields and Bayesian networks.

FACTOR GRAPHS

Basically, a factor graph encodes a global function that may be factorized into factors [27, 51]. For instance, the factor graph shown in Fig. 2.6A expresses the factorization of the function

$$f(X_1, X_2, X_3, X_4, X_5) = f_A(X_1) f_B(X_2) f_C(X_2, X_3, X_4) f_D(X_3, X_5) f_E(X_5).$$

The factors represent the local functions (e.g., f_A, f_C, and f_D), and their product f is called the global function. A factor graph is therefore a bipartite undirected graph whose vertices can be divided between factor vertices and variable vertices. A factor graph exhibits a variable vertex for each variable X_i, a factor vertex for each local function f_j, and an edge connecting variable vertex X_i to factor vertex f_j if and only if X_i is an argument of f_j. Importantly, factor graphs can model both Markov random fields and Bayesian networks (see Figs. 2.6B and 2.6C). Typically, factor graphs are used for probabilistic inference in PGMs: they support message passing algorithms to efficiently compute characteristics of the global function encoded, such as marginal distributions.

CHAIN GRAPHS

Chain graphs represent a class of graphical models that includes both Markov networks and Bayesian networks as special cases. Chain graphs are most appropriate when both response-explanatory and symmetric relationships exist among variables, whereas Bayesian networks more peculiarly focus on the former class of associations and Markov random fields specifically address

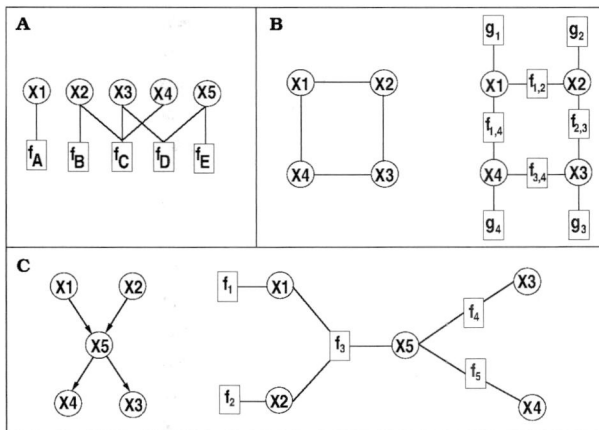

Fig 2.6 Factor graphs. A) Representation of a factorized function with a factor graph. The factorized function is $f(X_1, X_2, X_3, X_4, X_5) = f_A(X_1) f_B(X_2) f_C(X_2, X_3, X_4) f_D(X_3, X_5) f_E(X_5)$. B) Conversion of a pairwise Markov random field ($f(X) = \prod_i g_i(X_i) \prod_{i,j} f_{i,j}(X_i, X_j)$) into a factor graph. C) Conversion of a Bayesian network into a factor graph ($f(X) = f_1(X_1) f_2(X_2) f_3(X_5 \mid X_1, X_2) f_4(X_3 \mid X_5) f_5(X_4 \mid X_5)$).

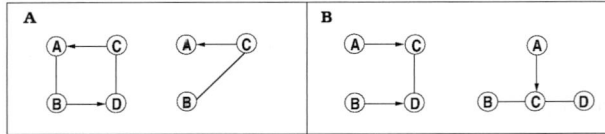

Fig 2.7 Mixed graphs. A) Mixed graphs containing semi-directed cycles. B) Chain graphs. In a mixed graph, a semi-directed cycle is a sequence of r distinct edges $E_1, \ldots, E_r (r \geq 3)$ with endpoints S_i and S_{i+1}, such that: (1) $S_1 = S_{r+1}$, (2) $\forall i (1 \leq i \leq r)$, edge E_i is either $S_i - S_{i+1}$ or $S_i \rightarrow S_{i+1}$, and (3) $\exists j (1 \leq j \leq r)$, edge $E_j = S_j \rightarrow S_{j+1}$.

the latter. As a mixed graph, a chain graph admits both directed and undirected edges; its characteristic lies in the absence of semi-directed (or partially directed) cycles (see Fig. 2.7).

2.4 Probabilistic Inference

In this section, we outline various mechanisms used for probabilistic inference, mostly focusing on Bayesian networks for the sake of conciseness. However, some of these methods are suitable for inference in both Bayesian networks and Markov random fields. Inference is the task of querying a graphical model. There are three standard inference queries to be solved.

Given $E \subset X$, $Q = \bar{E}$ its complement set, $Q' \subset Q$, and e, q, and q' three assignments of values to E, Q, and Q', respectively, Bayesian networks allow to answer the three following canonical queries:

- probability of evidence (PE): $\mathbb{P}(E = e)$ (and the related node marginals query $\mathbb{P}(X_i = x_i \mid E = e)$, $\forall X_i \in X$ and $\forall$ value $x_i \in D_{X_i}$),
- most probable explanation (MPE): $MPE(e) = \arg\max_q \mathbb{P}(Q = q \mid E = e)$, and
- maximum a posteriori hypothesis (MAP): $\mathbb{P}(Q' = q' \mid E = e)$.

More complex queries can be built from the previous ones. Algorithms for inference on Bayesian networks fall into two main categories: exact and approximate. An exact inference query is evaluated by marginalizing out the irrelevant variables. In general, the full summation required is not tractable. The decision version of PE is PP-complete, where PP stands for "probabilistic polynomial time": the problem can be resolved to any specified degree of accuracy by running a randomized, polynomial time algorithm a sufficient (but bounded) number of times. The decision version of MPE is NP-complete, meaning that it cannot be solved in polynomial time in any known way. The decision version of MAP is still more difficult. However, tractable exact algorithms could be developed.

2.4.1 Exact Inference

In this section, we give a flavor of the various classes of methods available for exact inference. Unless specified, the methods below address inference in Bayesian networks. However, Bayesian networks and Markov random fields can be both represented by factor graphs. Thus, common solutions for inference may be shared by both classes of graphical models.

All exact methods compute marginal probabilities by systematically exploiting the conditional independences encoded in the graph of a Bayesian network. Two categories of exact methods were proposed. The methods in the first category implement the computation of the probability of interest by propagating messages along the arrows of the acyclic graph. This category includes **variable elimination** and **message passing for trees**, later extended to polytrees. A polytree is a DAG which does not admit undirected cycles either. The principle of message propagation was generalized to graphs by adding **cycle-cutset conditioning**. The second category of methods builds a novel structure—a **junction tree**—from the original graph through **moralization** and **triangulation**; an adapted message propagation scheme is then applied to this novel representation of the graph.

VARIABLE ELIMINATION

In the variable elimination method, some steps in the marginalization process are simplified due to the factorized form of the joint probability distribution: the factorization dictates the ordering of the variables successively marginalized out (that is, *eliminated*) [25, 56, 99]. The marginalization then amounts to a series of local products and local marginalizations. The specific method for eliminating a variable depends on the query at hand. In particular, if the aim is to solve probability of evidence, then the variables are eliminated by summing them out. In an MPE query, variables are eliminated by maximizing them out. To solve MAP, both types of elimination need to be performed. We illustrate the elimination process with a toy example:

Example 2.1 (Variable elimination)
The DAG is $A \to B \to C$. Let us consider the probability of evidence query $\mathbb{P}(C = c)$:

$$\mathbb{P}(C = c) = \sum_b \mathbb{P}(C = c \mid B = b) \sum_a \mathbb{P}(B = b \mid A = a)\,\mathbb{P}(A = a)$$

$$= \sum_b \mathbb{P}(C = c \mid B = b)\,\mathbb{P}(B = b).$$

MESSAGE PASSING ALGORITHMS

Regarding the computation of marginals in the case of trees, a significant breakthrough was established by the pioneering message-passing algorithm proposed by [68]. In this scheme, two types of messages are conveyed through the tree, λ and π, where λ and π are set for both leaves and instantiated variables (forming evidence e); otherwise, λ and π messages are computed using the hierarchical structure. The following sketch is commented below.

- if X is instantiated: $\lambda(X) = [010\ldots 0]$ with 1 indicating the value of X
- if X is a leaf: $\lambda(X) = [1\ldots 1]$
- otherwise:

  $\lambda(P = p) = \prod_{C \in children(P)} \lambda_C(P = p)$
  with $\forall C \in children(P)$, $\lambda_C(P = p) = \sum_c \mathbb{P}(C = c \mid P = p)\,\lambda(C = c)$
- if X is instantiated: $\pi(X) = [010\ldots 0]$ with 1 indicating the value of X
- if X is the root: $\pi(X) = \mathbb{P}(X)$
- otherwise:

  $\pi(C = c) = \sum_p \mathbb{P}(C = c \mid P = p)\,\pi_C(P = p)$
  with $\pi_C(P = p) = \pi(P = p) \prod_{C_1 \in children(P)\backslash\{C\}} \lambda_{C_1}(P = p)$

A first pass starts from the leaves of the tree; therein each vertex C sends one message $\lambda_C(p)$ to its parent, for each possible value p of the parent; this message $\lambda_C(p)$ is compiled as a function of messages $\lambda(c)$, over the set of values c allowed for the child C. Then, for a given value p of the parent, the message $\lambda(p)$ of the parent is compiled as a function of the $\lambda_C(p)$ messages over the children. The second pass is initiated from the root; for each possible value p of its unique parent, each child C receives a message $\pi_C(p)$ from this parent; the latter message is a function of the $\lambda_C(p)$ messages in the sibling vertices of C (that were obtained in the first phase) and of the $\pi(p)$ message in the parent. Then for child C, message $\pi(c)$ is calculated as a function of the $\pi_C(p)$, over the set of values p allowed for the parent. In the end of the process, $\mathbb{P}(X_i = x_i \mid E = e)$ is computed as a function of $\lambda(X_i = x_i)$ and $\pi(X_i = x_i)$.

We now present a specific kind of message-passing algorithm, the *sum-product algorithm*, also named *belief propagation*, which has been the basis for many variants. The seminal algorithm was also provided by [69]. This algorithm operates on factor graphs that are trees. The message-passing algorithm described above for Bayesian networks [68] translates immediately into an instance of the sum-product algorithm operating on a factor graph. We recall that any Bayesian network or Markov random field can be converted into a factor graph (see Subsection 2.3.5, page 45).

The belief propagation procedure exploits the way in which the joint distribution factors, using the distributive law to simplify the summations. In addition, this procedure reuses intermediate values (partial sums). In computer science, arithmetic expressions are often represented by ordered rooted trees—or expression trees—in which internal vertices represent arithmetic operators (e.g., addition, negation, etc.) and leaf vertices represent variables or constants. Expression trees are extended so that the leaf vertices may also represent functions. As shown in Fig. 2.8, any cycle-free factor graph can be mapped into an expression tree. In this example, the marginalization developed is

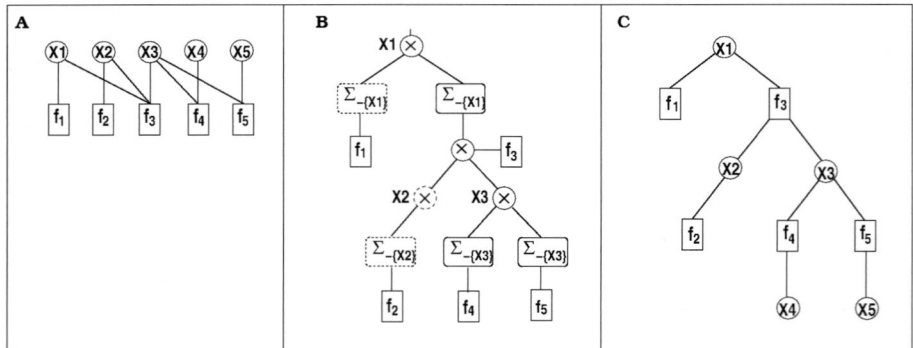

Fig 2.8 Factor graphs and the sum-product message-passing algorithm (belief propagation) - Example extracted from [51] - A) Cycle-free factor graph representing the product $f(X_1, X_2, X_3, X_4, X_5) = f_1(X_1) f_2(X_2) f_3(X_1, X_2, X_3)$ $f_4(X_3, X_4) f_5(X_3, X_5)$. B) Tree expression corresponding to the marginalization $g_1(X_1)$ of function f, described in the text. C) Factor graph rooted in X_1 and corresponding to B). The rules to map the cycle-free factor graph with joint probability distribution $f(X_1, \ldots, X_n)$ into the expression tree associated with marginal distribution $g_i(X_i)$ are the following: (1) root the factor graph in X_1, (2) replace each variable node with the product operator, (3) replace each factor node with the "product operator pointing to the factor node", and (4) insert $\sum_{-\{X_j\}}$ between a factor node and its parent X_j. Trivial products are omitted (leaf nodes, one or no operand).

$$g_1(X_1) = f_1(X_1) \left(\sum_{X_2} f_2(X_2) \left(\sum_{X_3} f_3(X_1, X_2, X_3) \right. \right.$$

$$\left. \left. \left(\sum_{X_4} f_4(X_3, X_4) \right) \left(\sum_{X_5} f_5(X_3, X_5) \right) \right) \right).$$

Thus, not only does the factor graph encode the factorization of the joint probability distribution $f(X_1, X_2, X_3, X_4, X_5)$, it also encodes arithmetic expressions by which the marginal distributions associated with the joint distribution may be computed. In Fig. 2.8, we use the following abbreviation: $g_i(X_i) = \sum_{-\{X_i\}} (X_1, \ldots, X_n)$, where "–" indicates the variables not being marginalized out.

Message passing is initiated at the leaves. A node v waits for the messages coming from its children, then computes the message to be sent to its parent, waits for a message returned from this latter node, and finally computes and sends a message to each of its children. The algorithm is completed when all leaves have received their messages. At a variable node X_i, the product of the two messages that passed through the node is the marginal function $g_i(X_i)$. The message computations are the following:

- variable to local function $\quad \mu_{X \to f}(X) = \prod_{v \in \mathcal{N}(X) \setminus \{f\}} \mu_{v \to X}(X)$
- local function to variable $\quad \mu_{f \to X}(X) = \sum_{-\{X\}} \left(f(X) \prod_{Y \in \mathcal{N}(f) \setminus \{X\}} \mu_{y \to f}(Y) \right)$

A detailed tutorial on the sum-product algorithm is provided in [51].

CONDITIONING

The main power of conditioning is its ability to decrease graph connectivity. In cut-set conditioning, the graph is reduced into a polytree for which inference is linear [18, 69]. A polytree (also referred to as singly connected network) is a DAG with at most one undirected path between any two vertices. The key to cut-set conditioning is the choice of a set of variables—the cycle-cutset Q—which when instantiated, will render the network singly connected. The conditional marginals $\mathbb{P}(x_i \mid x_Q)$ are computed by iterating over all the instantiations x_Q of the cut-set. Finally, the marginals are computed as $\mathbb{P}(x_i) = \sum_{x_Q} \mathbb{P}(x_i \mid x_Q) \mathbb{P}(x_Q)$. The complexity of this approach is exponential with the number of variables in the cut-set. Recursive conditioning drives the conditioning process through an additional graphical structure—a dtree. This tree determines several cut-sets, with the purpose of using each one at a different level of the conditioning process, to decompose the graph into smaller, independent networks. The worst case time complexity of recursive conditioning scales in $O(n \exp(w \log n))$ where w denotes the width of a given elimination order, a precursor in obtaining a dtree of width less than w (for details, see [20]).

MESSAGE PROPAGATION IN JUNCTION TREES

The second category of methods relies on message propagation in junction trees [41, 56, 72]. In such trees, the nodes are clusters of vertices in the original graph; hence the designation of these approaches as clustering methods. Therein, message propagation relies on the concept of potentials and on the factorization of the joint probability distribution into potentials of cliques and separators. A first step builds a junction tree for the original graph, applying moralization then triangulation. We define these notions below.

Definition 2.15 (Moral graph)
The moral graph is obtained by linking the parents of each vertex in the initial directed graph and dropping the directionality of the edges.

Definition 2.16 (Triangulated graph)
A triangulated (or chordal or decomposable) undirected graph does not admit a cycle of length strictly greater than three.

Definition 2.17 (Clique tree)
A clique tree is a tree-structured graph whose nodes are maximal cliques in the original graph. A separator set is the set of intersections of the corresponding cliques.

Definition 2.18 (Junction tree, treewidth)
A junction tree is a clique tree such that the running intersection property holds: if a variable is present in two distinct nodes, then the variable must also be present in every node on the path between the two nodes.

An undirected graph is triangulated if and only if it admits a junction tree.

The width of a junction tree is the size of the largest clique minus one. A graph generally has several junction trees. The treewidth of a graph is then the minimum of the widths over all possible junction trees of the (possibly triangulated) original graph.

The moralization step is time polynomial. This step produces an undirected graph. The generation of an optimized triangulation is an NP-hard problem [15]. In this respect, it is important to note that an optimal triangulation should minimize the size of the cliques, thus allowing tractable local computations. Several heuristics have been proposed for this purpose [46]. We describe a standard algorithm presented in [50]. It relies on a vertex elimination process used to triangulate an undirected graph and to simultaneously identify the nodes of a junction tree. The principle is the following: a graph is triangulated if and only if its vertices may be "eliminated". A vertex may be eliminated if all its neighbors are pairwise connected. Thus, a node may be eliminated if it belongs to a clique. Such a clique is a node of the junction tree. If no vertex can be eliminated, one has to choose a vertex amongst the remaining ones with respect to some criterion aiming at minimizing the state space relative to the future clique. Once such vertex is identified, edges are added between the neighbors of the vertex such that this vertex may be eliminated. Finally, the construction of the junction tree is achieved by implementing the running intersection property. For this purpose, for all $i < j$, the edge between cliques C_i and C_j is weighted with the size of the separator $(C_i \cap C_j)$ and the standard Prim's or Kruskal's algorithm is run to find a maximal weight spanning tree. This latter tree is a junction tree.

In a triangulated graph, the factorization of the joint probability distribution into potentials of cliques and separators writes

$$\mathbb{P}(X) = \frac{\pi_{i=1}^{k} \phi_{C_i}}{\pi_{i=1}^{k} \phi_{S_i}},$$

where C_i denotes a maximal clique (i.e., a node in the junction tree) and S_i denotes a clique separator (with S_k being always equal to $\emptyset$). Message propagation is implemented to guarantee that at the termination of the algorithm, the potential of each clique will correspond to the a posteriori joint distribution for the variables in the clique. The computational cost of tree clustering algorithms grows exponentially with the initial graph's treewidth. The treewidth of the initial graph is the minimum width over the junction trees of the moralized graph (see Definition 2.18, page 50). Naturally, the undirected graph of a Markov random field may be triangulated so that message propagation in a junction tree may be applied. For some classes of graphs such as chains and trees, the treewidth is small and the junction tree algorithm yields an efficient solution to the inference issue. To recapitulate, the message passing- and junction tree- algorithms are essentially dynamic programming algorithms that implement a calculus to share intermediate terms. The messages are exactly these shared intermediate terms. However, in the previous approaches presented, a specific instance of the algorithm has to be rerun for each novel query. In contrast, once a junction tree is built, it can be used to perform multiple queries.

All the methods described above have a complexity that is exponential in the network's treewidth.

On the other hand, the inference efficiency can be increased if algorithms also exploit the structure encoded by the parameters, including for instance independence which is not detectable by d-separation. Advances have been obtained through such methods, for Bayesian networks of quite large treewidths. However, these methods are still in their infancy (see [19] for a survey). A typical example is a technique that exploits parametric structure by compiling the Bayesian network into an arithmetic circuit. Both PE and MPE (see page 46) can be answered by traversing arithmetic circuits in time scaling linearly with the size of these circuits.

However, nowadays technologies frequently yield Bayesian networks whose prohibitive treewidth bans the use of exact algorithms and whose poor parametric structure discards any hope regarding methods based on transformation into arithmetic circuits. Coping with such models requires turning to approximate inference algorithms.

2.4.2 Approximate Inference

The most common approximate inference algorithms include **loopy belief propagation, generalized belief propagation, stochastic Markov chain Monte Carlo simulation** and **variational methods**.

LOOPY AND GENERALIZED BELIEF PROPAGATION

It has to be noted that the sum-product algorithm may also be applied to factor graphs with cycles since all updates are local. More generally, loopy belief propagation refers to using a message-passing scheme such as the well-known Pearl polytree algorithm [69] on BNs containing loops, that is, undirected cycles. Because of these loops, an iterative algorithm with no natural termination will result, with messages passed multiple times on a given edge. Though the results of the loopy belief propagation algorithm cannot be interpreted as exact marginal functions, this approximation scheme was shown to achieve astonishing performance for major applications such as error-correcting codes [60]. The intuition is that if the loops are sufficiently long, then the effect of the loops fades out as the messages propagate: all messages tend to be brought to some stable equilibrium as time elapses. The conditions under which loopy belief propagation is ensured to converge are still not well understood; for instance, it is known that convergence is guaranteed for graphs containing a single loop, but the probabilities obtained might be incorrect [64, 88].

Generalized belief propagation was proposed in [92] to cope with message passing in general graphs. In a nutshell, it was shown that belief propagation can only converge to a stationary point of an approximate of the free energy function, known as the Bethe free energy in statistical physics: namely, the marginal posterior probabilities are fixed-points of the belief propagation algorithm if and only if they are points where the gradient of the Bethe free energy is zero. Generalized belief propagation versions were developed based on more accurate free energy approximations, known as the Kikuchi approximations [45]. These new algorithms can be significantly more accurate than the original one, at a user-adjustable increase in complexity.

STOCHASTIC SAMPLING

Let us first indicate how to achieve sampling from the joint distribution of a Bayesian network in the case when we can easily draw samples. The principle is to sample variables one at a time, following a topological ordering σ such that if one has already sampled $X_{\sigma(1)}, \ldots, X_{\sigma(k)}$, then $X_{\sigma(k+1)}$ necessarily verifies $descendants(X_{\sigma(k+1)}) \cap \{X_{\sigma(1)}, \ldots, X_{\sigma(k)}\} = \emptyset$, and one can therefore sample $X_{\sigma(k+1)}$ given the evidence of its parents: $\mathbb{P}(X_{\sigma(k+1)} \mid X_{\sigma(1)}, \ldots, X_{\sigma(k)}) = \mathbb{P}(X_{\sigma(k+1)} \mid Pa_{X_{\sigma(k+1)}})$.

A prominent class of methods for approximate inference relies on stochastic sampling approaches, and offers many variants. Among them, Markov chain Monte Carlo (MCMC) methods are designed to sample from a probability distribution of interest by constructing a Markov chain that has the desired distribution as its equilibrium distribution. Sampling from the chain after it was run a large number of steps (burn-in phase) is then used to approximate the target marginal distribution of one of the variables or of some subset of the variables. Gibbs sampling and the Metropolis-Hastings algorithm are among the most popular algorithms of this class [31]. Importance sampling represents another category of methods.

Gibbs Sampling   Let us consider for illustration the joint distribution of some set of variables, $\mathbb{P}(X_1, \ldots, X_r \mid e_1, \ldots, e_q)$, conditioned on a set of evidence variables. When the above joint distribution is not known explicitly, but the conditional distribution of each variable is known, the Gibbs sampling algorithm offers an applicable solution to inference. The principle of Gibbs sampling lies in generating an instance from the distribution of each variable in turn, conditional on the current values of the other variables. It can be shown that the sequence of samples describes a Markov chain, and that the stationary distribution of that Markov chain is just the target joint distribution.

The following method indicates one possible way to create a Gibbs sampler and run a Markov chain on $X_1, \ldots, X_r$. For convenience of notation, we denote $e = (e_1, \ldots, e_q)$. The initialization step fixes $(X_1, \ldots, X_r)$ to one of its possible instantiations, say $(x_1, \ldots, x_r)$. Then, at each step, one of the variables X_i is selected (at random in the basic algorithm), the conditional distribution $\mathbb{P}(X_i \mid X \setminus \{X_i\}, e)$ is computed, a value x_i is sampled from this novel distribution, and the value of X_i is updated with x_i. The distribution of X_i conditional on all other variables only depends on the variables that belong to the Markov blanket of X_i. Finally, after the burn-in phase, that is when the chain reaches the stationary distribution, drawing samples from the Markov chain provides an approximation of the target distribution.

Proposition 2.1 (Conditioning on the Markov blanket)
We recall that in a Bayesian network, the Markov blanket $MB(X_i)$ of a variable vertex consists of its children, its parents, and its spouse nodes, that is, the parents of its children (see Definition 2.13, page 42).

$$\mathbb{P}(X_i \mid X \setminus \{X_i\},\ e) = \mathbb{P}(X_i \mid MB(X_i),\ e)$$

$$\propto \mathbb{P}(X_i \mid Pa_{X_i},\ e) \prod_{X_j \in\ children(X_i)} \mathbb{P}(X_j \mid Pa_{X_j},\ e).$$

Metropolis-Hastings Algorithm Another way of constructing a random walk to extract a sample of the target distribution $\mathbb{P}(X_1,\ \ldots,\ X_r \mid e)$ is to use the Metropolis-Hastings (MH) algorithm. The two ingredients of the MH strategy are the proposal distribution and the acceptance probability. Given a current instantiation $x_o = (x_{o_1},\ \ldots,\ x_{o_r})$, the proposal distribution $\mathbb{Q}(.\ \mid x_o)$ allows to sample a new incumbent state x_n. The acceptance probability α is computed as

$$\alpha = \min\left(1,\ \frac{\mathbb{P}(x_n \mid e)}{\mathbb{P}(x_o \mid e)} \frac{\mathbb{Q}(x_o \mid x_n)}{\mathbb{Q}(x_n \mid x_o)}\right).$$

With probability α, the algorithm accepts the proposal x_n and moves to x_n. In case of rejection, the chain remains in state x_o. As for the Gibbs sampler, the relative frequency of state x in the samples picked from the stationary phase provides an estimate for $\mathbb{P}((X_1,\ \ldots,\ X_r) = x \mid e)$.

Importance Sampling In the simplest sampling schemes, relative frequencies allow to compute estimates, such as $\mathbb{P}(Y = y) = \frac{\#(Y=y)}{m}$ and $\mathbb{P}(Y = y \mid E = e) = \frac{\#(Y=y,E=e)}{\#(E=e)}$, where m is the total number of samples generated by the simulation and $\#A$ counts the number of occurrences of event A. However, cases arise where samples cannot be easily drawn from the distribution needed for the estimate. Besides, if the probability of evidence (e) is small, the above acceptance–rejection scheme mentioned above performs poorly. The importance sampling technique offers an efficient alternative to the above stochastic algorithms [75, 93, 94].

First, let us sketch the principle underlying importance sampling. Given a function f of X to integrate over the domain of integration D_X

$$R = \int_{D_X} f(X)\ dX, \tag{2.2}$$

importance sampling tackles the problem by considering a probability distribution $I(x)$, called the importance function, such that $f(x) \neq 0 \implies I(x) \neq 0$. Then

$$R = \int_{D_X} \frac{f(X)}{I(X)} I(X)\ dX.$$

As I is the sampling distribution that samples will be drawn from, I should be chosen to be easy to sample from; that is, I can be used in importance sampling if there exists an algorithm for generating samples from I. Notably, the case where I is uniform coincides with general Monte Carlo sampling. In this sampling scheme, the strong law of large numbers allows to approximate the integration of f through

$$\hat{R} = \frac{1}{m} \sum_1^m \frac{f(x_i)}{I(x_i)},$$

where m is the number of samples.

The idea behind importance sampling is that, to estimate R, more weight is assigned to regions where $f(X) > I(X)$ and less weight to regions characterized by $f(X) < I(X)$: some values of the input random variables in a simulation have more impact on the function being estimated than others. If these "important" values are emphasized by sampling more frequently, then the estimator variance is expected to be reduced.

For instance, in this framework, computing $\mathbb{P}(E = e)$ requires the following summation:

$$\mathbb{P}(E = e) = \sum_{X \setminus E} \mathbb{P}(X \setminus E, E = e). \tag{2.3}$$

Since equation (2.3) is similar to equation (2.2), the same process as explained above may be applied to compute the target probability distribution. Typically, a generic framework for importance sampling is forward sampling (FS). FS generates samples following the topological order of the vertices in the BN and periodically updates the importance function $\mathbb{P}(X \setminus E)$ based on the samples generated so far. Thus, the initialization step of FS requires the production of a topological order, in which the parents of each vertex appear before the vertex; this step also initializes the importance function according to the following scheme

$$\mathbb{P}^{(0)}(X \setminus E) = \prod_{x_i \notin e} \mathbb{P}(x_i \mid Pa_{X_i}, E = e).$$

At step $k + 1$, a sample is generated based on $\mathbb{P}^{(k)}(X \setminus E)$: each evidence vertex (i.e., in E) is instantiated to its observed state and is further omitted from the sample generation; each root variable is sampled from the importance prior probability distribution of this variable which is derived from $\mathbb{P}^{(k)}(X \setminus E)$; following the topological order, each variable with instantiated parent variables is sampled from the importance conditional probability distribution, also derived from $\mathbb{P}^{(k)}(X \setminus E)$.

To compute the posterior probability $\mathbb{P}(a \mid e)$, one can first compute $\mathbb{P}(a, e)$ and $\mathbb{P}(e)$ and then rely on the definition of conditional probability

$$\mathbb{P}(a \mid e) = \frac{\mathbb{P}(a, e)}{\mathbb{P}(e)}.$$

VARIATIONAL METHODS

Another influential class of approximate inference algorithms relies on reducing the inference problem to an optimization problem. Variational methods are such methods. They provide a functional approximation to the posterior distribution of interest with a functional form that makes calculations tractable. This approach has a similar flavor to importance sampling. However, whereas importance sampling chooses a single distribution $\mathbb{Q}$, variational methods consider a family of approximating distributions $\{\mathbb{Q}\}$, from which the optimization process selects a particular member.

To give the intuition of the principle employed in variational inference, let us consider the following example, taken from [43], where the logarithm function is expressed *variationally*: $ln(x) = min_x (\lambda x - ln(\lambda) - 1)$. The functions $\lambda x - ln(\lambda) - 1$ form a family of upper bounds for the logarithm function. For each particular value of x, one of these upper bounds coincides exactly with $ln(x)$: $ln(x) \leq \lambda x - ln(\lambda) - 1, \forall \lambda$. In this transformation, λ represents the variational parameter. Thus, the variational transformation provides a family of upper bounds. Any value of λ provides an upper bound for the function. The tightness of the bound depends on the choice of λ. To obtain a target probability distribution in a graphical model, a product over local conditional probabilities must be computed. The variational transformation of each local function will allow to set a bound on the target distribution, provided that the transformation has algebraic properties.

We now explain how these variational approximations translate into approximations of target probability distributions, such as a conditional distribution $\mathbb{P}(L \mid E)$ or a marginal distribution $\mathbb{P}(E)$. To fix ideas, we consider a Bayesian network whose each local conditional distribution $\mathbb{P}(X_i \mid Pa_{X_i})$ is bounded by upper bound $(\mathbb{P}^U(X_i \mid Pa_{X_i}, \lambda_i^U))$ and lower bound $(\mathbb{P}^L(X_i \mid Pa_{X_i}, \lambda_i^L))$, respectively. Therein, λ_i^U and λ_i^L are variational parameterizations appropriately set for the upper and lower bounds. We choose to reason with the upper bound; the product of upper bounds is an upper bound:

$$\mathbb{P}(X) = \prod_i \mathbb{P}(X_i \mid Pa_{X_i}) \leq \prod_i \mathbb{P}^U(X_i \mid Pa_{X_i}, \lambda_i^U). \tag{2.4}$$

Equation (2.4) still holds for any subset of X whenever some other subset is held instantiated. Therefore, it is possible to derive

$$\mathbb{P}(E) = \sum_L \mathbb{P}(L, E) \leq \sum_L \prod_i \mathbb{P}^U(X_i \mid Pa_{X_i}, \lambda_i^U). \tag{2.5}$$

Care must be taken in the choice of the variational form $\mathbb{P}^U(X_i \mid Pa_{X_i}, \lambda_i^U)$ so that the summation over L may be performed efficiently. Whatever the equation considered ((2.4) or (2.5)), the aim is to minimize the right-hand side with respect to λ_i^U. The reasoning above readily extends to marginal probabilities obtained from lower bounds on the joint distribution. Finally, the conditional distribution $\mathbb{P}(L \mid E)$ is the ratio of two marginal distributions:

$$\mathbb{P}(L \mid E) = \frac{\mathbb{P}(L, E)}{\mathbb{P}(E)}.$$

Obtaining upper and lower bounds for this ratio requires that upper and lower bounds can be computed for both numerator and denominator. However, much of the work concerns the denominator, since the numerator involves fewer sums.

To summarize, the form of the joint probability distribution is simplified through an appropriate choice of variational transformation, which thereby simplifies the inference problem. Besides, some or all of the variables (i.e., vertices) may be transformed, which leads to two categories of variational approaches, the *sequential* and *block* methods. In the sequential approach, variables are transformed one at a time, in an order that is determined during the inference process. This order depends on the particular pattern of evidence. In some cases, however, it can be advantageous to determine in advance the variables to be transformed: the block approach therefore applies when some peculiar substructure is known in the graph.

The block approach, which transforms only some of the variables, implies that exact methods are used as subroutines within the variational approximation framework. In this scheme, partial transformations of the graph may keep some of the original graphical structure unchanged; or a new graphical structure may be introduced, which supports the application of exact inference methods; or both modalities may be considered. It is beneficial to use variational approximations in a limited way, to transform the graph into a simplified graph to which exact methods can be applied. This generally provides closer bounds than an algorithm that transforms the whole graph without regard for tractable substructure.

We now go a bit deeper into details in the case of the block approach, to show how the bounds are computed. We suppose that the target distribution is the conditional distribution $\mathbb{P}(L \mid E)$, where E are the observed variables and L are the latent (unobserved) variables. The block approach relies on the identification of a substructure in the graph of interest, such as a tree or a set of chains, known to be amenable to exact inference methods (or at least to efficient approximate inference methods). To approximate $\mathbb{P}(L \mid E)$, a family of approximating conditional distributions, $\{\mathbb{Q}(L \mid E, \lambda)\}$, is used. The graph representing $\mathbb{Q}$ is a subgraph of that representing $\mathbb{P}$. The selection of a particular distribution from the family $\{\mathbb{Q}(L \mid E, \lambda)\}$ of distributions approximating $\mathbb{P}(L \mid E)$ is performed through the minimization of the Kullback-Leibler divergence $D_{KL}(\mathbb{Q}(L \mid E, \lambda) \| \mathbb{P}(L \mid E))$, with respect to the variational parameters λ. Thus, the inference problem is transformed into the following optimization problem:

$$\lambda^* = \arg\min_{\lambda} D_{KL}\left(\mathbb{Q}(L \mid E, \lambda) \| \mathbb{P}(L \mid E)\right). \tag{2.6}$$

We recall the following definition:

Definition 2.19 (Kullback-Leibler divergence)
For any probability distributions $\mathbb{P}(X)$ and $\mathbb{Q}(X)$, the Kullback-Leibler (KL) divergence, which measures the closeness of the two distributions, is:

$$D_{KL}(\mathbb{Q} \| \mathbb{P}) = \sum_{X} \mathbb{Q}(X) \ln \frac{\mathbb{Q}(X)}{\mathbb{P}(X)}.$$

Notably, a motivation for using the KL divergence as a measure of approximation accuracy lies in that it provides the best lower bound on $\mathbb{P}(E)$ in the family of approximations $\mathbb{Q}(L \mid E, \lambda)$. Using the so-called Jensen's inequality (if function f is concave, then $f(E[X]) \geq E[f(X)]$), it is possible to bound the logarithm of $\mathbb{P}(E)$:

$$ln\, \mathbb{P}(E) = ln \sum_{L} \mathbb{P}(L, E)$$

$$= ln \sum_{L} \mathbb{Q}(L, E, \lambda) \frac{\mathbb{P}(L, E)}{\mathbb{Q}(L, E, \lambda)}$$

$$\geq \sum_{L} \mathbb{Q}(L, E, \lambda)\, ln \left[\frac{\mathbb{P}(L, E)}{\mathbb{Q}(L, E, \lambda)} \right].$$

The difference between the left- and right-hand sides of the above equation is exactly the Kullback-Leibler divergence $D_{KL}(\mathbb{Q}\|\mathbb{P})$. As the KL divergence is positive, the right-hand side is a lower bound for $ln\,\mathbb{P}(E)$. The tightest lower bound is obtained when $\lambda = \lambda^*$ (see equation (2.6)).

Variations on variational inference in graphical models are numerous. A wide literature referencing in particular the seminal works allows a more in-depth study on the subject (see [43, 57, 84, 85] to name but a few). For full details, worked examples, and additional insight on other inference strategies, the reader is referred to the compilation of [67].

2.5 Learning Bayesian networks

In most applications, the Bayesian network (BN) is unknown and one has to learn it from the data, namely a matrix D with p observations as rows and n variables as columns (e.g., individuals of a cohort that are described by genetic markers in a biomedical research framework). A BN is composed of a graph and parameters. Therefore, the taxonomy of BN learning tasks distinguishes whether the structure is known. In addition, when learning parameters given a known structure, one has to distinguish between generative learning and discrimination learning. The former attempts to maximize $\mathbb{P}(X)$ (or a variant) while the latter focuses on the maximization of $\mathbb{P}(Y \mid X)$, where a target variable Y will be used to classify further observations. In this chapter, we restrict our focus on generative learning. Another fact that influences the complexity of the learning task is the handling of **incomplete data**, due to data acquisition protocols proned to errors or ambiguities. Finally, in some applications, the presence of **latent—or unobserved—variables** in the model complicates the learning task all the more as they have to be learned as a part of the graphical structure. Table 2.7 recapitulates this taxonomy. We recall that in this chapter, we deal with discrete variables.

The following sections furnish a broad-brush picture of the range of generative techniques encompassed by this taxonomy. We will use $x_1, \ldots, x_n$ to denote the values respectively observed for $X_1, \ldots, X_n$. The symbol $x_{\ell i}$ denotes the value observed for ℓ^{th} data example and i^{th} variable. We recall that $\theta_i = \mathbb{P}(X_i \mid Pa_{X_i})$ designates the conditional probability table of X_i knowing the

Table 2.7 A taxonomy of the learning tasks for Bayesian networks.

Structure	Incomplete Data	Latent Variables	Main Representatives of Learning Approaches
Known	No	No	Maximum likelihood estimation, maximum a posteriori, expectation a posteriori
	Yes	Yes	Expectation-maximization, sampling methods
Unknown	No	No	Search through model space (hill-climbing, simulated annealing, K2, best-first search)
	Yes	No	Structural expectation-maximization
		Yes	Adapted structural expectation-maximization

state of its parents Pa_{X_i}. In particular, $\theta_{i,j,k} = \mathbb{P}(X_i = k \mid Pa_{X_i} = j)$. We recall that for the sake of conciseness, we use identifiers of instantiations $(X_i = k, Pa_{X_i} = j)$ instead of explicit instantiations $(X_i = x_k \mid Pa_{X_i} = pa_{ij})$.

2.5.1 Parameter Learning

Three standard methods allow us to handle parameter learning in the simple case of complete data: standard maximum likelihood estimation (MLE), maximum a posteriori (MAP) and expectation a posteriori (EAP). MLE is a statistical approach which estimates the probability of an event $(X_i = k \mid Pa_{X_i} = j)$ as its relative frequency in the data. In contrast, Bayesian approaches estimate the most probable parameters, acknowledging the fact that the data have been observed, which implies the use of a prior on the parameters. By virtue of Bayes theorem, hence the designation of "Bayesian" for these approaches, specifying a prior allows the derivation of the posterior distribution, which is the criterion to maximize.

COMPLETE DATA

Various scoring functions may be used to navigate in the space of parameters, given a known BN structure and the training data. The simplest approach to learn the parameters θ consists in maximizing in θ^{ML} the likelihood that the data D be generated by the model described by θ. Instead of the likelihood

$$L(\theta) = \mathbb{P}(D \mid \theta) = \prod_{i=1}^{n} \mathbb{P}(x_i \mid \theta_i) = \prod_{\ell=1}^{p} \prod_{i=1}^{n} \mathbb{P}(x_{\ell i} \mid \theta_i),$$

the log-likelihood scoring function allows local computations:

$$\log L(\theta) = \sum_{\ell=1}^{p} \sum_{i=1}^{n} \log \mathbb{P}(x_{\ell i} \mid \theta_i).$$

Such a decomposition thus enables us to maximize the contribution of each vertex independently. This statistical approach readily estimates probabilities through counts of occurrences $(X_i = k, Pa_{X_i} = j)$ in the data D.

Alternatively, Bayesian learning is a process that relies on the assignment of a prior probability distribution to each parameter vector θ_i. In the maximum a posteriori estimation, the aim is to maximize in θ^{MAP} the following score:

$$\log \mathbb{P}(\theta \mid D) \propto \mathbb{P}(D \mid \theta)\, \mathbb{P}(\theta),$$

derived from Bayes theorem. The motivation for the use of priors is two-fold. First, the absence of some occurrences $(X_i = k, Pa_{X_i} = j)$ in the training data set would bias further inference; therefore a compensation is required to account for missing occurrences. In addition, Dirichlet distributions or similar *conjugate* priors are chosen to enable a straightforward algebraic calculation for the posterior.

For instance, in the multinomial case, the likelihood writes

$$L(D \mid \theta) = \prod_{i=1}^{n} \prod_{j=1}^{q_i} \prod_{k=1}^{r_i} \theta_{i,j,k}^{N_{i,j,k}},$$

and the log-likelihood writes

$$LL(D \mid \theta) = \sum_{i=1}^{n} \sum_{j=1}^{q_i} \sum_{k=1}^{r_i} N_{i,j,k} \log \theta_{i,j,k}, \tag{2.7}$$

where $N_{i,j,k}$, the number of occurrences of $(X_i = k,\ Pa_{X_i} = j)$, is computed from the data based on counts relative to the r_i possible values of X_i and the q_i possible configurations of Pa_{X_i}.

When using a Dirichlet prior with coefficients $\alpha_{i,j,1}, \ldots, \alpha_{i,j,r_i}$ for the probability distribution $\mathbb{P}(X_i = k \mid Pa_{X_i} = j)$ $(1 \leq k \leq r_i)$ pseudo-counts are assigned to non-represented occurrences:

$$Dir(\theta_{i,j,1}, \ldots, \theta_{i,j,r_i-1} \mid \alpha_{i,j,1}, \ldots, \alpha_{i,j,r_i}) = \frac{1}{B(\alpha_{i,j})} \prod_{k=1}^{r_i} \theta_{i,j,k}^{\alpha_{i,j,k}-1},$$

with $B(\alpha_{i,j}) = \frac{\prod_{k=1}^{r_i} \Gamma(\alpha_{i,j,k})}{\Gamma\left(\sum_{k=1}^{r_i} \alpha_{i,j,k}\right)}$ and Γ the standard Gamma function.

$$\mathbb{P}(\theta_{i,j}) = Dir(\theta_{i,j} \mid \alpha_{i,j}) \implies \mathbb{P}(\theta_{i,j} \mid D) = Dir(\theta_{i,j} \mid \alpha_{i,j,1} + N_{i,j,1}, \ldots, \alpha_{i,j,r_i} + N_{i,j,r_i}).$$

Thus,

$$\mathbb{P}(\theta \mid D) = \prod_{i=1}^{n} \prod_{j=1}^{q_i} \prod_{k=1}^{r_i} \theta_{i,j,k}^{N_{i,j,k}+\alpha_{i,j,k}-1}.$$

Besides, such priors help to learn a generic model, smoothing the cases when the training data set is too specific; in such cases, ML would entail overfitting, reflected by configurations having very weak or null probabilities while others show overestimated probabilities.

In contrast with ML and MAP which each produce a point estimate for θ, *full Bayesian* parameter learning considers parameters as latent variables. The principle there lies in marginalizing over the unknown parameters (θ). Isolating an incomplete example observation $x = (x_1, \ldots, x_n)$, it is possible to derive the following (see Appendix 2.A, page 77):

$$\mathbb{P}(x \mid D) = \int_{\theta} \mathbb{P}(x \mid \theta)\, \mathbb{P}(\theta \mid D)\, d\theta$$

and thus

$$\mathbb{P}(x_1, x_2, \ldots, x_n \mid D) = \mathbb{E}_{\mathbb{P}(\theta \mid D)} \left[\prod_{i=1}^{n} \theta_{i,j,k} \right].$$

Table 2.8 Parameter learning for Bayesian networks. Parameter estimates obtained with maximum likelihood, maximum a posteriori* and expectation a posteriori* methods.

Method	Category	Principle	Estimate
Maximum likelihood	statistical	$\hat{\theta}^{ML} = \underset{\theta}{\arg\max} \log \mathbb{P}(D \mid \theta)$	$\hat{\theta}_{i,j,k}^{ML} = \dfrac{N_{i,j,k}}{\sum_{k=1}^{r_i} N_{i,j,k}}$
Maximum a posteriori	Bayesian	$\hat{\theta}^{MAP} = \underset{\theta}{\arg\max} \log \mathbb{P}(\theta \mid D)$	$\hat{\theta}_{i,j,k}^{MAP} = \dfrac{N_{i,j,k}+\alpha_{i,j,k}-1}{\sum_{k=1}^{r_i} (N_{i,j,k}+\alpha_{i,j,k}-1)}$
Expectation a posteriori		$\mathbb{P}(x \mid D) = \mathbb{E}_{\mathbb{P}(\theta \mid D)}\left[\prod_{i=1}^{n} \theta_{i,j,k}\right]$	$\hat{\theta}_{i,j,k}^{EAP} = \dfrac{N_{i,j,k}+\alpha_{i,j,k}}{N_{i,j}+\alpha_{i,j}}$

* The prior on parameters θ follows a Dirichlet distribution.

Not surprisingly in a model averaging approach, the above formalization exhibits an expectation term. In the case of a Dirichlet prior on θ, the calculus of the expectation a posteriori is tractable. However, depending on the prior, the computation of the full Bayesian posterior distribution is often intractable. Approximation techniques such as *point estimates* or *sampling* are therefore required.

The parameter estimates obtained with the three above classical methods are shown in Table 2.8, where $N_{i,j} = \sum_{k=1}^{r_i} N_{i,j,k}$ and $\alpha_{i,j} = \sum_{k=1}^{r_i} \alpha_{i,j,k}$.

INCOMPLETE DATA

The standard expectation-maximization (EM) procedure readily applies to parameter learning in the case of incomplete data. Note that this situation encompasses the case of latent variables, provided that the whole structure is known (including the parents and children of the latent variables) as well as the cardinalities of the latent variables: such variables are merely considered as variables whose observations across examples are all missing. Starting from a peculiar parameter instantiation θ^0 (generally chosen at random), the EM procedure iterates rounds of expectation (E) and maximization (M) steps until convergence. Expectation completes the data set based on $\theta^{(t)}$. Maximization re-estimates parameters using the completed data set, thus yielding $\theta^{(t+1)}$.

Formally, with D split into D_o (observed data) and D_m (missing data), we are interested in maximizing the log-likelihood $\log \mathbb{P}(D_o, D_m \mid \theta)$. This log-likelihood is per se a random variable which is a function of D_m. If we now assume that reference parameters are available, θ^*, that allow us to estimate $\mathbb{P}(D_m \mid \theta^*)$, then it is possible to compute the *expectation* of the above log-likelihood in the multinomial case:

$$Q(\theta : \theta^*) = \mathbb{E}_{\theta^*}\left[\log \mathbb{P}(D_o, D_m \mid \theta)\right]$$

$$= \sum_{i=1}^{n} \sum_{j=1}^{q_i} \sum_{k=1}^{r_i} N_{i,j,k}^* \log \theta_{i,j,k} \text{ (see equation (2.7), page 59)},$$

where $N_{i,j,k}^* = \mathbb{E}_{\theta^*}\left[N_{i,j,k}\right] = \sum_{\ell=1}^{p} \mathbb{P}(X_i = k \mid Pa_{X_i} = j, \theta_{i,j,k}^*, D_o)$. $N_{i,j,k}^*$ is either obtained from counts if occurrences (X_i, Pa_{X_i}) are fully observed, or from inference based on $\theta_{i,j,k}^*$ otherwise.

Then, $\hat{\theta}$ is estimated through maximum likelihood, maximum a posteriori, or expectation a posteriori.

In the standard case (ML), the EM procedure repeatedly implements the following tasks:

- E-step
  - compute $\mathbb{P}(X_i = k \mid Pa_{X_i} = j, \theta_{i,j,k}^{(t)}, D_o)$,
  - obtain $N_{i,j,k}^{(t)} = \sum_{\ell=1}^{P} \mathbb{P}(X_i = k \mid Pa_{X_i} = j, \theta_{i,j,k}^{(t)}, D_o)$
- M-step
  - compute $\theta_{i,j,k}^{(t+1)} = \dfrac{N_{i,j,k}^{(t)}}{\sum_{k=1}^{r_i} N_{i,j,k}^{(t)}}$.

Multiple variants of the EM scheme have been proposed. As they provide a point estimate of the parameters, they are proned to getting trapped in local optima. To circumvent this issue, sampling strategies such as Markov chain Monte Carlo methods may be used to estimate the full posterior distribution $\mathbb{P}(\theta \mid D)$.

2.5.2 Structure Learning

This section dedicated to structure learning in Bayesian networks is organized as follows. First, we present **constraint-based** approaches, which learn the structure based on the identification of conditional independences in the data. Then, we deal with **score-based** approaches that browse the space of structures while optimizing a score. The next subsection briefly mentions hybrid methods. Fourth, we mention how score-based methods may be adapted to handle **incomplete data**. Finally, a subsection addresses the more specific case of learning Bayesian networks with **latent variables**.

CONSTRAINT-BASED ALGORITHMS

Constraint-based algorithms perform statistical tests to learn conditional independences. To present the two main strategies, we need the following definitions:

Definition 2.20 (I-map, D-map, perfect map)
An independence map (I-map) for the probability distribution $\mathbb{P}$ is any directed graph verifying the local Markov property (see Definition 2.12, page 41), which can be written as follows:

$$X_i \perp\!\!\!\perp_{\mathbb{P}} X_j \mid S \Leftarrow X_i \perp\!\!\!\perp_{d-sep} X_j \mid S,$$

where S denotes a set of variables in X. In other words, any conditional independence that is encoded through d-separation in the graph is also encoded by the distribution. However, some conditional independences derived from $\mathbb{P}$ may not be encoded in the map.

Similarly, a dependence map (D-map) for distribution $\mathbb{P}$ is a graph in which all d-connected nodes correspond to conditional dependent variables according to $\mathbb{P}$:

$$X_i \perp\!\!\!\perp_{\mathbb{P}} X_j \mid S \Rightarrow X_i \perp\!\!\!\perp_{d-sep} X_j \mid S.$$

If a graph is both an independence map and a dependence map, then it is a perfect map of $\mathbb{P}$.

Definition 2.21 (Minimal I-map, maximal D-map)

A minimal independence map is an independence map that fails to be an independence map if any of its arcs are removed.

A maximal dependence map fails to be a dependence map if any arc is added.

The PC algorithm is an illustrative algorithm of the class of constraint-based methods [78]. Starting from the complete (undirected) graph, the PC algorithm successively tests conditional independences with increasing orders (order 0: $X_i \perp\!\!\!\perp X_j$; order 1: $X_i \perp\!\!\!\perp X_j \mid X_k$; order 2: $X_i \perp\!\!\!\perp X_j \mid \{X_k, X_\ell\}$... For the order k, assessing conditional independence for any subset S_k of neighbors of X_i (X_j excluded), $X_i \perp\!\!\!\perp X_j \mid S_k$, is the condition to remove the edge between X_i and X_j. Thus, starting from a trivial I-map (the complete graph contains no independence relationship), the PC algorithm produces successive I-maps until a minimal I-map is obtained. A final step handles the orientation of the I-map: first, the v-structures (see Definition 2.14) are identified, and then, the orientation is propagated from the v-structures. The order according which the variables are processed exerts a prominent influence on the result [21]. In contrast, the IC algorithm initiates the construction with an empty graph [71], and traverses an embedded family of D-maps until a maximal D-map is reached. In both schemes, additional independence tests allow the detection of v-structures, from which edge orientation is propagated. Once the structure is learned, standard parameter learning may be performed. The FCI [78] and IC* [70] algorithms are variants of PC and IC, respectively, that allow the detection of latent variables (see Subsection 2.5.2).

SCORE-BASED ALGORITHMS

Score-based methods are divided into two categories: some methods navigate in the search space of structures, optimizing the degree to which the current candidate structure explains the data, until an optimum score is obtained; other approaches identify the best structures and combine these structures' features. As the number of possible structures is super-exponential in the number of vertices [32], exhaustive enumeration of the candidates is not feasible. When the number n of variables is strictly greater than 1, this number is computed as: $NS(n) = \sum_{i=1}^{n} (-1)^{i+1} C_n^i 2^{i(n-1)} NS(n - i)$; $NS(1) = NS(0) = 1$. For instance, $NS(10) = 4.2 \times 10^{18}$. This has led to an extensive use of heuristic optimization strategies, encompassing local search and its variants, genetic algorithms, Monte Carlo methods, and variational methods.

A common feature to these score-based methods is the necessity to define the neighborhood for an incumbent solution. Typically, a novel candidate is obtained from the incumbent graph through operations such as addition, removal, or reversal of an edge (under the constraint of acyclicity). Second, the corner stone of these methods is the choice of a scoring criterion. Two important concepts are those of decomposable score and equivalent score.

Definition 2.22 (Decomposable score)

A scoring function is called decomposable if it can be expressed as a sum or a product of local scores which each only depends on a vertex and its parents.

Score decomposability is the condition to reduce the computational burden of the learning task: thus, the variation of the score between two structure neighbors needs only be computed, instead of the global score of the novel structure candidate.

Definition 2.23 (Markov equivalent DAGs, equivalent score)

Two directed acyclic graphs (DAGs) are Markov equivalent if they (equivalently) check the following conditions:

- *they entail the same conditional independences,*
- *they encode the same decomposition of the joint distribution,*
- *they have the same skeleton* and the same v-structures.*

* The skeleton of a DAG is the undirected graph that results from ignoring the directionality of every arc.

A score that returns the same value for two Markov equivalent DAGs is called an *equivalent score*.

In this section, we will further refer to the concept of completed partial DAG. We define this notion now as it sheds light on what equivalent DAGs share, which therefore characterizes their Markov equivalence class:

Definition 2.24 (Markov equivalence class)
An arc in a DAG $\mathcal{G}$ is compelled if this arc is shared by all DAGs belonging to the same equivalence class as $\mathcal{G}$. An arc which is not compelled is reversible.

Definition 2.25 (Partially directed acyclic graph (PDAG))
A partially directed acyclic graph is a graph which contains both directed and undirected edges, with no directed cycle in its directed subgraph.

Definition 2.26 (Completed PDAG))
As all DAGs in the same Markov equivalence class have the same set of compelled and reversible arcs, a canonical representation of this class is a PDAG having arcs for compelled arcs and edges for reversible arcs. This representation is called a completed PDAG.

Table 2.9 (page 65) recapitulates the most popular Markov equivalent scores used for structure learning. All the scores mentioned in Table 2.9 are decomposable.

Understanding the necessity of using equivalent scores to navigate the DAG search space is fundamental. Given the data, the only knowledge one has access to is the set of conditional dependences and independences the data encode. However, a given data set, which provides a unique such set of conditional dependences and independences, may be associated with several DAGs, which all belong to the same Markov equivalent class. There is no reason to advantage any of these equivalent DAG candidates when learning the structure of the BN. Therefore, it is essential that such candidate DAGs return the same value when assessed by a scoring function. In particular, this rationale explains the efforts to refine the Bayesian Dirichlet (BD) score, which is not an equivalent score, into the BDe score, which is an equivalent score. Judiciously tuning the prior (though keeping it enough general) was the key to this adaptation.

The Bayesian scoring criterion computes the posterior distribution $\mathbb{P}(\mathcal{G} \mid D)$, relying on a prior probability distribution $\mathbb{P}(\mathcal{G})$ on the possible DAGs. The derivation of the posterior probability writes:

$$\mathbb{P}(\mathcal{G} \mid D) = \int_{\theta} L(D \mid \theta, \mathcal{G})\, \mathbb{P}(\theta \mid \mathcal{G})\, \mathbb{P}(\mathcal{G})\, d\theta$$
$$= \mathbb{P}(\mathcal{G}) \int_{\theta} L(D \mid \theta, \mathcal{G})\, \mathbb{P}(\theta \mid \mathcal{G})\, d\theta.$$

Marginalizing over the parameters (θ) is generally not tractable. However, for a large data set size p, a Laplace approximation together with a further simplification allow us to obtain the Bayesian information criterion (BIC) score. On the other hand, the computation of a specific Bayesian score—the BD score—is practicable, due to a prior Dirichlet distribution with hyperparameters $\alpha_{i,j,k}$, assumed for each unknown parameter.

The BD was further refined to satisfy conservation for Markov equivalent structures, thus providing the BDe designed for a specific prior on parameters (see Table 2.9), where $\mathscr{G}_c$ is the complete graph and p' is the so-called equivalent size (size in connection with the number of examples in the data set) defined by the user. The variable p' expresses the strength of the user's belief in the prior distribution. Still more specifically, a BDeu (Bayesian Dirichlet equivalent uniform) score is defined when the conditional distribution for each variable is assumed *uniform*: $\mathbb{P}(X_i = k, Pa_{X_i} = j \mid \mathscr{G}) = \frac{1}{r_i \, q_i}$. The Bayesian score does not explicitly include a penalty for structure complexity. However, a penalty is implicitly determined by the hyperparameters $\alpha_{i,j,k}$: for instance, when using the BDeu score, the penalty decreases as p' increases [76].

The Minimum Description Length (MDL) principle allows us to trade off the accuracy of the learned network against its practical usefulness. The MDL principle assesses that the model best explaining the data is that which minimizes the sum of two terms: the number of bits sufficient to encode the DAG model, and the number of bits sufficient to encode the data given the model. In the variant developed in [34] (see Table 2.9), d_i is the number of parameters required to represent the conditional distributions associated with variable X_i and p is the number of observations. In this information-theoretic framework, other MDL variants assign different structure penalties, but encode the data in the same way.

Together with the definitions of a scoring function and of neighborhood, the third ingredient necessary to implement structure learning is the search strategy. To prevent the usual drawback of standard local search (or hill climbing) [36], that is, getting trapped in a local optimum, iterated hill climbing performs the search starting from multiple solutions (random restarts). Alternatively, other heuristic search methods are used, such as simulated annealing [36], tabu search [8], genetic algorithms [54, 90], variable neighborhood search [24], ant colony optimization [22], greedy randomized adaptive search procedures (GRASP) [23], and estimation of distribution algorithms [7].

On the other hand, some approaches focus on a smaller search space than the DAG space. The K2 algorithm exploits an order on the vertices such that only vertices preceding X_i in this order, $pred(X_i)$, may be allowed as parents of X_i [16]. In addition, an upper bound on the number of parents allowed is specified. The choice of the vertex ordering is crucial. Many solutions have been proposed to cope with optimization of vertex ordering: tests of conditional independence [77], MCMC methods [30], and genetic algorithms [38], to name but a few. To reduce the structure search space, some other methods limit it to tree-shaped structures. In this line, a maximum weight spanning tree is built, based on distances between vertices. Any locally decomposable score [36] may be used to define the distance, as well as the mutual information criterion [14]. With a decomposable score s, the weight of edge (X_i, X_j) is set to $s(X_i, X_j) - s(X_i, \emptyset)$ where $s(X_i, X_j)$ is the local score in X_i having X_j as its parent, and $s(X_i, \emptyset)$ is the local score in X_i assuming that X_i has no parents. After an undirected tree has been provided by an algorithm such as the standard Kruskal's algorithm, it has to be oriented (see for instance [26]).

Most learning algorithms employ the DAG search space. Possible alternatives consist in browsing through the space of orderings on the variables, the space of completed PDAGs (see Definition 2.26) or the space of the so-called RPDAGs (restricted PDAGs). In [24] and [53], a main search process explores the space of orderings on the variables. Then, for each candidate ordering, a scoring function evaluates the best Bayesian network obtained by means of a secondary search (typically the K2 algorithm) performed in the subspace of DAGs compatible with this

Table 2.9 Information theory-based scores and Bayesian scores p: number of observations, n: number of variables. For the definition of $N_{ij,k}$, N_{ij}, q_i, r_i and $\alpha_{ij,k}$, see Subsection 2.5.1.

Information Theory-based Scores

Acronym	Designation	Definition	Equiv.[1]	Reference
MDL	Minimum description length	$score_{MDL}(\mathscr{G},D) = \log L(D \mid \theta^{ML}, \mathscr{G}) - \lvert\mathscr{A}_{\mathscr{G}}\rvert \log p - c\, Dim(\mathscr{G})$ [2]	YES	variant of [52]
		$score_{MDL\_suzuky}(\mathscr{G},D)$ $$= \sum_{i=1}^{n} \frac{d_i}{2} \log_2 p$$ $$-p \sum_{i=1}^{n} \sum_{j=1}^{q_i} \sum_{k=1}^{r_i} \mathbb{P}(X_i = k,\, Pa_{X_i} = j)\, \log \frac{\mathbb{P}(X_i=k,Pa_{X_i}=j)}{\mathbb{P}(X_i=k)\,\mathbb{P}(Pa_{X_i}=j)}$$ [3]		variant of [34]
AIC	Akaike information criterion	$score_{AIC}(\mathscr{G},D) = \log L(D \mid \theta^{ML}, \mathscr{G}) - Dim(\mathscr{G})$		[2]
BIC	Bayesian information criterion	$score_{BIC}(\mathscr{G},D) = \log L(D \mid \theta^{ML}, \mathscr{G}) - \frac{1}{2} Dim(\mathscr{G}) \log p$		[74]

Bayesian scores

Acronym	Designation	Definition	Equiv.[1]	Reference
BD	Bayesian Dirichlet	$score_{BD}(\mathscr{G},D) = \mathbb{P}(\mathscr{G}) \prod_{i=1}^{n} \prod_{j=1}^{q_i} \frac{\Gamma(\alpha_{ij})}{\Gamma(N_{ij}+\alpha_{ij})} \prod_{k=1}^{r_i} \frac{\Gamma(N_{ij,k}+\alpha_{ij,k})}{\Gamma(\alpha_{ij,k})}$	NO	[16]
BDe	Bayesian Dirichlet equivalent	$\alpha_{ij,k} = p'\, \mathbb{P}(X_i = k, Pa_{X_i} = j \mid \mathscr{G}_c)$ [4]	YES	[36]
BDeu	Bayesian Dirichlet equivalent uniform	$\alpha_{ij,k} = \frac{p'}{r_i\, q_i}$		[9]

[1] Equiv.: the score is Markov equivalent.

[2] $\mathscr{A}_{\mathscr{G}}$ denotes the number of arcs in graph $\mathscr{G}$. c is the number of bits used to store each parameter and $Dim(\mathscr{G}) = \sum_{i=1}^{n} (r_i - 1)\, q_i$.

[3] d_i is the number of parameters needed to represent the conditional probability distributions associated with variable X_i, p is the number of observations.

[4] $\mathscr{G}_c$ denotes the complete graph (encoding no conditional independence) and p', the so-called equivalent sample size, is specified by the user.

topological ordering. On the other hand, greedy search in the space of Markov equivalents was proposed to spare time otherwise wasted in generating structures with equal scores. The seminal works in [61] and [12] laid the foundations for *Greedy equivalent search* (GES), which only relies on two operations (insertion or deletion of arcs under specific conditions of validity), and consists of two phases. In a nutshell, a first greedy stage iteratively augments the completed PDAG structure until a local maximum score is attained; symmetrically, the second stage simplifies the completed PDAG structure until a local maximum score is reached (see for instance [13]). Again, the use of an equivalent score is imperative. Similarly to completed PDAGs, RPDAGs also allow navigation in Markov equivalence classes of DAGs; however, for efficiency, two different RPDAGs may represent the same equivalence class (see for example [1]).

Finally, when the posterior distribution $\mathbb{P}(\mathcal{G}, \theta \mid D)$ is diffuse, model averaging is recommended and is performed using sampling methods, principally Markov chain Monte Carlo methods (see for instance [47]). Variational methods offer a faster alternative to MCMC but their implementation is more complex.

HYBRID METHODS

To achieve structure learning, several methods combine testing conditional independences and using a score. For example, in [77] and [81], a topological ordering is produced through tests of conditional independences; then the K2 algorithm is run. In [91], the algorithm run downstream of the dependence analysis is a genetic algorithm. In contrast to the previous scheme, other approaches conduct heuristic searches that rely on conventional constraint-based techniques. In [21], the algorithm searches the space of completed PDAGs using a heuristic based on dependence analysis; then each PDAG is converted into a DAG which is scored with a Bayesian score.

ADAPTING SCORE-BASED ALGORITHMS TO INCOMPLETE DATA

The EM algorithm was adapted to navigate in the structure-parameter space $(\mathcal{G}, \theta)$. This approach is designated as **structural EM**. However, it is not feasible to perform the maximization step directly in this latter space. In practice, given a reference Bayesian network $(\mathcal{G}^{(t-1)}, \theta^{(t-1)})$, two maximizations are performed, respectively in the space of DAGs and in the space of parameters:

- $\mathcal{G}^{(t)} = \arg \max_{\mathcal{G}} Q(\mathcal{G}, - : \mathcal{G}^{(t-1)}, \theta^{(t-1)})$

- $\theta^{(t)} = \arg \max_{\theta} Q(\mathcal{G}^{(t)}, \theta : \mathcal{G}^{(t-1)}, \theta^{(t-1)})$,

where $Q(\mathcal{G}, \theta : \mathcal{G}^*, \theta^*)$ is the expected score of a Bayesian network $(\mathcal{G}, \theta))$ computed using the distribution $\mathbb{P}(D_m \mid \mathcal{G}^*, \theta^*)$.

In the Bayesian framework, $Q(\mathcal{G}, - : \mathcal{G}^*, \theta^*)$ stands for $\mathbb{E}_{\theta}[Q(\mathcal{G}, \theta : \mathcal{G}^*, \theta^* \mid \mathbf{D_m})]$. For a maximum likelihood-based approach, this expression corresponds to $Q(\mathcal{G}, \theta^{ML} : \mathcal{G}^*, \theta^*)$. A generic version of the structural learning algorithm is described in Algorithm 1.

Structural EM may be interpreted as hill climbing through the structure space, embedding EM on parameters at each move in the DAG space. The pioneering works of [28] and [29] rely on an information-theoretic score (BIC, MDL) and a Bayesian framework (Bayesian score), respectively. Alternatively, evolutionary algorithms have also been proposed [65].

The EM algorithm provides a natural framework to estimate latent variables. In this line, variational EM approaches approximate posterior distributions to simpler ones, thus allowing

Algorithm 1 Structural learning

1: $i \leftarrow 0$
2: $(\mathcal{G}^{(0)}, \theta^{(0)}) \leftarrow initialization()$
3: **while true**
4: $\quad \mathcal{G}^{(i+1)} = \arg\max_{\mathcal{G} \in \mathcal{N}(\mathcal{G}^{(i)})} Q(\mathcal{G}, - : \mathcal{G}^{(i)}, \theta^{(i)})$
5: $\quad \theta^{(i+1)} = \arg\max_{\theta} Q(\mathcal{G}^{(i+1)}, \theta : \mathcal{G}^{(i)}, \theta^{(i)})$
6: $\quad$ **if** convergence **then** exit **end if**
7: $\quad$ incr(i)
8: **end while**
$\quad$ convergence: $|Q(\mathcal{G}^{(i+1)}, \theta^{(i+1)} : \mathcal{G}^{(i)}, \theta^{(i)}) - Q(\mathcal{G}^{(i)}, \theta^{(i)} : \mathcal{G}^{(i)}, \theta^{(i)})| \leq \epsilon$

the use of a lower bound on the marginal likelihood. This scheme generalizes the EM algorithm by maintaining posterior distributions over both latent variables and parameters [4].

LATENT VARIABLES

The issue of dealing with latent variables is two-fold: one has to discover these variables, and one has to adjust their cardinalities (in the case of discrete variables). We mentioned above that the EM algorithm provides a natural framework to estimate latent variables. This explains that the structural EM-based scheme is the general answer to address structure learning under these conditions.

An exception to this trend is the proposal of adapted constraint-based algorithms. The algorithms PC and IC (see Subsection 2.5.2) were adapted into FCI [79] and IC* [70], respectively, to allow the detection of latent variables. Central to these algorithms are several concepts around causality (in an acception different from the standard concept (see Section 2.7)): genuine causality $(X_i \rightarrow X_j)$, artificial causality $(X_i \leftrightarrow X_j)$ reflecting the existence of a latent variable H $(X_i \leftarrow H \rightarrow X_j)$, potential causality encompassing the two previous cases $(X_i \rightarrow X_j$ or $X_i \leftrightarrow X_j)$, undetermined causality stating that it is impossible to know whether X_i is the genuine cause of X_j or conversely, or if X_i and X_j are the consequences of a latent variable. Later, the FCI algorithm was corrected: the set of rules used to determine edge orientations was augmented into the complete set presented in [96].

In addition, in the case of latent variables, the quest for efficiency may compel us to tackle the learning task in a subclass of Bayesian networks. In this line, latent tree models (LTMs)—formerly called hierarchical latent class models—were subject to extensive investigations initiated with the seminal work in [100]. The structure of an LTM is constrained to layers where the only links allowed connect latent variables to their child nodes. Moreover, most often, the bottom layer is composed of all the observed variables—and solely them. In this case, the LTM's structure may be interpreted as a fractal topology, based on the elementary structure of the latent class model (LCM) (see Fig. 2.9, page 68). The methods used to learn LTMs fall into two categories: conventional search-based methods and approaches based on the clustering of variables. In the first category, various hill climbing approaches were proposed to search the space of regular LTMs.

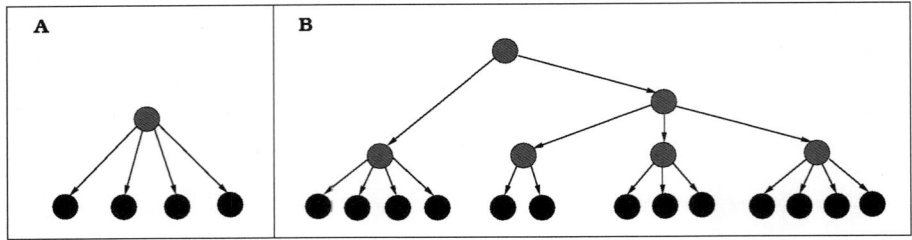

Fig 2.9 Hierarchical Bayesian networks. A) latent class model. B) latent tree model. The observed variables are colored in black, and the latent variables are colored in gray.

Definition 2.27 (Regular latent tree model (LTM))

We denote by $|A|$ the cardinality of a discrete variable A.

A latent tree model is regular if, for any latent variable H and its neighbors $Z_1, \ldots, Z_k$, $|H| \leq \frac{\prod_{i=1}^{k} |Z_i|}{\max_{i=1}^{k} |Z_i|}$, and the strict inequality holds when H has only two neighbors, at least one of which being a latent variable.

Definition 2.28 (Parsimonious LTM)

A parsimonious LTM does not contain any redundant latent variables or any redundant latent classes.[2]

An LTM is parsimonious if there does not exist another LTM that is Markov equivalent and has a smaller dimension.

A parsimonious LTM is necessarily regular.

To search the space of regular LTMs, a possible start may be an LCM. The operations involved to explore the neighborhood of a given LTM include the addition or the removal of a latent vertex, and neighbor relocation. Complementary operations aim at optimizing the cardinality of the latent variables by adding or removing a state for a given latent vertex. Such operations, which may be performed in sequence or simultaneously [11, 97, 98], allow a fine-grained exploration, given a current structure.

Efficient alternatives exploit the hierarchical structure to develop ascending construction strategies, based on the clustering of the variables. All the variables in a given cluster represent in essence the children of a latent variable to be created. The most simple methods are those that deal with binary trees [33, 39], possibly also restraining latent variables to have binary cardinalities [39]. It has to be noted that in [39], the binary tree-based LTM may be augmented with connections between sibling nodes. However, binary trees are not always easy to interpret. Neither do they optimize the number of latent variables. To relax this structural constraint, several strategies were used: for example, starting from a binary tree, pairwise independence tests between neighboring latent variables allow us to detect redundancy, and to dismiss in consequence one of the two neighbors and regraft its children as children of the other vertex variable [87]; other learning

---

[2] In the discrete framework, the different values a latent variable can take are also called the latent classes.

algorithms build the current layer of the hierarchy constructing LCMs from observed variables or imputed latent variables of the previous layer: in [86], at each step of the ascending construction, LCMs initialized with two child variables are enriched through the greedy addition of other variables until some validity criterion is met; in [59] and [62], at each step, a partition into cliques is obtained for the variables of the previous layer. Each such clique groups pairwise-dependent variables which represent as many candidates to subsumption by a common latent variable. A comprehensive survey of latent tree models is provided in [63].

2.6 Learning Markov random fields

Except in the case of triangulated Markov random fields, the presence of the normalizing constant renders the learning task intractable for Markov random fields. Regarding **parameter learning**, we mainly consider the case of complete data. We first lay the bases to understand how **gradient-based** and **second-order methods** address the problem, transforming it into an **unconstrained optimization problem**. Alternatively, a tractable solution may be achieved through the use of an approximation of the likelihood, the so-called **pseudo-likelihood**. Finally, to illustrate **structure learning** in Markov random fields, we choose to sketch a **score-based approach** that relies on the least absolute shrinkage and selection operator (LASSO) technique.

2.6.1 Parameter Learning

The ability to estimate likelihood is central to parameter learning. In Bayesian networks, where the joint distribution writes

$$\mathbb{P}(X) = \prod_{i=1}^{n} \mathbb{P}(X_i \mid Pa_{X_i}),$$

each probability distribution may be estimated separately. As concerns Markov random fields, the general case writes

$$\mathbb{P}(X = x) = \frac{1}{Z} \exp\left(-\sum_{C \in cl(UG)} \phi_C(x_{|C})\right),$$

(see Definition 2.10, page 40; for notation $x_{|C}$, see Notation 2.1, page 39). Therein,

$$Z = \sum_{x} \exp\left(-\sum_{C \in cl(UG)} \phi_C(x_{|C})\right),$$

the normalizing constant, thwarts local estimation. A notable exception is the case of triangulated Markov networks. We will mainly present parameter learning in the case of complete data, for both general and triangulated graphs.

COMPLETE DATA

General Graphs The likelihood is formulated as

$$L(\phi, D) = \prod_{i=1}^{p} \frac{1}{Z} \exp \left(- \sum_{C \in cl(UG)} \phi_C(x^i_{|C}) \right),$$

where $x^i_{|C}$ denotes the values taken by the variables in clique C, for observation i.

The normalization constant rules out the possibility of a closed-form solution. Computing this normalization constant would require us to sum over a number of states which grows exponentially with the number of variables. This computation is often unfeasible. However, this problem of unconstrained maximization of the likelihood $L(\phi, D) = L(\theta)$, where the set of parameters is $\theta = \{\phi_{C,v_C}\}_{v_C \in D_{X_C}, C \in cl(UG)}$, can be solved by a simple gradient ascent or second-order methods. Such iterative methods tackle the unconstrained (possibly non-concave) maximization problem starting from some initial guess θ_0 and generating a sequence $\theta_1, \theta_2, \ldots$ to reach a local maximum of an objective function f. Using a gradient ascent, one reaches a local maximum through steps proportional to the gradient: the principle lies in that $f(\theta)$ increases fastest if one moves from a current solution θ^* in the direction of the positive gradient of f at θ^* (denoted $\nabla f(\theta^*)$). Iteratively stepping from θ_m to $\theta_{m+1} = \theta_m + \gamma_m \nabla f(\theta_m)$, one obtains a sequence which converges to a local maximum. In a gradient ascent, the local maximum coincides with the global optimum if the function f is concave, which is the case of the likelihood. More demanding in terms of computation of derivatives of the objective function, second-order methods evaluate the objective function, its gradient, and its Hessian or an approximation thereof (involving terms of the form $\frac{\partial^2 f}{\partial \theta_i \partial \theta_j}$) at each iteration.

The derivative of the log-likelihood with respect to ϕ_{C,v_C} is shown to be

$$\frac{\partial \log L(\phi, D)}{\partial \phi_{C,v_C}} = \sum_{i=1}^{p} \left(\mathbb{P}(v_C \mid \phi - \mathbb{I}(x^i_{|C} = v_C)) \right) = p\, \mathbb{P}(v_C \mid \phi) - N_{v_C},$$

where N_{v_C} is the number of occurrences of v_C in the data set. It has to be noted that the gradient is zero when the counts from the data (N_{v_C}) correspond with the expected counts. In practice, a prior on the parameters is used to avoid overfitting. The evaluation of the above gradient requires the inference of probability $\mathbb{P}(v_C \mid \phi)$ in the Markov network, which adds complexity to the task. Inference in Markov random fields is intractable, in the general case. However, probabilistic inference can be performed by triangulating the graph associated with the Markov random field and then applying one of the standard algorithms for probabilistic inference for this specific case, such as the junction tree algorithm [41]. Alternatively, Gibbs sampling may be used.

In addition to the gradient and second-order methods, an alternative is to choose an objective function approximating the joint probability and which is tractable, in contrast to the likelihood. The so-called pseudo-likelihood measure provides a simple workaround for intractability [6].

Definition 2.29 (Pseudo log-likelihood for a Markov random field)

The pseudo log-likelihood of any instantiation $(x_1, \ldots x_n)$ of the variables $X_1, \ldots, X_n$ supposed to arise from one distribution among the family $\{\mathbb{P}^\theta, \theta \in \Theta\}$ is defined as

$$PLL(\theta, x_1, \ldots x_n) = \sum_{i=1}^{n} \log \mathbb{P}^\theta(x_i \mid x_{|\mathcal{N}(X_i)}),$$

where $x_{|\mathcal{N}(X_i)}$ denotes the state of the Markov blanket of X_i (see Definition 2.9, page 40).

Thus, the global log-likelihood is replaced by a sum of local indicators through the pseudo log-likelihood function. This quantity is computed in time linear with the number of variables. The maximum pseudo-likelihood estimate is then

$$\hat{\theta} = \arg\max_\theta \sum_{i=1}^{n} \left(\log Z_i(x_{|\mathcal{N}(X_i)}, \theta) + \sum_{C \ni i} \phi_C^\theta(x_{|C}) \right),$$

where the local partition functions $Z_i(x_{|\mathcal{N}(X_i)}, \theta)$ are usually accessible. Moreover, if one can compute the gradient of the pseudo log-likelihood, one can maximize this objective function by applying any gradient-based optimization technique.

Triangulated Graphs   In a Markov random field whose graph $\mathcal{G}$ is triangulated, the factorization more specifically writes as a product of the distributions of the maximal cliques divided by a product of the distributions of their intersections (or separators):

$$\mathbb{P}(X) = \prod_{i=1}^{k} \frac{\mathbb{P}(C_i)}{\mathbb{P}(C_i \cap C_{pred(i)})}$$

where $C_1, C_2, \ldots, C_k$ is a collection of maximal cliques of $\mathcal{G}$, and $C_{pred(i)}$ is the predecessor node of clique node C_i in an ordering of the nodes of some junction tree of $\mathcal{G}$.

Therefore, the clique parameters of Markov random fields defined on triangulated graphs can be determined in a straightforward manner from the low-order joint probabilities associated with the maximal cliques.

INCOMPLETE DATA

In the case of incomplete data, the likelihood loses the concavity property. As for Bayesian networks, expectation-maximization remains the panacea, and will provide a local maximum. The reader in quest of details may consult [10], for instance, where a procedure called Gibbsian EM generalizes the EM algorithm to the Markov random field context.

2.6.2 Structure Learning

The problem of structure learning for Markov random fields (MRFs) has received considerable attention from several communities. As for Bayesian networks, the approaches to this issue are either constraint-based or score-based. Constraint-based methods estimate conditional independences from data using dependence tests and then determine a graph that represents these independences [78]. To score a candidate graph, score-based approaches combine a metric to measure goodness-of-fit with a metric to measure graph complexity. The hill climbing process involved in the standard score-based approach makes the problem NP-hard due to the number of possible graphs, which is super exponential in the number of variables. Besides, structure learning for MRFs necessarily includes parameter estimation, a task which is much more complex for general undirected models. Again, the pseudo-likelihood represents the corner stone for developing efficient score-based methods aiming at browsing the space of possible structures. To avoid overfitting, a simple approach would consider regularizing the score, such as in the following MDL-like score:

$$MDL^*(\mathcal{M}) = \log \mathbb{P}^*(D \mid \mathcal{M}) - \lambda \, \frac{DL(\mathcal{M})}{2} \, \log p,$$

where $\mathcal{M}$ denotes the Markov random field, $DL(\mathcal{M})$ stands for the number of edges in $\mathcal{M}$, and λ is a regularization parameter. However, any heuristic search method only provides a local optimum for some penalized likelihood score. Rather than viewing MRF structure learning as a combinatorial search problem, several works have considered convex approximations to both the goodness-of-fit metric and the model complexity metric [37, 58, 73, 83]. Among the latter works, those methods based on L1-regularization offer strong theoretical properties: they are consistent in both parameters and structure, which means that increasing the number of samples ensures that the true model is discovered; and their statistical efficiency is high, that is, discovering the true model only requires a small number of samples. L1-regularization converts the structure learning problem into a convex optimization problem. Besides, formalizing the problem into a convex optimization problem in a continuous space allows us to use efficient gradient methods. L1-regularization was initially examined for the class of Gaussian graphical models [3], but these techniques were then extended to pairwise log-linear models for discrete data. When problems involve a large number of features, many of which may be irrelevant, L1-regularization is a recognized technique that suitably biases the model toward sparsity [82]. In linear regression, LASSO addresses the issue of finding a linear predictor by minimizing the sum of squared errors. For this purpose, an L1-penalty is used to shrink the regression coefficients and thus lead to sparse predictors. The shrinking is implemented via a bound on the sum of the absolute values of the coefficients.

We illustrate the L1-regularization approach for MRF structure learning considering the formulation of an MRF into a Markov log-linear model. Therein, pairwise log-linear potentials are considered. For expositional conciseness, we consider multinomial random variables X all defined over the same discrete domain of size k.

Definition 2.30 (Markov log-linear model)
The Markov log-linear model is defined as $(X, \mathcal{G}, \phi, \psi)$ where $(X, \mathcal{G})$ is an undirected graph defined on the variables X, and ϕ and ψ are respectively a set of vertex potentials and a set of edge potentials, chosen to be log-linear functions as illustrated below:

$$\phi_i(X_i) = (e^{v_{i_1}} \; e^{v_{i_2}} \; \ldots \; e^{v_{i_k}}), \qquad \psi_{i,j}(X_i, X_j) = \left\{ \begin{matrix} e^{w_{i,j_{11}}} & e^{w_{i,j_{12}}} & \ldots & e^{w_{i,j_{1k}}} \\ e^{w_{i,j_{21}}} & e^{w_{i,j_{22}}} & \ldots & e^{w_{i,j_{2k}}} \\ & \ldots & \\ e^{w_{i,j_{k1}}} & e^{w_{i,j_{k2}}} & \ldots & e^{w_{i,j_{kk}}} \end{matrix} \right\},$$

where $v = \{v_i\}_{i=1,\ldots,n}$ and $w = \{w_{i,j}\}_{(i,j)\in\mathcal{G}}$ are respectively vertices and edge weights. The joint probability writes

$$\mathbb{P}(X) = \frac{1}{Z} \prod_{X_i \in X} \phi_i(X_i) \prod_{(i,j)\in\mathcal{G}} \psi_{i,j}\,(X_i, X_j).$$

As usual, the normalizing constant Z is defined as a sum over all possible instantiations x of X:

$$Z = \sum_{x} \prod_{X_i \in X} \phi_i(X_i) \prod_{(i,j)\in\mathcal{G}} \psi_{i,j}\,(X_i, X_j).$$

Let $\theta = (\mathcal{G}, v, w)$, the log-likelihood is then expressed as;

$$LL(\theta) = \frac{1}{p} \sum_{o=1}^{p} \left(\sum_{X_i \in X} \log \phi_i(X_i^o) \sum_{(i,j)\in\mathcal{G}} \log \psi_{i,j}\,(X_i^o, X_j^o) \right) - \log Z,$$

where o denotes the oth observation in the data set.

Block-regularization is similar in spirit with group LASSO [95]. Group LASSO extends the LASSO, at the group level, by grouping the weights of some features of the predictor: then, depending on the regularization parameter, an entire group of predictors may drop out of the model, thus leading to sparse selection of groups. The adaptation of group LASSO to MRF structure learning is two-fold: (1) the sum of squared errors of the predictor is replaced with an approximation to the log-likelihood; and (2) all the feature weights of an edge are grouped. To avoid both overfitting and densely connected structures, the log-likelihood is optimized using block-L1 regularization:

$$\max_{\theta} LL(\theta) - R(\theta), \tag{2.8}$$

with $R(\theta) = \lambda_{vertex} \; || \; v \; ||_2 + \lambda_{edge} \sum_{1 \le i < j \le n} || \; w_{i,j} \; ||_\alpha$.
Various norms may be used in equation (2.8):

Definition 2.31 (L1-, L2-, and L∞-norms)
Given a vector v with m components, the three norms are defined as:

$$\text{L1-norm:} \qquad || \; v \; ||_1 = \sum_{i=1}^{m} | \; v_i \; |,$$

$$\text{L2-norm:} \qquad || \; v \; ||_2 = \sqrt{\sum_{i=1}^{m} | \; v_i \; |^2},$$

$$\text{L∞-norm:} \qquad || \; v \; ||_\infty = \max_{1 \le i \le m} | \; v_i \; | \, .$$

In equation (2.8) (page 73), the regularization parameters λ_{vertex} and λ_{edge} determine how strongly higher weights are penalized. These parameters control the trade-off between the log-likelihood term and the regularization term. An L2 regularizer is placed on the parameters v (which do not affect the graph structure directly). Regarding the edge (structural) regularizer, block-L1 regularization is considered for each block—or group—of parameters associated with an edge. Each edge is in fact associated to many parameters; in the log-linear model, there is approximately one such parameter for every state the two vertices the edge connects can take. The purpose is sparsifying the graph by driving to zeroes entire groups, that is, removing the corresponding edges. The choice of norms is usually selected from $\alpha \in \{1, 2, \infty\}$ and affects the nature of sparsity. Using $\alpha = 1$ degenerates into the standard LASSO regularizer and encourages sparsity in the parameters: in this case, the goodness-of-fit is penalized by the sum of the L1-norms of the individual parameters; thus, individual parameters are encouraged to be zeroes. However, even if the method aims at having as many entries to zero as possible, some entries may be non-zeroes. In contrast, using $\alpha \in \{2, \infty\}$ aims at getting rid of as many edges as possible: the absence of an edge is equivalent to having the entire group of parameters associated with this edge filled with zeroes. Encouraging individual parameters to be zeroes and encouraging a whole group to be filled with zeroes represents all the difference between regular and structural sparsities. The reason why the above scheme (see equation (2.8), page 73) is called block-**L1** regularization even in the case of $\alpha \in \{2, \infty\}$ is the following: the second term in this equation is equivalent to the L1-norm of a vector where each component is the L2-norm (or L∞-norm) of a group of parameters (corresponding to some edge).

Again, the intractability of the partition function advocates the use of the pseudo log-likelihood, which, in this case, writes

$$PLL(\theta) = \frac{1}{P} \sum_{o=1}^{P} \sum_{X_i \in X} \log \mathbb{P}(X_i^o \mid \mathcal{N}^o(X_i)) \tag{2.9}$$

$$= \frac{1}{P} \sum_{o=1}^{P} \sum_{X_i \in X} \left(\log \phi_i(X_i^o) + \sum_{X_j \in \mathcal{N}(X_i)} \log \psi_{i,j}(X_i^o, X_j^o) - Z_i \right).$$

In equation (2.9), the local normalization constants Z_i's defined for the vertices ensure the tractability of the approximation function (kept concave as the initial function). It remains to identify the set of weights that maximizes the pseudo log-likelihood under the block-regularization specification of equation (2.8) (page 73). However, the objective function associated with block-L1 regularization is not smooth. To overcome this obstacle, a standard solution consists in converting the non-differentiable part of the objective into a constraint making the new objective function differentiable:

$$\max_{\theta, \gamma} PLL(\theta) - \lambda_{vertex} \parallel v \parallel_2 - \lambda_{edge} \sum_{1 \leq i < j \leq n} \gamma_{i,j}$$

$$\text{subject to: } \forall 1 \leq i < j \leq n \quad \gamma_{i,j} \geq \parallel w_{i,j} \parallel_{\alpha}.$$

Typically, the algorithms for resolving this reformulated problem fall in the class of gradient-descent methods [73].

2.7 Causal Networks

We end this chapter emphasizing the difference between Bayesian networks and causal networks. A causal network is a Bayesian network with the added property that the parents of each vertex are its direct causes. In particular, we furnish a broad brush description of the principles used to learn causal models from observational and interventional data.

The reader must be aware of the fact that a Bayesian network encodes conditional dependences and independences, but no causality relations. For instance, an arc $X_i \rightarrow X_j$ must not be interpreted as a relation of causality between X_i and X_j and even less as X_i being a cause of X_j. Though, it has to be recognized that a DAG built by an expert will often reflect relations of causality since this corresponds to a prevailing mode of thinking. In contrast, there is no reason that such relations of causality be exhibited or verified in a DAG learned from data through an automatic process. A telling example is that of a Bayesian network showing a dependence between the indications of a barometer and the onset of thunderstorms. A distinction must be made between the notions of dependence and causality.

In Subsection 2.5.2 (page 61), we explained how, for the purpose of structure learning, score-based methods have to rely on equivalent scores (see Definition 2.23, page 62) to navigate the DAG search space: given the data, the only knowledge one has access to is the set of conditional dependences and independences they encode. However, a given data set that provides a unique set of conditional dependences and independences may be associated with several DAGs, which all belong to the same Markov equivalent class. The interpretations of these DAGs, in terms of causality, would be pairwise different. These *observational* data can only encode the causality relations that are common to all DAGs in the same Markov equivalence class, for example, the v-structures. These reasons explain that scores imperatively have to be constant within the Markov equivalence class. To further discriminate between Markov equivalent DAGs and identify relations of causality, relying on observational data is generally insufficient.

Passive learning (that is, only using observational data) only looks at whether two variables are statistically dependent, which can be explained by many different underlying causal structures. A remarkable exception lies in that all v-structures in a causal network may be discovered from observational data. Moreover, several works have been devoted to identifying conditions under which certain causal hypotheses can be reliably assessed considering observational data alone [70, 79]. In an *active* learning framework, interventional data allow us to infer relations of causality through perturbations of the observed system [80]. Formally, an intervention on a single variable can be modeled by inserting an extraneous variable as an additional parent of this former variable. This "surgery" on the network may break the Markov equivalence between Bayesian networks. However, except in simple cases, it is not likely that an intervention would be maximally effective for identifying the true causal structure (maximal effectiveness meaning that this intervention would allow us to distinguish between the different DAGs in the Markov equivalence class). Instead, typically, a Markov equivalence class of DAGs is sequentially refined into some smaller subclasses via randomized experiments. The edge orientation can then be performed separately in each so-called chain component of the PDAG (see Definition 2.25, page 63) representing the Markov equivalence class.

Definition 2.32 (Chain component)
In a graph G, a path $(X_0, X_{i+1}, \ldots, X_{r-1}, X_r)$ is directed (semi-directed) if $X_i \rightarrow X_{i+1} \in G$ for all (at least one) $i = 1, \ldots, r$.

The absence of semi-directed cycles (see Fig. 2.7, page 46) in the PDAG representing a Markov equivalence class implies that the vertex set of this chain graph can be partitioned into chain components verifying: (1) the edges within a chain component are undirected; and (2) the edges between two chain components are directed and they are oriented in the same way.

It has been shown that there are neither new v-structures nor cycles introduced by edge orientation in the causal network under construction as long as there are no v-structures or cycles introduced in any chain component [34]. Thus, in the orientation process, instead of checking the whole graph, it is only necessary to check for v-structures and cycles in each component. The variables in chain components can be manipulated iteratively until the relevant Markov equivalence subclass is reduced to a DAG.

Notation 2.2 (Refined Markov equivalence class)

Let $\mathscr{C}$ denote the PDAG representing the Markov equivalence class and $e(X_i)$ the set of edges whose orientation is fixed by manipulating variable X_i. The post-intervention Markov equivalence class is denoted $\mathscr{C}_{e(X_i)}$.

The previous definition readily extends to $e(S)$, where S is a sequence of manipulated variables.

In a randomized experiment, some variables are manipulated from external interventions by assigning some levels of these variables to p individuals in a probabilistic way [34]. Thus, the pre-intervention conditional probability $\mathbb{P}(X_i \mid Pa_{X_i})$ is replaced with the post-intervention probability $\mathbb{P}'(X_i)$, whereas the other conditional probabilities are kept unchanged. Then, the orientation of an edge $X_i - X_j$ only requires testing independence for the marginal distribution of variables X_i and X_j in the post-intervention distribution. In this context, detecting independence between X_i and X_j allows us to assign the orientation $X_j \rightarrow X_i$, whereas the orientation $X_i \rightarrow X_j$ is assigned in the case of dependence. Given the PDAG representing the Markov equivalence class, if it is possible to orient all undirected edges after a sequence of variables has been manipulated no matter which DAG in the class is the genuine causal structure, then the sequence is said to be *sufficient*. Any permutation of a sufficient sequence is also sufficient. An optimal *batch intervention* design attempts to find a sufficient set with the smallest number of manipulated variables. By definition, in the batch intervention design, no benefit can be drawn from newly inferred orientations to enhance the remaining inference process. In contrast, *sequential interventions* select the next variable to be tested, based on some criterion. This time, the optimal design aims at lowering the remaining uncertainty in the refined Markov equivalent classes, after intervention. For instance, at iteration i in the ordering $X_{(1)}, X_{(2)}, \ldots, X_{(i-1)}$, the next variable $X_{(i)}$ can be selected to minimize the maximum size of refined subclasses $\mathscr{C}_{e(X_{(1)}, \ldots, X_{(i)})}$ for all possible orientations encompassed by $e(X_{(1)}, \ldots, X_{(i)})$. An alternative is maximizing the following entropy criterion:

$$H_{X_i} = -\sum_{j=1}^{r} \frac{\ell_j}{L} \log \frac{\ell_j}{L},$$

where ℓ_j denotes the number of possible DAGs restrained to the chain component (containing X_i) when this chain is subject to orientation $e(X_i)_j$ obtained by manipulating X_i, $L = \sum_j \ell_j$ and r is the number of possible orientations $e(X_i)_1, \ldots, e(X_i)_r$ [34].

2.8 List of General Monographs and Focused Chapter Books

This chapter presented a tutorial on probabilistic graphical models, encompassing the presentation of the most popular classes of models, inference algorithms, and learning algorithms. A final section was devoted to causal networks. The author's choice was to favor sufficient in-depth presentation over topic coverage. Thus, this chapter naturally ends with a list—inevitably non-exhaustive—of several notable monographs written about PGMs as well as of books more specifically focused on either inference or learning, for such models.

GENERAL MONOGRAPHS

- Pearl (1988) [69]
- Whittaker (1990) [89]
- Lauritzen (1996) [55]
- Jensen (1997) [40]
- Cowell, Dawid, Lauritzen, and Spiegelhalter (1999) [17]
- Jordan (1999) [42]
- Neapolitan (2003) [67]
- Wainwright and Jordan (2008) [85]
- Darwiche (2009) [19]
- Koller and Friedman (2009) [48]
- Spirtes, Glymour, and Scheines (2009) [79]
- Nagarajan, Scutari, and Lèbre (2013) [66]

CHAPTER BOOKS AND REPORTS

- Learning in Bayesian networks: Heckerman (1996) [35]
- Probabilistic inference in graphical models: Jordan and Weiss (2002) [44]
- General survey: Koller, Friedman, Getoor, and Taskar (2007) [49]

APPENDIX 2.A BAYESIAN PARAMETER LEARNING FOR A BAYESIAN NETWORK

In this framework, the uncertainty relative to the parameter values is acknowledged through the modeling of a probability distribution of the parameters, considered as latent variables. Given the parameter prior, $\mathbb{P}(\theta)$, the joint probability of the data D, a new observation x and the parameter θ, writes

$$\mathbb{P}(D, x, \theta) = \mathbb{P}(x \mid \theta) \, \mathbb{P}(D \mid \theta) \, \mathbb{P}(\theta). \tag{2.10}$$

Marginalizing out parameter θ means performing the following integration:

$$\mathbb{P}(x, D) = \int_\theta \mathbb{P}(D, x, \theta) \, d\theta. \tag{2.11}$$

Applying the product rule (see Section 2.1, page 32) to the above left-hand side and substituting the joint probability developed in equation (2.10) for the corresponding term in the above right-hand side, we obtain

$$\mathbb{P}(x \mid D) \, \mathbb{P}(D) = \int_\theta \mathbb{P}(x \mid \theta) \, \mathbb{P}(D \mid \theta) \, \mathbb{P}(\theta) \, d\theta. \tag{2.12}$$

By application of the Bayes theorem $\mathbb{P}(\theta \mid D) \, \mathbb{P}(D) = \mathbb{P}(D \mid \theta) \, \mathbb{P}(\theta)$, it follows

$$\mathbb{P}(x \mid D) = \int_\theta \mathbb{P}(x \mid \theta) \, \mathbb{P}(\theta \mid D) \, d\theta. \tag{2.13}$$

Exploiting the above full Bayesian posterior requires that a prior distribution, $\mathbb{P}(\theta)$, be specified. When no practicable algebraic derivation of the posterior distribution can be obtained, approximate methods need to be applied, such as point estimates or sampling techniques.

REFERENCES

[1] S. Acid, L.M. Campos, and J.G. Castellano. Searching for Bayesian network structures in the space of restricted acyclic partially directed graphs. *Journal of Artificial Intelligence Research*, 18:445–490, 2003.

[2] H. Akaike. Statistical predictor identification. *Annals of the Institute of Statistical Mathematics*, 22:203–217, 1970.

[3] O. Banerjee, L.E. Ghaoui, and A. d'Aspremont. Model selection through sparse maximum likelihood estimation for multivariate Gaussian or binary data. *Journal of Machine Learning Research*, 9:485–516, 2008.

[4] M.J. Beal and Z. Ghahramani. The variational Bayesian EM algorithm for incomplete data: with application to scoring graphical model structures. In J.M. Bernardo, M.J. Bayarri, J.O. Berger, A.P. Dawid, D. Heckerman, A.F.M. Smith, and M. West, editors, *Bayesian Statistics 7*, pages 453–463. Oxford University Press, 2003.

[5] J. Besag. Spatial interaction and the statistical analysis of lattice systems. *Journal of the Royal Statistical Society. Series B (Methodological)*, 36(2):192–236, 1974.

[6] J. Besag. Statistical analysis of non-lattice data. *The Statistician*, 24(3):179–195, 1975.

[7] R. Blanco, I. Inza, and P. Larrañaga. Learning Bayesian networks in the space of structures by estimation of distribution algorithms. *International Journal of Intelligent Systems*, 18:205–220, 2003.

[8] R.R. Bouckaert. *Bayesian Belief Networks: From Construction to Inference*. PhD thesis, University of Utrecht, 1995.

[9] W. Buntine. Theory refinement on Bayesian networks. In B. d'Ambrosio, P. Smets, and P. Bonissone, editors, *Proceedings of the Seventh Conference on Uncertainty in Artificial Intelligence (UAI 91)*, pages 52–60. Morgan Kaufmann Publishers, 1991.

[10] B. Chalmond. An iterative Gibbsian technique for reconstruction of m-ary images. *Pattern Recognition*, 22(6):747–761, 1989.

[11] T. Chen, N.L. Zhang, T. Liu, K.M. Poon, and Y. Wang. Model-based multidimensional clustering of categorical data. *Artificial Intelligence*, 176(1):2246–2269, 2012.

[12] D.M. Chickering. Learning equivalence classes of Bayesian network structures. *Journal of Machine Learning Research*, 2:445–498, 2002.

[13] D.M. Chickering. Optimal structure identification with greedy search. *Journal of Machine Learning Research*, 3:507–554, 2002.

[14] C.K. Chow and C.N. Liu. Approximating discrete probability distributions with dependence trees. *IEEE Transactions on Information Theory*, 3(14):462–467, 1995.

[15] G.F. Cooper. The computational complexity of probabilistic inference using Bayesian belief networks. *Artificial Intelligence*, 42 (Issues 2-3):393–405, 1990.

[16] G.F. Cooper and E. Herskovits. A Bayesian method for the induction of probabilistic networks from data. *Machine Learning*, 9:309–347, 1992.

[17] R.G. Cowell, P. Dawid, S.L. Lauritzen, and D.J. Spiegelhalter. *Probabilistic Networks and Expert Systems, Exact Computational Methods for Bayesian Networks*. Springer, Series: Information Science and Statistics, 1st edition, 1999.

[18] A. Darwiche. Recursive conditioning. *Artificial Intelligence*, 126(1-2):5–41, 2001.

[19] A. Darwiche. *Modeling and Reasoning with Bayesian Networks*. Cambridge University Press, 1st edition, 2009.

[20] A. Darwiche and M. Hopkins. Using recursive decomposition to construct elimination orders, jointrees, and dtrees. In *Sixth European Conference on Symbolic and Quantitative Approaches to Reasoning with Uncertainty*, pages 180–191, 2001.

[21] D.H. Dash and M.J. Druzdzel. A hybrid anytime algorithm for the construction of causal models from sparse data. In *Proceedings of the Fifteenth Annual Conference on Uncertainty in Artificial Intelligence (UAI 99)*, pages 142–149. Morgan Kaufmann Publishers, 1999.

[22] L.M. de Campos, J.M. Fernández-Luna, J.A. Gámez, and J.M. Puerta. Ant colony optimization for learning Bayesian networks. *International Journal of Approximate Reasoning*, 31:291–311, 2002.

[23] L.M. de Campos and J.F. Huete. In B. Bouchon-Menieur, J. Gutiérrez-Rios, L. Magdalena, R. R. Yager, editors, *Technologies for Constructing Intelligent Systems 2—Tools*, Stochastic algorithms for searching causal orderings in Bayesian networks pages 327–340. Physica-Verlag, 2002.

[24] L.M. de Campos and J.M. Puerta. Stochastic local and distributed search algorithms for learning belief networks. In *Third International Symposium on Adaptive Systems: Evolutionary Computation and Probabilistic Graphical Model*, pages 109–115, 2001.

[25] R. Dechter. Bucket elimination: A unifying framework for probabilistic inference. In E. Horvitz and F. Jensen, editors, *Proceedings of the Twelfth Conference on Uncertainty in Artificial Intelligence (UAI 96)*, pages 211–219. Morgan Kaufmann Publishers, San Francisco, 1996.

[26] D. Dor and M. Tarsi. A simple algorithm to construct a consistent extension of a partially oriented graph. Technical report, Cognitive Systems Laboratory, Technical Report (R-185), UCLA Computer Science Department, 1992.

[27] G.D. Forney Jr. Codes on graphs: normal realizations. *IEEE Transactions, Information Theory*, 47(2):520–548, 2001.

[28] N. Friedman. Learning belief networks in the presence of missing values and hidden variables. In *Fourteenth International Conference on Machine Learning*, pages 125–133. Morgan Kaufmann Publishers, 1997.

[29] N. Friedman. The Bayesian structural EM algorithm. In G.F. Cooper and S. Moral, editors, *Proceedings of the Fourteenth Conference on Uncertainty in Artificial Intelligence (UAI 98)*, pages 129–138. Morgan Kaufmann Publishers, 1998.

[30] N. Friedman and D. Koller. Being Bayesian about network structure: a Bayesian approach to structure discovery in Bayesian networks. *Machine Learning*, 50:95–126, 2003.

[31] T. L. Griffiths and A. Yuille. A primer on probabilistic inference. In M. Oaksford and N. Chater, editors, *The Probabilistic Mind: Prospects for Rational Models of Cognition*, pages 33–57. Oxford University Press, 2008.

[32] F. Harary and E.M. Palmer. *Graphical Enumeration*. Academic Press, 1973.

[33] S. Harmeling and C.K.I. Williams. Greedy learning of binary latent trees. *IEEE Transactions on Pattern Analysis and Machine Intelligence*, 33(6):1087–1097, 2011.

[34] Y.-B. He and Z. Geng. Active learning of causal networks with intervention experiments and optimal designs. *Journal of Machine Learning Research*, 9:2523–2547, 2008.

[35] D. Heckerman. A tutorial on learning with Bayesian networks. Technical report, MSR-TR-95-06, Microsoft Research, 1996.

[36] D. Heckerman, D. Geiger, and D. Chickering. Learning Bayesian networks: the combination of knowledge and statistical data. *Machine Learning*, 20(3):197–243, 1995.

[37] H. Höfling and R. Tibshirani. Estimation of sparse binary pairwise Markov networks using pseudo-likelihoods. *The Journal of Machine Learning Research*, 10:883–906, 2009.

[38] W.H. Hsu, H. Guo, B.B. Perry, and J.A. Stilson. A permutation genetic algorithm for variable ordering in learning Bayesian networks from data. In W.B. Langdon, E. Cantú-Paz, K.E. Mathias, R. Roy, D. Davis, R. Poli, K. Balakrishnan, V. Honavar, G. Rudolph, J. Wegener, L. Bull, M.A. Potter, A.C. Schultz, J.F. Miller, E.K. Burke, and N. Jonoska, editors, *The Genetic and Evolutionary Computation Conference (GECCO)*, pages 383–390. Morgan Kaufmann Publishers, 2002.

[39] K.-B. Hwang, B.-H. Kim, and B.-T. Zhang. Learning hierarchical Bayesian networks for large-scale data analysis. In I. King, J. Wang, L.-W. Chan, and D. L. Wang, editors, *Neural Information Processing, Lecture Notes in Computer Science*, volume 4232, pages 670–679. Springer, 2006.

[40] F.V. Jensen. *An Introduction to Bayesian Networks*. Springer Verlag, 1st edition, 1997.

[41] F.V. Jensen, S.L. Lauritzen, and K.G. Olesen. Bayesian updating in causal probabilistic networks by local computations. *Computational Statistics Quarterly*, 4(4):269–282, 1990.

[42] M.I. Jordan, editor. *Learning in Graphical Models*. The MIT Press, 1999.

[43] M.I. Jordan, Z. Ghahramani, T.S. Jaakkola, and L.K. Saul. An introduction to variational methods for graphical models. *Machine Learning*, 37:183–233, 1999.

[44] M.I. Jordan and Y. Weiss. Graphical models: probabilistic inference. In M. Arbib, editor, *Handbook of Neural Networks and Brain Theory*, pages 490–495. The MIT Press, 2nd edition, 2002.

[45] R. Kikuchi. A theory of cooperative phenomena. *Physical Review*, 81:988–1003, 1951.

[46] U. Kjaerulff. Triangulation of graphs—algorithms giving small total state space. Technical report, R-90-09, Department of Mathematical and Computer Science, Aalborg University, 1990.

[47] T. Kocka and R. Castelo. Improved learning of Bayesian networks. In J. Breese, D. Koller, editors, *Proceedings of the Seventeenth Conference on Uncertainty in Artificial Intelligence (UAI 2001)*, pages 269–276. Morgan Kaufman Publishers, 2001.

[48] D. Koller and N. Friedman. *Probabilistic Graphical Models: Principles and Techniques*. The MIT Press, 1st edition, 2009.

[49] D. Koller, N. Friedman, L. Getoor, and B. Taskar. Graphical models in a nutshell. In L. Getoor and B. Taskar, editors, *Introduction to Statistical Relational Learning*, pages 13–55. The MIT Press, 2007.

[50] A. Kong. *Multivariate Belief Functions and Graphical models*. PhD thesis, Department of Statistics, Harvard University, 1986.

[51] F.R. Kschischang, B.J. Frey, and H.-A. Loeliger. Factor graphs and the sum-product algorithm. *IEEE Transactions on Information Theory*, 47(2):498–519, 2001.

[52] W. Lam and F. Bacchus. Using causal information and local measures to learn Bayesian networks. In D. Heckerman and E. H. Marndani, editors, *Proceedings of the Ninth Annual Conference on Uncertainty in Artificial Intelligence (UAI 93)*, pages 243–250. Morgan Kaufmann Publishers, 1993.

[53] P. Larrañaga, C.M.H. Kuijpers, R.H. Murga, and Y. Yurramendi. Learning Bayesian network structures by searching for the best ordering with genetic algorithms. *IEEE Transactions on Systems, Man, and Cybernetics*, 26:487–493, 1996.

[54] P. Larrañaga, M. Poza, Y. Yurramendi, R. Murga, and C. Kuijpers. Structure learning of Bayesian networks by genetic algorithms: a performance analysis of control parameters. *IEEE Transactions on Pattern Analysis and Machine Intelligence*, 18:912–926, 1996.

[55] S.L. Lauritzen. *Graphical Models*. Oxford University Press, 1996.

[56] S.L. Lauritzen and D.J. Spiegelhalter. Local computations with probabilities on graphical structures and their application to expert systems. *Journal of the Royal Statistical Society. Series B (Methodological)*, 50(2):157–224, 1988.

[57] N.D. Lawrence. *Variational Inference in Probabilistic Models*. PhD thesis, University of Cambridge, 2000.

[58] S.-I. Lee, V. Ganapathi, and D. Koller. Efficient structure learning of Markov networks using L1-regularization. In B. Schölkopf, J. Platt, and T. Hoffman, editors, *Advances in Neural Information Processing Systems*, volume 19, pages 817–824. MIT Press, 2007.

[59] J. Martin and K. Vanlehn. Discrete factor analysis: learning hidden variables in Bayesian networks. Technical report, Department of Computer Science, University of Pittsburgh, 1995.

[60] R.J. McEliece, D.J.C. MacKay, and J.-F. Cheng. Turbo decoding as an instance of Pearl's "belief propagation" algorithm. *IEEE Journal on Selected Areas in Communications*, 16(2):140–152, 1998.

[61] C. Meek. *Graphical Models: Selecting Causal and Statistical Models*. PhD thesis, Carnegie Mellon University, 1997.

[62] R. Mourad, C. Sinoquet, and P. Leray. A hierarchical Bayesian network approach for linkage disequilibrium modeling and data-dimensionality reduction prior to genome-wide association studies. *BMC Bioinformatics*, 12(16), 2011.

[63] R. Mourad, C. Sinoquet, N.L. Zhang, T. Liu, and P. Leray. A survey on latent tree models and applications. *Journal of Artificial Intelligence Research*, 47:157–203, 2013.

[64] K.P. Murphy, Y. Weiss, and M.I. Jordan. Loopy belief propagation for approximate inference: an empirical study. In K.B. Laskey and H. Prade, editors, *Proceedings of the Fifteenth Conference on Uncertainty in Artificial Intelligence (UAI 99)*, pages 467–475. Morgan Kaufmann Publishers, 1999.

[65] J.W. Myers, K.B. Laskey, and T. Levitt. Learning Bayesian networks from incomplete data with stochastic search algorithms. In K. B. Laskey and H. Prade, editors, *Proceedings of the Fifteenth Conference on Uncertainty in Artificial Intelligence (UAI 99)*, pages 476–485. Morgan Kaufman Publishers, 1999.

[66] R. Nagarajan, M. Scutari, and S. Lèbre. *Bayesian Networks in R with Applications in Systems Biology*. Springer-Verlag, US, Use R series, 1st edition, 2013.

[67] R.E. Neapolitan. *Learning Bayesian Networks*. Artificial Intelligence, Prentice Hall, 1st edition, 2003.

[68] J. Pearl. Fusion, propagation, and structuring in belief networks. *Journal of Artificial Intelligence*, 29(3):241–288, 1986.

[69] J. Pearl. *Probabilistic reasoning in intelligent systems: networks of plausible inference*. Morgan Kaufmann Publishers, San Francisco, 1988.

[70] J. Pearl. *Causality: models, reasoning, and inference*. Cambridge University Press, 2nd edition, 2009.

[71] J. Pearl and T. Verma. A theory of inferred causation. In J.A. Allen, R. Fikes, and E. Sandewall, editors, *Second International Conference on Principles of Knowledge Representation and Reasoning*, pages 441–452. Morgan Kaufmann Publishers, 1991.

[72] P.P. Prakash. A valuation-based language for expert systems. *International Journal of Approximate Reasoning*, 3(5):383–411, 1989.

[73] M. Schmidt, K. Murphy, G. Fung, and R. Rosales. Structure learning in random fields for heart motion abnormality detection. In *IEEE Computer Society Conference on Computer Vision and Pattern Recognition (CVPR 2008)*, pages 1–8, 2008.

[74] G. Schwarz. Estimating the dimension of a model. *Annals of Statistics*, 6(2):461–464, 1978.

[75] R.D. Shachter and M.A. Peot. Simulation approaches to general probabilistic inference on belief networks. In M. Henrion, R.D. Shachter, D. Ross, L.N. Kanal, and J.F. Lemmer, editors, *Proceedings of the Fifth Annual Conference on Uncertainty in Artificial Intelligence (UAI 89)*, pages 311–318. Elsevier Science, 1989.

[76] T. Silander, P. Kontkanen, and P. Myllymäki. On sensitivity of the map Bayesian network structure to the equivalent sample size parameter. In R. Parr and L. van der Gaag, editors, *Proceedings of the Twenty-Third Conference on Uncertainty in Artificial Intelligence (UAI 2002)*, pages 360–367. AUAI Press 2002.

[77] M. Singh and M. Valtorta. Construction of Bayesian network structures from data: a brief survey and an efficient algorithm. *International Journal of Approximate Reasoning*, 12(2):111–131, 1995.

[78] P. Spirtes, C. Glymour, and R. Scheines. *Causation, Prediction, and Search*. Springer Verlag, 1993.

[79] P. Spirtes, C. Glymour, and R. Scheines. *Causation, Prediction, and Search*. The MIT Press, 2nd edition, 2009.

[80] M. Steyvers, J.B. Tenenbaum, E.J. Wagenmakers, and B. Blum. Inferring causal networks from observations and interventions. *Cognitive Science*, 27(3):453–489, 2003.

[81] S. Storari, F. Riguzzi, and E. Lamma. Exploiting association and correlation rules parameters for learning Bayesian networks. *Intelligent Data Analysis*, 13(5):689–701, 2009.

[82] R. Tibshirani. Regression shrinkage and selection via the Lasso. *Journal of the Royal Statistical Society. Series B (Methodological)*, 58(1):267–288, 1996.

[83] M. Wainwright, P. Ravikumar, and J. Lafferty. High-dimensional graphical model selection using l1-regularized logistic regression. In B. Schölkopf, J. Platt, and T. Hoffman, editors, *Advances in Neural Information Processing Systems*, volume 19, pages 1465–1472. MIT Press, 2007.

[84] M.J. Wainwright and M.I. Jordan. A variational principle for graphical models. In S. Haykin, J. Principe, T. Sejnowski, and J. McWhirter, editors, *New Directions in Statistical Signal Processing: From Systems to Brain*, pages 21–93. MIT Press, 2005.

[85] M.J. Wainwright and M.I. Jordan. Graphical models, exponential families, and variational inference. *Foundations and Trends in Machine Learning*, 1(1-2):1–305, 2008.

[86] Y. Wang. *Latent Tree Models for Multivariate Density Estimation: Algorithms and Applications*. PhD thesis, Department of Computer Science and Engineering, The Hong Kong University of Science and Technology, 2009.

[87] Y. Wang, N. L. Zhang, and T. Chen. Latent tree models and approximate inference in Bayesian networks. *Journal of Artificial Intelligence Research*, 32:879–900, 2008.

[88] Y. Weiss. Correctness of local probability propagation in graphical models with loops. *Neural Computation*, 12(1):1–41, 2000.

[89] J. Whittaker. *Graphical Models in Applied Multivariate Statistics*. Wiley, 1st edition, 1990.

[90] M.L. Wong, W. Lam, and K.S. Leung. Using evolutionary computation and minimum description length principle for data mining of probabilistic knowledge. *IEEE Transactions on Pattern Analysis and Machine Intelligence*, 21:174–178, 1999.

[91] M.L. Wong, S.Y. Lee, and K.S. Leung. Data mining of Bayesian networks using cooperative coevolution. *Decision Support Systems*, 38(3):451–472, 2004.

[92] J.S. Yedidia, W.T. Freeman, and Y. Weiss. Generalized belief propagation. *Advances in Neural Information Processing Systems*, 13:689–695, 2000.

[93] C. Yuan. *Importance Sampling for Bayesian Networks: Principles, Algorithms and Performance*. PhD thesis, University of Pittsburg, 2000.

[94] C. Yuan and M.J. Druzdzel. Theoretical analysis and practical insights on importance sampling in Bayesian networks. *International Journal of Approximate Reasoning*, 46(2):320–333, 2007.

[95] M. Yuan and Y. Lin. Model selection and estimation in regression with grouped variables. *Journal of the Royal Statistical Society. Series B (Methodological)*, 68:49–67, 2006.

[96] J. Zhang. *Causal Inference and Reasoning in Causally Insufficient Systems*. PhD thesis, Canergie Mellon University, 2006.

[97] N. L. Zhang and T. Kocka. Efficient learning of hierarchical latent class models. In *Sixteenth IEEE International Conference on Tools with Artificial Intelligence (ICTAI)*, pages 585–593, 2004.

[98] N. L. Zhang, T. D. Nielsen, and F. V. Jensen. Latent variable discovery in classification models. *Artificial Intelligence in Medicine*, 30(3):283–299, 2004.

[99] N. L. Zhang and D. Poole. Exploiting causal independence in Bayesian network inference. *Journal of Artificial Intelligence Research*, 5:301–328, 1996.

[100] N.L. Zhang. Structural EM for hierarchical latent class models. Technical report, HKUST-CS03-06, Department of Computer Science, Hong Kong University of Science and Technology, 2003.

PART II

Gene Expression

CHAPTER 3

Graphical Models and Multivariate Analysis of Microarray Data

HARRI KIIVERI

The usual analysis of gene expression data ignores the correlation between gene expression values. Biologically, this assumption is unreasonable. The approach presented in this chapter allows for correlation between genes through a sparse Gaussian graphical model: sparse inverse covariance matrices and their associated graphical representations are used to capture the notion of gene networks. Existing methods find their limitations in the issue posed by the identification of the pattern of zeroes in such inverse covariance matrices. A workable solution for determining the zero pattern is provided in this chapter. Two other important contributions of this chapter are a method for very high dimensional model fitting and a distribution-free approach to hypothesis testing. Such tests address assessment of differential expression and of differential connection, a novel notion introduced in this chapter. An example dealing with real data is presented.

3.1 Introduction

A typical gene expression data set consists of measurements of a large number of genes for a set of subjects who have been given a pre-specified treatment or have a known genotype. This data is arranged into an $n \times p$ matrix where n, the sample size, is typically of the order of tens to hundreds, and p is of the order of tens of thousands or more. A simple and common treatment structure is one in which each individual is either exposed to a treatment (treatment) or not (control). The usual analysis of this data [21] attempts to identify differentially expressed genes i.e., genes

Probabilistic Graphical Models for Genetics, Genomics, and Postgenomics. First Edition. Christine Sinoquet & Raphaël Mourad (Eds). © Oxford University Press 2014. Published in 2014 by Oxford University Press.

whose mean expression over the treatment group is significantly higher or lower than the mean expression over the control group. The usual analysis makes no attempt to model the correlation structure between genes and so effectively assumes that gene expression measurements are uncorrelated. Biologically this assumption is unreasonable as genes are known to be connected in pathways or networks.

In this chapter, we present a model for gene expression data which explicitly allows for correlation between genes. However, in attempting to model the correlation or covariance matrix for gene expression data sets we encounter several problems which need to be overcome. Firstly, the sample covariance matrix is singular (not full rank or invertible) due to the small sample size n relative to the number of variables (genes) p, i.e., there are more variables than samples. This raises issues about the ability to estimate the covariance matrix (or equivalently its inverse) and the quality of the estimate, since the usual theory assumes that there are many more samples than variables. Secondly, the number of entries or parameters in the covariance or correlation matrix is astronomical, with numbers of the order of hundreds of millions quite common. It is apparent that we must reduce the number of these parameters somehow in order to make progress.

An attractive class of models that can be used in this context are Gaussian graphical models [14, 22] which are defined by patterns of zeroes in the *inverse* covariance matrix. Application of these models to genomic data has become popular, see for example [5, 20] and the references therein. At first glance it may appear unusual to consider the inverse covariance matrix, but as shown below, it is an interesting object because it encodes all the regression relationships between each gene and the remaining genes. In addition, zeroes in the inverse covariance matrix specify conditional independence constraints amongst the genes [14, 22]. Gaussian graphical models effectively address the issues mentioned above provided they are sufficiently sparse, i.e., they have large numbers of off-diagonal elements of the inverse covariance matrix equal to zero. These models can also be conveniently represented by graphs and, when applied to gene expression data, provide a representation of a gene network as a context in which analysis, including differential expression, can take place. An argument for the sparsity of biological networks can be found in [15].

This work differs from existing work on graphical models for gene expression data in that it does not just attempt to identify gene networks but also analyzes the effects of treatments in the context of the network. Other new elements are an algorithm for very high-dimensional model fitting and a distribution-free approach to hypothesis testing. The discussion below is a slightly simplified version of [12], but with the addition of the calculation of signal-to-noise ratios for all the relationships between genes, and an alternative analysis strategy which identifies interesting genes from a multivariate perspective. In this framework, genes which are not differentially expressed or differentially connected (see below) can still appear to be interesting.

The structure of this chapter is as follows. Section 3.2 presents the mean and covariance model specified for the data. In Section 3.3, we discuss the calculation of maximum likelihood estimates of the mean and inverse covariance parameters in the model, including the determination of the pattern of zeroes in the inverse covariance matrix. Hypothesis testing is covered in Section 3.4, where we consider an overall test of the hypothesis of no difference between treatment and control as well as tests for differences in components. We also consider an adjustment for multiple testing. An example illustrating the ideas presented in this chapter is given in Section 3.5, and we conclude with a discussion in Section 3.6.

3.2 The Model

Consider the $n \times p$ expression data matrix X. Associated with this matrix we have an $n \times r$ design matrix D, the rows of which describe the treatments which were applied to the n individuals/samples in the data set. An example of design matrix in the case of an experiment with a treatment and a control is given in equation (3.10) below (see also Appendix 3.A, page 101). For simplicity, in the following we do not distinguish between variables and their realizations, relying instead on the context to determine the distinction. Writing B for an $r \times p$ matrix of regression parameters, $X_{i.}$ for the ith row of X and X_j for the jth column of X, for the usual model we have

$$X_j \sim N(DB_j, \sigma_{jj}I). \tag{3.1}$$

i.e., we have a regression model with independent and identically distributed errors for the variables associated with each column of X. In the case of an experiment with a treatment and a control group, each column of the matrix B represents the effect of treatment and the effect of control on the mean expression of the associated gene. For the rows we have

$$\text{vec}\{X_{i.}\} \sim N(\text{vec}\{D_{i.}B\}, \text{diag}(\sigma)) \tag{3.2}$$

where σ is a $p \times 1$ vector of variances with elements σ_{ii} and the vec{} operator takes a matrix or row vector and forms it into a column vector row by row. We would like to keep the structure in equation (3.1) above in the usual model while expanding the variance model in equation (3.2) to allow for correlations. Let Σ be a $p \times p$ covariance matrix. Then the following model simultaneously achieves both aims:

$$\text{vec}\{X\} \sim N(\text{vec}\{DB\}, I \otimes \Sigma) \tag{3.3}$$

where I denotes the $n \times n$ identity matrix and $\otimes$ denotes the tensor product of matrices.

When the design matrix contains indicator $(0, 1)$ variables, as will be the case here, equation (3.3) is an example of a high-dimensional mixed graphical model (see [4] and [12]). This representation makes it easy to see and parameterize generalizations of the model, such as allowing the covariance matrix to be factor/treatment dependent.

As mentioned previously, we will specify our model for Σ indirectly through patterns of zeroes in Σ^{-1}, and we will estimate Σ indirectly by estimating Σ^{-1}. To see why this might be an interesting class of models, note that there is a one-to-one relationship between the elements of the inverse covariance matrix and regression coefficients according to the following. For Gaussian variables with covariance matrix Σ, writing X_i for the ith variable (gene), X_{-i} for the vector of the remaining variables, μ_i for the mean of variable i, and σ^{ij} for the ijth element of the inverse covariance matrix Σ^{-1}, we have

$$E\{X_i|X_{-i}\} = \mu_i + \sum_{j \neq i} \beta_{ij}(X_j - \mu_j),$$

$$\tag{3.4}$$

$$V\{X_i|X_{-i}\} = (\sigma^{ii})^{-1}$$

Fig 3.1 Example of a zero pattern (* denotes non-zero) in an inverse covariance matrix and the corresponding graphical representation for a five-variable model.

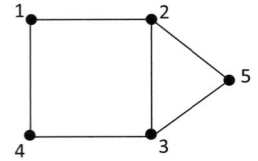

where the regression coefficients β_{ij} are related to the elements of the inverse covariance matrix by the equation $\beta_{ij} = -\sigma^{ij}/\sigma^{ii}$, E denotes conditional expectation, and V denotes conditional variance. (As a matter of notation, in this chapter we will use *superscripts*, e.g., σ^{ij} to denote elements of the inverse covariance matrix and *subscripts*, e.g., σ_{ij} to denote elements of the covariance matrix.) Equation (3.4) is easy to show from the properties of the multivariate normal distribution and the inverse of a partitioned matrix (see Appendix 3.B, page 102). Hence, inverting a covariance matrix is equivalent to simultaneously performing the p regressions of each variable on the remaining variables. Zeroes in the inverse covariance matrix now take on an interesting interpretation in that they imply that not all variables are required in these regressions. Equivalently, there are conditional independence constraints among the variables [22].

The pattern of zeroes in the inverse covariance matrix can be used to define a graphical representation of the model as follows. Each variable represents a vertex of the graph, and there is an edge in the graph between vertex i and j if and only if $\sigma^{ij} \neq 0$. For any given vertex i, the set of neighbors of i, $n(i) = \{j : \sigma^{ij} \neq 0\}$ is the set of vertices with an edge connected to vertex i. An illustration of this is given in Fig. 3.1.

For example, from this graph we can see that the neighbors of variable/vertex 5 are 2 and 3 and the regression of variable 5 on the rest has non-zero regression coefficients only for variables 2 and 3. More details can be found in [14] and [22]. Concerning the distributional assumptions, for Fig. 3.1 we have a five-dimensional multivariate normal distribution, and thus each node/vertex has a univariate Gaussian distribution.

Finally, in Appendix 3.C (page 103) it is shown that the signal-to-noise ratio (*snr*) for the regression of the ith variable on the rest (i.e., the ratio of the expected length of the regression-fitted value vector to the error variance) can be computed from the diagonals of the covariance matrix and its inverse as

$$snr_i = \sigma^{ii}\sigma_{ii} - 1. \tag{3.5}$$

This provides an indication of the strength of the relationship, with larger values indicating stronger relationships. The sparse Takahashi algorithm [6, 13] can be used to compute the diagonals of Σ without the need to compute or store all of Σ.

Having described the model, in the next section we consider the problem of estimating the parameters B and Σ^{-1} in order to fit the model given a data matrix X.

3.3 Model Fitting

In this section, we consider the estimation of the treatment and control effects in the parameter matrix B as well as the estimation of the inverse covariance matrix Σ^{-1} in the model (3.3). Note that the reader should keep in mind the distinction between the regression parameters in B and the regression interpretation of the elements of Σ^{-1}. The parameters in B are treatment and control effects while the regression parameters defined by (3.4) determine linear relationships between gene expression values. One set of parameters is associated with the mean structure of the model and the other set with the variance structure, i.e., the network.

In the following we first consider the case when the zero pattern in Σ^{-1} is known. We then present a strategy for determining the zero pattern given the expression data $\mathbf{X}$, and finally combine these strategies to handle the general case when the zero pattern is not known beforehand.

3.3.1 Maximum Likelihood Estimation when the Zero Pattern is Known

We use the maximum likelihood method to estimate the parameters B and Σ^{-1} in the model (3.3) given the n by p data matrix X, and we assume that the pattern of zeroes in Σ^{-1} is known. Apart from a constant, twice the log-likelihood function for model (3.3) is

$$2 * L = n \log(\det(\Sigma^{-1})) - \text{vec}\{X - DB\}^T (I \otimes \Sigma^{-1}) \text{vec}\{\mathbf{X} - \mathbf{DB}\}$$
$$= n \log(\det(\Sigma^{-1})) - \text{Tr}\{(X - DB)\Sigma^{-1}(X - DB)^T\} \qquad (3.6)$$

Equations for computing the maximum likelihood estimates can be obtained by setting the first derivatives of the log-likelihood function (3.6) to zero. We simply state the results here. For the reader interested in the details of these derivations see [12] and [3]. It can be shown [12] that the regression parameters B can be estimated without knowledge of Σ^{-1}. The maximum likelihood estimate of the regression parameters in the mean model is

$$\hat{B} = (D^T D)^{-1} D^T X. \qquad (3.7)$$

For an experiment with a treatment and a control group, $\hat{B}$ is a $2 \times p$ matrix with rows giving the mean gene expression values for each gene in the treatment and control groups.

The likelihood equations for estimating the non-zero elements of Σ^{-1} are as follows. For $1 \leq i, j \leq p$ let $NZ = \{(i,j), i \leq j : \sigma^{ij} \neq 0\}$ and $Z = \{(i,j), i < j : \sigma^{ij} = 0\}$. It is easily seen that NZ and Z are disjoint sets whose union is the set of all pairs (i,j) with $i \leq j$. The likelihood equations specify constraints on the disjoint sets NZ and Z as follows:

$$\hat{\sigma}_{ij} = s_{ij} \text{ for } (i,j) \in NZ,$$
$$\hat{\sigma}^{ij} = 0 \text{ for } (i,j) \in Z \qquad (3.8)$$

where $S = (s_{ij}) = R^T R / n$ and $R = X - D\hat{B}$. Note the distinction between subscripts and superscripts in (3.8). From equation (3.7) we see that the parameters in B are estimated by individual regression of each gene's expression data on the design matrix D. Equation (3.8) implies that the maximum likelihood estimate of Σ satisfies simultaneous constraints in disjoint parts of $\hat{\Sigma}^{-1}$ and $\hat{\Sigma}$. In particular, whenever $\hat{\sigma}^{ij} \neq 0$ (equivalently there is an edge in the model graph) the fitted *covariance* between variables i and j is equal to the ij^{th} element of the residual covariance matrix S. For the remaining elements, $\hat{\sigma}^{ij}$ is equal to 0, a constraint on elements of $\hat{\Sigma}^{-1}$.

In general, the solution of the likelihood equations (3.8) is obtained by maximizing (3.6) and requires an iterative process such as the limited memory quasi-Newton method [16]. A straight-forward approach is difficult if not impossible. It is necessary to avoid computing and storing all the elements of Σ (most of which are not needed) both from the point of view of memory and CPU requirements. For example if p equals 20 000 there are roughly 200 million unique elements in Σ, namely $\{\sigma_{ij} : 1 < i \leq j < p\}$. Sparse matrix representations and careful organization of calculations are required due to the high dimension of the matrix Σ^{-1}. We give a brief outline of the steps below. Details can be found in [13], where it is also pointed out that reasonable estimates can be expected if the sample size is sufficiently large compared to the size of the mean corrected sample covariance matrices restricted to the cliques of the model graph.

To solve the likelihood equations we maximize the log-likelihood function using the limited memory quasi-Newton algorithm [16] mentioned above. This method requires calculation of values of the likelihood function and its first derivative with respect to the non-zero elements of Σ^{-1}.

The key calculations involve computing a (sparse) Cholesky factor [2] in order to compute the determinant in (3.6) and the use of the sparse Takahasi algorithm [6] to compute only the small subset of the elements of Σ required in the first derivative, from Σ^{-1}. The calculations can be described as follows.

1 Reorder the variables so as to minimize fill-in (additional non-zero entries) when the sparse Cholesky factor is computed. This step only needs to be done once as the ordering stays the same throughout the iterations.

2 Only compute and store the elements of the sample covariance matrix which correspond to vertices and edges in the model graph.

3 Compute the sparse Cholesky factor in order to evaluate determinants.

4 When evaluating gradients, use the sparse Takahashi algorithm to compute only a (small) subset of the elements of Σ.

We refer the interested reader to [13] for more details.

3.3.2 Determining the Pattern of Zeroes in the Inverse Covariance Matrix

Clearly, identifying the pattern of zeroes in the inverse covariance matrix given the data matrix X is an important fundamental problem whose solution effectively determines the interconnections between gene expression values. Note that determining the pattern of zeroes is equivalent to the problem of determining the pattern of non-zeroes, and we approach the problem by determining the pattern of non-zeroes below.

Given the relationship (3.4) between regression coefficients and elements of the inverse covariance matrix, the following intuitively appealing method suggests itself:

1 For each column i of the matrix $R = X - D\hat{B}$, use some form of forward stepwise linear regression to produce an increasing number of predictors for column i up to some maximum number

$k_{max} < n$. After this step we can formulate a sequence of regression models involving the "best" k predictors for $(1 < k \leq k_{max})$ as determined by forward stepwise regression. The next step will be to choose one of these models, i.e., sets of predictors;

2 To choose a specific model for each variable from the given sequence of models from step (1), use the modified Bayesian information criterion (BIC) criterion of [1],

$$BIC_\gamma = n \log(\hat{\sigma}^2(k)) + k \log(n) + 2\gamma \left(\log\binom{p}{k}\right) \tag{3.9}$$

where $\hat{\sigma}^2(k)$ is the maximum likelihood residual variance estimate for a linear regression model with k predictors, $0 \leq \gamma \leq 1$ and $\binom{p}{k} = p!/((p-k)!k!)$ denotes the number of subsets of size k when there are p variables to choose from. Note that when γ equals zero, (3.9) corresponds to the usual BIC. When there are many more variables than observations the recommended value of γ is one [1]. This is the value we use.

Each of the regressions mentioned above contributes to a sparse p by p neighbor matrix A defined by

$$A_{ij} = \begin{cases} 1 & \text{if variable } j \text{ is chosen as a predictor of variable } i, \\ 0 & \text{otherwise.} \end{cases}$$

Finally the zero pattern is determined by computing $N = A + A^T$ and setting all diagonals and non-zero entries to 1 in the resulting matrix. This last step ensures that the pattern of zeroes is symmetric, as it must be in an inverse covariance matrix.

The above method is similar to that of [18], which has a tendency to overfit, but without the need to set a data-dependent tuning parameter for determining the sparsity of the estimated inverse covariance matrix. The method is appealing because of its simplicity and the ability to distribute large problems over multiple processors. Some arguments for preferring this type of method for very high-dimensional data sets are given in [13]. A high-dimensional simulation study [11] suggests using least absolute shrinkage and selection operator (LASSO) regression as a variable selector in step (1) above and demonstrates that the strategy can be quite effective. As might be expected, relationships with high signal-to-noise ratios also appear to be easier to detect for a given sample size. On the basis of simulations, a suggested size for *kmax* is approximately $n/20$, corresponding to approximately 20 observations per estimated regression parameter.

In Subsection 3.3.1, we presented the likelihood equations and briefly described an algorithm for estimating Σ^{-1} given the pattern of zeroes. We now suggest a two-step strategy for estimating Σ^{-1} when the pattern is not known. First, apply the strategy above to identify the zero pattern, and then fit the inverse covariance matrix using the methods described in Subsection 3.3.1 and in more detail in [13]. This two-step strategy can be applied by using the R library *sparse.inv.cov* available at http://www.bioinformatics.csiro.au/sparse.inv.cov.

Armed with estimates of the parameters in our model, we are now in a position to test hypotheses about treatment effects. This is the topic of the next section.

3.4 Hypothesis Testing

For simplicity, in this section we restrict ourselves to the simple treatment versus control structure for the matrix D in equation (3.3). Here we have

$$D = \begin{bmatrix} 1_q & 0 \\ 0 & 1_v \end{bmatrix} \tag{3.10}$$

where for example 1_q is an q by 1 vector of ones and q is the number of samples in the treatment group with $n = q + v$. The more general case is discussed in [12]. The usual null hypothesis in this situation is that there is no difference between the treatment and control groups for any of the genes. This hypothesis can be expressed in terms of a vector $c^T = [1, -1]/\sqrt{2}$ of unit length and be written in the form

$$\gamma = c^T B = 0 \tag{3.11}$$

where B denotes the matrix of true mean values for the treatment and control groups for each gene.

In deriving significance tests, our basic building block will be generating realizations from the null distribution of the estimated contrast vector $\hat{\gamma} = c^T \hat{B}$, and we describe the process for doing this in Subsection 3.4.1. To perform significance tests, intuitively, we generate realizations from the null distribution of $\hat{\gamma}$ by permutation, form a histogram for each component, and compute the quantiles (tails) of the histogram to determine critical values beyond which values of each component of $\hat{\gamma}$ are deemed to be significant, i.e., we reject the null hypothesis that for a given gene the means of the treatment and control groups are the same. A similar process applies for other test statistics which are a function of $\hat{\gamma}$.

A multivariate test statistic, T, which is a quadratic function of $\hat{\gamma}$ and which provides a single overall test of the null hypothesis. is presented in Subsection 3.4.2. It is shown how realizations from its null distribution can be obtained from realizations from the null distribution of $\hat{\gamma}$.

To enable a detailed examination of significant values of the multivariate test statistic T, we decompose it into p components T_i one for each gene, in Subsection 3.4.3. In addition, each of these components T_i consists of the product of two terms: one corresponds to the notion of differential expression and the other corresponds to a new concept termed differential connection. As all of these components are functions of $\hat{\gamma}$, realizations from their null distributions can be obtained from realizations from the null distribution of $\hat{\gamma}$ as before, and significance tests can be constructed to identify interesting genes.

3.4.1 Null Distributions by Permutation

Since we have estimated the inverse covariance matrix and are likely to have errors in the pattern of zeroes, rather than assume multivariate normality, we will use permutations to get null distributions of the parameter estimates and various functions of them. Following [7], we use the strategy below.

1 Fit the mean model under the null hypothesis that $c^T B = 0$. The fitted values are given by

$$\hat{F} = (11^T/n)X$$

where 1 is a $n \times 1$ vector of ones, since under the null hypotheses the means of the two groups are the same.

2 Compute the residuals R under the null hypothesis

$$R = X - \hat{F}.$$

3 For $k = 1, \ldots, m$ generate new data sets $X(k)$ according to

$$X(k) = \hat{F} + P_k R$$

where each P_k is a randomly chosen permutation matrix.

4 Compute the contrast $\tilde{\gamma}(k)$ using (3.7), namely

$$\tilde{\gamma}(k) = c^T (D^T D)^{-1} D^T X(k).$$

5 Build up an empirical null distribution of $\hat{\gamma}$ in (3.11) above using the m $\tilde{\gamma}(k)$ values from step 4.

Note that step 3 is equivalent to permuting the rows of X, so the procedure boils down to simply computing estimated contrasts from row permuted versions of the matrix X.

3.4.2 A Multivariate Test Statistic

Let $\hat{\gamma} = c^T \hat{B}$ be the observed contrast value for the data set, obtained by using (3.7) and (3.11). A simple calculation shows that under the null hypothesis

$$\hat{\gamma} \sim N(\gamma, \kappa \Sigma) \tag{3.12}$$

where $\kappa = c^T (D^T D)^{-1} c$ is a constant. This follows from (3.7), (3.11) and properties of linear transformations of moments of random variables. If we knew Σ, an overall test for the hypotheses (3.11) could be based on the Euclidean norm of the uncorrelated vector $(\kappa \Sigma)^{-1/2} \hat{\gamma}$ which we write as

$$T = \kappa \hat{\gamma}^T \Sigma^{-1} \hat{\gamma}. \tag{3.13}$$

In practice we ignore the constant κ and replace Σ^{-1} by its estimate $\hat{\Sigma}^{-1}$. Note that under the normality assumption, if $\hat{\Sigma}$ is diagonal, (3.13) is proportional to the sum of squares of p univariate t-statistics. Thus (3.13) could be thought of as a multivariate generalization of the t-test [19, p. 541]. Given m realizations $\tilde{\gamma}$ from the null distribution of $\hat{\gamma}$ as outlined above, we can generate a null distribution for (3.13) from

$$T(k) = \tilde{\gamma}(k) \hat{\Sigma}^{-1} \tilde{\gamma}(k)$$

for $k = 1, \ldots, m$ and compare the observed value to specified quantiles of this null distribution to define an overall test of the hypothesis (3.11).

3.4.3 Partitioning of the Test Statistic

The test statistic (3.13) gives an overall test of the null hypothesis; however an explanation of a significant value in terms of the behavior of individual genes is of interest. To see that (3.13) can be partitioned into components for each variable/gene, note that we can write

$$\hat{\gamma}^T \hat{\Sigma}^{-1} \hat{\gamma} = \sum_{i=1}^{p} \hat{\gamma}_i (\hat{\Sigma}^{-1} \hat{\gamma})_i$$

$$= \sum_{i=1}^{p} T_i \qquad (3.14)$$

where $T_i = \hat{\gamma}_i (\hat{\Sigma}^{-1} \hat{\gamma})_i$. Observed values of T_i can be calculated from $\hat{\gamma}_i$ and $\hat{\Sigma}^{-1}$. Realizations from the null distribution of T_i can be calculated from realizations from the null distribution of $\hat{\gamma}_i$. Estimated quantiles from this null distribution can be used to define tests to identify significant values, and thus, which components have made most contribution to the overall value. Bearing in mind the graph of the model, identifying significant components can be visualized as highlighting vertices of the graph (network) where potentially interesting things are happening.

Each T_i itself has two components $\hat{\gamma}_i$, $(\hat{\gamma}_N)_i$, which we can see from

$$T_i = \hat{\gamma}_i (\hat{\Sigma}^{-1} \hat{\gamma})_i$$

$$= \hat{\gamma}_i \hat{\sigma}^{ii} \left(\hat{\gamma}_i - \sum_{j \in n(i)} \hat{\beta}_{ij} \hat{\gamma}_j \right)$$

$$= \hat{\sigma}^{ii} \hat{\gamma}_i (\hat{\gamma}_N)_i$$

where $n(i) = \{j : \sigma^{ij} \neq 0\}$ is the set of neighbors of i, $\hat{\beta}_{ij} = -\hat{\sigma}^{ij}/\hat{\sigma}^{ii}$ and $(\hat{\gamma}_N)_i = \hat{\gamma}_i - \sum_{j \in n(i)} \hat{\beta}_{ij} \hat{\gamma}_j$ is a neighbor-corrected contrast value. Hence we can see that, apart from the scale factor $\hat{\sigma}^{ii}$, each T_i has two components, the first one being the difference in mean expression between treatment and control, $\hat{\gamma}_i$, i.e., related to differential expression, and the second term involving a neighbor-corrected contrast value which we term differential connection. Observed values of these individual components can be calculated and null distributions obtained by permutation as before for T and T_i.

To understand these components let us assume the ideal situation in which we know Σ^{-1}. Note that from (3.12) we have $\hat{\gamma}_i \sim N(\gamma_i, \sigma_{ii})$, so the hypothesis $\gamma_i = 0$ is the hypothesis of no differential expression and is a hypothesis in a marginal distribution. From (3.12) we also have

$$\Sigma^{-1} \hat{\gamma} \sim N(\Sigma^{-1} \gamma, \Sigma^{-1}) \qquad (3.15)$$

and hence $(\hat{\gamma}_N)_i \sim N((\gamma_N)_i, 1/\sigma^{ii})$. From (3.12), and the structure in $\hat{\Sigma}^{-1}$, we also have

$$
\begin{aligned}
E\{\hat{\gamma}_i|\hat{\gamma}_{-i}\} &= \gamma_i + \sum_{j\in n(i)} \beta_{ij}(\hat{\gamma}_j - \gamma_j) \\
&= \left(\gamma_i - \sum_{j\in n(i)} \beta_{ij}\gamma_j\right) + \sum_{j\in n(i)} \beta_{ij}\hat{\gamma}_j \\
&= (\gamma_N)_i + \sum_{j\in n(i)} \beta_{ij}\hat{\gamma}_j
\end{aligned}
\tag{3.16}
$$

(see equation (3.4)). Hence the hypothesis $(\gamma_N)_i = 0$ is a hypothesis in the conditional distribution of $\hat{\gamma}_i$ given $\hat{\gamma}_{-i}$ or, equivalently, $\hat{\gamma}_{n(i)}$. Under the null hypothesis the expected contrast value for gene i is a weighted linear combination of the contrast values for its neighboring genes (see equation (3.15)). Differential connection means there is a pattern of differential expression involving gene i and its neighbors which causes the weighted linear combination of neighboring contrast values to over- or underestimate the expected contrast value for gene i, thus highlighting gene i and its connections to its neighboring genes for further investigation. This is new information (i.e., the gene, the identity of its neighbors, and an interesting pattern involving them all) not available from the traditional analysis. Biologically, this may suggest that the gene and its neighbors belong to one or more pathways which have been affected by the treatment. A gene can be identified with both properties, i.e., differential expression and differential connection, neither property, or simply one of them. The case of differential connection without differential expression may point to some form of post-transcriptional regulation.

3.4.4 Testing Strategies

In the previous subsections, we presented some test statistics and the calculation of associated null distributions for use in significance testing. We now present two different strategies for organizing the above information into an analysis of a data set.

Our first testing strategy is simply to simultaneously test for differential expression and differential connection. This will involve $2p$ significance tests and we require (quantiles of) the null distributions for all the $\hat{\gamma}_i$ and $(\hat{\gamma}_N)_i$. To adjust for multiple testing, we use the modified Bonferroni method of [8] where the significance level is $\alpha/(2p)$, and where α is the expected number of false positives, a number typically greater than one.

Our second strategy is a sequential one and begins with the overall test using T, then p tests for each of the T_i in (3.14). We then only test for differential expression and differential connection for the significant T_i. For multiple testing, here we could use significance level $\alpha/(p + u)$, where $u \le p$ is an a priori upper bound on the number of differentially expressed and differentially connected genes.

In computing null distributions, the number of permutations m needs to be chosen so that there are at least a few permuted statistic values greater than the selected quantile.

Note that the null distributions described above are distribution-free in the sense that they only depend on first-and second-order moments and not on the Gaussian assumption. Unlike other methods, such as the t-test, no divisor needs to be estimated or regularized for each individual test. Hence we are not relying on model assumptions to provide regularized estimates of variances and tests. It should not be surprising that (particularly with small sample sizes) there will be some differences in the results of testing compared to the usual model-based methods.

3.5 Example

To illustrate some of the ideas presented above, we use the data in [23]. The data comprise 29 samples of gene expression measurements obtained with Affymetrix U133A chips. The "treatment group" here consists of 14 samples of apparently normal breast epithelial cells from women *with* breast cancer, and the control group consists of 15 samples of apparently normal breast epithelial cells from women *without* breast cancer who were undergoing breast reduction surgery. The study sought to determine abnormalities in the normal appearing epithelium of breast cancer patients which might, among other things, improve cancer risk assessment. Following [23], we removed genes with little variation between the two groups, which resulted in 14 681 genes remaining in the study.

Due to the small sample size, to determine the pattern of zeroes in the inverse covariance matrix, we restricted the search for the neighbors of each gene to a maximum of two genes, corresponding to roughly 15 observations per regression coefficient. Note that this does not imply that the maximum number of connections that a gene has is two, as will be seen below. Clearly, we cannot hope to reliably fit a comprehensive network structure to this data set but we can hope to estimate some strongly connected (sub) components of it. Applying the zero pattern detection strategy described in Subsection 3.3.2, we obtained a sparse inverse covariance matrix with 24 313

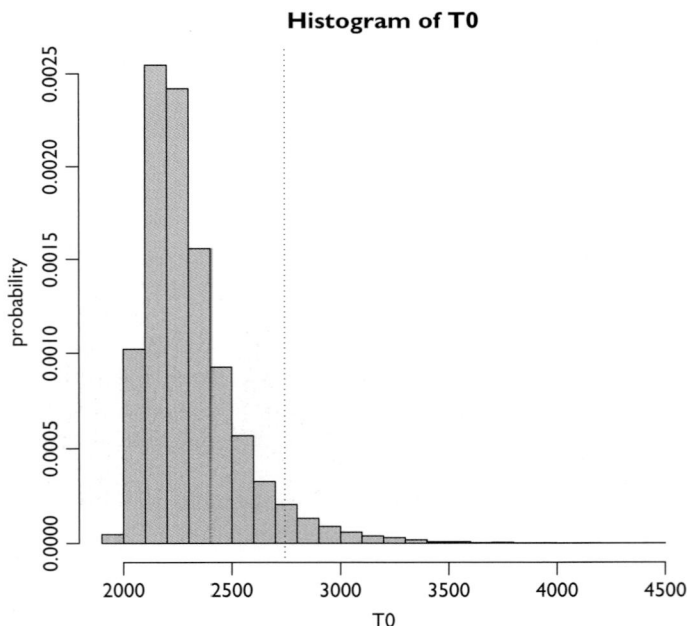

Fig 3.2 Histogram of null distribution of *T*.

non-zero elements. The graph of the model was relatively simple with 3395 cliques of size 1, i.e., apparently independent genes, and 9632 cliques of size two. This implies that the graph of the model is a simple tree-like structure. The signal-to-noise ratios of the fitted regression relationships (3.5) had a median value of approximately 1.5.

We could produce lists of differentially expressed and differentially connected genes as in [12]. However, to illustrate something different, we will use the components of the overall test (3.14) to identify interesting genes and interesting places in the network instead. This corresponds to using the second strategy for testing mentioned in Subsection 3.4.4.

For testing, we used $m = 40\,000$ permutations to generate the null distributions. The significance levels were determined conservatively by a priori assuming up to half of the genes could be differentially expressed, and up to half differentially connected, so we might expect to do roughly $2p$ tests. In other words, to determine significant components of T we do p tests and then to identify differential expression and connection on the selected genes, we would do at most an additional p tests. The expected number of false positives is set to 14.681, giving a p-value of 5×10^{-4} for use in multiple testing.

The overall test of the hypothesis that the mean gene expression of the two groups is the same gave a T statistic value of 5771.52 with p-value strictly less than 2.5×10^{-5}. Hence there is strong evidence for expression differences between the two groups. A plot of the null distribution for this test is given in Fig. 3.2. The vertical line is at $T = 2744.16$, which is the quantile corresponding to a p-value of 0.05.

The tests for significant components of the T statistic are illustrated in Fig. 3.3. In Fig. 3.3, the components of T and the significance levels adjusted for multiple testing have been transformed so that the upper and lower critical levels are 1 and –1 respectively. The lower edge at –1 is

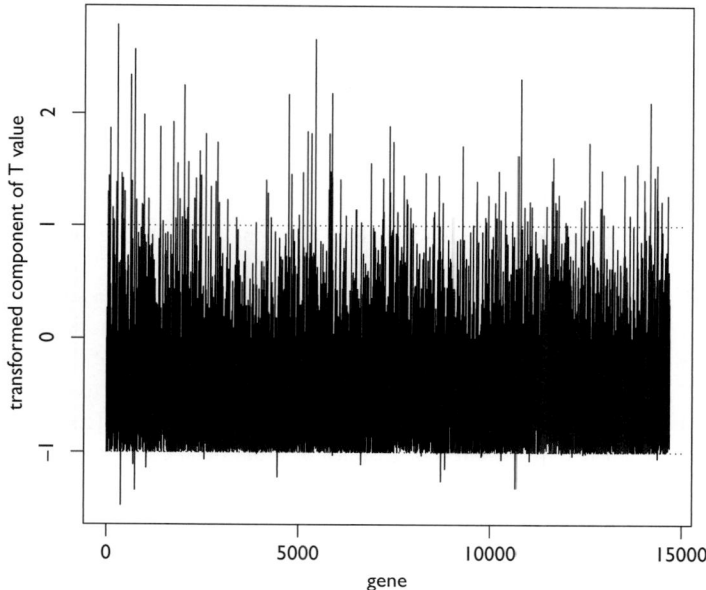

Fig 3.3 Transformed components of T.

EXAMPLE | 97

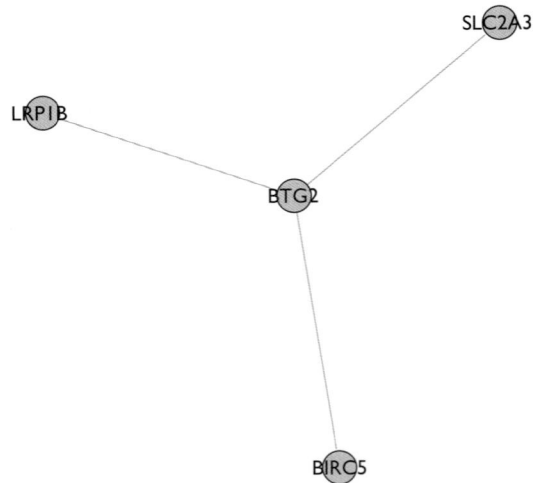

Fig 3.4 Plot of connections to gene (BTG2) with the largest T component.

in fact much more ragged than shown. However the large number of genes plotted in the relatively small space of the x axis obscures this fact.

A total of 167 components were identified as significant. Interesting functions associated with some of these genes are immune response, intracellular signaling, cell morphogenesis, cell cycle regulation, and proteolysis. The list also contains genes known to be associated with breast cancer or breast cancer susceptibility. The largest component of T corresponded to the gene BTG2, which is placed in the class of transcription factors and regulators in [23]. We extracted the neighbors of this gene, which are shown in Fig. 3.4. The estimated signal-to-noise ratio of this relationship was 4.74. A cursory functional analysis [9, 10] of these genes reveals descriptive words such as "inhibition of apoptosis," "B cell survival pathway," and a protein product known to be deleted in tumors. This illustrates that the genes and relationships in Fig. 3.4. are biologically interesting and bear further investigation.

Testing further, of the 167 genes with significant T components, 47 were differentially expressed, and 136 were differentially connected. The number that were both differentially expressed and connected was 41, and, interestingly, 25 were neither significantly differentially expressed nor connected, indicating the possibility of finding interesting genes which are neither differentially expressed nor differentially connected.

We also built a local network of size 3 (i.e., at most three edges from a specified gene to any other connected gene) around each of these 167 genes. Of these, the local network with the largest number of edges is given in Fig. 3.5. The estimated signal-to-noise ratio of the regression for the gene RPS11 is 21.2, suggesting a very strong relationship. It is interesting to note that there is an over-representation of genes associated with KEGG pathway "ribosome" (e.g., those with names beginning with RP). Clearly we have fertile ground for more detailed checking of the 167 lists of genes and their relationships, known pathways, and functions in relation to the development and detection of breast cancer. These lists may also contain new yet to be documented functions and pathways.

None of the genes in Fig. 3.5 appears to be differentially expressed. Normally we might expect to explain differential connection through patterns of differential expression involving a gene and

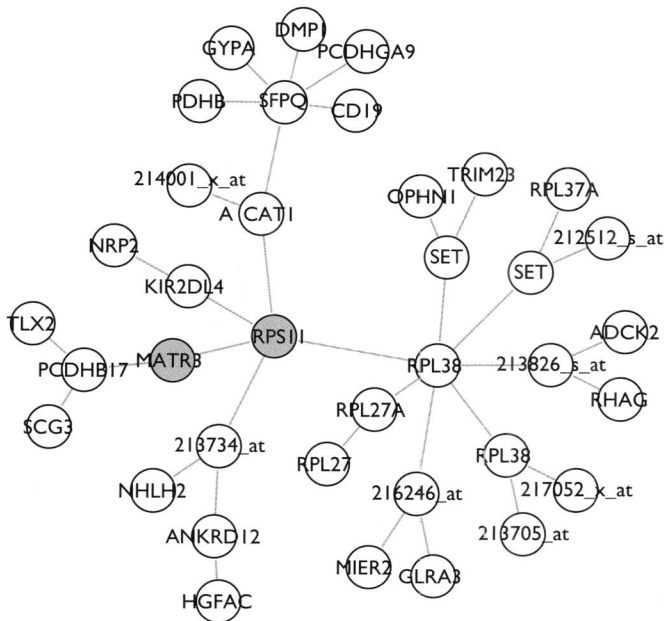

Fig 3.5 Graph of local network about the gene RPS11. Gray denotes genes the differentially connected genes.

its neighbors. Here this does not appear to be the case, and, for example, we might hypothesize the possibility of post-transcriptional regulation. In any case, to explore this further, we plot the expression pattern of the gene RPS11 in relation to its immediate neighbors in Fig. 3.6 (page 100). In the figure, the black line identifies the observed contrast values for each of the genes. The box plot is related to the null distribution of the contrast values, and the inter-quartile range and critical quantiles are plotted for each gene. Values between the extremes of the whiskers are not deemed differentially expressed. The expected value of the contrast for RPS11 as a weighted sum of its neighboring contrasts is plotted as a gray dot. Estimated regression coefficients, except the value for RPS11, which is the estimated residual variance, are also displayed on the plot.

On the basis of this plot, the location of the gray dot appears to be due mostly to the contrast values of the genes RPL38 and ACAT1. The discrepancy from the contrast value for RPS11 is intriguing and requires further investigation. We could also draw a similar plot for the other differentially connected gene, MATR3.

3.6 Discussion and Conclusions

In this chapter, we have presented a model for microarray data which specifically allows for correlation between genes. We showed how to estimate the mean parameters in the model as well as the parameters in the inverse covariance matrix, including its pattern of zeroes. This involved determining regression relationships among each gene and the remaining genes, a computationally demanding step not performed in the usual analysis of differential expression. We also introduced

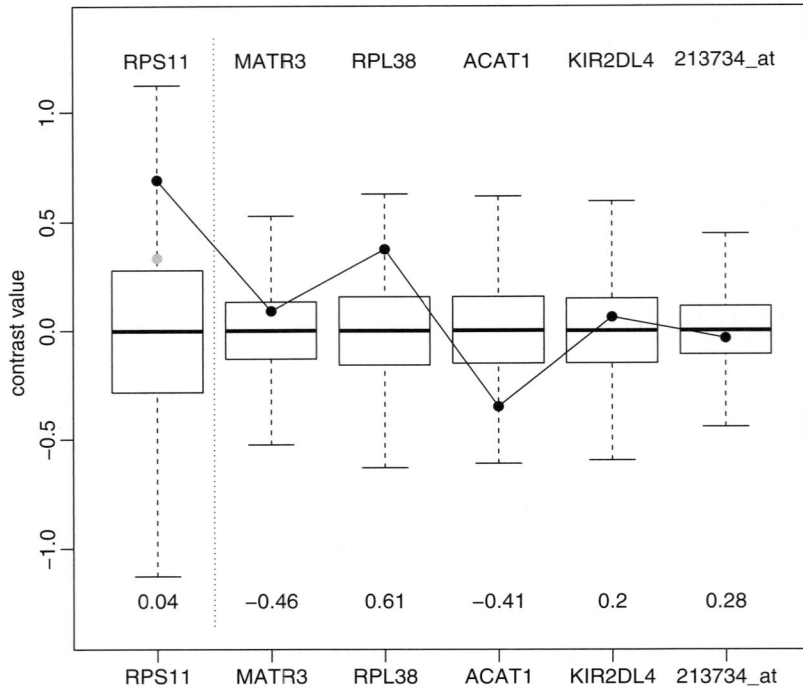

Fig 3.6 Expression pattern for gene RPS11 and its neighbors.

the notion of differential connection which involves a gene and its immediate neighbors. Given the parameter estimates, we then constructed distribution-free tests of hypotheses through the use of permutations. These tests are also useful for the analysis of differential expression. We derived an overall test of hypothesis as well as tests for individual components with an adjustment for multiple testing. It was shown how the model could be represented graphically and the analysis of the data done in the context of the model graph (network). In the example, we showed that there is an alternative way to analyze the data which goes beyond the production of lists of differentially expressed genes, and respects the idea that genes can belong to biological pathways.

If we consider the possibility that each non-zero entry in the inverse covariance matrix could be the basis of an experiment to determine the biological reasons for a relationship between two genes, it becomes clear that there is the potential of an overwhelming amount of information in a microarray data set and that this information has typically been overlooked. A graph is a convenient way to summarize this information, though more work needs to be done to enable the manipulation and display of these very large sparse graphs.

We now turn to a discussion of some new ideas and extensions of the work presented above.

It is interesting to consider the possibility of data transformations to try to make the data look more multivariate normal, as this may improve the linear predictive relationships between genes (see [17] for example).

Graphical queries such as the shortest path between two genes, local connectivity, enumerating cliques of genes, and identifying connected components of genes are easy to do and may also provide useful information.

Any regression relationship between a gene and its neighbors can be subject to further statistical analysis by cross validation, model checking, and examining fitted values and residuals.

It would also be interesting to check the linearity assumption. At the biological level, we can search for known relationships in the literature which could help explain the regression results. However we should remain open to the possibility of discovering new relationships, as existing knowledge is incomplete.

In this chapter, we have assumed the same inverse covariance matrix for the treatment and control groups; however, models which allowed a different inverse covariance matrix for the treatment and control groups could also be fitted and tested. Such models would allow for the possibility of a treatment activating different biological pathways or the same pathways in different ways. To be feasible, this would require each group to have a reasonably large number of observations.

We have used undirected graphical models in this chapter; however, given an ordering of the variables, we could also have used models with directed graphs (Bayesian networks (BN)) to model the covariance matrix. In this case, the sparse Cholesky factor of the inverse covariance matrix becomes the object of interest as it encodes the regression relationships between each variable and its predecessors or parents. An analysis similar to the one above could be done in this case also.

The results in this chapter were produced by using the R packages *sparse.inv.cov* and *mvama*. These packages along with simple tutorials are freely available at http://www.bioinformatics.csiro.au.

In conclusion, there is a lot more information in a microarray data set than the usual analysis extracts. In this chapter, we have presented some strategies for extracting some of this additional information, and we have shown the usefulness of graphical models for representing this information, and guiding the analysis.

APPENDIX 3.A DESIGN MATRICES

A design matrix is a matrix of explanatory variables (which may be indicator variables) used to define the linear relationship between the mean value of a set of observations and the explanatory variables. For example, suppose we have observations $y^T = (y_1, y_2, y_3, y_4, y_5, y_6, y_7, y_8)$ with associated values of a covariate $z^T = (z_1, z_2, z_3, z_4, z_5, z_6, z_7, z_8)$. If observations 1 to 4 come from group 1 and observations 5 to 8 come from group 2, then a linear model in which the mean (expected value) of y depends on the group and the covariate in a linear manner can be written as

$$y = E\{y\} + \varepsilon$$
$$= Db + \varepsilon$$

where D is the design matrix and E denotes expectation. In expanded form, this equation is

$$\begin{bmatrix} y_1 \\ y_2 \\ y_3 \\ y_4 \\ y_5 \\ y_6 \\ y_7 \\ y_8 \end{bmatrix} = \begin{bmatrix} 1 & 0 & z_1 \\ 1 & 0 & z_2 \\ 1 & 0 & z_3 \\ 1 & 0 & z_4 \\ 0 & 1 & z_5 \\ 0 & 1 & z_6 \\ 0 & 1 & z_7 \\ 0 & 1 & z_8 \end{bmatrix} \begin{bmatrix} b_1 \\ b_2 \\ b_3 \end{bmatrix} + \begin{bmatrix} \varepsilon_1 \\ \varepsilon_2 \\ \varepsilon_3 \\ \varepsilon_4 \\ \varepsilon_5 \\ \varepsilon_6 \\ \varepsilon_7 \\ \varepsilon_8 \end{bmatrix}.$$

Here b_1 is a mean parameter associated with group 1, b_2 is a mean parameter associated with group 2, and b_3 is the parameter associated with the covariate.

APPENDIX 3.B THE RELATIONSHIP BETWEEN REGRESSION COEFFICIENTS AND THE ELEMENTS OF THE INVERSE COVARIANCE MATRIX

In this section, we derive expressions for regression parameters in terms of the elements of sigma inverse. Suppose that $x = (x_1, x_2, \ldots, x_p)^T$ has a multivariate Gaussian distribution with mean μ and covariance matrix Σ. We write I for the set $\{1, 2, \ldots, p\} \setminus \{i\}$ and partitioning the covariance matrix as

$$\Sigma = \begin{pmatrix} \Sigma_{ii} & \Sigma_{iI} \\ \Sigma_{Ii} & \Sigma_{II} \end{pmatrix}$$

where, for example, $\Sigma_{ii} = \sigma_{ii}$ and Σ_{iI} denote the $1 \times (p-1)$ matrix of covariances between variable i and variables indexed by I. It is well known [19, p. 522] that the conditional mean and variance of variable i given the rest is given by

$$E\{x_i | x_I\} = \mu_i + \Sigma_{iI} \Sigma_{II}^{-1} (x_I - \mu_I)$$
$$= \mu_i + \beta_I^T (x_I - \mu_I)$$
$$V\{x_i | x_I\} = \Sigma_{ii} - \Sigma_{Ii} \Sigma_{II}^{-1} \Sigma_{Ii}$$

where μ_I denotes the mean vector of the variables indexed by I and $\beta_I^T = \Sigma_{iI} \Sigma_{II}^{-1}$. Partitioning the inverse covariance matrix in the same way, but with superscripts, note that we have

$$\begin{pmatrix} \Sigma_{ii} & \Sigma_{iI} \\ \Sigma_{Ii} & \Sigma_{II} \end{pmatrix} \begin{pmatrix} \Sigma^{ii} & \Sigma^{iI} \\ \Sigma^{Ii} & \Sigma^{II} \end{pmatrix} = \begin{pmatrix} 1 & 0 \\ 0 & J_{p-1} \end{pmatrix} \tag{3.17}$$

where J_{p-1} is the $(p-1) \times (p-1)$ identity matrix. From the $(2, 1)$ element of the matrix product in (3.17), we obtain the equation

$$\Sigma_{Ii} \Sigma^{ii} + \Sigma_{II} \Sigma^{Ii} = 0$$

from which we obtain

$$\beta_I = \Sigma_{II}^{-1} \Sigma_{Ii}$$
$$= -\Sigma^{Ii} (\Sigma^{ii})^{-1} \tag{3.18}$$
$$= -\Sigma^{Ii} / \sigma^{ii}.$$

Hence it follows that the conditional expectation can be expressed in terms of elements of the inverse covariance matrix. Similarly from the $(1, 1)$ element of the matrix product in (3.17), we obtain

$$\Sigma_{ii} \Sigma^{ii} + \Sigma_{iI} \Sigma^{Ii} = 1. \tag{3.19}$$

Rearranging (3.19), we get

$$(\Sigma^{ii})^{-1} = \Sigma_{ii} + \Sigma_{iI}\Sigma^{Ii}(\Sigma^{ii})^{-1}$$

and from (3.18),

$$(\sigma^{ii})^{-1} = (\Sigma^{ii})^{-1} = \Sigma_{ii} - \Sigma_{iI}(\Sigma_{II})^{-1}\Sigma_{Ii}.$$

APPENDIX 3.C SIGNAL-TO-NOISE RATIOS

Consider the typical regression model

$$y = X\beta + \varepsilon \tag{3.20}$$

where y is $n \times 1$, the columns of X ($n \times p$) are mean corrected, and the components of ε are independent and identically distributed with variance σ^2. We can define the signal-to-noise ratio as the ratio of the squared norm of the fitted value in the regression to the squared norm of the residual in (3.20), i.e.,

$$snr = \frac{\|X\beta\|^2}{\|\varepsilon\|^2}. \tag{3.21}$$

Taking the expectation of the numerator and denominator of (3.21) we can define an expected *snr* as

$$snr = \frac{\beta^T\Sigma_{xx}\beta}{\sigma^2} \tag{3.22}$$

where $\Sigma_{xx} = E\{X^TX\}/n$. Thus, the *snr* is a measure of the (expected) relative lengths of the fitted values and residuals in the regression. Bearing in mind the regression interpretation of the elements in the inverse covariance matrix, using the above ideas we can characterize the signal-to-noise ratios of the regressions encoded in the inverse covariance matrix as follows. Let $I = \{1, 2, \ldots, p\}\setminus\{i\}$. Then, writing Σ_{Ii} for the ith column of Σ with the ith row removed, and Σ_{II} for the submatrix of Σ with ith row and ith column removed, with $\Sigma_{iI} = (\Sigma_{Ii})^T$ and $\beta = (\Sigma_{II})^{-1}\Sigma_{Ii}$ we can calculate the (expected) signal-to-noise ratio as

$$\begin{aligned} snr_i &= \frac{\beta_i^T\Sigma_{II}\beta_i}{(1/\sigma^{ii})} \\ &= \Sigma_{iI}(\Sigma_{II})^{-1}\Sigma_{Ii}\,\sigma^{ii} \\ &= (\sigma_{ii} - 1/\sigma^{ii})\sigma^{ii} \\ &= \sigma_{ii}\sigma^{ii} - 1 \end{aligned} \tag{3.23}$$

where σ_{ii} denotes the ith diagonal element of Σ and σ^{ii} denotes the ith diagonal element of Σ^{-1}. Equation (3.23) gives a simple expression for the signal-to-noise ratios of the regressions defined by the inverse covariance matrix. Note that the calculation of (3.23) requires the diagonal elements of both Σ and Σ^{-1}.

REFERENCES

[1] J. Chen and Z. Chen. Extended Bayesian information criteria for model selection with large model spaces. *Biometrika*, 95(3):759–771, 2008.

[2] Y. Chen, T. Davis, W. Hager, and S. Rajamanickam. Algorithm 887 CHOLMOD, supernodal sparse Cholesky factorization and update/downdate. *ACM Transactions on Mathematical Software*, 35(3):1–14, 2008.

[3] A. P. Dempster. Covariance selection. *Biometrics*, 28(1):157–175, 1972.

[4] D. Edwards. *Introduction to Graphical Modelling*. Springer Texts in Statistics. Springer, 2000.

[5] D. Edwards, G. de Abreu, and R. Labouria. Selecting high-dimensional mixed graphical models using minimal AIC or BIC forests. *BMC Bioinformatics*, 11(1):18, 2010.

[6] A. Erisman and W. Tinney. On computing certain elements of the inverse of a sparse matrix. *Communications of the ACM*, 18(3):177–179, 1975.

[7] D. Freedman and D. Lane. A nonstochastic interpretation of reported significance levels. *Journal of Business & Economic Statistics*, 1(4):292–298, 1983.

[8] A. Gordon, G. Glazko, X. Qiu, and A. Yakovlev. Control of the mean number of false discoveries, Bonferroni and stability of multiple testing. *The Annals of Applied Statistics*, 1(1):179–190, 2007.

[9] D. W. Huang, B. T. Sherman, and R. A. Lempicki. Systematic and integrative analysis of large gene lists using DAVID bioinformatics resources. *Nature Protocols*, 4(1):44–57, 2008.

[10] D. W. Huang, B. T. Sherman, and R. A. Lempicki. Bioinformatics enrichment tools: paths toward the comprehensive functional analysis of large gene lists. *Nucleic Acids Research*, 37(1):1–13, 2009.

[11] H. T. Kiiveri and R. Dunne. Determining the pattern of zeroes in very large sparse inverse covariance matrices. *CMIS technical report*, 2011.

[12] H.T. Kiiveri. Multivariate analysis of microarray data: differential expression and differential connection. *BMC Bioinformatics*, 12(42), 2011.

[13] H.T. Kiiveri and F. de Hoog. Fitting very large sparse Gaussian graphical models. *Computational Statistics & Data Analysis*, 56(9):2626–2636, 2012.

[14] S.L. Lauritzen. *Graphical Models*. Oxford Science Publications. Clarendon Press, 1996.

[15] R.D. Leclerc. Survival of the sparsest: robust gene networks are parsimonious. *Molecular Systems Biology*, 4:Article 213, 2008.

[16] D. Liu and J. Nocedal. On the limited memory method for large scale optimization. *Mathematical Programming B*, 45(3):503–528, 1989.

[17] H. Liu, J. Lafferty, and L. Wasserman. The nonparanormal: semiparametric estimation of high dimensional undirected graphs. *Journal of Machine Learning Research*, 10:1–37, 2009.

[18] N. Meinshausen and P. Buhlman. High-dimensional graphs and variable selection with the Lasso. *The Annals of Statistics*, 34(3):1436–1462, 2006.

[19] C.R. Rao. *Linear Statistical Inference and its Applications*. Wiley Series in Probability and Mathematical Statistics. Probability and Mathematical Statistics. Wiley, 2001.

[20] J. Schafer and K. Strimmer. Learning large-scale graphical Gaussian models from genomic data. *AIP Conference Proceedings*, 776:263–276, 2005.

[21] G.K. Smyth. Linear models and empirical Bayes methods for assessing differential expression in microarray experiments. *Statistical Applications in Genetics and Molecular Biology*, 3(1):Article 3, 2004.

[22] T.P. Speed and H.T. Kiiveri. Gaussian Markov distributions over finite graphs. *Annals of Statistics*, 14(1):138–150, 1986.

[23] A. Tripathi, C. King, A. de la Morenas, V.K. Perry, B. Burke, G.A. Antoine, E. F. Hirsch, M. Kavanah, J. Mendez, M. Stone, N. P. Gerry, M. E. Lenburg, and C. L. Rosenberg. Gene expression abnormalities in histologically normal breast epithelium of breast cancer patients. *International Journal of Cancer*, 122(7):1557–1566, 2008.

Comparison of Mixture Bayesian and Mixture Regression Approaches to Infer Gene Networks

SANDRA L. RODRIGUEZ-ZAS AND BRUCE R. SOUTHEY

Most Bayesian network applications to gene network reconstruction assume a single distributional model across all the samples and treatments analyzed. This assumption is likely to be unrealistic, especially when describing the relationship between genes across a range of treatments with potentially different impacts on the networks. To address this limitation, a mixture Bayesian network approach has been developed. In addition, the equivalence between Bayesian networks and regression approaches has been demonstrated. The goal of this chapter is to compare the two strategies, the mixture Bayesian network approach and the mixture regression approach, when used for the purpose of gene network inference. The finite mixture model that is integrated into both strategies allows the characterization of gene relationships unique to particular conditions as well as the identification of interactions shared across conditions. The chapter reviews both approaches, comparing their performances on real data describing a pathway analyzed under up to nine different experimental conditions, and highlights the strengths of the approaches evaluated. An interpretation of the mixtures estimated as best describing the data is provided.

Probabilistic Graphical Models for Genetics, Genomics, and Postgenomics. First Edition. Christine Sinoquet & Raphaël Mourad (Eds). © Oxford University Press 2014. Published in 2014 by Oxford University Press.

4.1 Introduction

Gene expression is measured in samples representing a range of treatments or conditions typically using microarray, RNA-sequencing (RNA-seq), or real-time quantitative PCR experiments. Several approaches, from simple pairwise correlation between gene expression profiles to Bayesian networks, have been proposed to infer gene networks using gene expression profiles. These gene networks can verify known pathways or serve as pilot work to discover new relationships among genes.

Bayesian networks offer an intuitive model of the relationships between genes using acyclic, directed graphs based on a probabilistic framework that can accommodate missing data and prior information [11]. Bayesian networks are well fitted to identify relationships between genes in a network [11, 16, 24]. Most Bayesian network applications to gene network reconstruction assume a single distributional model across all the samples and treatments analyzed. This assumption is likely to be unrealistic, especially when describing the relationship between genes across a range of treatments with potentially different impacts on the network. To address this limitation, we developed a mixture Bayesian network approach and demonstrated the superiority of this approach on a wide range of pathways and data sets [17, 18, 23]. These results showcased the flexibility of mixture Bayesian networks to infer gene relationships that vary between treatments. This is a valuable feature when exploiting information from large gene expression data sets that span across many treatments.

Bayesian network learning is NP-hard and can impose an intense computational challenge [2, 26, 32]. This is due to the presence of large and densely connected networks that are intractable for exact computations [32]. The number of possible networks increases super-exponentially with the number of genes considered [26]. For instance, there are 1018 possible network topologies in a small network of 10 genes. Greedy search [17], genetic algorithm [34], bootstrap [5], or Markov chain Monte Carlo (MCMC) [14, 23] methods have been proposed to learn Bayesian networks. Thus, Bayesian network applications require the availability of adequate computational environments and implementation expertise that is seldom available to researchers in the life sciences. In addition, the long time required to infer a network could discourage the application of Bayesian networks to many gene expression experiments.

The goal of this chapter is to compare two strategies, a mixture Bayesian network approach and a mixture regression approach, to infer gene networks. The latter strategy is easy to implement and offers a computationally efficient approach to infer gene networks. The equivalence between Bayesian network and regression approaches has been demonstrated [25]. The finite mixture model that was integrated into both strategies allows both the characterization of gene relationships unique to particular conditions or shared across conditions. The chapter is organized in the following sections. Section 4.2 reviews the mixture Bayesian approach, introduces the mixture regression approach, and describes the data used to evaluate the network approaches. Section 4.3 includes a comparison of the performances of both approaches applied to the genes in the adherens junction pathway, a demonstration of the insights provided by the mixture model on the variability of the relationships between genes across conditions, and the application of mixture regression to a network with a greater number of genes. Finally, Section 4.4 summarizes the findings and highlights the strengths of the approaches evaluated.

4.2 Methods

4.2.1 Mixture Bayesian Network

In gene network applications Bayesian networks are represented by directed acyclic graphs that include nodes denoting genes and directed edges denoting the relationships between the genes [17, 18, 23]. These graphs can be extended to include a finite mixture model to accommodate for potential changes in the relationship between genes across the samples or conditions studied as follows. Let G denote a Bayesian network and $\mathbb{P}(G)$ represent the joint probability distribution over all M genes in the network $(g_1, g_2, \ldots, g_j, \ldots, g_M)$. Conditional on its parent genes, each child gene is independent of all other genes in the network. This conditional independence allows the factorization of the joint probability distribution into the product of the conditional probabilities of the genes in the network. Thus, the complete network is dealt with as a set of M gene subnetworks, each including a child gene g_j and the parent gene(s) $A(g_j)$.

The likelihood of the model is:

$$
\mathbb{P}(G) = \prod_{j=1}^{M} \frac{\mathbb{P}\left(g_j, A\left(g_j\right)\right)}{\mathbb{P}\left(A\left(g_j\right)\right)} = \prod_{j=1}^{M} \frac{\prod_{i=1}^{N}\left[\sum_{k=1}^{K_j} w_{ij}\phi_{ik}\left(y_{ji}\right)\right]}{\prod_{i=1}^{N}\left[\sum_{k=1}^{K_j} w_{ij}\phi_{ik}\left(y_{ji}^*\right)\right]}
$$

$$
= \prod_{j=1}^{M} \frac{\prod_{i=1}^{N}\left[\sum_{k=1}^{K_j} w_{ij}\left(\frac{2\pi}{V_{ik}}\right)^{-\frac{1}{2}} \exp\left[-\frac{1}{2}\left(y_{ji} - \hat{y}_{jk}\right)'\left(V_{jk}\right)^{-1}\left(y_{ji} - \hat{y}_{jk}\right)\right]\right]}{\prod_{i=1}^{N}\left[\sum_{k=1}^{K_j} w_{ij}\left(\frac{2\pi}{V_{ik}^*}\right)^{-\frac{1}{2}} \exp\left[-\frac{1}{2}\left(y_{ji}^* - \hat{y}_{jk}^*\right)'\left(V_{jk}^*\right)^{-1}\left(y_{ji}^* - \hat{y}_{jk}^*\right)\right]\right]}
$$

In the previous equation, the ratio between the joint probability $\mathbb{P}\left(g_j, A\left(g_j\right)\right)$ of the parent $A\left(g_j\right)$ and child gene g_j and the marginal probability of the parent genes $A\left(g_j\right)$ is the conditional probability $\mathbb{P}\left(g_j \mid A\left(g_j\right)\right)$ of the child gene given the parent gene(s). The normalized expression values are considered to be continuous and exhibit a Gaussian distribution. In this context, the joint and marginal probability density functions are assumed to follow a finite mixture of K_j multivariate Gaussian distributions (ϕ_{jk}), each receiving a weight w_{jk}. For an individual gene, the sum of the weights over all the components of the finite mixture is equal to 1 ($\sum_{k=1}^{K_j} w_{jk} = 1$). The maximum number of mixture components can be set to any limit and can vary between subnetworks. One-fifth of the number of samples available can be used as a guideline for the maximum number of components that would allow the precise estimation of the mixture parameters. Let p_j denote the number of child and parent genes in the subnetwork of the jth gene. Let y_{ji} denote a p_j dimensional vector of the observations where the ith subscript denotes the ith gene expression observation ($i = 1, \ldots, N$) of child gene g_j and parent genes $A\left(g_j\right)$ in the jth subnetwork. Similarly, Y_{ji}^* is a p_{j-1} dimensional vector including the ith gene expression measurements of the parent genes $A\left(g_j\right)$. For the joint probability $\mathbb{P}\left(g_j, A\left(g_j\right)\right)$, the kth component of the mixture is described with an expected vector $\widehat{Y}_{ij}$ and an p_j-by-p_j variance-covariance matrix V_{jk}. For the marginal probability of the parent genes $\mathbb{P}\left(A\left(g_j\right)\right)$, the kth component of the mixture is described with an expected $(p_j - 1)$ vector and an $(p_j - 1)$-by-$(p_j - 1)$ variance-covariance matrix.

The inference of the network topology encompasses two major steps, 1) the inference of each child-gene subnetwork and 2) the integration of these networks into an overall network [3, 17]. The parent genes of each child gene can be identified using the sparse candidate algorithm [11, 17] or the simulated annealing method [18, 30]. In the sparse candidate algorithm, the parent genes are added to the subnetwork one at a time in order of association to the child-gene expression profile. The process is iterated until the Bayesian information criterion (BIC) [27] indicates that the addition of the next parent gene does not significantly improve the prediction of the child-gene expression profile [17]. The simulated annealing approach can resolve potential local maxima detected by the sparse candidate approach [17, 18]. At each iteration, the simulated annealing algorithm randomly moves from a set of parent genes in the current subnetwork structure to a similar structure by the addition, deletion, or reversion of an edge. The probability of moving to another structure depends on the annealing temperature that is empirically adjusted based on pilot runs. This procedure is repeated iteratively until convergence. Once each child-gene subnetwork is obtained, the subnetworks are combined into one overall network using the conditional independence of the child gene given the parent genes, and any cyclic relationships between genes are removed. A schematic representation of this process is provided in [17].

Once the parent genes have been identified for each child-gene subnetwork, the expectation-maximization (EM) algorithm is used to estimate the mixture parameters within each gene subnetwork [4]. The EM formulae were described in detailed in [17]. Briefly, the expectation step provides the probability that a gene expression observation pertains to each mixture component, conditional on the parameter estimates from the maximization step (i.e., the mean and variance). The maximization step provides the estimated mixture component weights and the parameter estimates that maximize the restricted-log likelihood of the observed data, given the distribution of the observations over the mixture components. The BIC is used to identify the number of mixtures for each gene subnetwork, and to evaluate the overall network [27]. An iterative process between the inference in individual subnetworks and the mixture model is implemented until there is no improvement in the likelihood-based criteria of the overall network.

Approaches that limit the superexponential number of gene subnetworks that are evaluated have been proposed [17, 18]. These approaches may result in incomplete total network topologies because a subnetwork that does not surpass the limiting criterion will not be considered although this subnetwork would have improved the total network. In addition to the computational complexity of inferring gene networks, the size of the microarray experiments available to infer networks is typically small compared to the potential dimension (number of genes and relationships) of the network. In the next subsection, we present an alternative strategy to infer gene networks based on regression analysis. The similarities between Bayesian networks and regression analysis have been demonstrated, and the regression approach addresses the computational complexity of the mixture Bayesian network approach.

4.2.2 Mixture Regression Approach

There are several strategies to implement a mixture Bayesian network approach that can infer a gene network and identify the parent genes of each child gene, including sparse candidate algorithm [17] and simulated annealing [17, 18]. These strategies are typically computationally intense. An alternative method to reconstruct gene networks draws on the equivalence between Bayesian networks and regression. This method integrates regression analysis with parent-gene selection and mixture model. This strategy can be applied using standard statistical or programming language software, making it accessible to the general research community. In this subsection, the relationship between Bayesian networks and regression is introduced; then parent gene selection approaches are discussed; and lastly, the integration with finite mixture models is considered.

In the context of continuous data analysis, Bayesian networks have been proposed as a tool for solving regression problems, and conversely, regression models can be used to describe Bayesian networks [8, 10, 21]. The relationship between Bayesian networks and regression problems has been demonstrated in [9, 15, 25]. This was accomplished by a reparameterization of the Bayesian network into a regression model. Briefly, viewing a Bayesian network as a regression and assuming that Y is the response variable (child gene) and $X_1, \ldots, X_n$ are the explanatory variables (parent genes), the conditional density $f(y \mid x_1, \ldots, x_n)$ is computed, and a prediction for Y is obtained. The conditional density is proportional to $f(y) * f(x_1, \ldots, x_n \mid y)$, and, thus, estimating the regression parameters necessitates the specification of an n dimensional density. Here, the factorization embedded in the Bayesian network allows the simplification of this scenario conditional of the structure of the network. In the simplest case, all explanatory variables are independent, given Y [9]. In [9], a mixture of truncated exponential distributions was used, whereas we used mixtures of Gaussian distributions [9]. Likewise, in [25] it was noted that a Bayesian network represented as a perfect directed acyclic graph is equivalent to a regression problem and that the global maximum of the conditional likelihood can be found by simple local optimization methods to estimate the parameters [25]. A strategy for pruning the number of parameters (parent genes) in regression models provides good results, with heavily pruned submodels containing many fewer parameters than the Bayesian network being compared [25]. This strategy can be implemented using one of the many feature or variable selection methods developed in the context of regression models.

A regression variable selection approach allows the detection of relationships between child and parent genes, and the elimination phase provides a parsimonious network. Applications of regression feature selection methods to gene networks have been reported [1, 6, 19, 31]. The present study is the first to consider the relationship between regression and Bayesian network approaches and integrate regression and mixture models. This development enhances the understanding of the relationship between genes and the potential changes in the relationships between genes across conditions. Three feature selection approaches are considered: forward stepwise, least absolute shrinkage and selection operator (LASSO) and least angle regression (LARS). The forward stepwise approach starts with an empty model with no parent gene associated with the child gene. All genes are evaluated as potential parents, and the one with the most significant association that also surpasses a minimum "entry" p-value threshold is included in the model and accepted as parent gene. The residual gene expression that remains after adjusting for the expression of the parent gene is subsequently modeled with each of the remaining potential parent genes, and the one with the most significant association is introduced in the model as an additional parent gene. The association of each previously introduced parent genes is re-evaluated, and parent genes are removed from the model when the statistical association does not surpass a minimum "keep" p-value threshold. The parent-gene p-value was obtained from a t-test that assesses the covariation between the child and parent expression profiles. The process is repeated until no additional parental gene fulfills the entry or keep thresholds. This method has two potential limitations. First, the levels for the threshold entry and keep p-value must be specified; and second, the selection strategy may discard parental genes that may be correlated to already selected parent genes. In the LASSO method [29], the parameters are estimated through the minimization of the sums of squares of residuals of $\|y - X\beta\|^2$ subject to the L1-penalty constraint $\sum_{q=1}^{Q} |\beta_q| \leq t$, where y denotes the gene expression of the child gene, β denotes the regression coefficients of the potential parent genes, $X\beta$ denotes the prediction based on the parent-gene expression profiles, q is the number of parent genes in the model, and t is the LASSO constraint parameter. For small values of t, many of the regression coefficients of the potential parent genes will be equal to zero. The corresponding parent candidates are removed from the model and are not considered parents of the child gene in the network structure.

Depending on the strength of the coassociation between the child gene and the parent genes and the value of the LASSO constraint selected, the LASSO approach can be computationally inefficient. The LARS method offers a more computationally efficient strategy to identify the parent genes [6]. The parent gene that is most correlated with the residual of the child-gene expression (expression adjusted by any parent gene already considered) is determined, and the regression estimate changes in the direction of this parent gene based on the covariation between the parent and child genes. The magnitude of the estimate change is such that the next potential parent gene and the current set of parent genes have the same sum-of-expression cross product and, thus, the correlation with the current residual. The criterion of this methodology differs from previous approaches in which the estimate changes to minimize the residuals at the current model selection stage. For an angle representing the degree of association between the parent and child genes, the LARS parameter estimates are modified in a direction equiangular to each potential parent-gene correlation with the residual. This process is repeated until all parent genes are in the model.

Once the parent genes for each child gene have been identified, a finite mixture model is used to describe each child-gene subnetwork following the methodology described in the mixture Bayesian network approach section. Using the expectation-maximization formulae described in [17], the weight of each mixture and the parameter estimates describing the relationship between the parent and child genes in each mixture are computed. Analogous to the mixture Bayesian network approach, the mixture regression network inference encompasses the inference of each child-gene subnetwork, mixture model, and the integration of these networks into an overall network. The gene subnetworks are combined into one overall network assuming conditional independence of the child genes given the parent genes, and any cyclic relationships between genes are removed. Thus, the mixture regression approach provides a gene network that is equivalent to the gene network provided by the mixture Bayesian network approach. The performance of the mixture Bayesian network and mixture regression approaches to infer gene networks was evaluated using a real-life experiment and a known pathway, the adherens junction pathway.

4.2.3 Data

The relationships between genes in the adherens junction pathway were characterized using the mixture Bayesian network and mixture regression approaches. The adherens junction is a type of cell–cell adhesion structure or junction that has a main role in the maintainance of the physical association between the cells [20]. The integrity of the structure relies on catenins and associated molecules, actin filaments, microtubules, and the recycling machinery. The relationship between several genes in the adherens junction pathway is available in the KEGG (Kyoto Encyclopedia of Genes and Genomes) pathway (http://www.genome.jp/) and serves as benchmark for the applications presented in this study.

The relationship between genes in the adherens junction pathway was inferred using gene expression profiles from nine mouse microarray experiments that evaluated multiple toxic or teratogenic agents that have the capacity to interfere with normal embryo development and lead to birth defects [24]. The expression profiles were obtained from the National Center for Biotechnology

Information Gene Expression Omnibus or GEO repository (http://www.ncbi.nlm.nih.gov/geo). The GEO experiments or series considered were GSE1068, GSE1069, GSE1070, GSE1072, GSE1074, GSE1075, GSE1076, GSE1077, and GSE1079. The number of samples in each experiment is provided in Table 4.1. Four teratogenic agents (ethanol, methylmercury, low oxygen, and the metabolic toxin 2-chloro analog of 2'-deoxyadenosine) and two intervention agents (PK11195 or Ro5-4864) were studied [13]. Each experiment used headfold and forebrain samples from mouse embryos. These samples were collected on day 8 after insemination and approximately 3 hours after the dams were injected with the teratogenic agents. The treatment doses and timing, sampling time, mouse strains, platform, RNA isolation, labeling, and hybridization are described in detail in the GEO database and in Nemeth et al., Singh et al., and Green et al. [13, 22, 28]. Briefly, GSE1068 studied 5.0, 2.5, 1.25, or 0.625 mg/kg of 2-chloro-2'deoxyadenosine at 3 hours after injection. GSE1069 studied 2.5 and 4.5 mg/kg of 2-chloro-2'deoxyadenosine at 3 and 6 hours after injection, respectively. GSE1070 studied 2-chloro-2'deoxyadenosine with PK11195 intervention at 3, 4.5, and 6 hours after injection. GSE1072 studied 2-chloro-2'deoxyadenosine, ethanol, and methylmercury plus intervention with PK11195 or Ro5-4864 at 4.5 hours after injection. GSE1074 studied ethanol at 2.9 g ethanol per kg body weight and intervention with PK11195, and GSE1075 studied 10, 5, 2.5, or 1.25 mg/kg of methylmercury. GSE1076 studied 5 mg/kg of methylmercury at 1.5, 3, 4.5, 6 or 12 hours after injection. GSE1077 studied 5 mg/kg of methylmercury at 3, 4.5, and 6 hours after injection and intervention with PK11195. GSE1079 studied 21% and 5% oxygen and intervention with PK11195 or Ro5-4864.

The nine experiments utilized the same spotted two-dye complementary DNA (cDNA) platform and fluorescence intensities (see Chapter 1, Section 1.3, page 16) available on 2382 sequence-verified human gene elements that hybridize to mouse target messenger RNA (mRNA). A microarray included two samples; all samples in each experiment were labeled with both dyes in a reverse labeling design and, thus, all condition or treatment levels were present in at least two microarrays. The experimental design of the studies included two types of direct comparisons; comparisons between samples that received different treatments (e.g., ethanol, methylmercury, low oxygen, the metabolic toxin 2-chloro analog of 2'-deoxyadenosine administered at different time-points or doses), and comparisons between treated and control samples. Ninety microarrays were used in the mixture Bayesian network and mixture regression analyses.

Data preprocessing included the removal of observations when the median background subtracted intensity across both dyes did not surpass a minimum threshold of 200. Subsequently, the background-subtracted intensities were log-2 transformed, a loess normalization was applied to remove channel-dependent biases, and the normalized intensities were adjusted for global dye or microarray technical noise [24]. Gene expression data processing was implemented in Beehive (http://stagbeetle.animal.uiuc.edu/Beehive).

Two networks of the mouse adherens junction pathway were studied that differed in the number of genes used. The first network included the same seven genes used in the application of mixture Bayesian network reported by [17]. Since then, improvements in the annotation of the platform allowed the inference of a second network that included 24 genes in the mouse adherens junction pathway. The networks inferred through the two strategies were compared; these networks were also compared to the mouse adherens junction pathway in the KEGG database (mmu04520, accessed on October 2011).

Several genes present in the KEGG adherens junction pathway do not have a corresponding probe in the microarray platform. Thus, two types of relationships between genes were inferred: a) direct relationships present in the KEGG pathway (denoted with a straight line or edge) and,

b) indirect relationships through one or more intermediate genes that are not present in the microarray platform (denoted with a dashed line).

Both the mixture Bayesian network and mixture regression approaches are able to predict directional relationships where the parent- and child-gene nodes are identified. In addition, the adherens-junction mouse pathway in the KEGG database includes non-directional gene relationships. Thus, non-directional gene relationships in the KEGG database that were inferred by the mixture approaches are depicted with lines without directionality or arrow heads.

4.3 Results

4.3.1 Comparison of Mixtures

The network topologies were obtained from the mixture Bayesian network and mixture regression approaches that were applied to seven genes in the adherens junction pathway. In the mixture regression approach, the network topologies obtained from the forward stepwise, LASSO, and LARS methods were consistent, and only the results from the LASSO implementation are presented. The mixture Bayesian network approach was able to detect all except one gene relationship in the KEGG pathway database [17]. The mixture regression approach was able to detect all the relationships between the genes in the KEGG pathway database. No false edges were inferred by either approach. The relationship between genes identified by the mixture Bayesian network and mixture regression approaches are denoted by the edges in Fig. 4.1. The gene relationship only detected by the regression mixture approach is denoted by a dash-dotted edge in Fig. 4.1. This result suggests that the performances of the mixture Bayesian network and mixture regression approaches were comparable.

4.3.2 Mixture Modeling of Changes in Gene Relationships

In the mixture Bayesian network approach, the estimates from the EM algorithm indicated that three mixture components described the subnetwork of the gene α-actinin 2. Each mixture component corresponded to low, intermediate, and high levels of gene expression, and the associated weights were 0.28, 0.57, and 0.16, respectively [17]. Similarly to the mixture Bayesian network approach, the mixture regression approach detected changes in the relationships between genes in the α-actinin 2 subnetwork across samples and conditions. The mixture component estimates are depicted in Fig. 4.2 (page 114). The relationship between α-actinin 2 and its four parent genes was best depicted by a mixture of three components. The results from mixture regression approaches were consistent with those results reported from the mixture Bayesian network implementation [17]. Specifically, the weights of the three mixture components were 0.28, 0.57, and 0.16 in the mixture Bayesian approach and 0.24, 0.55, and 0.21 in the mixture regression approach, respectively.

An additional insight gained from the mixture implementation of the network inference methodology is that whereas the relationships between α-actinin 2 and some parent genes were consistent across mixtures for some genes (all positive associations with tight junction protein 1 (TJP1) or all negative associations with afadin (MLLT4)), the relationships with other parent genes, e.g., ras homolog gene family, member A (RHOA) and catenin (CTNNA1), varied (positive to negative) across mixture components. In the context of this chapter, positive and negative associations between parent and child genes suggested enhanced and depressed child-gene expression,

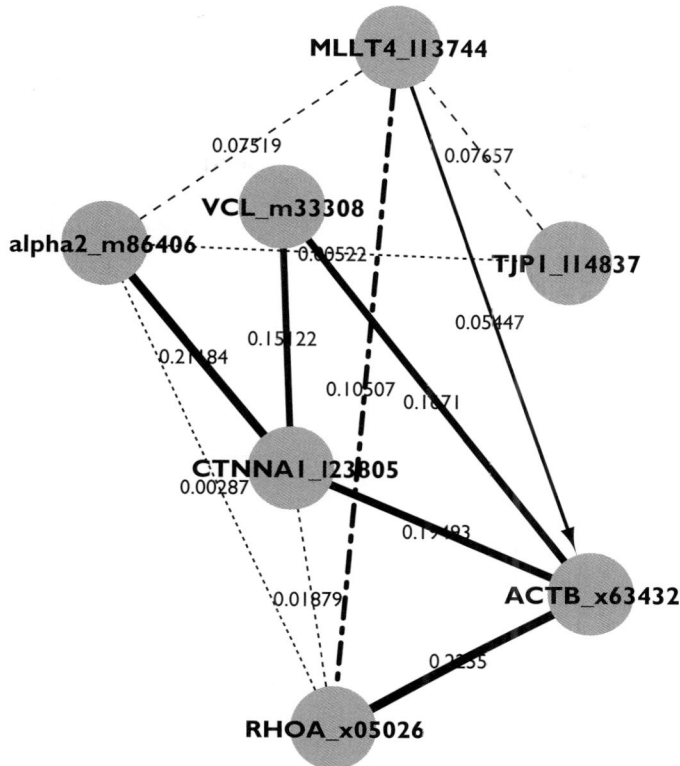

Fig 4.1 Adherens junction network inferred using the mixture Bayesian network and mixture regression approach on seven genes. Nodes represent genes. MLLT4: afadin, VCL: vinculin, TJP1: tight junction protein 1 or ZO1, ACTB: β-actin, RHOA: ras homolog gene family, member A, CTNNA1: catenin or cadherin associated protein α1, alpha2: α-actinin 2. Lines denote edges representing the relationship between genes. Straight and dashed lines denote direct and indirect relationships detected by the mixture Bayesian and mixture regression network approaches and confirmed in the KEGG pathway, respectively. The dashed-dotted line denotes an indirect relationship between genes only detected by the mixture regression approach and confirmed in the KEGG pathway. Numbers associated with each edge denote the partial correlation between the genes. The thickness of the edge indicates the strength of the partial correlation. The arrow denotes directionality of the edge consistent with the KEGG pathway.

respectively. Even for the genes that exhibited the same relationship (positive or negative) with α-actinin 2 across mixtures, the mixture differed on the strength of these relationships. For example, the relationship between MLLT4 and α-actinin 2 across all nine experiments was estimated to be described by a mixture of three components each with weights 0.24, 0.55, and 0.21 and correlations −0.14, −0.05, and −0.56, for mixture components one, two and three respectively (see Fig. 4.2). From the regression estimate and standard error, a measurement of the statistical significance of the associations between the parent genes and α-actinin 2 was computed. Accounting for multiple testing, the parent genes that exhibited significant relationships with α-actinin 2 at p-value < 0.005 were CTNNA1 and RHOA (mixture component one), RHOA (mixture component two), MLLT4, TJP1, CTNNA1, and RHOA (mixture component three).

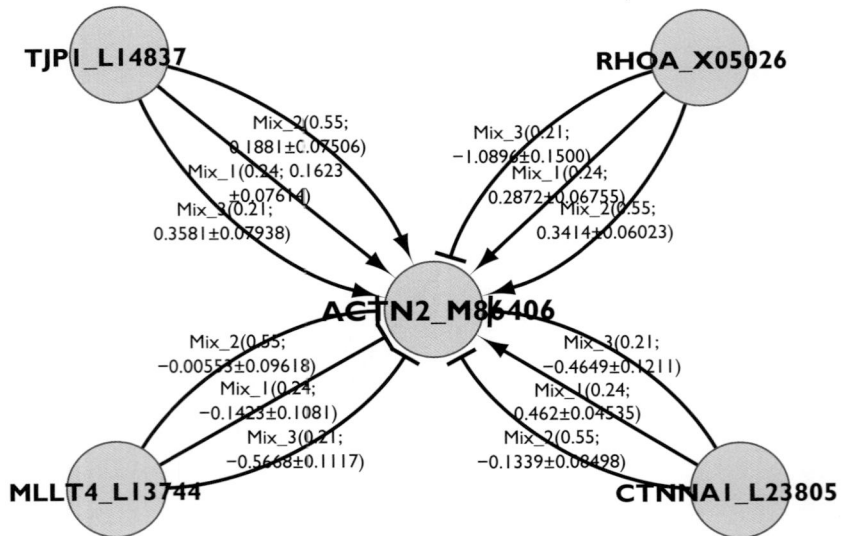

Fig 4.2 Depiction of the different relationships between the child gene α-actinin and the parent genes across the three mixture components. Nodes represent genes. Child gene: ACTN2: α-actinin 2; parent genes, MLLT4: afadin, CTNNA1: catenin α 1, TJP1: tight junction protein 1 or ZO1, and RHOA: ras homolog gene family, member A. Lines denote edges or relationship between genes; arrowheads denote positive associations, and flat heads denote negative associations between the pair of genes connected by the edge. Edges are labeled with the mixture component weight (mixture component one: Mix_1, mixture component two: Mix_2, mixture component 3: Mix_3) followed by the regression coefficient ± standard error.

4.3.3 Interpretation of Mixtures

The mixture Bayesian network approach estimated that a mixture of two Gaussian components provided the best description of the β-actin gene subnetwork [17]. In this subnetwork, most samples were assigned to the first mixture component. This result suggests that the conditions studied across experiments do not have a major effect on the relationship between genes in the β-actin subnetwork [17]. Only the samples receiving 2-chloro-2′deoxyadenosine before (three hours) or after (six hours) the critical event of p53 protein induction and samples treated with Ro5-4864 were assigned to mixture component two.

Most of the adherens junction pathway genes studied were assigned to two mixtures, with one mixture assigned to CTNNA1 and three mixtures assigned to ACTN2. For the genes with two mixtures, the number of samples assigned to each mixture within experiment and within common treatment are provided in Tables 4.1 and 4.2, respectively. Consistent with the mixture Bayesian network, in the actin subnetwork, all the samples within an experiment were assigned to one mixture component. In other gene subnetworks, samples were assigned to both mixture components. Consideration of mixture of distributions confirms that, within a gene network (in our case, the adherens junction pathway), there are some relationships between genes that do not vary across condition and, thus, are assigned to a single mixture component. Other gene relationships that are more flexible and vary across conditions can result in multiple mixture components.

Table 4.1 Number of samples assigned to each mixture component by experiment and gene sub-network.

Gene[1]	Mixture[2]	Experiment[3]								
		68	69	70	72	74	75	76	77	79
MLLT4	1	4	2	3	4	7	5	5	1	2
MLLT4	2	1	0	1	3	6	0	2	6	4
RHOA	1	3	0	2	5	9	0	3	7	3
RHOA	2	2	4	2	2	4	5	4	0	3
VCL	1	3	3	2	7	4	1	6	4	3
VCL	2	2	1	2	0	9	4	1	3	3
ACTB	1	0	4	4	7	13	5	0	7	6
ACTB	2	5	0	0	0	0	0	7	0	0
Total samples		5	4	4	7	13	5	7	7	6

[1] Genes: MLLT4: afadin, VCL: vinculin, ACTB: β-actin, and RHOA: ras homolog gene family, member A.
[2] Mixture: mixture component one or two.
[3] Experiments: Gene Expression Omnibus series
GSE1068, GSE1069, GSE1060, GSE1072, GSE1074, GSE1075, GSE1076, GSE1077, GSE1079.

The distribution of samples across mixture components was consistent between the mixture Bayesian network [17] and the mixture regression approaches. The sample were more evenly distributed across mixture components within the most common treatments (see Table 4.2) than within experiments for the genes studied in the adherens junction network (see Table 4.1). This result suggests that the different teratogenic agents studied caused similar disruptions in the gene relationships within experiments.

The correlation of the number of samples assigned to each mixture between pairs of genes with two mixtures across all studies listed in Table 4.1 was computed. The correlations of genes with CTNNA1 were not evaluated because this gene was represented by a mixture of a single component and has no variation. The correlations of genes with ACTN2 were not evaluated because this gene was represented by a mixture of three components and differed from all other genes that were represented by a mixture of two components. The purpose of the correlation was to identify consistent assignment of samples across genes to mixtures. The correlation between assignment to the same mixture component (one to one and two to two) and to different mixture components (one to two and two to one) were computed, and the maximum correlation was considered. The correlation of the assignment of samples to mixtures was very high between MLLT4 and RHOA, MLLT4 and ACTB, RHOA and ACTB, and VCL and ACTB, with an average correlation equal to 0.68. Likewise, the correlation between the number of samples assigned to each mixture between pairs of genes with two mixtures across treatments was computed (see Table 4.2). The correlation of the assignment of samples to mixtures between all pairs of genes across treatments was

Table 4.2 Number of samples assigned to each mixture component by treatment and gene subnetwork.

Gene[1]	Mixture[2]	Treatment[3]				
		Control	tCdA	Mercury	Ethanol	Oxygen
MLLT4	1	6	10	11	5	2
MLLT4	2	3	3	7	5	4
RHOA	1	5	6	9	7	3
RHOA	2	4	7	9	3	3
VCL	1	6	8	12	2	3
VCL	2	3	5	6	8	3
ACTB	1	7	11	16	10	6
ACTB	2	2	2	2	0	0
Total samples		9	13	18	10	6

[1] Genes: MLLT4: afadin, VCL: vinculin, ACTB: β-actin, and RHOA: ras homolog gene family, member A.
[2] Mixture: mixture component one or two.
[3] Treatments, control: untreated mice, ethanol: mice treated with ethanol, mercury: mice treated with methylmercury, oxygen: mice treated with low oxygen, and tCdA: mice treated with the metabolic toxin 2-chloro analog of 2′-deoxyadenosine.

even higher than the correlation across experiments, with an average correlation equal to 0.91. These results are supported by the strong relationship between these genes observed in Fig. 4.1.

4.3.4 Inference of Large Networks

Since the publication of the first inferred adherens junction pathway [17], additional probes that correspond to additional genes in the pathway have been identified. The mixture regression approach was applied to infer this more complete pathway (see Fig. 4.3). All the relationships between the genes were confirmed in the KEGG pathway with the exception of the relationship between transforming growth factor β receptor 2 (TGFBR2), which corresponds to the weak adhesion pathway component, and genes on the strong adhesion pathway. Also some of the relationships between the Wiskott-Aldrich syndrome (WAS) genes and other members of the strong adhesion components of the adherens junction pathway could not be confirmed in the KEGG pathway. Although not currently included in the KEGG pathway, the identified relationships between TGFBR2 or WAS and strong adhesion genes such as α-actinin have been reported [7, 12, 33].

4.4 Conclusions

Gene expression experiments offer a rich resource to infer gene networks. The large number of genes in a network and the limited number of gene expression observations typically available

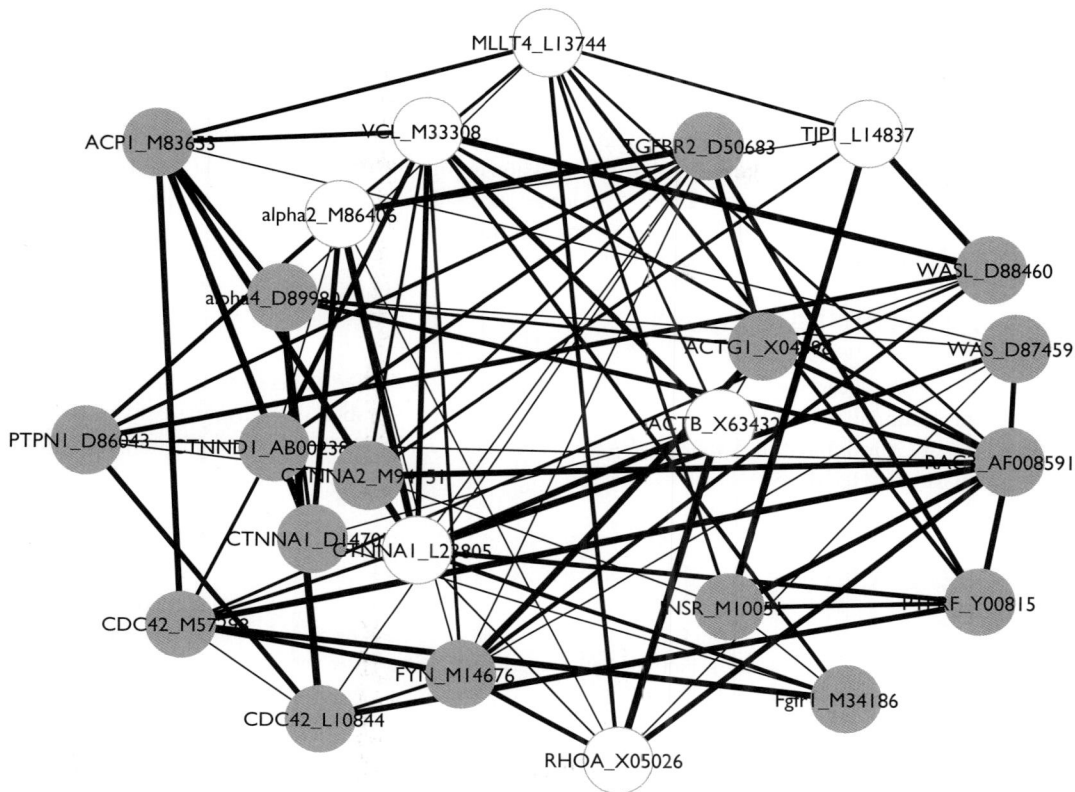

Fig 4.3 Reconstruction of the adherens junction pathway based on 24 genes. White nodes: genes in the previously inferred adherens junction pathway using seven genes; MLLT4: afadin, VCL: vinculin, TJP1: tight junction protein 1 or ZO1, ACTB: β-actin, RHOA: ras homolog gene family, member A, CTNNA1: catenin or cadherin associated protein α 1, and alpha2: α-actinin 2. Gray nodes: genes that were included in the subsequent pathway inference; ACP1: acid phosphatase 1 or LMW-PTP, PTPN1: protein tyrosine phosphatase, non-receptor type 1, alpha4: α-actinin 4, CTNND1: catenin δ 1 or P120, CTNNA2: Catenin α 2, CDC42: cell division cycle 42 homolog, FYN: Fyn proto-oncogene tyrosine-protein kinase, INSR: insulin receptor, Fgfr1: fibroblast growth factor receptor 1 or Hspy, PTPRF: protein tyrosine phosphatase, receptor type F or LAR, RAC3: RAS-related C3 botulinum toxin substrate 3, WAS: Wiskott-Aldrich syndrome protein or Wasp, WASL: WAS-like, ACTG1: actin γ 1, TGFBR2: transforming growth factor β receptor 2. Lines denote edges. The thickness of the edge indicates the strength of the partial correlation between genes.

in an experiment impose challenges in the inference of networks. Furthermore, few network inference approaches allow for changes in the gene relationships across the multiple treatments or samples being analyzed. Mixture Bayesian network is a flexible and powerful framework to infer gene network topologies and accommodate for changes in the gene relationships across treatments. The mixture regression approach provided an alternative method to infer a gene network. The relationship between the Bayesian network and regression methodologies was demonstrated by the similarity of gene networks using different data sets. Therefore, the mixture regression approach presented can be viewed as a strategy to learn the mixture-based Bayesian network model.

The mixture Bayesian and regression approaches inferred consistent gene network topologies that were supported by empirical evidence. Likewise, the parameter estimates from both approaches were consistent, and insights into the relationships between genes were obtained from the mixture component estimates. The mixture component accommodated changes in the relationships between genes across treatments or conditions considered. This study demonstrates that the mixture Bayesian and mixture regression approaches have comparable performance supported by experimental evidence. The latter approach is computationally simpler and can be easily implemented by the biology research community.

ACKNOWLEDGMENTS

The project described was supported by Award Numbers R21DA027548 and P30DA018310 from the National Institute on Drug Abuse (NIDA), Award Number 1R03CA143975 from the National Cancer Institute (NCI), and Award Number R21 MH096030 from the National Institute on Mental Health (NIMH).

REFERENCES

[1] S. Zeke, H. Chan, I. Havukkala, V. Jain, Y. Hu, and N. K. Kasabov. Soft computing methods to predict gene regulatory networks: an integrative approach on time-series gene expression data. *Applied Soft Computing*, 8(3):1189–1199, 2008.

[2] P. Dagum and M. Luby. Approximating probabilistic inference in Bayesian belief networks is NP-hard. *Artificial Intelligence*, 60(1):141–153, 1993.

[3] S. Davies and A.W. Moore. Mix-nets: factored mixtures of Gaussians in Bayesian networks with mixed continuous and discrete variables. In Craig Boutilier and Moisés Goldszmidt, editors, *Proceedings of the Sixteenth Conference in Uncertainty in Artificial Intelligence (UAI00)*, pages 168–175. Morgan Kaufmann Publishers, 2000.

[4] A. Dempster, N. Laird, and D. Rubin. Maximum likelihood from incomplete data via the EM algorithm. *Journal of the Royal Statistical Society. Series B (Methodological)*, 39(1):1–38, 1977.

[5] A. Djebbari and J. Quackenbush. Seeded Bayesian Networks: constructing genetic networks from microarray data. *BMC Systems Biology*, 2(1):57, 2008.

[6] B. Efron, T. Hastie, I. Johnstone, and R. Tibshirani. Least angle regression. *The Annals of Statistics*, 32(2):407–499, 2004.

[7] F. Facchetti, L. Blanzuoli, W. Vermi, L.D. Notarangelo, S. Giliani, M. Fiorini, A. Fasth, D.M. Stewart, and D.L. Nelson. Defective actin polymerization in EBV-transformed B-cell lines from patients with the Wiskott–Aldrich syndrome. *The Journal of Pathology*, 185(1):99–107, 1998.

[8] A. Fernández, M. Morales, and A. Salmerón. Tree augmented naive Bayes for regression using mixtures of truncated exponentials: application to higher education management. In M.R. Berthold, J. Shawe-Taylor, and N. Lavrac, editors, *Intelligent Data Analysis (IDA)*, volume 4723 of *Lecture Notes in Computer Science*, pages 59–69. Springer, 2007.

[9] A. Fernández and A. Salmerón. Extension of Bayesian network classifiers to regression problems. In H. Geffner, R. Prada, I.M. Alexandre, and N. David, editors, *IBERAMIA*, volume 5290 of *Lecture Notes in Computer Science*, pages 83–92. Springer, 2008.

[10] E. Frank, L. Trigg, G. Holmes, and I.H. Witten. Technical note: Naive Bayes for regression. *Machine Learning*, 41:5–25, 2000.

[11] N. Friedman, M. Linial, I. Nachman, and D. Pe'er. Using Bayesian networks to analyze expression data. *Journal of Computational Biology*, 7(3-4):601–620, 2000.

[12] R. Fuchshofer, S. Ullmann, L.F. Zeilbeck, M. Baumann, B. Junglas, and E.R. Tamm. Connective tissue growth factor modulates podocyte actin cytoskeleton and extracellular matrix synthesis and is induced in podocytes upon injury. *Histochemistry and Cell Biology*, 136:301–319, 2011.

[13] M.L. Green, A.V. Singh, Y. Zhang, K.A. Nemeth, K.K. Sulik, and T.B. Knudsen. Reprogramming of genetic networks during initiation of the fetal alcohol syndrome. *Developmental Dynamics: An Official Publication of the American Association of Anatomists*, 23:6613–631, 2007.

[14] M. Grzegorczyk, D. Husmeier, K.D. Edwards, P. Ghazal, and A.J. Millar. Modelling non-stationary gene regulatory processes with a non-homogeneous Bayesian network and the allocation sampler. *Bioinformatics*, 24(18):2071–2078, 2008.

[15] D. Hecker and C. Meek. Embedded Bayesian network classifiers. Technical Report MSR-TR-97-06, Microsoft Research, 1997.

[16] S. Imoto, T. Goto, and S. Miyano. Estimation of genetic networks and functional structures between genes by using Bayesian networks and nonparametric regression. *Pacific Symposium on Biocomputing*, 175–186, 2002.

[17] Y. Ko, C. Zhai, and S.L. Rodriguez-Zas. Inference of gene pathways using mixture Bayesian networks. *BMC Systems Biology*, 3:54, 2009.

[18] Y. Ko, C. Zhai, and S.L. Rodriguez-Zas. Discovery of gene network variability across samples representing multiple classes. *International Journal of Bioinformatics Research and Applications*, 6:402–417, 2010.

[19] F. Li and Y. Yang. Recovering genetic regulatory networks from micro-array data and location analysis data. *Genome Information*, 15:131–140, 2004.

[20] C. Meng and M. Takeichi. Adherens junction: molecular architecture and regulation. *Cold Spring Harb Perspectives in Biology*, 136:a002899, 2009.

[21] M. Morales, C. Rodriguez, and A. Salmerón. Selective naive Bayes for regression using mixtures of truncated exponentials. *International Journal of Uncertainty, Fuzziness and Knowledge Based Systems*, 15:697–716, 2007.

[22] K.A. Nemeth, A.V. Singh, and T.B. Knudsen. Searching for biomarkers of developmental toxicity with microarrays: normal eye morphogenesis in rodent embryos. *Toxicology and Applied Pharmacology*, 20:6219–6228, 2005.

[23] S.L. Rodriguez-Zas and Y. Ko. Elucidation of general and condition-dependent gene pathways using mixture models and Bayesian networks. In M. Dehmer, M. Emmert-Streib, A. Graber, and A. Salvador, editors, *Applied Statistics for Network Biology: Methods in Systems Biology*, pages 91–103. Wiley-Blackwell, 2011.

[24] S.L. Rodriguez-Zas, Y. Ko, H.A. Adams, and B.R. Southey. Advancing the understanding of the embryo transcriptome co-regulation using meta-, functional, and gene network analysis tools. *Reproduction*, 135:213–224, 2008.

[25] T. Roos, H. Wettig, P. Grünwald, P. Myllymäki, and H. Tirri. On discriminative Bayesian network classifiers and logistic regression. *Machine Learning*, 59(3):267–296, 2005.

[26] E.E. Schadt, M.D. Linderman, J. Sorenson, L. Lee, and G.P. Nolan. Computational solutions to large-scale data management and analysis. *Nature Reviews Genetics*, 11:647–657, 2010.

[27] G. Schwarz. Estimating the dimension of a model. *Annals of Statistics*, 6:461–464, 1978.

[28] A.V. Singh, K.B. Knudsen, and T.B. Knudsen. Computational systems analysis of developmental toxicity: design, development and implementation of a Birth Defects Systems Manager (BDSM). *Reproductive Toxicology*, 19:421–439, 2005.

[29] R. Tibshirani. Regression shrinkage and selection via the Lasso. *Journal of the Royal Statistical Society. Series B (Methodological)*, 58:267–288, 1996.

[30] J. Tomshine and Y.N. Kaznessis. Optimization of a stochastically simulated gene network model via simulated annealing. *Biophysics Journal*, 91:3196–3205, 2006.

[31] E. P. van Someren, L. F. A. Wessels, E. Backer, and M. J. T. Reinders. Multi-criterion optimization for genetic network modeling. *Signal Processing*, 83(4):763–775, 2003.

[32] W. Wiegerinck and B. Kappen. Approximations of Bayesian networks through KL minimisation. *New Generation Computing*, 18(2):167–176, 2000.

[33] L. You and F.E. Kruse. Differential effect of activin A and BMP-7 on myofibroblast differentiation and the role of the Smad signaling pathway. *Investigative Ophthalmology and Visual Science*, 43:72–81, 2002.

[34] J. Yu, V.A. Smith, P.P. Wang, A.J. Hartemink, and E.D. Jarvis. Advances to Bayesian network inference for generating causal networks from observational biological data. *Bioinformatics*, 20(18):3594–3603, 2004.

CHAPTER 5

Network Inference in Breast Cancer with Gaussian Graphical Models and Extensions

MARINE JEANMOUGIN, CAMILLE CHARBONNIER,
MICKAËL GUEDJ, AND JULIEN CHIQUET

Clustering genes with high correlations will group genes with close expression profiles, defining clusters of co-expressed genes. However, such correlations do not provide any clues about the chain of information going from gene to gene. Partial correlation consists in quantifying the correlation between two genes after excluding the effects of the other genes. Partial correlation thus allows us to distinguish between the correlation of two genes due to direct causal relationships from the correlation that originates via intermediate genes. Gaussian graphical models (GGMs) offer a well-studied framework to spot direct relationships. In this chapter, to model gene networks, Gaussian graphical model learning is set up as a covariate selection problem. Two LASSO-type techniques[1] are described, the Graphical LASSO approach and the neighborhood selection. Then two extensions to the classical GGM are presented. Prior knowledge on the latent structure is expected to help find the correct graph in a more robust way. For this purpose, GGMs are extended in structured GGMs, to account for modularity, and more generally heterogeneity in the gene connection features. Structure adaptive penalty parameters are considered. A second extension is proposed to merge different experimental conditions and thus enlarge the number of observations available to infer interactions. To remedy the problem

---

[1] Least absolute shrinkage and selection operator.

Probabilistic Graphical Models for Genetics, Genomics, and Postgenomics. First Edition. Christine Sinoquet
& Raphaël Mourad (Eds). © Oxford University Press 2014. Published in 2014 by Oxford University Press.

of heterogeneity among microarray experiments, multiple GGMs are estimated simultaneously instead of merging the various conditions as is usually done. A structured penalizer encourages networks to share similar up- or down-regulations across conditions. Finally, an analogous way to handle time-varying data is presented. The extension using a biological prior on the network structure is illustrated on real data.

5.1 Introduction

Correlations between gene expression levels provide information on whether genes seem to be expressed or inhibited at the same time. Therefore, clustering genes with high correlations will group genes with close expression profiles, defining clusters of co-expressed genes. However, if correlations highlight genes expressed in similar conditions or time points and thereby likely to interact together, they do not provide any clue as to how the chain of information goes from gene to gene. Thus, we need to evaluate the correlation between genes conditioning on other genes. This concept is known as the partial correlation and consists in quantifying the correlation between two genes after excluding the effects of other genes. Partial correlation coefficient allows us to distinguish between the correlation of two genes due to direct causal relationships from the correlation that originates via intermediate genes.

Through the notion of partial correlation, Gaussian graphical models (GGMs) provide a well-studied framework to spot those direct relationships. When adapted to biological network (or graph) modeling, GGMs consider genes as vertices and their levels of expression as Gaussian random variables. In a Gaussian world, independence between variables is equivalent to absence of correlation. As a consequence, an edge will be drawn between two genes in the graph if they have a non-zero partial correlation coefficient, meaning that their expressions are dependent conditional on all other gene expressions in the data set.

In our framework, GGMs are used to model gene regulatory networks. However, we encourage the reader to bear in mind that the only accurate definition of what we are estimating is the statistical, that is to say, graphical representation of the conditional dependence structure between RNA measurements. Gene regulation is a very complex phenomenon due to the number of actors involved but which cannot be accounted for (proteins, microRNAs, chromatin modifications, . . .) and to the complexity of RNA level dynamics (production, storing and degradation). Therefore, the only reasonable aim of such networks resides in highlighting potential regulations or chains of regulations. It is then the role of the biologists to validate them or elucidate the exact paths followed by regulatory chains.

This chapter is outlined as follows: first, classical modeling through GGMs will be briefly introduced; the learning problem, understood here as the recovery of the network associated to the partial correlation structure between variables, will be discussed and set up as a problem of covariate selection. Then we will underline two natural extensions of these techniques, in the development of which the authors have been involved. These refinements were always motivated by biological considerations. The use of these techniques will be illustrated in the second part of this chapter, devoted to the study of a novel breast cancer data set with a large cohort of patients: this application involves the recent GGM techniques depicted here and illustrates what kind of insights can be reasonably expected from these methods.

5.2 Modeling of Gene Networks by Gaussian Graphical Networks

Notation 5.1

For every size p vector u, we denote respectively by $\|u\|_{L_0}$, $\|u\|_{L_1}$, and $\|u\|_{L_2}$ its L0-, L1- and L2-norms, defined as follows:

$$\|u\|_{L_0} = \sum_{k=1}^{p} \mathbb{I}_{\{u_k \neq 0\}}, \qquad \|u\|_{L_1} = \sum_{k=1}^{p} |u_k|, \qquad \|u\|_{L_2} = \sqrt{\sum_{k=1}^{p} u_k^2}.$$

In the sequel, when applied to matrices, L0-, L1- and L2-norms are implicitly applied entrywise. We denote the trace operator by Tr. Then, for any pair (A, B) of square, real-valued matrices, $\langle A, B \rangle = Tr(B^\mathsf{T} A)$ defines the matrix inner product associated with the Frobenius norm, which is nothing more than the usual L2-norm applied entrywise. For every size $n \times p$ matrix A, the Frobenius norm writes

$$\|A\|_F = \sqrt{\sum_{i=1}^{n} \sum_{j=1}^{p} A_{ij}^2}.$$

5.2.1 Simple Gaussian graphical network

Let us represent the expressions of the p genes in the set $\mathcal{G} = \{1, \ldots, p\}$ by a Gaussian random vector $X = (X_1, \ldots, X_p)^\mathsf{T} \in \mathbb{R}^p$ that follows a multivariate Gaussian distribution with unknown mean and unknown covariance matrix Σ. No loss of generality is involved when centering X, so we may assume that $X \sim \mathcal{N}(0, \Sigma)$. The covariance matrix, equal to $\mathbb{E}(XX^\mathsf{T})$ under the assumption that X is centered, belongs to the set $\mathcal{S}_p^+$ of positive definite symmetric matrices of size $p \times p$. Positive definite matrices are the matrix analogs of positive real numbers: for every positive vector u, $u^\mathsf{T} \Sigma u$ is also positive.

GRAPH OF CONDITIONAL DEPENDENCES

GGMs endow Gaussian random vectors with a graphical representation Γ of their *conditional dependence structure*. Each gene g is represented by a vertex in the graph. Two genes g and h are linked by an undirected edge (g, h) if, conditional on all other gene expressions indexed by $\mathcal{G}\backslash\{g, h\}$, random variables X_g and X_h remain or become dependent. Thanks to the Gaussian assumption, conditional independence actually boils down to a zero conditional covariance $\mathrm{cov}(X_g, X_h | X_{\mathcal{G}\backslash\{g,h\}})^2$, or equivalently to a zero partial correlation $r_{gh|\mathcal{G}\backslash\{g,h\}}$. The latter is a useful normalized expression of the former, as recalled in (5.1).

---

[2] Let X, Y, Z be random variables on respective probabilistic spaces $\mathcal{X}, \mathcal{Y}, \mathcal{Z}$ with joint density probability $f_{(X,Y,Z)}$. For every x in $\mathcal{X}$ and y in $\mathcal{Y}$, the conditional distributions of X and Y with respect to Z are defined, for every z in $\mathcal{Z}$ such that $f_Z(z) > 0$, by

$$f_{X|Z=z}(x) = \frac{f_{(X,Z)}(x, z)}{f_Z(z)}, \quad f_{Y|Z=z}(y) = \frac{f_{(Y,Z)}(y, z)}{f_Z(z)}, \quad f_{(X,Y)|Z=z}(x, y) = \frac{f_{(X,Y,Z)}(x, y, z)}{f_Z(z)}.$$

$$r_{gh|\mathcal{G}\backslash\{g,h\}} = \frac{\text{cov}(X_g, X_h|X_{\mathcal{G}\backslash\{g,h\}})}{\sqrt{\text{var}(X_g|X_{\mathcal{G}\backslash\{g,h\}})\text{var}(X_h|X_{\mathcal{G}\backslash\{g,h\}})}}. \tag{5.1}$$

Concretely, the inference of a GGM is based upon a classical result originally emphasized in [13] claiming that partial correlations $r_{gh|\mathcal{G}\backslash\{g,h\}}$ are actually proportional to the corresponding entries in the *inverse* of the covariance matrix $(\Sigma^{-1})_{gh}$. Denoted by Θ in the following, the covariance inverse, also known as the *concentration matrix*, thus directly describes the conditional dependence structure of X. Indeed, after a simple rescaling, Θ can be interpreted as the adjacency matrix of an undirected weighted graph Γ representing the partial correlation structure between variables $X_1, \ldots, X_p$. This graph has no self-loop and contains all edges (g, h) such that Θ_{gh} is non-zero. Therefore, recovering non-zero entries of Θ is equivalent to inferring the graph of conditional dependences Γ, and the correct selection of non-zero entries is the main issue in this framework.

MAXIMUM LIKELIHOOD INFERENCE

We observe n identically and independently distributed (hereafter i.i.d.) observations from such a multivariate Gaussian distribution with covariance Σ, which are stored once centered in a matrix $X \in \mathbb{R}^{n \times p}$. For each observation $i, (i = 1, \ldots, n)$ and gene $g, (g = 1, \ldots, p)$, entry x_{ig} contains the expression level observed for gene g in the ith sample. These n observations must be collected in close enough conditions so that we can assume that they follow the exact same distribution once centered. Independence of the observations also implies that time-course measurements do not fit this model. We devote Subsection 5.2.2 to models designed for time-course, i.e., longitudinal, data.

GGMs fall in the family of exponential models, for which the whole range of classical statistical tools apply. As soon as n is greater than p, the model likelihood admits a unique maximum over the set $\mathcal{S}_p^+$, defining a Maximum Likelihood Estimator (MLE). Following the assumption that X is Gaussian, the MLE of Θ is defined by:

$$\widehat{\Theta}^{\text{MLE}} = \underset{\Theta \in \mathcal{S}_p^-}{\arg \max} \quad (2\pi)^{-p/2} \det(\Theta) \exp\left(-\frac{1}{2}X^\top \Theta X\right). \tag{5.2}$$

Let us denote the empirical covariance matrix by $\widehat{S} = X^\top X/n$. After log-transformation and use of the Trace operator property $\text{Tr}(a^\top b) = \text{Tr}(ba^\top)$, for every compatible vectors a and b, problem (5.2) becomes

$$\widehat{\Theta}^{\text{MLE}} = \underset{\Theta \in \mathcal{S}_p^+}{\arg \max} \quad \log \det(\Theta) - \langle \Theta, \widehat{S} \rangle. \tag{5.3}$$

When n is larger than p, problem (5.3) admits a unique solution equal to $\widehat{S}^{-1}$. As the square product of a centered and scaled Gaussian vector, $\widehat{S}$ follows a Wishart distribution, which is the

---

The conditional moments are defined with respect to these distributions:

$$\mathbb{E}(X|Z = z) = \int_{\mathcal{X}} x f_{X|Z=z}(x)\, dx, \quad \text{cov}(X, Y|Z = z) = \mathbb{E}(XY|Z = z) - \mathbb{E}(X|Z = z)\,\mathbb{E}(Y|Z = z).$$

multivariate generalization of the chi-square distribution. As a result, its inverse $\widehat{S}^{-1}$ naturally follows an inverted Wishart distribution with computable parameters.

There are two major limitations with the MLE regarding the objective of graph reconstruction by recovering zeroes in the estimate of Θ. First, it provides an estimate of the saturated graph: all genes are connected to each other; and second, we need n to be larger than p to be able even to define this estimator, which is never the case in microarray studies.

What saves us here is a common property of biological networks, namely sparsity: among all $p(p-1)/2$ possible interactions between genes, only a few actually take place. The true number of edges to look for in the graph is much more on the order of p rather than the order of p^2. Sparsity makes the estimation feasible in the case where n is smaller than p, since we can concentrate on sparse or shrinkage estimators with fewer degrees of freedom than in the original problem. Henceforth, the question of selecting the correct set of edges in the graph is treated as a question of model (or covariate) selection.

BACKGROUND ON HIGH-DIMENSIONAL INFERENCE OF GAUSSIAN GRAPHICAL MODEL

The different methods for model selection/estimation in GGMs roughly fall into three categories. The first contains constraint-based methods, performing statistical tests. We mention that the procedure in [17, 18] relies on asymptotic considerations, a regime never attained in real situations. The forward selection method combined with permutation tests suggested in [28] would fall into this category. Limited-order partial correlations were also considered in [7, 64]. The second of these categories is composed of Bayesian approaches (see for instance [14, 26, 48]). However, constructing priors on the set of concentration matrices is not a trivial task, and the use of Markov chain Monte Carlo (MCMC) procedures limits the range of applications to moderate-sized networks. The third category contains regularized estimators, which add a penalty term to the likelihood in order to reduce the complexity or remove degrees of freedom from the estimator. A first shrinkage estimator was proposed by [53]. This approach consists in using a weighted average of two different estimators, the first being unconstrained (thus having small bias but large variance), and the second being low dimensional (and thus exhibiting small variance but large bias).

Let us now introduce $L1$-regularized procedures. In the context of linear regression, the LASSO technique was introduced by [59]. The idea underlying this procedure is that the ordinary least square criterion may be improved in a sparse context, using an $L1$-norm penalty. The $L1$-norm penalty is used as a convex relaxation of the ideal but too computationally intensive $L0$-regularized problem (5.4).

$$\arg\max_{\Theta \in \mathcal{S}_p^+} \quad \log\det(\Theta) - \langle \Theta, \widehat{S} \rangle - \rho \, \|\Theta\|_{L0}. \tag{5.4}$$

Problem (5.4) looks for a trade-off between the maximization of the likelihood and the sparsity of the graph in one single optimization problem. BIC[3] or AIC[4] criteria, are special cases of such $L0$-regularized problems, except that the maximization is made upon a restricted subset of candidate estimates $\{\tilde{\Theta}_1, \ldots, \tilde{\Theta}_m\}$ and the choice of ρ is fixed, namely $\log(n)$ for BIC and 0.5 for AIC. Actually solving (5.4) would require the exploration of all 2^p possible graphs. On the contrary, by preserving the convexity of the optimization problem, $L1$-regularization opens the way

---

[3] Bayesian information criterion
[4] Akaike information criterion

to extremely fast algorithms. For the price of a little bias on all coefficients, we get to shrink some coefficients to exactly 0, operating selection and estimation in one single step as hoped in problem (5.4).

In [40], a first attempt was made to apply LASSO techniques to the inference of a concentration matrix in a GGM framework, under the name of *neighborhood selection*. This approach solves p different LASSO regression problems, where p is the number of genes in the network. Subsequently, two other articles, [5] and [66], independently provided an improvement of the initial work of [40]. In both works, the problem is seen as a penalized maximum likelihood problem and is solved as a recursive "LASSO-like" problem. The next improvement in this vein comes with the *Graphical* LASSO, or gLASSO, of [20], which makes this penalized likelihood approach highly attractive in terms of computational cost, with very recent improvements for high dimension developed in [38]. Still, the neighborhood selection approach remains a lot cheaper, computationally speaking.

HIGHLIGHTS ON L1-REGULARIZERS FOR GAUSSIAN GRAPHICAL MODELS

Let us review in a little more detail the two LASSO-type techniques upon which we build, namely the *neighborhood selection* and the *Graphical* LASSO approaches.

On the one hand, the $L1$-penalized estimator, proposed in [5] and advantageously solved by the gLASSO algorithm, directly considers the original penalized likelihood problem:

$$\widehat{\Theta}^\rho = \arg \max_{\Theta \in \mathcal{S}_p^+} \log \det(\Theta) - \langle \Theta, \widehat{S} \rangle - \rho \, \|\Theta\|_{L1}. \tag{5.5}$$

In this regularized problem, the $L1$-norm on the entries of the concentration matrix drives some coefficients to zero: it enforces sparsity. The non-negative parameter ρ tunes the global amount of sparsity: the larger the parameter ρ, the fewer edges in the graph. A large enough penalty level produces an empty graph. As ρ decreases toward zero, the estimated graph tends toward the saturated graph, and the estimated concentration matrix tends toward the usual unpenalized maximum likelihood estimator $\widehat{\Theta}^{ML}$. By construction, this approach guarantees a well-behaved estimator of the concentration matrix, that is to say sparse, symmetric, and positive definite.

On the other hand, the more naïve neighborhood selection procedure has been reported to be more accurate in terms of edge detection. The reader is referred to [62] and [51]. This approach determines the graph of conditional dependences Γ by solving a series of p independent $L1$-penalized regression problems, successively estimating each gene neighborhood. Recall that X is the $n \times p$ matrix of observations, with column g containing the vector X_g of n observations for gene g. Matrix $X_{\backslash g}$ contains all columns $\mathbf{X}$ except its gth column, that is to say, observations on all genes except expression levels of gene g. Concretely, for each gene g, expression levels are "explained" by the expression levels of remaining genes. Neighbors of gene g in the graph Γ are estimated by the non-zero elements of $\hat{\beta}_g$ solving problem (5.6).

$$\hat{\beta}_g = \arg \min_{\beta \in \mathbb{R}^{p-1}} \frac{1}{n} \left\| X_g - X_{\backslash g} \beta \right\|_{L2}^2 + \rho \, \|\beta\|_{L1}. \tag{5.6}$$

Indeed, if ne(g) denotes the set of neighbors of gene g in the graph of conditional dependences Γ associated with the concentration matrix Θ, then the best linear approximation of the random vector X_g by the remaining gene expressions $X_{\backslash g}$ is given by

$$X_g = \sum_{h \in ne(g)} \beta_{gh} X_h = - \sum_{h \in ne(g)} \frac{\Theta_{gh}}{\Theta_{gg}} X_h.$$

As a result, problem (5.6) aims to estimate coefficients β_{gh} proportional to the concentration matrix entries of interest Θ_{gh}.

Actually, solving the p regression problems defined by (5.6) may be interpreted as inferring the concentration matrix in a penalized maximum pseudo-likelihood framework, as depicted in [1, 50, 51]: the joint distribution of X is approximated by the product of the p distributions of the p variables, conditional on the other ones, as if these distributions were independent, that is

$$\mathcal{L}(\Theta; X) = \sum_{g=1}^{p} \sum_{i=1}^{n} \log \mathbb{P}(x_{ig} | X_{i \backslash g}; \Theta_g),$$

where $X_{i \backslash g}$ is the ith observation of the vector X deprived of the gth coordinate. This pseudo-likelihood is based upon the (false) assumption that conditional distributions of expression levels are independent. Particularly, the distribution of gene g expression levels conditional on gene h is assumed independent from the distribution of gene h conditional on gene g, ignoring the symmetry condition on concentration matrices. Because the neighborhoods of the p genes are selected separately, a post-symmetrization must be applied to manage inconsistencies between edge selections; [40] suggests AND or OR rules.

Even though there is no reason for the *Graphical* LASSO and the neighborhood selection approach to result in identical estimates at fixed n, they are both shown to be asymptotically consistent in terms of edge detection (as n goes to infinity) under their respective strong but necessary assumptions on the true covariance matrix Σ, most often referred to as irrepresentability conditions. The reader is referred to [49] for a detailed comparison of those results.

In the sequel, for the sake of clarity, we present our suggested extensions in terms of penalized maximum likelihood. However, in practice, all extensions are independent from the formulation used (either likelihood or pseudo-likelihood), and the final choice between those two strategies is left to the user.

5.2.2 Extensions Motivated by Regulatory Network Modeling

While sparsity is necessary to solve the problem when few observations are available, biasing the estimation of the network toward a given topology (what we call "structure" herein) can help us find the correct graph in a more robust way, preventing the algorithms from looking for solutions in regions we know the correct graph is less likely to reside. Introduction of knowledge about structure in the network inference is discussed in the next section, as a first extension to the classical GGM framework. As a second extension, we propose to remedy the problem of heterogeneity among microarray experiments by estimating multiple GGMs, each of which matching different modalities of the same set of variables, which correspond here to different experimental conditions. Finally, we conclude this part on an adaptation to time-course data, which can be combined with both previous extensions.

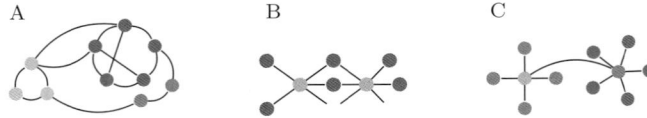

Fig 5.1 Examples of typical network structures. A) Community (or affiliation). B) Star (or hub). C) Mixed community and star.

STRUCTURED GAUSSIAN GRAPHICAL MODELS

Modularity and more generally heterogeneity in the gene connection behaviors are an important property of biological networks (see [23]). Typical network structures which can all be expected in biological networks are illustrated in Fig. 5.1. For instance, the so-called "hubs," as depicted in Fig. 5.1B, are highly connected genes, showing a different behavior from the rest of the graph. To model this heterogeneity, we assume that the graph Γ we wish to infer is endowed with a latent structure Z which clusters the genes into a set $\mathcal{Q}$ of Q groups. This partition is represented by a set of size Q random vectors $\{Z_i\}_{i=1,\ldots,p}$ such that Z_{gq} equals 1 if gene g belongs to cluster q, and 0 otherwise. The latent structure Z should come up with parameters to define how genes spread among the clusters $\mathcal{Q}$ and how a gene from cluster q connects to a gene from cluster L. These parameters could either be considered as prior knowledge, motivated by biological expert knowledge or bibliographical references, or be estimated by various algorithms based on an initial estimate of the graph. As our main point is to show how such a knowledge can be easily integrated into the GGM inference process, we first consider the hidden structure Z as known. This latent structure should help us refine the regularization term in (5.5) by adding entrywise *structure adaptive* penalty parameters to the single parameter ρ governing the overall amount of regularization.

High-dimensional Inference with Latent Network Structure   This generalization of (5.5) was suggested in [1]. The idea is to have a different penalty parameter for each entry Θ_{ij}:

$$\widehat{\Theta} = \arg \max_{\Theta \in \mathcal{S}_p^+} \log \det(\Theta) - \langle \Theta, \widehat{S} \rangle - \rho \, \|P^Z \star \Theta\|_{L1}, \qquad (5.7)$$

where P^Z is a $p \times p$ matrix of penalty terms whose entries depend on the latent structure Z and $\star$ denotes the term-by-term product. Note the decomposition of the penalty term into a common part ρ used to tune the overall amount of penalty, which will govern the overall number of edges in the graph as in (5.5), and a new structured part P^Z used to tune the strength of prior information, which will encourage the edge structure to adopt more or less strongly the prior structure Z.

Indeed, we wish to penalize the elements of the concentration matrix according to the unobserved clusters to which the genes belong. For instance, let us imagine a graph endowed with a community or affiliation structure as presented in Fig. 5.1A: genes belonging to the same community are highly connected to each other, in contrast to genes belonging to different communities. This canonical structure typically arises when clusters are defined by biological functions. If two genes belong to the same unobserved community, we wish to lower the penalty parameter acting on the corresponding entry in the concentration matrix. Conversely, we want to increase the penalty parameters on entries corresponding to genes belonging to different communities with low connectivity probability, so that we can shrink the estimated partial correlation to zero.

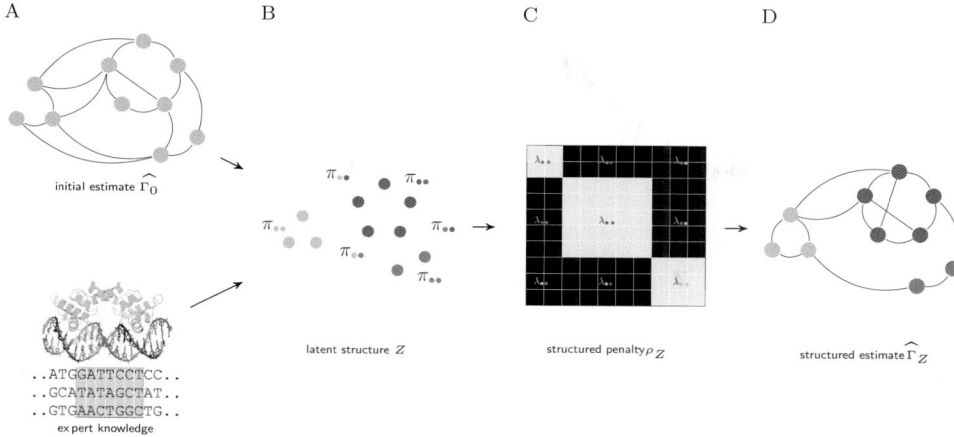

Fig 5.2 Overall inference strategy in two steps. Step 1: A) Collect prior information on the structure either by initial inference of the network via a usual $L1$-regularized Gaussian graphical model or expert knowledge, B) derive the latent structure, and C) design the structured penalty matrix. Step 2: D) Infer the structured Gaussian graphical model.

In the following, we adopt a simple two-step approach illustrated in Fig. 5.2: step 1, definition of the latent structure Z and corresponding penalty matrix P^Z; and step 2, structure adaptive inference of the network. We refer to two articles for a detailed description of possible all-encompassing frameworks justifying mathematically the introduction of this structure adaptive penalty. In [1], an elegant complete likelihood framework is developed, solved by an expectation-maximization (EM) algorithm. Another work, [37], provides a refined Bayesian algorithm to implement this approach.

We now focus on two options to realize step 1. The first one is to infer the latent structure using mixture models for graphs on an initial graph estimate. The second is to find biologically grounded elements of structure based upon various bioinformatics or bibliographic tools.

Inference of the Latent Structure   An interesting graph modeling which captures the features of biological networks is the Stochastic Block Model (SBM) framework, providing mixture models for random graphs. This model has been rediscovered many times in the literature and a non-exhaustive bibliography includes [12, 19, 36, 43, 57, 58]. The most important parameter, allowing us to describe a large panel of network topologies, is the connectivity matrix $\pi = (\pi_{q\ell})_{q,\ell \in \mathcal{Q}}$, describing $\mathbb{P}(i \leftrightarrow j | i \in q, j \in \ell)$, that is, how genes from each cluster connect to each others. Note that even though SBM models describe how to generate edges conditional on the clustering of genes, the use of SBM models follows the reverse path: the objective is to recover the clustering and connectivity coefficients which best fit the observed network.

Inference of such models has been implemented in various R packages, for instance, **mixer**, which is of straightforward use. Details about a large panel of methods to infer SBM can be found for instance in [12, 29].

SBM structures are integrated within step 1 by first inferring an initial graph estimate $\widehat{\Gamma}_0$ based upon a usual unweighted $L1$-penalty as in problem (5.5) (Fig. 5.2A); second, inferring the latent structure via an SBM algorithm on $\widehat{\Gamma}_0$ (Fig. 5.2B); and, finally, deriving the structured penalty matrix from SBM parameters (Fig. 5.2C). Various penalty values can be defined as decreasing functions of the estimated connectivity matrix $\hat{\pi}$. Suppose genes g and h are assigned to their most probable clusters q and L; then an efficient penalty weight for edge Θ_{gh} is $\lambda_{q\ell} = 1 - \pi_{q\ell}$.

Of course, it is tempting to interpret this partition of genes as a model for biological functional modules, genes being more likely to connect to each other if they are linked to the same main biological function, in the spirit of community structures. The SBM framework embraces a larger variety of network topologies. In that respect, Fig. 5.1 illustrates only a small subset of all possible structures. In such structures, clusters gather genes which share similar connectivity patterns to genes in other clusters. Hence, this model is able to capture not only functional modules in the spirit of community structures but also other major topological properties of biological networks, such as star-shaped models isolating transcription factors. This large range of potential topologies represents an equal number of opportunities to introduce biological information via an expert-based definition of the latent structure.

In this case, the use of an initial estimate $\widehat{\Gamma}_0$ to refine the estimation in a second step is close to the idea of the adaptive LASSO developed by [67]. The adaptive LASSO principle is to lower the bias of the LASSO on large coefficients by adapting the penalty parameter of each coefficient so that it automatically scales with the initially inferred value $(\widehat{\Gamma}_0)_{gh}$. Initially small values of $\widehat{\Gamma}_0$ are more strongly penalized in the second step, whereas initially large values are attributed lower penalties. Instead of focusing on edges individually and from a pure statistical point of view, our procedure rather adapts to the underlying and global structure of the graph.

Examples of Where to Find Biological Information about Latent Structure   Many sources can be used as prior biological structures, as long as they provide information on the pattern of regulations. We only provide here some hints at what could be used. However, these expert-based topologies highly depend on the biological model under study and the extent of expert knowledge available at the time of research.

A first source of information lies within metabolic pathways as available from the KEGG or BioCarta databases.[5] Genes belonging to the same pathway are more likely to interact together and be connected in the regulatory network. When possible, information on which genes in the data set code for transcription factors is highly relevant. When they exist, computational predictions of the number of potential binding sites for every known transcription factor in the data set is of even greater use to indicate where to look for potential edges. Such information is for instance available for *Saccharomyces cerevisiae* in the YEASTRACT database.

MULTIPLE GAUSSIAN GRAPHICAL MODELS

Merging different experimental conditions from wet lab data as in [53, 60] is a common practice in GGM-based inference methods. This process enlarges the number of observations available to infer interactions. However, GGMs assume that the observed data form an independent and identically distributed sample. In the aforementioned paradigm, assuming that the merged data are drawn from a single Gaussian component is obviously wrong and is likely to have detrimental effects on the estimation process. In this section, we propose to remedy this problem by estimating multiple GGMs.

From a statistical viewpoint, we have n observations belonging to C different subpopulations (or "tasks"), hence with different distributions. Assuming that each sample was drawn independently from a Gaussian distribution $X^{(c)} \sim \mathcal{N}(\mathbf{0}_p, \Sigma^{(c)})$, the C samples may be processed

---

[5] Kyoto Encyclopedia of Genes and Genomes

separately by following the approach described in Subsection 5.2.1. Denoting by $\mathcal{L}(\Theta^{(c)}; \widehat{S}^{(c)})$ the Gaussian likelihood in condition c, the objective function is expressed compactly as a sum:

$$\underset{\{\Theta_{ij}^{(c)}: i \neq j\}_{c=1}^{C}}{\arg\max} \sum_{c=1}^{C} \left(\mathcal{L}(\Theta^{(c)}; \widehat{S}^{(c)}) - \rho \, \|\Theta^{(c)}\|_{L1} \right). \tag{5.8}$$

Note that it is sensible to apply the same penalty parameter ρ for all samples since we assume normalized data.

Problem (5.8) ignores the relationships between regulation networks. When subpopulation networks are assumed to share a large common core of edges and only differ by a small subset of edges, the multitask learning framework presented in Fig. 5.3 is well adapted, especially for small sample sizes. On the one hand, sharing information may considerably improve estimation accuracy. On the other hand, keeping the opportunity to identify differences between the networks is the key to understanding the regulatory system up to its subpopulation variations. Starting from problem (5.8), coupling the estimation of $\Theta^{(1)}, \ldots, \Theta^{(C)}$ may be achieved by either modifying the data-fitting term or the penalizer. Several proposals were developed in [9] to address this issue. We focus here on the use of a structured penalizer encouraging networks to share similar up- or down-regulations across conditions.

This kind of setting has received much attention in the statistics and machine learning communities, mainly on variants and applications of the group-LASSO proposed in [65], which has already inspired some multitask learning strategies as in [3, 4, 34, 39]. We shortly describe how the group-LASSO may be used to infer multiple graphical models before introducing a slightly more complex penalty that was directly inspired by the application to gene regulatory networks.

As in the independent and identically distributed (i.i.d.) case, sparsity of the concentration matrices is obtained via $L1$ penalization of their entries. The common penalty structure imposes similarities between the concentration matrices. The group-LASSO penalty is based upon a mixed norm which encourages sparse solutions with respect to groups, where groups form a predefined

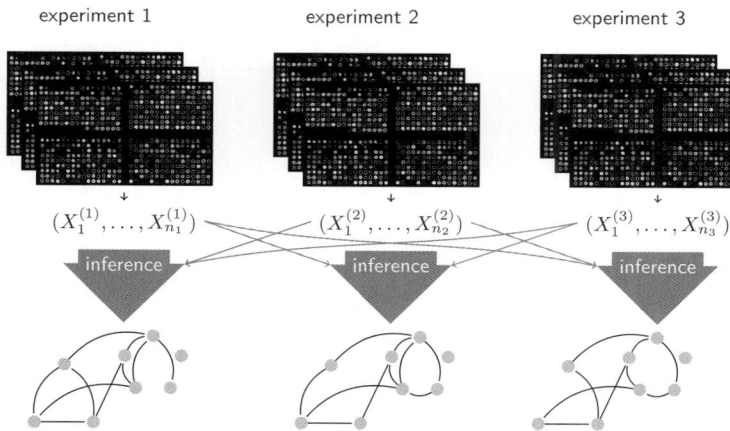

Fig 5.3 Multitask learning framework.

partition of variables. Contrary to the previous section, the partition no longer acts at the gene level but at the edge level, by grouping each partial correlation coefficient across the C conditions. It is therefore useful to define vectors $\{\theta_{gh}\}_{g \neq h}$ containing all partial correlations between genes g and h across the C conditions. Such a penalty will favor graphs $\Gamma^{(1)}, \dots, \Gamma^{(C)}$ with common regulations, not necessarily with the same strength but present or absent together across the conditions. The group-LASSO learning problem is then

$$\underset{\{\Theta_{gh}^{(c)} : g \neq h\}_{c=1}^{C}}{\arg\max} \sum_{c=1}^{C} \mathcal{L}(\Theta^{(c)}; \widehat{S}^{(c)}) - \rho \sum_{g \neq h} \|\theta_{gh}\|_{L2}. \qquad (5.9)$$

Though this formalization expresses some of our expectations regarding the common core between subpopulation networks, it is not really satisfying here since the nature of the group norm enforces all graphs to share exactly the same set of edges. However, we may cure this problem by considering a regularization term that better suits our needs. Specifically, regulations encompass up- and down-regulations and the type of regulation is not likely to be inverted across assays: in terms of partial correlations, sign swaps are very unlikely. This additional constraint is formalized in the following cooperative-LASSO learning problem (5.10), whose properties in the linear regression context have been studied in [10]:

$$\underset{\{\Theta_{gh}^{(c)} : g \neq h\}_{c=1}^{C}}{\arg\max} \sum_{c=1}^{C} \mathcal{L}(\Theta^{(c)}; \widehat{S}^{(c)}) - \lambda \sum_{g \neq h} \left(\|\theta_{gh}^{+}\|_{L2} + \|\theta_{gh}^{-}\|_{L2} \right), \qquad (5.10)$$

where $u^{+} = \max(0, u)$ and $u^{-} = \max(0, -u)$.

Fig. 5.4 illustrates the construction of the two grouped penalties. The group-LASSO switches on or off all edges between gene g and h across all conditions, whereas the cooperative-LASSO

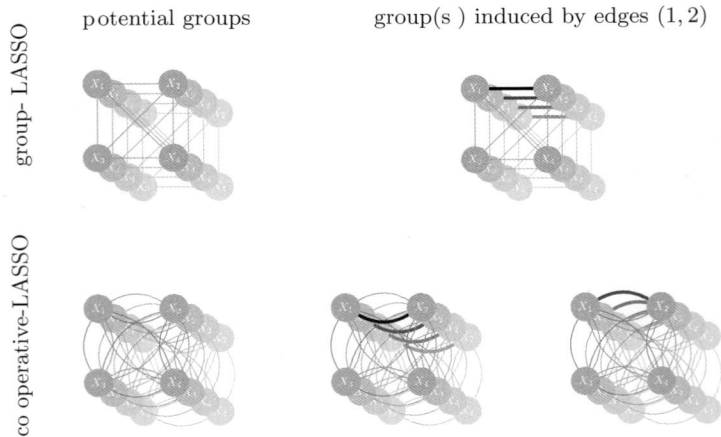

Fig 5.4 Grouping edges without or with sign effect for $C = 4$ conditions.

disconnects the activation of up- and down-regulation. In this way, the cooperative-LASSO allows, for instance, the activation of an up-regulation in a subset of conditions while this regulation disappears in the remaining conditions.

ADAPTATION TO TIME-COURSE DATA

Finally, [30, 45, 55] underline that transcriptomic data sets are not i.i.d. when considering time course expression data. Most learning strategies rely on first-order vector autoregressive (VAR(1)) models, which we define in the next paragraph. Many inference methods have been suggested to handle high-dimensional settings in this VAR(1) framework: [45] depicts a shrinkage estimate, whereas in [30], statistical tests on limited-order partial correlations are performed to select significant edges. In [55], the VAR(1) setup is dealt with by combining ideas from two major developments of the LASSO, thus defining the recursive elastic net. As the elastic-net penalty [68], this method adds an $L2$ term to the original $L1$-regularization, thus encouraging the simultaneous selection of highly correlated covariates on top of the automatic selection process due to the $L1$ norm. As in the adaptive-LASSO [67], weights are corrected on the basis of a former estimate so as to adapt the regularization parameter to the relative importance of coefficients.

Graph of Conditional Dependences in VAR(1) Modeling Here the purpose is to illustrate how to adapt previous developments to time-course settings, following [8]. The dynamics of RNA measurements $X_0, \ldots, X_T$ at regular time points are represented by a VAR(1) model as in equation (5.11). Each measurement X_t is a size p vector containing the expression levels of the p genes at time t:

$$X_t = X_{t-1}A + \varepsilon_t, \quad \text{for all } t \geq 1, \tag{5.11}$$

where matrix $A = (A_{gh})_{g,h \in \mathcal{G}}$ is a $p \times p$ matrix governing the dynamics of expression levels over time. Variations from these dynamics are captured by the white Gaussian process $\{\varepsilon_t\}_{t=1,\ldots,T}$, with zero mean and no correlation between different time points. Under technical assumptions on ε_t, $\{X_t\}_{t \geq 0}$ follows a first-order Markov process homogeneous in time: if expression levels vary over time according to equation (5.11), the regulatory structure among these expression levels is assumed constant over time.

In this setting, matrix A plays the role of the concentration matrix Θ in the i.i.d. framework presented in the previous sections. Indeed, each entry A_{gh} is proportional to the partial correlation coefficient between variables $X_{g,t}$ and $X_{h,t-1}$, that is to say, between the expression of gene g at time t and the expression of gene h at the previous time point, with respect to all other gene expressions at time $t - 1$, as expressed in equation (5.12):

$$A_{gh} = \frac{\text{cov}\left(X_{g,t}, X_{h,t-1} | X_{\backslash h, t-1}\right)}{\text{var}\left(X_{h,t-1} | X_{\backslash h, t-1}\right)}. \tag{5.12}$$

As in the i.i.d. setting, non-zero entries of A code for a directed graph describing the conditional dependences between the elements of $\mathcal{G}$. An edge from h to g is added to the graph if, conditional on all gene expressions except gene h at time $t - 1$, the covariance between $X_{g,t}$ and $X_{h,t-1}$ is non-zero. Inferring A is again equivalent to reconstructing the graph of conditional dependences. However, there are two main differences between this dynamic version of partial correlation and the notion of partial correlation expressed in the previous section. First, the conditioning is made upon all gene expressions from the previous time point; therefore self-loops are allowed. Second, the correlation considered between the two genes is asymmetric: we consider the correlation between the past expression levels of gene h and the present expression levels of gene g, leading naturally to an asymmetric matrix of partial correlations and a directed graph of conditional dependences.

The Penalized Likelihood Similarly to i.i.d. settings, matrix A is given a prior distribution depending on a latent structure Z. Our aim is to maximize the posterior probability of A, given the data X. For a fixed structure Z, if $\widehat{V} = X_{\backslash 0}^{\mathrm{T}} X_{\backslash T}/n$ denotes the across-time empirical covariance matrix, the estimate of A is the solution of

$$\widehat{A} = \arg \max_{A \in \mathbb{R}^{p \times p}} = \langle A, \widehat{V} \rangle - \frac{1}{2} \langle \widehat{S}A, A \rangle - \rho \, \|P^Z \star A\|_{L1}.$$

The structure adaptive penalty matrix P^Z can be adjusted as in the i.i.d. settings either by statistical inference of the latent structure or prior biological knowledge. Since the network is now directed, this structure needs to take the direction of edges into account.

Discussion of VAR(1) Modeling for Time-course Data Caution must be taken when applying VAR(1) models to time-course data. Indeed VAR(1) models apply to dynamic measurements but do not provide dynamic networks. Regulations are assumed to be constant over time. Therefore, this model is better suited to draw a picture of short-term regulation dynamics based upon measurements taken at close time points and over a short period of time. Models taking into account possible evolutions of the networks over time and better suited for life cycle data sets were, for instance, developed in [31].

5.3 Application to Estrogen Receptor Status in Breast Cancer

In this section, we illustrate on real data the practical use of GGM covariance selection with biologically grounded latent structure in a multitask framework. We tested our method on a gene expression data set provided in [22], including 466 patients with breast cancer.

5.3.1 Context

Breast cancer is a heterogeneous disease comprising various subtypes defined by their amplification status of the epidermal growth factor receptor 2 gene ERBB2 and the presence of hormone receptors. In this work, we are particularly interested in the estrogen receptor (ER) status which is an essential parameter of the pathological analysis of breast cancer. Estrogen is a female sex hormone that may stimulate the growth of cancer by triggering particular proteins (receptors) in the damaged cells. If breast cancer cells have estrogen receptors, the cancer is said to be ER positive (ER$^+$). The corresponding ER$^+$ tumors are routinely treated using antihormonal therapies (such as tamoxifen) that can stop estrogen from stimulating the cells to divide and grow. ER$^+$ tumors generally have a more favorable prognosis than tumors with estrogen negative (ER$^-$) status that are usually more clinically aggressive. Understanding the molecular mechanisms underlying the ER status could help better characterize tumors and give appropriate treatment and prognostic information to the patient. Therefore, in this study we aimed to identify changes in the structures of gene networks inferred under ER^+ and ER^- conditions. Based on the GGM framework described in the previous sections, we developed an approach to infer gene regulation networks under each condition using a biological prior on their structures. The biological a priori information is provided from a pathway enrichment analysis conducted on the genes that are differentially expressed between ER^+ and ER^- patients. Thus, the general idea is that the presence of an edge between two genes of the network is promoted or penalized depending on whether the genes belong to the same pathway or not.

In Subsection 5.3.2, we give details about the generation of the biological prior information. This part is structured in two steps: (1) we will perform a differential analysis to reduce data dimensionality by selecting the most informative genes, and (2) we conduct a pathway enrichment analysis to define the latent structure. In Subsection 5.3.3, this structure is used as a prior knowledge to drive the network inference in a multitask framework, where the experimental conditions are the ER^+ or ER^- statuses. This three-step analysis pipeline has been proposed in [25].

5.3.2 Biological Prior Definition

Identification of Differentially Expressed Genes Looking for differentially expressed genes is a widely used approach in microarray data analysis to identify genes associated with a phenotype of interest and to reduce data dimensionality.

For each gene, differential analysis consists in testing the null hypothesis (H_0) that the levels of expression are the same across the conditions of interest, against the alternative hypothesis (H_1) that they differ. Under the assumption of homoscedasticity (i.e., homogeneity of variance) between conditions, the general model is given by

$$\mathbb{E}(X_{ig}^{(c)}) = \mu_g^{(c)} \quad \text{and} \quad \mathbb{V}(X_{ig}^{(c)}) = \sigma_g^2,$$

where $X_{ig}^{(c)}$ is the expression level of the ith sample for gene g under condition c and σ_g^2 is the variance of gene g expression levels, constant across conditions. For two conditions, the null hypothesis to test comes down to

$$\begin{cases} H_0: \ \mu_g^{(1)} = \mu_g^{(2)}, \\ H_1: \ \mu_g^{(1)} \neq \mu_g^{(2)}. \end{cases}$$

We arbitrarily set the ER^- status as Condition 1 and ER^+ as Condition 2.

If the t-test is certainly the most natural and popular test to assess differential expression, its efficiency in terms of variance modeling and power on small sample sizes has been seriously questioned. In [24], the authors report that the Bayesian framework proposed by **limma** in [56] outperforms all the usual methods to test for differential expression. Briefly, the statistic used in **limma** is very similar to an ordinary t-statistic except that the estimations of gene expression variances $(S_g^{\text{limma}})^2$ are moderated across genes:

$$t_g^{\text{limma}} = \frac{\bar{x}_{\cdot g}^{(1)} - \bar{x}_{\cdot g}^{(2)}}{(S_g^{\text{limma}})^2 \sqrt{\frac{1}{n_1} + \frac{1}{n_2}}},$$

where n_1 and n_2 are the number of samples relative to the ER$^-$ and ER$^+$ conditions and $\bar{x}_{\cdot g}^{(1)}$ is the natural estimator of $\mu_g^{(1)}$, i.e., the average expression level for gene g under Condition 1.

In the **limma** statistic, a Bayesian estimator of the variance $(S_g^{\text{limma}})^2$ has been substituted for the usual empirical variance S_g^2, in the classical t-statistic. Assuming a common prior distribution centered around the same value S_0^2 for all gene expression variances σ_g^2, this Bayesian framework results in a posterior estimator shrinking each empirical gene-by-gene estimate S_g^2 toward the common prior value S_0^2; for each gene g, $(S_g^{\text{limma}})^2$ is a weighted average of S_g^2 and S_0^2:

$$(S_g^{\text{limma}})^2 = \frac{d_0 S_0^2 + d_g S_g^2}{d_0 + d_g},$$

where d_0 and d_g are, respectively, the residual degrees of freedom for the prior estimator S_0^2 and for the linear model for gene g.

An empirical Bayes approach is adopted in **limma**, estimating the hyperparameter S_0^2 from the data. We refer to [56] for details about this estimation. It appears that the estimated value of S_0^2 is usually a little less than the mean of the S_g^2's.

Such a strategy is motivated by the fact that the accuracy of the gene variance estimation is a key step in the identification of differentially expressed genes. Indeed, with few samples, it is difficult to assess robustly the variability of each gene individually. To gather as much information as possible and improve the variance of the estimator, one alternative would be to model a common variance σ^2 for all genes. However, the assumption of homoscedasticity across genes is generally not valid. By introducing a common prior distribution on gene variances, the **limma** statistic adopts an intermediate approach which uses information from the whole set of genes to stabilize the estimation of individual gene variances. The Bayesian approach provides robustness to the estimation of specific gene variances, moderating possibly extreme empirical variances that one could expect from so small numbers of samples. This approach is implemented in the R package **limma**.

The differential analysis between the ER$^+$ and ER$^-$ conditions yielded more than 12 000 genes with statistically significant differences at a 5% false discovery rate (FDR) level. In order to ease the interpretation, we focus on the top 200 genes. In the following, we will refer to this set of genes as the signature. The most significant gene is unsurprisingly ESR1, which encodes estrogen receptor alpha. The top gene list also includes GATA3, which is known to have a strong association with the estrogen receptor and CA12, a gene regulated by estrogen via estrogen receptor alpha, as shown in [6, 63].

Pathway Analysis   A common challenge faced by researchers is to interpret gene signatures in biologically meaningful terms. Group testing for pathway analysis, as mentioned in [16], is getting especially popular in the field of gene expression data analysis. It consists in determining whether the signature is enriched in pathway key actors. By conducting an over-representation analysis as done in [35], we search for pathways that are represented more than expected by chance in the set of genes of the signature.

Given p genes measured on a microarray, the signature is defined as a subset of s genes, and a given pathway is defined as another subset of length t of these p genes. Let assume that we observe y of these t genes that are differentially expressed. The probability of having y genes of a given pathway in the list of differentially expressed genes is modeled by the hypergeometric distribution

$$\mathbb{P}(Y = y) = \frac{\binom{s}{y}\binom{p-s}{t-y}}{\binom{p}{t}}.$$

Under the null hypothesis of no over-representation, the probability of observing at least y genes of a pathway of size t in the signature can be calculated by

$$\mathbb{P}(Y \geq y) = 1 - \mathbb{P}(Y \leq y)$$
$$= 1 - \sum_{i=0}^{y} \frac{\binom{s}{i}\binom{p-s}{t-i}}{\binom{p}{t}}.$$

The probability $\mathbb{P}(Y \geq y)$ corresponds to the p-value of a one-sided test. A pathway is said to be "significant" if the null hypothesis of no over-representation is rejected. The significance level of the test is set at 5%.

This approach first requires a pre-defined set of pathways to analyze. The enrichment test was conducted using the KEGG [27], and the BioCarta database. The corresponding sets of pathways are available on the MSigDB website,[6] which contains a collection of annotated gene sets. The enrichment of the signature in KEGG and BioCarta pathways has been measured and led to the identification of 28 significant pathways at a 5% level (see Fig. 5.5, page 138).

The problem with the analysis of pathways is that they do not clearly represent distinct entities. Thus, two pathways can involve common genes and hence share common biological information. For instance, an identical set of genes is responsible for the positive results of three pathways: "G1," "Cell Cycle", and "RacCycD." Therefore, we propose to summarize the set of pathways found significant because of the same genes into a reduced set of "core pathways" (each core pathway is represented by a set of pathways). In practice, we apply a hierarchical clustering algorithm on a binary matrix, denoted by $M = (m_{u,v})_{1 \leq u \leq y,\ 1 \leq v \leq k}$, where y is the length of the signature and k the number of significant pathways, such that

$$m_{u,v} = \begin{cases} 1 & \text{if the gene } u \text{ belongs to the pathway } v, \\ 0 & \text{otherwise.} \end{cases}$$

Dissimilarity between pathways, which accounts for pairwise differences between two given pathways (denoted v_1 and v_2 in the following), is assessed by using a binary metric, also known as the Jaccard distance:

---

[6] http://www.broadinstitute.org/gsea/msigdb/index.jsp

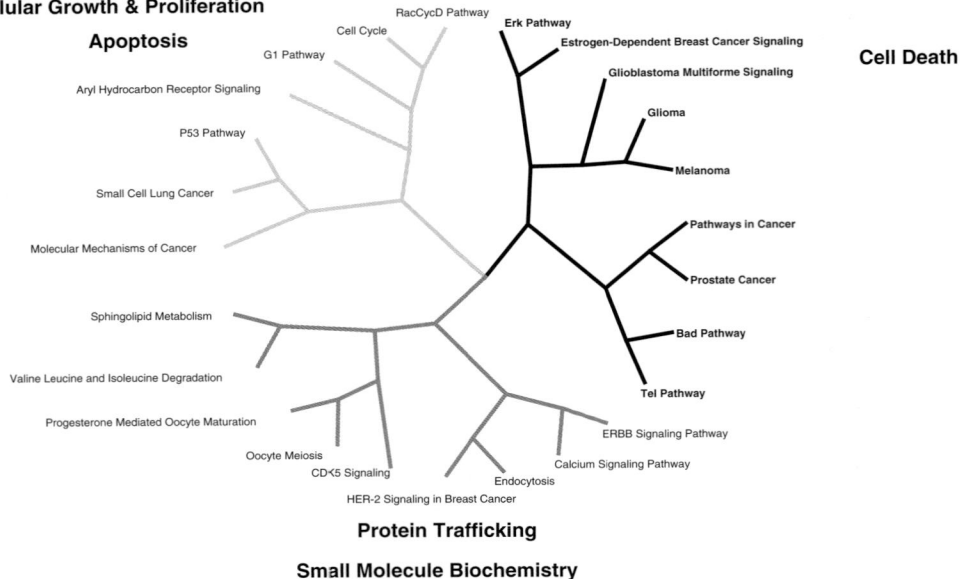

Fig 5.5 Core pathways. From the hierarchical clustering we identify three core pathways related to several molecular and cellular functions. The core pathway associated with apoptosis and cellular growth and proliferation is displayed in light gray. The one related to cell death is colored in black. Finally, the last core pathway (in dark gray) targets protein trafficking and small molecule biochemistry.

CDK5: cyclin-dependent Kinase 5, HER-2: human epidermal growth factor receptor 2.

$$J_\delta = 1 - \frac{\sum_{u=1}^{y} \mathbb{I}_{\{m_{u,v_1}=1,\, m_{u,v_2}=1\}}}{y - \sum_{u=1}^{y} \mathbb{I}_{\{m_{u,v_1}=0,\, m_{u,v_2}=0\}}}.$$

This metric measures the percentage of non-zero elements of two binary vectors that differ. In our case, it corresponds to the percentage of genes that belong exclusively to either one of the two pathways of interest. Finally, from this dissimilarity matrix we perform a hierarchical agglomerative clustering using Ward's criterion.

Three core pathways, displayed in Fig. 5.5, were identified from the hierarchical clustering. The first core pathway is related to cellular growth and proliferation. It is also associated with apoptosis via the gene BLC2, which is involved in blocking apoptotic death in some cells. Apoptosis is one of the main types of programmed cell death, which basically leads to the suicide of the cell. The second core pathway is related to cell death, and is particularly involved in anti-apoptotic functions by enhancing cell survival via the IGF1R gene. The last core pathway is more heterogeneous. We identify two main categories of functions: the protein trafficking (transport or signaling of proteins for instance) and the biochemistry of small molecules (principally fatty acid and lipid metabolism). Each of these core pathways will be used in the next step as a biological prior information to drive the network inference.

5.3.3 Network Inference from Biological Prior: Application and Interpretation

Based on the cooperative strategy detailed in Section 2.2 and implemented in the R package SIMoNe from [11], we conduct network inference from the gene signature. We thus consider two networks (one per condition) containing 200 nodes, each representing a gene from the signature. The ER⁻ network has 649 edges, and the ER⁺ network has 647 edges. Both networks are summarized in Fig. 5.6, which shows a subpart of the entire network that we will take as an example in the following. Both networks have a strong relationship, which means that they have a large number of common edges, symbolized in gray in Fig. 5.6.

In order to illustrate the relevance of the inferred network, we focus on a set of edges (displayed in bold in Fig. 5.6) that are involved in the anti-apoptotic mechanisms. The inactivation of apoptosis plays a key role in the development of cancer, and a wide range of genes are involved in this cellular suicide. Among them, the BCL2 gene is known to mediate anti-apoptotic signals in various human cell systems. In our study, we observe an edge between BCL2 and ESR1, suggesting that there is a regulation between the anti-apoptotic gene and the ESR1 estrogen receptor. This was corroborated by [46], in which it was observed that ESR1 targets BCL2 in the MCF7 breast cancer cell line. The ESR1 receptor has been a focus of attention in the literature,

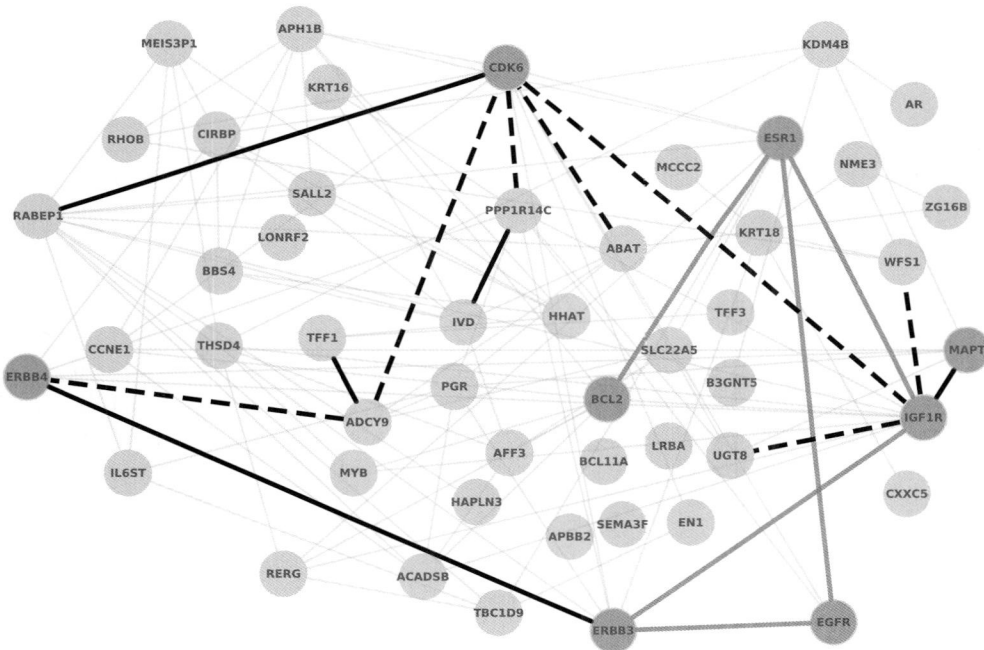

Fig 5.6 Subnetwork inferred from the ER status signature. This figure displays an enlarged view of the network subpart. The common edges of both ER⁺ and ER⁻ networks are symbolized in gray. The dashed black edges are inferred only under the ER⁻ condition, and the solid black edges are only predicted under the ER⁺ condition. The edges and nodes discussed in the main text are in bold.

as it regulates several key genes in cancer development. For instance, in our study we identify a potential regulation between ESR1 and the receptors EGFR and IGF1R. Both regulations are known in the literature. Indeed [52] showed that ESR1 increases the transcription of EGFR, and [44] demonstrated that the expression of IGF1R is regulated by ESR1. Eventually, the edges inferred from our study and validated from the literature allow us to reconstruct the map presented in Fig. 5.7.

In addition to such a result, our approach provides additional information about gene regulations. Indeed, it enables us to specify under which condition the regulation occurs. For instance, [32] highlights that ERBB3 regulates the ERBB4 protein. Our analysis suggests that it operates only in ER$^+$ tumors. On the contrary, under the ER$^-$ condition we note the presence of an edge between CDK6 and IGF1R that does not appear in the ER$^+$ network.

This application to the ER status of breast tumors highlights key mechanisms in cancer progression (cellular growth and proliferation, cell death . . .). To illustrate our approach, we focused on the anti-apoptotic processes. We show the relevance of edges inferred by comparing the ER$^+$ and ER$^-$ networks to what is known in the literature. In addition, the framework we provide

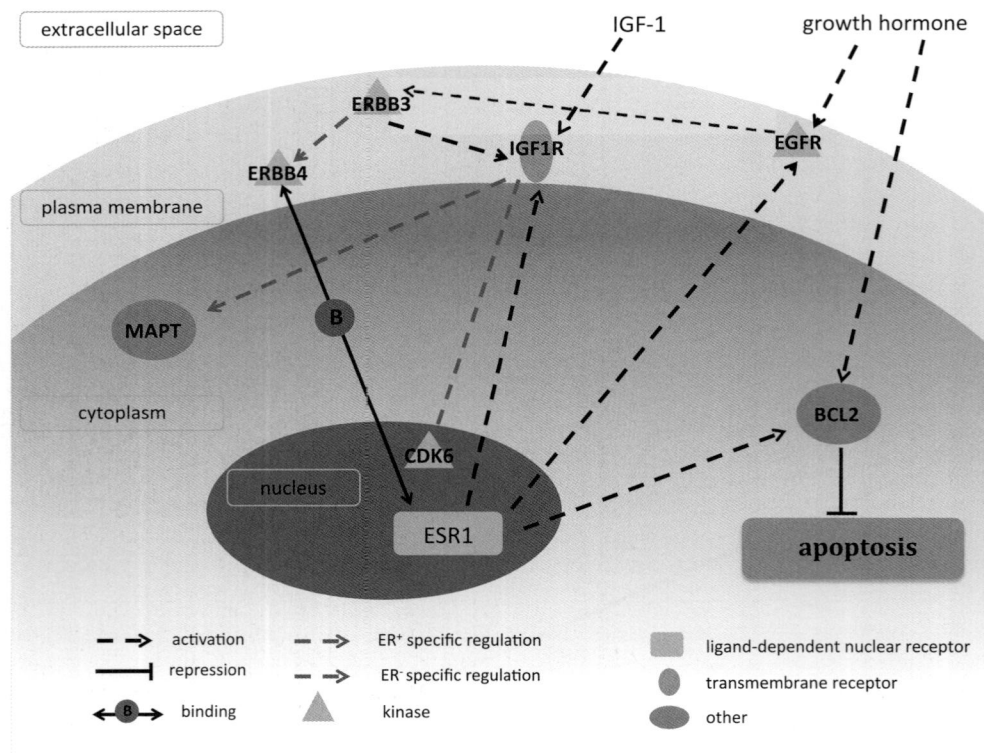

Fig 5.7 Anti-apoptotic molecular mechanisms. Illustration of the results obtained from our network inference approach. By focusing on anti-apoptotic mechanisms, we identify regulations that occur under both ER$^+$ and ER$^-$ conditions, that is, condition-specific regulations.

also allows us to specify condition-specific regulations. Thus, we are able to target the molecular mechanisms occurring specifically in ER^+ or ER^- tumors. Such results can be a starting point for applications at the clinical level to customize treatments according to the ER status of patients.

5.4 Conclusions and Discussion

Inference of regulatory networks is a very challenging issue in systems biology for which GGMs constitute a promising framework. In this chapter, we discussed several models and their extensions motivated by biological applications. First, we described an approach taking into account a latent structure on the concentration matrix that enables us to model heterogeneity in gene connection features. It is particularly relevant for inferring gene networks, since organized modularity is ubiquitous in biological systems. We then introduced various methods intended to infer networks under different experimental conditions. These approaches refer to multiple GGM inference in the chapter and are motivated by the need to analyze transcriptomic data across multiple conditions where many samples measuring the expression of the same molecules are available. Rather than merging the various conditions as it is usually done, we proposed to estimate networks by learning the estimation problems simultaneously. It consists in grouping the partial correlations between variables across the conditions to favor graphs with common edges. Finally, we presented an analogous way of tackling time-varying data. Since all the previous approaches have an impact on different parts of the likelihood function, they can be combined in many different ways: multiple GGMs based upon time-course measurements or multiple GGMs combined with latent topological structures, for instance. We therefore provide a complete toolbox for network inference. The application on breast cancer data is a perfect example of the combination of multiple inference with expert-based latent structure.

From a practical point of view, the GGMs and their extensions can provide useful insights into the mutual influence existing between genes. The study about ER status in breast tumors is a good illustration of the potential of our method to highlight relevant biological phenomena from microarray data. However, as edges are defined conditional on all other genes present in the data set, the relevance of the inferred network greatly depends on the inclusion of all potential covariates in the analysis. Thus, results should be treated with caution because some missing covariates may lead to highlighting direct causal relationships between two genes when an intermediate gene is actually responsible for this regulation. Therefore, as a future study, we aim to integrate heterogeneous "omics" data coming from various technologies, as gene expression data enable us to understand only a limited part of the whole system. Indeed, the regulation of gene expression is governed by an intricate combination of transcription factors, microRNAs, splicing factors and other complex processes occurring at transcriptomic, proteomic, or metabolomic levels. Despite the scarcity of large-scale data sets measuring protein or metabolic levels, other sources of information such as protein–protein interactions, interactions that map between transcription factors and their DNA binding locations or microRNAs could be integrated in such studies. The reader should be aware that the use of partial correlation in biological applications implies a linear association between two genes. In consequence, such a strategy is not suitable for modeling saturation effects, for instance.

From a methodological point of view, the crucial issue of how to tune the overall amount of penalty is still a matter of discussion. Cross-validation is a popular solution. However, its own construction makes it more suited to prediction problems than selection problems. On top of that, it is highly unstable on small samples such as the ones we have to cope with in microarray experiments. BIC adaptations [15, 69] lose their asymptotic justification [47, 54] in the "large p, small n" setting but tend to present extremely good behavior in terms of model selection in simulated experiments. Some methods were developed to meet this need in penalties, guaranteeing robust and stable model selection in high dimension. We should mention at least three interesting approaches in that respect. The GGMselect procedure [21], along with its corresponding R package, GGMselect, was specifically designed to handle the case in which the sample size is smaller than the number of variables. Two bootstrap-based procedures were designed in [41] on the one hand and in [33] on the other hand, the latter being available within the R package **huge**.

As future work, many points draw our attention. First of all, there is an essential need to increase the robustness of the learning process, which remains quite unstable due to the high-dimensional setting and the high level of noise. Theoretical insights like [2, 49] might help to understand the real limits of the number of observations required to obtain fully trustworthy estimates. Development of bootstrap-based methods could also improve the robustness of the networks. Obviously, this also stresses the need for high-dimensional testing frameworks in order to derive confidence intervals on estimated networks and statistically validate the differences observed between inferred networks. High-dimensional testing schemes like [42, 61] pave the way toward a solution but have not yet answered the exact question of how to compare high-dimensional estimates based upon two different populations.

REFERENCES

[1] C. Ambroise, J. Chiquet, and C. Matias. Inferring sparse Gaussian graphical models with latent structure. *Electronic Journal of Statistics*, 3:205–238, 2009.

[2] A. Anandkumar, V. Tan, and A. Willsky. High-dimensional Gaussian graphical model selection: tractable graph families, arXiv: 1203. 0697, 2011.

[3] A. Argyriou, T. Evgeniou, and M. Pontil. Convex multi-task feature learning. *Machine Learning*, 73(3):243–272, 2008.

[4] F. Bach. Consistency of the group LASSO and multiple kernel learning. *Journal of Machine Learning Research*, 9:1179–1225, 2008.

[5] O. Banerjee, L. El Ghaoui, and A. d'Aspremont. Model selection through sparse maximum likelihood estimation for multivariate Gaussian or binary data. *Journal of Machine Learning Research*, 9:485–516, 2008.

[6] D. Barnett, S. Sheng, T. Howe Charn, A. Waheed, W. Sly, C. Lin, E. Liu, and B. Katzenellenbogen. Estrogen receptor regulation of carbonic anhydrase XII through a distal enhancer in breast cancer. *Cancer Research*, 68(9):3505–3515, 2008.

[7] R. Castelo and A. Roverato. A robust procedure for Gaussian graphical model search from microarray data with p larger than n. *Journal of Machine Learning Research*, 7:2621–2650, 2006.

[8] C. Charbonnier, J. Chiquet, and C. Ambroise. Weighted-LASSO for structured network inference from time-course data. *Statistical Applications in Genetics and Molecular Biology*, 9(1):15, 2010.

[9] J. Chiquet, Y. Grandvalet, and C. Ambroise. Inferring multiple graphical structures. *Statistics and Computing*, 21(4):537–553, 2011.

[10] J. Chiquet, Y. Grandvalet, and C. Charbonnier. Sparsity with sign-coherent groups of variables via the cooperative-LASSO. *The Annals of Applied Statistics*, 6(2):795–830, 2012.

[11] J. Chiquet, A. Smith, G. Grasseau, C. Matias, and C. Ambroise. SiMoNe: Statistical Inference for Modular Networks. *Bioinformatics*, 25(3):417–418, 2009.

[12] J.-J. Daudin, F. Picard, and S. Robin. A mixture model for random graphs. *Statistics and Computing*, 18(2):173–183, 2008.

[13] A.P. Dempster. Covariance selection. *Biometrics, Special Multivariate Issue*, 28:157–175, 1972.

[14] A. Dobra, C. Hans, B. Jones, J. R. Nevins, G. Yao, and M. West. Sparse graphical models for exploring gene expression data. *Journal of Multivariate Analysis*, 90(1):196–212, 2004.

[15] C. Dossal, M. Kachour, J. Fadili, G. Peyré, and C. Chesneau. The "degrees of freedom" of the LASSO for underdetermined systems of linear equations. In *SPARS 11*, page 56 2011.

[16] S. Draghici, P. Khatri, R. Martins, G. Ostermeier, and S. Krawetz. Global functional profiling of gene expression. *Genomics*, 81(2):98–104, 2003.

[17] M. Drton and M.D. Perlman. Multiple testing and error control in Gaussian graphical model selection. *Statistical Science*, 22:430, 2007.

[18] M. Drton and M.D. Perlman. A SINful approach to Gaussian graphical model selection. *Journal of Statistical Planning and Inference*, 138(4):1179–1200, 2008.

[19] O. Frank and F. Harary. Cluster inference by using transitivity indices in empirical graphs. *Journal of the American Statistical Association*, 77(380):835–840, 1982.

[20] J. Friedman, T. Hastie, and R. Tibshirani. Sparse inverse covariance estimation with the graphical LASSO. *Biostatistics*, 9(3):432–441, 2008.

[21] C. Giraud, S. Huet, and N. Verzelen. Graph selection with GGMselect. arXiv:0907.0619, 2009.

[22] M. Guedj, L. Marisa, A. de Reynies, B. Orsetti, R. Schiappa, F. Bibeau, G. MacGrogan, F. Lerebours, P. Finetti, M. Longy, P. Bertheau, F. Bertrand, F. Bonnet, A.L. Martin, J.P. Feugeas, I. Bieche, J. Lehmann-Che, R. Lidereau, D. Birnbaum, F. Bertucci, H. de The, and C. Theillet. A refined molecular taxonomy of breast cancer. *Oncogene*, 31(9):1196–1206, 2012.

[23] J. Ihmels, G. Friedlander, S. Bergmann, O. Sarig, Y. Ziv, and N. Barkai. Revealing modular organization in the yeast transcriptional network. *Nature Genetics*, 31(4):370–377, 2002.

[24] M. Jeanmougin, A. de Reynies, L. Marisa, C. Paccard, G. Nuel, and M. Guedj. Should we abandon the t-test in the analysis of gene expression microarray data: A comparison of variance modeling strategies. *PLOS ONE*, 5(9):e12336, 2010.

[25] M. Jeanmougin, M. Guedj, and C. Ambroise. Defining a robust biological prior from pathway analysis to drive network inference. *Journal de la Société Française de Statistique*, 152:97–110, 2011.

[26] B. Jones, C. Carvalho, A. Dobra, C. Hans, C. Carter, and M. West. Experiments in stochastic computation for high-dimensional graphical models. *Statistical Science*, 20(4):388–400, 2005.

[27] M. Kanehisa, S. Goto, M. Hattori, K.F. Aoki-Kinoshita, M. Itoh, S. Kawashima, T. Katayama, M. Araki, and M. Hirakawa. From genomics to chemical genomics: new developments in KEGG. *Nucleic Acids Research*, 34:D354–D357, 2006.

[28] H. Kiiveri. Multivariate analysis of microarray data: differential expression and differential connection. *BMC Bioinformatics*, 12(1):42, 2011.

[29] P. Latouche, E. Birmele, and C. Ambroise. Variational Bayesian inference and complexity control for stochastic block models. *Statistical Modelling*, 12(1):93–115, 2012.

[30] S. Lèbre. Inferring dynamic genetic networks with low order independencies. *Statistical Applications in Genetics and Molecular Biology*, 8(1):9, 2009.

[31] S. Lèbre, J. Becq, F. Devaux, M. P. H. Stumpf, and G. Lelandais. Statistical inference of the time-varying structure of gene-regulation networks. *BMC Systems Biology*, 4(130):1–16, 2010.

[32] H. Lee, R.W. Akita, M.X. Sliwkowski, and N.J. Maihle. A naturally occurring secreted human ERBB3 receptor isoform inhibits Heregulin-stimulated activation of ERBB2, ERBB3, and ERBB4. *Cancer Research*, 61(11):4467–4473, 2001.

[33] H. Liu, K. Roeder, and L. Wasserman. Stability approach to regularization selection (StARS) for high dimensional graphical models. In J. Lafferty, C. K. I. Williams, J. Shawe-Taylor, R. S. Zemel, and A. Culotta, editors *Neural Information Processing Systems (NIPS 2010)*, pages 1432–1440. NIPS, 2010.

[34] K. Lounici, M. Pontil, A.B. Tsybakov, and S. van de Geer. Taking advantage of sparsity for multi-task learning. In *Twenty-second Annual Conference on Learning Theory (COLT 2009)*. Omnipress, 2009.

[35] T. Manoli, N. Gretz, H.-J. Grone, M. Kenzelmann, R. Eils, and B. Brors. Group testing for pathway analysis improves comparability of different microarray datasets. *Bioinformatics*, 22(20):2500–2506, 2006.

[36] M. Mariadassou and S. Robin. Uncovering latent structure in valued graphs: a variational approach. Technical Report 10, Statistics for Systems Biology, 2007.

[37] B. Marlin, M. Schmidt, and K. Murphy. Group sparse priors for covariance estimation. In J. Bilmes and A. Y. Ng, editors, *Proceedings of the Twenty-Fifth Conference on Uncertainty in Artificial Intelligence* pages 383–392. Morgan Kauffman Publishers, 2009.

[38] R. Mazumder and D.K. Agarwal. A flexible, scalable and efficient algorithmic framework for primal graphical LASSO. Technical report, arXiv: 1110. 5508, 2011.

[39] L. Meier, S. Van De Geer, and P. Bühlmann. The group LASSO for logistic regression. *Journal of the Royal Statistical Society, Series B*, 70:53–71, 2008.

[40] N. Meinshausen and P. Bühlmann. High-dimensional graphs and variable selection with the LASSO. *The Annals of Statistics*, 34(3):1436–1462, 2006.

[41] N. Meinshausen and P. Bühlmann. Stability selection. *Journal of the Royal Statistical Society, Series B*, 72:417–473, 2010.

[42] N. Meinshausen, L. Meier, and P. Buhlmann. P-values for high-dimensional regression. *Journal of the American Statistical Association*, 104:1671–1681, 2009.

[43] K. Nowicki and T.A.B. Snijders. Estimation and prediction for stochastic block-structures. *Journal of the American Statistical Association*, 96(455):1077–1087, 2001.

[44] S. Oesterreich, P. Zhang, R.L. Guler, X. Sun, E.M. Curran, W.V. Welshons, C.K. Osborne, and A.V. Lee. Re-expression of estrogen receptor alpha in estrogen receptor alpha- negative MCF-7 cells restores both estrogen and insulin-like growth factor-mediated signaling and growth. *Cancer Research*, 61(15):5771–7, 2001.

[45] R. Opgen-Rhein and K. Strimmer. Learning causal networks from systems biology time course data: an effective model selection procedure for the vector autoregressive model. *BMC Bioinformatics*, 8 (Suppl 2): S3, 2007.

[46] T.J. Peterson, S. Karmakar, M.C. Pace, T. Gao, and C.L. Smith. The silencing mediator of retinoic acid and thyroid hormone receptor (SMRT) corepressor is required for full estrogen receptor alpha transcriptional activity. *Molecular and Cellular Biology*, 27(17):5933–48, 2007.

[47] A.E. Raftery. Bayesian model selection in social research. *Sociological Methodology*, 25:111–164, 1995.

[48] A. Rau, F. Jaffrézic, J.-L. Foulley, and R. W. Doerge. Reverse engineering gene regulatory networks using approximate Bayesian computation. *Statistics and Computing*, 22(6): 1257–1271, 2012.

[49] P. Ravikumar, M. Wainwright, G. Raskutti, and B. Yu. High-dimensional covariance estimation by minimizing ℓ_1-penalized log-determinant divergence. *Electronic Journal of Statistics*, 5:935–980, 2011.

[50] P. Ravikumar, M. J. Wainwright, and J. Lafferty. High-dimensional Ising model selection using ℓ_1-regularized logistic regression. *The Annals of Statistics*, 38:1287–1319, 2010.

[51] G.V. Rocha, P. Zhao, and B. Yu. A path following algorithm for sparse pseudo-likelihood inverse covariance estimation (SPLICE), arXiV: 0807. 3734, 2008.

[52] L Salvatori, L Ravenna, MP Felli, MR Cardillo, MA Russo, L Frati, A Gulino, and E. Petrangeli. Identification of an estrogen-mediated deoxyribonucleic acid-binding independent transactivation pathway on the epidermal growth factor receptor gene promoter. *Endocrinology*, 141(6):2266–2274, 2000.

[53] J. Schäfer and K. Strimmer. A shrinkage approach to large-scale covariance matrix estimation and implications for functional genomics. *Statistical Applications in Genetics and Molecular Biology*, 4(1):32, 2005.

[54] G. Schwarz. Estimating the dimension of a model. *The Annals of Statistics*, 6(2):461–464, 1978.

[55] T. Shimamura, S. Imoto, R. Yamaguchi, A. Fujita, M. Nagasaki, and S. Miyano. Recursive regularization for inferring gene networks from time-course gene expression profiles. *BMC Systems Biology*, 3(41), 2009.

[56] G.K. Smyth. Linear models and empirical Bayes methods for assessing differential expression in microarray experiments. *Statistical Applications in Genetics and Molecular Biology*, 3(1), 2004.

[57] T.A.B. Snijders and K. Nowicki. Estimation and prediction for stochastic blockmodels for graphs with latent block structure. *Journal of Classification*, 14(1):75–100, 1997.

[58] C. Tallberg. A Bayesian approach to modeling stochastic block structures with covariates. *Journal of Mathematical Sociology*, 29(1):1–23, 2005.

[59] R. Tibshirani. Regression shrinkage and selection via the LASSO. *Journal of the Royal Statistical Society, Series B*, 58(1):267–288, 1996.

[60] H. Toh and K. Horimoto. Inference of a genetic network by a combined approach of cluster analysis and graphical Gaussian modeling. *Bioinformatics*, 18:287–297, 2002.

[61] N. Verzelen and F. Villers. Tests for Gaussian graphical models. *Computational Statistics & Data Analysis*, 53(5):1894–1905, 2009.

[62] F. Villers, B. Schaeffer, C. Bertin, and S. Huet. Assessing the validity domains of graphical Gaussian models in order to infer relationships among components of complex biological systems. *Statistical Applications in Genetics and Molecular Biology*, 7(2):1–34, 2008.

[63] D. Voduc, M. Cheang, and T. Nielsen. GATA-3 expression in breast cancer has a strong association with estrogen receptor but lacks independent prognostic value. *Cancer Epidemiology Biomarkers and Prevention*, 17(2):365–373, 2008.

[64] A. Wille and P. Bühlmann. Low-order conditional independence graphs for inferring genetic networks. *Statistical Applications in Genetics and Molecular Biology*, 5(1), 2006.

[65] M. Yuan and Y. Lin. Model selection and estimation in regression with grouped variables. *Journal of the Royal Statistical Society, Series B*, 68(1):49–67, 2006.

[66] M. Yuan and Y. Lin. Model selection and estimation in the Gaussian graphical model. *Biometrika*, 94(1):19–35, 2007.

[67] H. Zou. The adaptive LASSO and its oracle properties. *Journal of the American Statistical Association*, 101(476):1418–1429, 2006.

[68] H. Zou and T. Hastie. Regularization and variable selection via the elastic net. *Journal of the Royal Statistical Society. Series B. Statistical Methodology*, 67(2):301–320, 2005.

[69] H. Zou, T. Hastie, and R. Tibshirani. On the degrees of freedom of the LASSO. *The Annals of Statistics*, 34(5):2173–2192, 2007.

PART III

Causality Discovery

CHAPTER 6

Utilizing Genotypic Information as a Prior for Learning Gene Networks

KYLE CHIPMAN AND AMBUJ SINGH

The combination of genotypic and genome-wide expression data arising from segregating populations offers an unprecedented opportunity to model and dissect complex phenotypes. Leading studies have augmented Bayesian networks with genotypic data, providing a powerful framework for learning and modeling causal relationships. However, one major drawback of these methods is that they are generally limited to resolving causal orderings for transcripts most proximal to the genomic loci. This chapter reviews two methods where all interactions between genotype and gene transcripts are considered collectively in order to better resolve causal relationships between gene transcripts. The likelihood-based causality model selection (LCMS) of Schadt and collaborators is first described. Then the stochastic causal tree (SCT) method is depicted. It allows the learning of causal hierarchies representing the propagation of influence that emanates from genomic loci and is transmitted through gene transcripts. The information provided by such methods is intended to be used as a prior for Bayesian network structure learning, resulting in enhanced performance for gene network reconstruction.

6.1 Introduction

This chapter addresses methodologies for probabilistic causal modeling of gene networks. In particular, we describe methodologies tailored for data sets where both gene expression levels and genotypic data are provided. Studies of this nature have been described as genetics of gene expression (GOGE) due to the fact that, when analyzed in tandem, genotypic and expression data

Probabilistic Graphical Models for Genetics, Genomics, and Postgenomics. First Edition. Christine Sinoquet
& Raphaël Mourad (Eds). © Oxford University Press 2014. Published in 2014 by Oxford University Press.

empower researchers to model the genetic basis of transcriptional variation. Since standard quantitative trait loci (QTL) mapping is applied in these studies to establish associations between genomic loci and gene transcripts, the corresponding genotypic and expression data are often referred to as expression QTL (eQTL) data sets. While the methodologies covered in this chapter are described in the context of GOGE studies modeling causal relationships between gene transcripts, we emphasize that they are readily applicable to other phenotypes, including disease states and physiological traits. Indeed, the prevalence of studies focused on genome-wide levels of gene expression reflects the relative ease with which measurements of transcript abundance are acquired, and the coming years will provide data sets representing general physiological phenotypes to which the methodologies presented herein can be applied.

The combination of genotypic and expression data, when collected at the population level or generated from an inbred cross, offers an unprecedented opportunity to model and dissect complex phenotypes. The immense potential offered by these data stems from the fact that genotypic variation can be assumed to directly influence phenotypes, which, as we explain in this chapter, has important consequences for structure learning in graphical models. The computational methodologies presented herein can be conceptually viewed as the synergy between the forerunning methodologies of QTL mapping and Bayesian networks [15]. eQTL mapping is a straightforward procedure based on univariate mapping for learning significant associations between transcripts and genetic variations, while Bayesian networks have long been recognized as a state-of-the-art method for causal modeling of random variables. While both methodologies are especially useful by themselves, ideally the benefits of both approaches could be fused into a single methodology. Since Bayesian networks are designed to model a set of random variables where no prior structural relationships are assumed, they are naturally suitable for biological studies featuring only gene expression data. For example, Bayesian network structure learning was first applied to biological networks by Friedman et al. [8], who modeled gene expression networks from microarray data on *Saccharomyces cerevisiae* cell-cycle measurements. Since this leading research was published, there have been many successful applications of Bayesian network modeling to gene expression data sets.

Opportunities for more advanced modeling methodologies arose upon the inception of eQTL data sets featuring both genotypic and expression data [1]. In one particular pioneering study, Brem et al. [1] published expression and genotypic data for over 100 yeast strains derived from an inbred cross. Immediately, such data sets are amenable to eQTL mapping, where significant associations between genotype and transcript levels allow for focused, point experiments to better elucidate the molecular basis underlying the associations. However, while eQTL associations are often useful for providing leads for experimentalists, they represent "point solutions" that are considered independent of each other. Toward the goal of producing more global, holistic models, it is potentially advantageous to consider all interactions between genotype and gene transcripts collectively in order to better resolve causal relationships between gene transcripts. This need was answered by Schadt and colleagues [16], who introduced novel probabilistic modeling techniques to integrate genotype–transcript interactions into modeling gene expression networks. In this chapter, we describe in detail the LCMS method of Schadt and colleagues, as well as recently published alternatives to the original methodology.

Before presenting the specifics of these methodologies, it is instructive to consider that methodologies belonging to this class do not by themselves learn causal networks. Instead, they are designed to learn the likelihoods of causal relationships between pairs of gene transcripts. The information gleaned from the LCMS and related techniques is incorporated as prior knowledge into Bayesian network structure learning, which ultimately learns causal gene networks. The integration of the LCMS method with Bayesian network structure learning is outlined in Subsection 6.2.4.

6.2 Methods

This section outlines steps required to construct Bayesian networks from eQTL data sets, where directed edges represent causal relationships between gene transcripts. We focus on modeling eQTL data sets derived from studies concerning GOGE, which feature data for both gene expression and DNA variation. We summarize the important aspects of eQTL data sets and detail how these data are especially amenable for causal modeling of gene transcripts. To convey the significance of this line of research, we describe in detail leading methods for causal modeling of gene transcripts, in particular the LCMS method of [16]. Additionally, we provide technical details of how the LCMS method is integrated with Bayesian network structure learning in a principled manner. Finally, we cover recently published methodologies for causal modeling that have been motivated by the LCMS method.

6.2.1 eQTL Data sets

For studies involving GOGE, the samples (columns) correspond to either individuals from an outbred population (association studies) or, alternatively, individuals from a segregating population arising from an inbred cross, termed RIAILs (recombinant inbred advanced intercross lines) [1]. In both cases, genotypic and transcript expression data are provided, which results in greater power to learn causal networks as compared to data sets featuring expression data alone. Thus, the advent of eQTL data sets produced by GOGE studies offers the potential for a new class of computational methods that leverage genotypic data to resolve causal relationships between transcripts. Fig. 6.1(page 152) depicts a typical eQTL data set, which is composed of both genotypic and transcript expression data for a cohort of samples of size P. The gene expression data comprise N transcripts: $T_1, \ldots, T_N$. Similarly, the genotypic data set consists of M loci: $L_1, \ldots, L_M$. The expression level of each transcript is generally represented as a continuous variable, produced by the log-transformed signals from microarrays [17] or the "digital" readouts from RNA sequencing technologies [18]. The genotypic data are binary or tertiary for haploid and diploid species, respectively. Since the number of gene transcripts is on the order of thousands, the expression data are commonly discretized for computational efficiency.

6.2.2 LCMS Method for Learning a Prior Matrix of Causal Relationships

This section details the steps required to construct a matrix of causal relationships between transcripts using the likelihood-based causality model selection (LCMS) method developed by Schadt and colleagues, which is the most widely utilized procedure for learning causal models from eQTL data sets [16, 21]. The core of the LCMS method consists of three likelihood models corresponding to three distinct configurations of one genomic locus and two gene transcripts. The three

	S_1	S_2	S_3	S_4	...	S_P
T_1	0.7	0.5	0.4	1.9	0.8	0.9
T_2	2.1	1.2	1.4	1.3	0.2	1.0
⋮	1.2	1.3	2.2	0.8	1.9	0.9
T_N	1.9	0.9	2.4	2.0	1.5	0.1
L_1	1	0	1	0	1	0
L_2	2	0	2	1	1	0
⋮	1	2	0	1	1	2
L_M	0	1	1	2	0	1

Fig 6.1 In this example eQTL data set, the set of P samples, $S_1, \ldots, S_P$ are represented as columns. The expression levels of all N gene transcripts, $T_1, \ldots, T_N$, and all L loci, $L_1, \ldots, L_M$, are in rows.

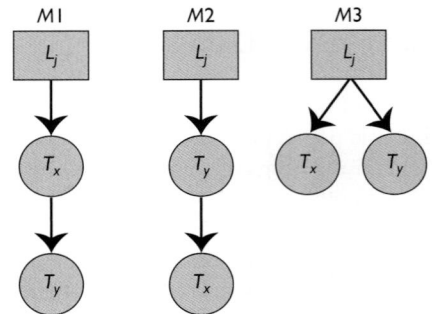

Fig 6.2 Models $M1$ and $M2$ imply a causal relationship between variables T_x and T_y. In model $M3$, the two transcripts are influenced by locus L_j in an independent manner.

alternative models are depicted in Fig. 6.2, in which models $M1$ and $M2$ indicate causal relationships of either T_x controlling T_y or T_y controlling T_x, respectively. By contrast, model $M3$ indicates that transcripts T_x and T_y are influenced by locus L_j in an independent manner.

It might initially seem that more than three models are possible, given that our example subnetwork consists of three nodes. However, biological reasoning allows us to assume that the influence of genomic loci on transcripts is unidirectional, and therefore genomic loci serve as causal anchors in the network. This has a very important practical implication for structure learning, as the two models $L_j \rightarrow T_x \rightarrow T_y$ and $T_y \rightarrow T_x \rightarrow L_j$ are no longer likelihood equivalent. In probabilistic terms, the three models can be expressed as:

$$M1 = \mathbb{P}(T_x \rightarrow T_y | T_x, T_y, L_j)$$
$$M2 = \mathbb{P}(T_y \rightarrow T_x | T_x, T_y, L_j)$$
$$M3 = \mathbb{P}(T_x \perp\!\!\!\perp T_y | T_x, T_y, L_j).$$

The probabilities of the three models highlight the fact that we are considering possible relationships between the two transcripts T_x and T_y. The model likelihoods as a function of the genotypic and expression data are expressed as:

$$M1 = \prod_{i=1}^{P} \sum_{j=1}^{K} \mathbb{P}(L_j)\, l(\theta_{T_{x_i}|L_j}; T_{x_i}|L_j)\, l(\theta_{T_{y_i}|T_{x_i}}; T_{y_i}|T_{x_i})$$

$$M2 = \prod_{i=1}^{P} \sum_{j=1}^{K} \mathbb{P}(L_j)\, l(\theta_{T_{y_i}|L_j}; T_{y_i}|L_j)\, l(\theta_{T_{x_i}|T_{y_i}}; T_{x_i}|T_{y_i})$$

$$M3 = \prod_{i=1}^{P} \sum_{j=1}^{K} \mathbb{P}(L_j)\, l(\theta_{T_{x_i}|L_j}; T_{x_i}|L_j)\, l(\theta_{T_{y_i}|T_{x_i},L_j}; T_{y_i}|T_{x_i}, L_j).$$

Each model applies to a single locus, L_j, and two transcripts, T_x and T_y. Each of the respective likelihood functions iterates over the P samples and the K possible states of L_j ($K = 2$ for haploid, $K = 3$ for diploid). Assuming the expression data are normally distributed, the likelihood functions of the form $l(\theta_{T_{x_i}|L_j}; T_{x_i}|L_j)$ are implemented using a Bivariate normal distribution. The likelihood function in M3, $l(\theta_{T_{y_i}|T_{x_i}, L_j}; T_{y_i}|T_{x_i}, L_j)$, is modeled as a conditional bivariate normal distribution [16]. Note that Markov equivalence between the two possible configurations for M3 assures that $\mathbb{P}(L_j)\mathbb{P}(T_x|L_j)\mathbb{P}(T_y|T_x, L_j) = \mathbb{P}(L_j)\mathbb{P}(T_y|L_j)\mathbb{P}(T_x|T_y, L_j)$.

After the respective likelihoods for the three models have been computed, it is necessary to account for model complexity. One option for adjusting for model complexity is the Akaike information criterion (AIC), where the penalized score is obtained from the model log-likelihood score by applying the following equation: $AIC_i = -2 \times LLS_i + \log(P) \times p_i$, where LLS_i is the likelihood of the ith model, P is the number of samples in the data set, and p_i is the number of parameters for the ith model. At this point, we have three penalized likelihood scores: AIC_1, AIC_2, and AIC_3, corresponding to models $M1$, $M2$, and $M3$, respectively. From these penalized likelihood scores, we wish to obtain probabilities for the causal relationships between pairs of gene transcripts. Toward this goal, bootstrapping is applied to estimate the stability of the relative likelihoods of each model. Statistical bootstrapping [6] consists of randomly selecting P indices from the full set of P samples. In practice, applying modeling methodologies to a large set of bootstrapped data sets (1000 or more) provides a measure of the consistency at which one model obtains the highest likelihood. To convey this point, we consider hypothetical results for the example triplet of L_j, T_x, and T_y depicted in Fig. 6.2. Assuming there are 1000 bootstrapped data sets, one might observe that models $M1$, $M2$, and $M3$ achieve the most optimal AIC score 700, 40, and 260 times. This hypothetical outcome equates to $\mathbb{P}(T_x \rightarrow T_y|T_x, T_y, L_j) = 0.7$, $\mathbb{P}(T_y \rightarrow T_x|T_x, T_y, L_j) = 0.04$, and $\mathbb{P}(T_x \perp\!\!\!\perp T_y|T_x, T_y, L_j) = 0.26$ for models $M1$, $M2$, and $M3$, respectively. In this example, model $M1$ clearly represents the most likely causal relationship between transcripts T_x and T_y. This procedure is applied on a genome-wide scale to all pairs of gene transcripts, as we detail next.

In order to generate a matrix of probabilities between all transcript pairs, the LCMS method is applied to all possible triplets consisting of a single locus and two transcripts, provided that both transcripts show significant association to a common locus. In practice, it is desirable to only

---

[0] We use the same notations as those of Schadt et al. [16].

consider triplets where the two transcripts are significantly associated to a common genomic locus in order to reduce computation time and limit noise. Therefore, for each locus, the set of significantly associated transcripts is determined by applying univariate mapping techniques, usually the t-test for normally distributed data or the Wilcoxon rank-sum test in cases when the data are non-normally distributed. Finally, the experiment-wise p-value threshold of significance is determined using the false discovery rate (FDR) [5]. The choice of FDR threshold is arbitrary, though we note that publications utilizing the LCMS method typically employ an FDR of 0.05. After the set of significantly associated transcripts has been calculated for each locus, the matrix of pairwise causal relationships, R, can be calculated as follows:

$$R_{x,y} = \sum_{j=1, x, y \in L_j}^{M} \mathbb{P}(T_x \rightarrow T_y | T_x, T_y, L_j)$$

where R is an $N \times N$ matrix, M is the total number of genomic loci, and $x, y \in L_j$ expresses that transcripts T_x and T_y are both associated with locus L_j. At this point, the LCMS method has extracted insight from the full eQTL data set and encapsulated it into matrix R. As we discuss in Subsection 6.2.4, this matrix of pairwise causal relationships will serve as a prior over the set of possible graph structures for Bayesian network structure learning.

6.2.3 Bayesian Network Structure Learning

Bayesian networks provide a graphical representation of the joint probability distribution for a set of random variables, allowing for efficient computation of the probability of graphical structures [15]. A signature aspect of Bayesian networks is that the global likelihood function that defines the probability of the structure is decomposable into the product of local likelihood functions. This is due to the fact that Bayesian networks carry Markov assumptions. This implies that each variable, e.g., T_x, is independent of all other ancestor variables given its parents, denoted by the set $Pa(T_x)$. Formally, Bayes' formula allows the posterior probability of a network, $\mathbb{P}(G|D)$, to be calculated as $\frac{\mathbb{P}(G) \times \mathbb{P}(D|G)}{\mathbb{P}(D)}$. $\mathbb{P}(G)$ represents the prior over the possible states of graph structures, which, as we detail in the next section, is useful for incorporating prior biological knowledge. $P(D)$ is the marginal likelihood of the data, and is the average probability of the data over all possible parameter values. Since it serves as a normalizing constant that cancels out when comparing the relative likelihoods of two distinct structures, it is omitted from further consideration in this chapter. We therefore focus on optimizing the likelihood of the data given a structure, $\mathbb{P}(D|G)$.

For a specific graph structure, G, the model parameters θ specify the distribution for the data set of random variables $T_1, T_2, \ldots T_N$. The chain rule allows the joint distribution over the data, $\mathbb{P}(T_1, T_2 \ldots T_N)$, to be expressed as the product of conditional probabilities: $\prod_{i=1}^{N} \mathbb{P}(T_i | T_1 \ldots T_{i-1})$.

Following from the aforementioned Markov assumptions, conditional independence implies that the probability of a variable, e.g., T_i, is independent of all other ancestor variables in the network given its parents, $Pa(T_i)$:

$$\mathbb{P}(T_1, T_2 \ldots T_N) = \prod_{i=1}^{N} \mathbb{P}(T_i | T_1 \ldots T_{i-1}) = \prod_{i=1}^{N} \mathbb{P}(T_i | Pa(T_i)).$$

For a data set composed of N genes and P samples, the likelihood of the data, given parameters θ is

$$L(\theta : D) = \prod_{j=1}^{P} \prod_{i=1}^{N} \mathbb{P}\left(T_i[j]|Pa(T_i), \theta_{T_{i|Pa(T_i)}}\right),$$

where $Pa(T_i)$ is the set of parents for transcript T_i. Again, conditional independence implies that the global likelihood can be decomposed into the product of local likelihoods, and the parametrization of transcript T_i is solely a function of the parent set $\theta_{T_{i|Pa(T_i)}}$.

This has very important implications for the computational complexity of structure learning. Namely, the local likelihood score for each node can be computed based on the sufficient statistics encoded by its parents.[1] For discretized data with three values, the computational complexity is at worst $N \times 3^k$, where N is the number of random variables and k indicates the maximum number of parents encountered for any node. For studies focused on modeling gene networks, it is very common to limit the maximum value of k to 3.

6.2.4 Integrating the Prior Matrix

Rapid advancements in technologies for profiling DNA variation, gene transcript levels, and protein abundance have fueled the respective fields of genomics, transcriptomics, and proteomics. Importantly, as the costs of profiling technologies decline, data sets featuring multiple data types are becoming ubiquitous. Such advancements provide ample opportunity for modeling heterogeneous data types with the goal of building more holistic biological models. Such approaches, however, mandate formal methods to integrate and weight the relative scores corresponding to the different data types. While this is potentially a broad area of computational research, the scope of this chapter focuses on methods for integrating knowledge from external data sources as prior information for Bayesian network structure learning [10, 11, 14]. One especially influential approach for integrating biological network data into structure learning was introduced by Imoto et al. [11], which, building on methodology from statistical physics, formulates the prior over the distribution of possible structures as

$$\mathbb{P}(G|\beta) = \frac{e^{-\beta E(G)}}{Z(\beta)}$$

where $E(G)$ is the energy function, and the partition function $Z(\beta)$ is the normalizing constant. The energy function is always dependent solely on the current graph structure, whereas the normalizing constant can be viewed as the sum of all possible states. The hyperparameter β is used to modulate the strength of the prior information as compared to the Bayesian network likelihood score. Since the value of β relative to the Bayesian network likelihood score will vary depending on the application and specific data set, its value is a heuristic choice that will depend on the practitioner's goals. However, we note that a recent study involving simulated eQTL data sets reported that network reconstruction accuracy is relatively stable across a broad range

---

[1] In other words, no additional data points or statistics from the data set can theoretically improve parameter estimation or model likelihood.

of values for β [3], indicating that performance gains should be observed for most non-zero values of β. Exact computation of $Z(\beta)$ entails enumerating over all possible network structures, which is not computationally feasible. An upper bound estimate for the partition function $Z(\beta)$ has been presented by Imoto et al. [11] and Husmeier et al. [10]; however, the partition function may be disregarded entirely for most purposes, since the structure learning procedure evaluates the ratio of two likelihood functions to identify whether a novel structure proposal improves on the incumbent structure. Consequently, when evaluating the ratio of two likelihood scores, $Z(\beta)$ occurs in both the numerator and denominator and thus cancels out, leaving a ratio of two exponentials. Therefore, given the $N \times N$ prior matrix R, the calculation of the graph prior, $\mathbb{P}(G|\beta)$, is as follows:

$$\mathbb{P}(G|\beta) = \exp\left\{ -\beta \sum_{i,j\in e(G)} R_{i,j} \right\},$$

where $e(G)$ denotes the set of edges in G.

We now revisit the posterior probability of the Bayesian network, which can be evaluated as the product of the graph prior and data likelihood: $\mathbb{P}(G) * \mathbb{P}(D|G)$. Having described how to score the probability of the graph structure, $\mathbb{P}(G)$, we now summarize the log-likelihood score of the entire system:

$$LLS_{\text{total}} = \log\left(\prod_{j=1}^{P} \prod_{i=1}^{N} \mathbb{P}\left(T_i[j]|Pa(T_i), \theta_{T_i|Pa(T_i)} \right) \right) + \left\{ \beta \sum_{i,j\in e(G)} R_{i,j} \right\}.$$

Since the total score is evaluated on a log scale, the second component of the total score, corresponding to the graph prior, $\mathbb{P}(G)$, sees the removal of the exponential term, simplifying to $\beta \sum_{i,j\in e(G)} R_{i,j}$.

6.2.5 Stochastic Causal Tree Method

Following the widespread success of the LCMS method, several alternative methods have been published [2, 3]. In this section, we describe the recently published stochastic causal tree (SCT) method of Chipman et al. [3]. Much like the LCMS method, the SCT method produces a prior matrix of dimension $N \times N$ representing causal relationships between pairs of gene transcripts. The SCT method is similar to the LCMS method in essence, since the primary aim is to properly align node triplets, and indeed it also utilizes combinations of pairwise and third-order correlations. However, the SCT method differs in that it does so in the context of a tree, resulting in a more global solution with greater coverage extending to transcripts located more distantly in the network with respect to genomic loci. The trees consist of genomic loci, which serve as roots for their respective trees, and an arbitrary number of transcripts that are stochastically added to the growing tree. Much like the LCMS method, the SCT method employs bivariate and conditional normal distributions to ensure the integrity of the alignment of a node with its grandparent and parent nodes.

Referring to Fig. 6.3, which illustrates the SCT method, one can see that the tree is initiated with a particular locus serving as a root, and given the current structure of the tree, the crux of the method involves including the node with the highest likelihood. In this figure, the current tree contains one locus (L_1, shaded square) and four transcripts (T_a, T_b, T_c, and T_d, shaded circles).

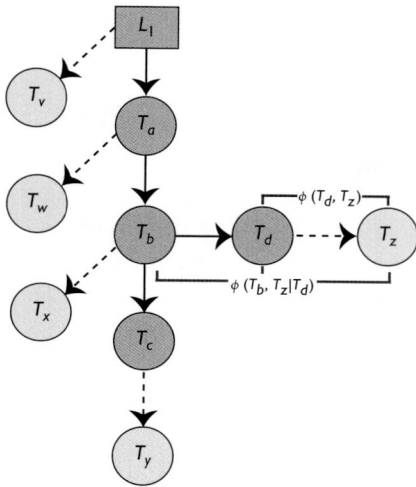

Fig 6.3 Schematic of the stochastic causal tree method. Shaded nodes represent the locus (square) and transcripts (circles) that are currently part of the tree; solid arrows represent causal relationships between nodes. The four transcripts currently included in the tree, T_a, T_b, T_c, and T_d, each have a respective best candidate node (unshaded circles) that can potentially be included. The associations between the incumbent nodes and their respective best candidate node are represented by dashed arrows. The likelihood of any candidate transcript being added to the tree depends on the candidate, its parent, and its grandparent. For example, the likelihood of the candidate transcript T_z being connected to T_d is a function of two potentials: $\phi(T_d, T_z)$ and $\phi(T_b, T_z|T_d)$.

Each of these five nodes has a corresponding best candidate, denoted by dashed arrows to unshaded circles. At this point, the SCT algorithm will stochastically choose one of the five best candidates, where the likelihood of any candidate transcript being added to the tree depends on the candidate, its parent, and its grandparent.

Again referring to Fig. 6.3, the likelihood of the candidate transcript T_z being connected to T_d is a function of two potentials: $\phi(T_d, T_z)$ and $\phi(T_b, T_z|T_d)$. The potential functions can be expressed abstractly as:

$$\phi(n_p, n_c), \tag{6.1}$$

$$\phi(n_g, n_c|n_p). \tag{6.2}$$

While several different functions are suitable for implementing the potentials, including Pearson's correlation or mutual information, the authors opted to use regression functions due to the fact that eQTL data sets consist of both binary (genomic loci) and continuous (expression) data. We next detail how the second-order potential, $\phi(n_p, n_c)$, is derived from the following regression function:

$$n_c = \hat{\beta}_0 + \hat{\beta}_1 \times n_p + \epsilon.$$

Specifically, the second order potential is a function of the regression sum of squares (SSR) and the total sum of squares (SSTO) from the above function where n_c is regressed against n_p:

$$\phi(n_p, n_c) = \sqrt{SSR_{n_p, n_c}/SSTO_{n_p, n_c}} = \sqrt{R^2_{n_p, n_c}}.$$

The regression sum of squares, SSR_{n_p, n_c}, is calculated as:

$$\sum_{i=1}^{P}(\hat{n}_{c_i} - \bar{n}_c)^2 = \sum_{i=1}^{P}(\hat{\beta}_0 + \hat{\beta}_1 \times n_{p_i} - \bar{n}_c)^2$$

where $\bar{n}_c$ is the mean value of node n_c over the P samples. The ordinary least squares method is used to estimate the β parameters [12].

The total sum of squares, $SSTO_{n_p,n_c}$, is expressed as: $\sum_{i=1}^{P}(\bar{n}_c - n_{c_i})^2$.

For the third-order potential, $\phi(n_g, n_c|n_p)$, a two-step regression procedure is employed:

1. Let $\mathbf{e}$ be the residuals from the linear regression model: $n_c = \hat{\beta}_0 + \hat{\beta}_1 \times n_p + \epsilon$.

2. Set $\phi(n_g, n_c|n_p) = \sqrt{SSR_{n_g,e}/SSTO_{n_g,e}} = \sqrt{R^2_{n_g,e}}$, where $SSR_{n_g,e}$ and $SSTO_{n_g,e}$ are the regression sum of squares and total sum of squares, respectively, from the linear regression model: $e = \hat{\beta}_0 + \hat{\beta}_1 \times n_g + \epsilon$.

On a practical level, the utility of the SCT method can be understood by conceptualizing the grandparent, parent, and child nodes. If n_p is an intermediate between n_g and n_c, then the residuals from the first step will not result in a high value for $SSR_{n_g,e}$ in the second linear regression function.

These regression functions form the concrete basis of the SCT method, and the likelihood of obtaining a particular value for the potentials is modeled as:

$$l(\phi(n_p, n_c)), \tag{6.3}$$

$$l(\phi(n_g, n_c|n_p)). \tag{6.4}$$

Given the regression functions to quantify interaction strengths between nodes, likelihood functions are utilized to model the likelihood of obtaining values from the regression functions. For example, the function $l(\phi(n_p, n_c))$ assesses the likelihood of obtaining a value for $\phi(n_p, n_c)$. This technique accounts for the fact that nodes vary in their background distributions of interaction strengths with other nodes. For example, if a particular node is naturally connected to a relatively large number of variables in the data set, then it will have a relatively high expected value for $\phi(n_p, n_{ran})$, given a randomly selected node n_{ran}. In order to estimate the "background" mean and standard deviation for each node, denoted as $\mu_{\phi(n_p,.)}$ and $\sigma_{\phi(n_p,.)}$, a sampling procedure is employed whereby a randomly generated series of interaction partners is created; then $\phi(n_p, n_{ran})$ is quantified for each of the n_{ran} random interaction partners.

Assuming that the values obtained from the regression functions are normally distributed, bivariate Gaussian functions are employed to approximate the distribution of interaction strengths. For example, the function $l(\phi(n_p, n_c))$ assesses the likelihood of obtaining a value for $\phi(n_p, n_c)$, and depends on the respective mean and standard deviation of nodes n_p and n_c. This likelihood is calculated as:

$$l(\phi(n_p, n_c)) = \frac{1}{Z} \exp\left(-\frac{v_{n_p}^2 - 2\rho v_{n_c} v_{n_p} + v_{n_c}^2}{2(1 - \rho^2)}\right)$$

$$Z = 2\pi \sigma_{\phi(n_p,.)}\sigma_{\phi(n_c,.)}\sqrt{1 - \rho},$$

$$v_{n_p} = \frac{\phi(n_p, n_c) - \mu_{\phi(n_{p,\cdot})}}{\sigma_{\phi(n_{p,\cdot})}},$$

$$v_{n_c} = \frac{\phi(n_p, n_c) - \mu_{\phi(n_{c,\cdot})}}{\sigma_{\phi(n_{c,\cdot})}}.$$

The covariance, ρ, is assumed to be zero in Chipman et al. [3]. However, note that this choice is not justified but rather is implemented for convenience. Note that for two continuous variables, the $\sqrt{R^2}$ value is symmetric and equal to the correlation coefficient.

For the conditional potential, $\phi(n_g, n_c | n_p)$, a mean of 0.0 is assumed. The value of $l(\phi(n_g, n_c | n_p))$ is obtained from the following univariate normal probability density function:

$$l(\phi(n_g, n_c | n_p)) = \frac{1}{\sqrt{2\pi \sigma_{\phi(n_*, n_* | n_*)}^2}} \exp\left(-\frac{\phi(n_g, n_c | n_p)^2}{2\sigma_{\phi(n_*, n_* | n_*)}^2}\right)$$

where the standard deviation for all triplets, denoted as $\sigma_{\phi(n_*, n_* | n_*)}$, is obtained from a sampling procedure whereby data for linear triplets are randomly generated and the conditional potential is computed [3].

Likelihoods $l(\phi(n_p, n_c))$ and $l(\phi(n_g, n_c | n_p))$ are used in tandem to model the likelihood that a transcript should be included as a leaf on a growing tree, and the log-likelihood score is expressed as follows:

$$LLS(n_c, n_p, n_g) = c_1 \times \log(l(\phi(n_g, n_c | n_p))) - \log(l(\phi(n_p, n_c))), \tag{6.5}$$

where c_1 is a constant used to modulate the extent to which the conditional potential is weighted against the pairwise potential. This constant is ultimately a heuristic choice; however, we emphasize that a recent study using simulated eQTL data sets demonstrated that the SCT method was effective for a broad range of non-zero values for parameter c_1 [3]. Finally, we point out that if c_1 is set to zero and $LLS(n_c, n_p, n_g) = -\log(l(\phi(n_p, n_c)))$, the SCT method can be viewed as a stochastic version of the Chow-Liu algorithm [4], which involves greedily adding nodes with the highest pairwise mutual information. In [3], the authors demonstrated that setting c_1 to 0 yielded relatively poor performance, reflecting the fact that spuriously correlated nodes may be included when not required to be aligned with the parent and grandparent nodes.

To describe the SCT method conceptually, the starting points of the algorithm are at the genomic loci, each of which serves as a root for their respective trees. As an example, a hypothetical tree is depicted in Fig. 6.3. The current state of the tree includes one locus, L_1 (shaded square), and four transcripts, T_a, T_b, T_c, and T_d (shaded circles). Each of the five nodes (1 locus, 4 transcripts) has a corresponding best candidate transcript (clear circles) that is stochastically chosen for inclusion into the growing tree. Each candidate has its own log-likelihood score (equation (6.5)), which is determined by the child, parent, and grandparent that factor into the second- and third-order potentials. Given the set of best candidates and their respective log-likelihood scores, the SCT method randomly chooses one candidate with a probability proportional to its likelihood score. Therefore, the candidates with the best and worst likelihood scores will have the greatest and least chances of being selected, respectively. In [3], the authors vary the extent to which the candidate with the best log-likelihood score is selected versus other suboptimal candidates. From their findings, it is clear that the best log-likelihood score should be highly prioritized; however, the option of stochastically choosing other, lower-scoring candidates is crucial for performance in that a broader range of tree structures (and therefore causal relationships) can be learned. This process is repeated until the maximum number of nodes allowed per tree is reached, represented by the parameter W. In [3], values for W in the range of 50–100 are used with consistently strong results.

The SCT method produces a set of trees which can be represented as adjacency matrices. The SCT output is converted into an $N \times N$ prior matrix by dividing the frequency at which node x is a parent of node y by the number of times the SCT method is run. Consider a hypothetical example where there are 50 loci (parameter M) and the SCT method is run 1000 times from each locus. In this example, if node x is a parent of node y on 100 occasions, then the prior matrix will have a value of $\frac{100}{50\,000}$ at entry x,y, since there are 50×1000 trees constructed.

We conclude this section by providing pseudocode for the three core functions of the SCT method. Formally, the SCT method accepts as input an eQTL data set D, which is composed of two components, gene expression and genotypic data. The expression data set E is a set of N transcripts $T_i, \ldots T_N$. Similarly, the genotypic data set G consists of a set of M loci $L_i, \ldots L_M$. If there are P samples, the two components of the data set D can be expressed as: $E = T[1] \ldots T[P]$ and $G = L[1] \ldots L[P]$. These data represent the input to the main algorithm (Algorithm 1). The SCT main procedure iterates procedure StochasticLeaf (Algorithm 2). In its turn, this latter algorithm iterates procedure GetBestCandidate (Algorithm 3).

Algorithm 1 SCT Main Procedure

INPUT: The data set, $D = \{d[1], \ldots, d[P]\}$
        W, the maximum number of leaves on a tree
        I, the number of iterations for the procedure (conducted over all loci)
        L, the set of M loci
OUTPUT: R, the $N \times N$ prior matrix representing the causal relationships


```
 1: for i = 1 to I do
 2:   for ℓ ∈ L do
 3:     S ← ∅; //S is the set nodes included in the tree
 4:     include node ℓ into the set S;
 5:     for w = 1 to W do
 6:       leaf ← StochasticLeaf(S);
 7:       S ← S ∪ {leaf}; //add leaf to the set, S
 8:       w ← w + 1;
 9:     end for
10:   end for
11:   i ← i + 1
12: end for
13: //Convert the trees into the prior matrix, R.
```
14: $\sum_{i,j}^{N} R_{i,j} = \frac{f(i,j)}{|L| \times I}$, where $f(i,j)$ represents the number of directed edges from node i to node j, summed over all trees.

Algorithm 2 StochasticLeaf (Node S)

INPUT: S, the set of nodes currently included in the tree

OUTPUT: the next leaf to be included in the tree

1: // B is the set of best candidates
2: $B \leftarrow \emptyset$;
3: **for** $s \in S$
4:     $b \leftarrow$ GetBestCandidates(s.id, s.parentId);
5:     $B \leftarrow B \cup \{b\}$;
6: **end for**
7: return $b \in B$, where b is stochastically chosen based on relative scores.

Algorithm 3 GetBestCandidate (p, g)

INPUT: p, the leaf node
        g, the parent of p

OUTPUT: the best candidate (child) node corresponding to p and g

1: // C is the set of candidates for node p
2: **for** $c \in C$
3:     $score_{n_g,n_p,n_c} = c_1 \log(\ell(\phi(n_g, n_c | n_p))) - \log(\ell(\phi(n_p, n_c)))$
4: **end for**
5: return $c_{best} \in C$, where c_{best} corresponds to $\max(score_{n_g,n_p,\star})$

6.3 Conclusion

In this chapter, we provided an introduction to studies focused on GOGE, including the fundamentals of an eQTL data set, which are instrumental in that they provide data for both genotype and gene expression. We then presented the LCMS method, which is the most widely deployed technique for learning causal models of gene regulation from eQTL data sets. After outlining the details of the LCMS method, we described how the matrix of causal relationships between transcripts produced by the LCMS method can be integrated into Bayesian network structure learning. Finally, we presented the recently published SCT method, which can be viewed as an alternative to the LCMS method.

To date, the LCMS method has be applied with remarkable success to several genomics data sets representing diverse systems, including yeast [21], mouse [9, 13, 16, 19], and humans [7]. In a pioneering study, Schadt et al. used the LCMS method to model the relationships between genotypic, transcriptomic, and obesity-related traits in an inbred mouse population [16].

Subsequently, Zhu et al. modeled causal gene networks with an eQTL data set representing a segregating population of *S. cerevisiae* derived from two parental strains [21]. This study was able to experimentally validate several causal regulators and the gene transcripts that they regulate. In a study related to bone mineral density traits, the LCMS method was applied to an inbred population of mice for which genotypic, transcriptomic and bone mineral density measurements are available [9]. Interestingly, by applying their LCMS method to mouse data consisting of genotype, expression, and clinical traits, the authors demonstrated that their mouse model is able to prioritize candidate genes identified in a genome-wide association study (GWAS) study related to bone mineral density in humans. In other words, the LCMS method is able to assign higher confidence to genotype-bone mineral density QTL that are supported by an intermediate transcript, which is quantitatively assessed by the likelihood of a causal model (M1, Fig. 6.2). Another especially important study utilizing the LCMS method involved a large cohort of human samples for which genotypic and transcriptomic measurements were acquired from adipose and blood cells [7]. This study was particularly successful at establishing the causal relationship between gene expression in adipose tissue and obesity-related traits [7]. Finally, we mention a follow-up study by Yang et al. [19] that employed extensive molecular genetics experimentation in the aforementioned inbred mouse population with the goal of identifying causal regulators of obesity-related traits. Using transgenic mice, the authors validated eight out of nine genes predicted to be causal to obesity-related traits by the LCMS method [19]. In summary, the LCMS method has been successfully utilized across a broad range of systems and tissue types, providing general validation for using integrative modeling techniques to guide experimentation.

Not surprisingly, this success has motivated several studies that aim to extend and improve upon this pioneering technique. In order to possibly help researchers who are new to this domain, whether they are experimentalists interested in modeling eQTL data sets, or computational scientists interested in making further improvements, we attempt to offer some objective opinions regarding the methodologies in this domain. The recently published SCT method produced larger performance gains over the LCMS method for simulated eQTL data sets [3]. However, this in and of itself does not indicate the SCT method would outperform the LCMS method on real data sets. As discussed earlier, the LCMS method has achieved remarkable success in application to real data sets across a broad range of biological systems. The probabilistic models driving the LCMS method offer very strong advantages, including efficient computation, a rigorous formal mathematical basis, and simplicity in terms of a limited number of parameters. This simplicity is beneficial for the purposes of comprehending and implementing the model. Additionally, although it is impossible to ascertain without empirical evidence, the LCMS method is more likely to be robust to complexities in biological systems underlying real data sets. Indeed, the idealized gene network model used in Chipman et al. [3], which was originally proposed by the LCMS authors in Zhu et al. [20], may differ from real biological networks in various ways. For example, the artificial gene network model does not contain cycles, though real gene networks clearly do. Also, real biological networks are almost certainly more modular than the artificial gene network model utilized in Chipman et al. [3]. The SCT method, which employs more heuristics and carries more parameters that require tuning, seems more likely to be sensitive to cases where real networks differ from the idealized model. Therefore, the large performance gains of the SCT method over the LCMS method reported in Chipman et al. [3] should be viewed with these caveats in mind.

However, this is not to say that the SCT method does not have merits, as it clearly addresses the greatest weakness of the LCMS method: lack of depth. The LCMS method is limited in the sense that causal relationships between transcripts can only be evaluated when two transcripts are both significantly associated with a genomic locus. At present, most eQTL data sets are composed of a limited number of samples on the order of hundreds and are characterized by limited statistical power to detect transcript-loci associations. This limitation may be alleviated as the sample sizes of genomics data sets increase, in which case greater statistical power in univariate mapping will lead to increased coverage for the LCMS method. Still, even with improvements to statistical power on the horizon, we believe that there is ample room for a probabilistic methodology that combines all the advantages of the two methods such that the underlying statistical models are formal, robust, and relatively parameter free like the LCMS method, but, at the same time, with the depth and resolution that the SCT method offers.

REFERENCES

[1] R.B. Brem, G. Yvert, R. Clinton, and L. Kruglyak. Genetic dissection of transcriptional regulation in budding yeast. *Science*, 296(5594):752–755, 2002.

[2] E. Chaibub Neto, C.T. Ferrara, A.D. Attie, and B.S. Yandell. Inferring causal phenotype networks from segregating populations. *Genetics*, 179:1089–1100, 2008.

[3] K.C. Chipman and A.K. Singh. Using stochastic causal trees to augment Bayesian networks for modeling eQTL data sets. *BMC Bioinformatics*, 12:7, 2011.

[4] C. Chow and C. Liu. Approximating discrete probability distributions with dependence trees. *IEEE Transactions on Information Theory*, 14(3):462–467, 1968.

[5] G.A. Churchill and R.W. Doerge. Empirical threshold values for quantitative trait mapping. *Genetics*, 138:963–971, 1994.

[6] B. Efron and R.J. Tibshirani. *An Introduction to the Bootstrap*. Chapman & Hall, 1993.

[7] V. Emilsson, G. Thorleifsson, B. Zhang, A.S. Leonardson, F. Zink, J. Zhu, S. Carlson, A. Helgason, G.B. Walters, S. Gunnarsdottir, M. Mouy, V. Steinthorsdottir, G.H. Eiriksdottir, G. Bjornsdottir, I. Reynisdottir, D. Gudbjartsson, A. Helgadottir, A. Jonasdottir, A. Jonasdottir, U. Styrkarsdottir, S. Gretarsdottir, K.P. Magnusson, H. Stefansson, R. Fossdal, K. Kristjansson, H.G. Gislason, T. Stefansson, B.G. Leifsson, U. Thorsteinsdottir, J.R. Lamb, J.R. Gulcher, M.L. Reitman, A. Kong, E.E. Schadt, and K. Stefansson. Genetics of gene expression and its effect on disease. *Nature*, 452(7186):423–428, March 2008.

[8] N. Friedman, M. Linial, and I. Nachman. Using Bayesian networks to analyze expression data. *Journal of Computational Biology*, 7:601–620, 2000.

[9] Y.-H. Hsu, M.C. Zillikens, S.G. Wilson, C.R. Farber, S. Demissie, N. Soranzo, E. N. Bianchi, E. Grundberg, L. Liang, J.B. Richards, K. Estrada, Y. Zhou, A. van Nas, M.F. Moffatt, G. Zhai, A. Hofman, J.B. van Meurs, H.A.P. Pols, R.I. Price, O. Nilsson, T. Pastinen, L.A. Cupples, A.J. Lusis, E.E. Schadt, S. Ferrari, A. G. Uitterlinden, F. Rivadeneira, T.D. Spector, D. Karasik, and D.P. Kiel. An integration of genome-wide association study and gene expression profiling to prioritize the discovery of novel susceptibility loci for osteoporosis-related traits. *PLOS Genetics*, 6(6):e1000977+, 2010.

[10] D. Husmeier and A. V. Werhli. Bayesian integration of biological prior knowledge into the reconstruction of gene regulatory networks with Bayesian networks. *Computational Systems Bioinformatics Conference*, 6:85–95, 2007.

[11] S. Imoto, T. Higuchi, T. Goto, K. Tashiro, S. Kuhara, and S. Miyano. Combining microarrays and biological knowledge for estimating gene networks via Bayesian networks. *IEEE Computer Society Bioinformatics Conference*, 2:104–113, 2003.

[12] M.H. Kutner, C.J. Nachtsheim, and J. Neter. *Applied Linear Regression Models*. McGraw-Hill/Irwin, 4th international edition, 2004.

[13] M. Mehrabian, H. Allayee, J. Stockton, P.Y. Lum, T.A. Drake, L.W. Castellani, M. Suh, C. Armour, S. Edwards, J. Lamb, A.J. Lusis, and E.E. Schadt. Integrating genotypic and expression data in a segregating mouse population to identify 5-lipoxygenase as a susceptibility gene for obesity and bone traits. *Nature Genetics*, 37(11):1224–1233, 2005.

[14] S. Mukherjee and T.P. Speed. Network inference using informative priors. *Proceedings of the National Academy of Sciences of the United States of America*, 105:14313–14318, 2008.

[15] J. Pearl. *Probabilistic Reasoning in Intelligent Systems: Networks of Plausible Inference*. Morgan Kaufmann Publishers, 1988.

[16] E.E. Schadt, J. Lamb, X. Yang, J. Zhu, S. Edwards, D. Guhathakurta, S.K. Sieberts, S. Monks, M. Reitman, C. Zhang, P.Y. Lum, A. Leonardson, R. Thieringer, J.M. Metzger, L. Yang, J. Castle, H. Zhu, S.F. Kash, T.A. Drake, A. Sachs, and A.J. Lusis. An integrative genomics approach to infer causal associations between gene expression and disease. *Nature Genetics*, 37(7):710–717, 2005.

[17] M. Schena, D. Shalon, R.W. Davis, and P.O. Brown. Quantitative monitoring of gene expression patterns with a complementary DNA microarray. *Science*, 270:467–470, 1995.

[18] Z. Wang, M. Gerstein, and M. Snyder. RNA-Seq: a revolutionary tool for transcriptomics. *Nature Review Genetics*, 10:57–63, 2009.

[19] X. Yang, J.L. Deignan, H. Qi, J. Zhu, S. Qian, J. Zhong, G. Torosyan, S. Majid, B. Falkard, R.R. Kleinhanz, J. Karlsson, L.W. Castellani, S. Mumick, K. Wang, T. Xie, M. Coon, C. Zhang, D. Estrada-Smith, C.R. Farber, S.S. Wang, A. van Nas, A. Ghazalpour, B. Zhang, D.J. Macneil, J.R. Lamb, K.M. Dipple, M.L. Reitman, M. Mehrabian, P.Y. Lum, E.E. Schadt, A.J. Lusis, and T.A. Drake. Validation of candidate causal genes for obesity that affect shared metabolic pathways and networks. *Nature Genetics*, 41:415–423, 2009.

[20] J. Zhu, M. C. Wiener, C. Zhang, A. Fridman, E. Minch, P.Y. Lum, J.R. Sachs, and E.E. Schadt. Increasing the power to detect causal associations by combining genotypic and expression data in segregating populations. *PLOS Computational Biology*, 3:e69, 2007.

[21] J. Zhu, B. Zhang, E.N. Smith, B. Drees, R.B. Brem, L. Kruglyak, R.E. Bumgarner, and E.E. Schadt. Integrating large-scale functional genomic data to dissect the complexity of yeast regulatory networks. *Nature Genetics*, 40:854–861, 2008.

Bayesian Causal Phenotype Network Incorporating Genetic Variation and Biological Knowledge

JEE YOUNG MOON, ELIAS CHAIBUB NETO,
XINWEI DENG, AND BRIAN S. YANDELL

I n a segregating population, quantitative trait loci (QTL) mapping can identify QTLs with a causal effect on a phenotype. Several approaches in the literature take advantage of QTLs identified by QTL mapping, to determine causal relations among phenotypes. A common feature of these methods is that QTL mapping and phenotype network reconstruction are conducted separately. As both tasks have to benefit from each other, this chapter presents an approach which jointly infers a causal phenotype network and causal QTLs. The joint network of causal phenotype relationships and causal QTLs is modeled as a Bayesian network. Besides, a prior distribution on phenotype network structures is adjusted by biological knowledge. This integrative approach can incorporate several sources of biological knowledge such as protein–protein interactions, gene ontology annotations, transcription factor and DNA binding information. This framework allows a flexible tuning on the confidence of the various sources of knowledge through priors on biological knowledge weights. A Metropolis–Hastings scheme is described that iterates between accepting a network structure and accepting k weights corresponding to the k types of biological knowledge. The way to encode biological knowledge is described in the case of protein–protein interactions, similarity-based measures derived from a gene ontology, and transcription factor bindings to DNA. The integrative method is then applied to reconstruct a network involved in the cell cycle in yeast, relying on transcription factor binding knowledge.

Probabilistic Graphical Models for Genetics, Genomics, and Postgenomics. First Edition. Christine Sinoquet & Raphaël Mourad (Eds). © Oxford University Press 2014. Published in 2014 by Oxford University Press.

7.1 Introduction

A key interest in molecular biology is to understand how DNA, RNA, proteins, and metabolic products regulate each other. In this regard, people have considered constructing regulatory networks from microarray expression data with time-series measurements or transcriptional perturbations [14, 15]. A regulatory network can also be constructed in a segregating population where genotypes perturb the gene expression, protein, and metabolite levels. The genetic variation information can decipher genetic effects on traits and help discover causal regulatory relationships between phenotypes. In addition, knowledge of regulatory relationships is available in various biological databases, which can improve the reconstruction of causal networks. This chapter focuses on combining genetic variations in a segregating population and biological knowledge to improve the inference of causal networks.

Given the quantitative nature of a gene expression phenotype, one can perform quantitative trait loci (QTL) mapping to detect the genomic locations affecting the phenotype [29]. The genotypes at a location are often coded as *AA*, *Aa*, or *aa*, where allele *A* and *a* are distinct variant forms of a genetic locus. A quantitative phenotype/trait is any observable physical or biochemical quantitative feature of an organism such as weight, blood pressure, gene expression, or protein levels. The basic idea of QTL mapping is to detect genomic regions, or QTLs, where variation in genotype is associated with quantitative variation in phenotype. For example, tall parents tend to have tall children, whereas short parents tend to have short children. Then, it appears that there are genetic factors to be associated with the height, and the genetic factors can be identified by QTL mapping. In an experimental population, where genotypes are randomly assigned, the genetic variation at QTLs can be interpreted as causing later changes in the phenotype of interest.

In a segregating population, QTL mapping can identify QTLs with a causal effect on a phenotype. The causal effect can be direct from QTL to phenotype, or indirect via other intermediate phenotypes. We only label the direct QTLs as "causal QTLs," recognizing that they have a more proximal effect on a phenotype than indirect QTLs. We also acknowledge that there may be many other molecular factors in a pathway between the QTL and the phenotype that were not measured in a particular study. Indirect and direct QTLs can be used to help determine the direction of the edges in a causal phenotype network (i.e., a directed graph composed of phenotype nodes, whose edges represent causal relations). Several approaches in the literature take advantage of QTLs identified by QTL mapping to determine causal relations among phenotypes including: structural equation modeling [2, 34, 35] score-based methods for Bayesian networks [56, 60, 62]; causal algorithms for Bayesian networks based on independence tests [8, 53]; and causality tests on pairs of phenotypes [10, 11, 32, 38, 48]. A common feature of the above approaches is that QTL mapping and phenotype network reconstruction are conducted separately. QTL mapping without consideration of a phenotype network may find indirect QTLs. As pointed out by [9], incorrect or indirect QTLs may compromise the inference of causal relationships among phenotypes. To address this issue, several researchers [9, 20] proposed to jointly infer causal phenotype networks and causal QTLs.

Various sources of biological knowledge have been incorporated with gene expression in the reconstruction of phenotype networks, because it is difficult to determine the causal direction of gene regulation using expression data only. Transcription factor binding information was leveraged by [52], whereas [40] used protein–protein interaction knowledge to construct phenotype networks. Methods integrating multiple sorts of biological knowledge have been proposed by [25], [55], and [12].

In this chapter, we propose a Bayesian approach to jointly infer a causal phenotype network and causal QTLs with a prior distribution on phenotype network structures adjusted by biological knowledge. The joint network of causal phenotype relationships and causal QTLs is modeled as a Bayesian network[1] adopted from [9], QTLnet. Causal QTLs can be inferred by QTL mapping conditional on the phenotype network. Since the phenotype network is unknown, QTLnet traverses the space of phenotype networks and updates causal QTLs using Markov chain Monte Carlo (MCMC). We extend the framework of QTLnet by incorporating biological knowledge into the prior distribution on phenotype network structures. The incorporation of biological knowledge is expected to increase the accuracy of the model estimation, enhancing the predictive power of the network [62]. The prior probability on phenotype network structures is based on the Gibbs distribution to integrate different sources of biological information, allowing for flexible tuning of the analyst's confidence on this knowledge [55]. The consideration of reliability of biological knowledge is necessary since biological knowledge can be incomplete and inaccurate. While [62] proposed a method to incorporate genetic variation and biological knowledge to phenotype networks, their method does not consider the reliability of biological knowledge. Our proposed approach (QTLnet-prior) can integrate phenotype data, genetic variation, and several sources of biological knowledge (protein–protein interaction, gene ontology annotation, and transcription factor and DNA binding information) with the consideration of the reliability of each source of biological knowledge in the network reconstruction algorithm.

The details of our integrated framework for the joint inference of causal phenotype network and causal QTLs are organized as follows. Section 7.2 describes the QTLnet method for the joint inference of causal network and causal QTLs. Section 7.3 presents the proposed QTLnet-prior, which incorporates biological knowledge into the prior probability distribution of phenotype network structures. A simulation study is conducted in Section 7.4 to compare the proposed method with several existing approaches. In Section 7.5, the proposed method is used to reconstruct a network of 26 genes involved in the yeast cell cycle. Finally, in Section 7.6, we discuss the strengths and caveats of our approach and point out future research directions.

7.2 Joint Inference of Causal Phenotype Network and Causal QTLs

In Subsection 7.2.1, we first present a standard Bayesian network for modeling phenotype data. Next, in Subsection 7.2.2, we present an extended model, QTLnet, based on the homogeneous conditional Gaussian regression (HCGR) model, to incorporate QTL nodes into the phenotype network. Directed edges in the standard Bayesian network can be interpreted as causal relationships. By extending the phenotype network with causal QTL nodes, we can further claim causal interpretations. In Subsection 7.2.3, we present a rationale for the joint inference of the

---

[1] Note that Bayesian networks can be inferred in a Bayesian framework or a frequentist framework. Here, we take a Bayesian approach to infer a Bayesian network. The reason that the term "Bayesian" is used in a Bayesian network is described in [43]. The following is an excerpt from page 14 of [43]: "Bayesian networks, a term coined in Pearl (1985) to emphasize three aspects: (1) the subjective nature of the input information; (2) the reliance on Bayes's conditioning as the basis for updating information; and (3) the distinction between causal and evidential modes of reasoning, a distinction that underscores Thomas Bayes's paper of 1763.

causal phenotype network and causal QTLs, and in Subsection 7.2.4, we describe the QTL mapping conditional on the phenotype network. Finally, we give an overview of our joint approach for phenotype network and causal QTL inference in Subsection 7.2.5.

7.2.1 Standard Bayesian Network Model

A standard Bayesian network is a probabilistic graphical model whose conditional independence is represented by a directed acyclic graph (DAG). A node t in a DAG G corresponds to a random variable Y_t in the Bayesian network. A directed edge from node u to node v can supposedly represent that Y_v is causally dependent on Y_u, though an edge truly represents the conditional dependence. The local directed Markov property of Bayesian networks states that each variable is independent of its non-descendant variables conditional on its parent variables:

$$Y_t \perp\!\!\!\perp Y_{V \setminus de(t)} | Y_{pa(t)} \quad \text{for all } t \in V,$$

where V is the set of all nodes in a DAG, $de(t)$ is the set of descendants of node t, $pa(t)$ is the set of parents of node t and $Y_{pa(t)}$ is a set of variables indexed by $pa(t)$, that is, $\{Y_i : i \in pa(t)\}$. Assume the node index is ordered such that the index of descendants is always bigger than the index of their parents. Since $\{t - 1, \ldots, 1\}$ is a set of non-descendants of node t and $pa(t)$ is included in the non-descendant set $\{t - 1, \ldots, 1\}$, Y_t is independent of $Y_{\{t-1,\ldots,1\}}$ conditional on $Y_{pa(t)}$; that is, $P(Y_t | Y_{pa(t)})$ is equivalent to $\mathbb{P}(Y_t | Y_{t-1}, \ldots, Y_1)$. The joint distribution can be written as:

$$\mathbb{P}(Y_1, \ldots, Y_T) = \prod_{t=1}^{T} \mathbb{P}(Y_t | Y_{t-1}, \ldots, Y_1)$$

$$= \prod_{t=1}^{T} \mathbb{P}(Y_t | Y_{pa(t)}), \tag{7.1}$$

where the first equality is satisfied by the chain rule in probability theory.[2]

---

[2] In probability theory, the chain rule permits that the joint probability of two variables X and Y can be written as

$$\mathbb{P}(X, Y) = \mathbb{P}(Y|X)\,\mathbb{P}(X) = \mathbb{P}(X|Y)\,\mathbb{P}(Y).$$

This can be extended to the joint probability of multiple variables:

$$\begin{aligned}
\mathbb{P}(Y_T, \ldots, Y_1) &= \mathbb{P}(Y_T | Y_{T-1}, \ldots, Y_1)\mathbb{P}(Y_{T-1}, \ldots, Y_1) \\
&= \mathbb{P}(Y_T | Y_{T-1}, \ldots, Y_1)\mathbb{P}(Y_{T-1} | Y_{T-2}, \ldots, Y_1)\mathbb{P}(Y_{T-2} | Y_{T-3}, \ldots, Y_1) \\
&= \ldots \\
&= \prod_{t=1}^{T} \mathbb{P}(Y_t | Y_{t-1}, \ldots, Y_1).
\end{aligned}$$

7.2.2 HCGR Model

The parametric family of a Bayesian network that jointly models phenotypes and QTL genotypes corresponds to an HCGR model. Conditional on the QTL genotypes and covariates, the phenotypes are distributed according to a multivariate normal distribution, where QTLs and covariates enter the model via the mean, and the correlation structure among the phenotypes is explicitly modeled according to the DAG representing the phenotype network structure [9]. Fig. 7.1 depicts one example of a joint Bayesian network of phenotypes and QTL genotypes.

The HCGR model is derived from a series of linear regression equations. For $i = 1, \ldots, n$ and $t = 1, \ldots, T$, let Y_{ti} be the value of phenotype for individual i and trait t. Then we assume for each phenotype that Y_{ti} can be modeled as follows:

$$Y_{ti} = \mu_{ti}^{\star} + \sum_{v \in pa(t)} \beta_{tv} Y_{vi} + \epsilon_{ti}, \quad \epsilon_{ti} \sim \mathcal{N}(0, \sigma_t^2). \tag{7.2}$$

The model can be decomposed into three parts: a genetic part ($\mu_{ti}^{\star}$), a phenotypic part ($\sum_{v \in pa(t)} \beta_{tv} Y_{vi}$), and an error term ($\epsilon_{ti}$). In the phenotypic part, β_{tv} is the effect of parent phenotype v on phenotype t. The error term, ϵ_{ti}, follows a normal distribution. The genetic part, $\mu_{ti}^{\star}$, corresponds to a model of QTL genotypes and possibly covariates:

$$\mu_{ti}^{\star} = \mu_t + \sum_{k=1}^{C} \vartheta_{tk} Z_{ki} + \sum_{k=1}^{K} \gamma_{tk} \theta_{tk} X_{ki},$$

where μ_t is the overall mean for trait t, Z_{ki} represents a covariate, ϑ_{tk} represents the effect of the covariate on the phenotype, and $\sum_{k=1}^{K} \gamma_{tk} \theta_{tk} X_{ki}$ is the overall effect of QTLs. For the simplicity, we will not consider the covariates Z later on. The parameter γ_{tk} is unknown, and represents the inclusion ($\gamma_{tk} = 1$) or exclusion ($\gamma_{tk} = 0$) of the QTL located at the genomic position k, into the model. The genetic effects of QTL can be partitioned into different types of genetic effects, e.g. additive and dominance effects, and hence the genotype of the QTL is coded into the variables to estimate the different genetic effects. The vector X_{ki} represents a column vector of coded variables of the genotype at the genomic location k for individual i, and the vector θ_{tk} is a row vector of several types of genetic effects of QTL at the location k on phenotype t. The coding of a genotype may follow Cockerham's genetic model [30]. For example, in an intercross, the segregating genotypes at a locus are denoted by AA, Aa, or aa, and we can code the genotype into an additive variable by the number of A alleles in the genotype minus 1 and a dominance variable by $1/2$ if it is Aa and $-1/2$ otherwise. In this case, the additive effect is the effect of substituting one allele a with another allele A and the dominance effect is the deviation of Aa from the mean of AA and aa. Accordingly, in an intercross, X_{ki} is a column vector of additive and dominance coding variables,

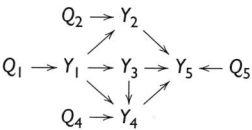

Fig 7.1 Example network with five phenotypes ($Y_1, \ldots, Y_5$) and five QTLs ($Q1, \ldots, Q5$).

and θ_{tk} is the row vector of additive and dominance effects on phenotype t. It was shown by [9] that these linear regression equations in equation (7.2) set a HCGR model for phenotypes and QTL genotypes.

7.2.3 Systems Genetics and Causal Inference

Systems genetics aims to understand the complex interrelations between genetic variations and phenotypes from large-scale genotype and phenotype data [39]. Here we explain how the systems genetics approach can infer causal networks. Causal relations from QTLs to phenotypes are justified by the unidirectional influence of the genotype on phenotype and the random allocation of genotypes to individuals. In contrast, causal relations among phenotypes are induced from conditional independence. The key idea of systems genetics is that by incorporating QTL nodes into phenotype networks, we create new sets of conditional independence relationships for distinguishing network structures that would, otherwise, belong to the same equivalence class (see Tables 7.1 and 7.2).

First, we give a more detailed description for the causal relations between QTLs and phenotypes. As stated in the central dogma of molecular biology, the hereditary DNA information is transferred to phenotypes. Thus a genotype influences phenotypes in general but not the other way around. A genotype is assumed to be randomized to other environmental factors by independent segregation of chromosomes in meiosis and random mating between gametes. These special characteristics enable us to infer causal effects of QTLs on phenotypes since, by analogy with a randomized experiment, we have that: (1) the treatment (genotype) to an experimental unit precedes the measured outcome (phenotype), and (2) random allocation of treatments to experimental units guarantees that other common causes get averaged out. Two loci on the same chromosome are highly correlated when their distance is small. But crossovers between two loci can still occur randomly in proportion to the distance. One can distinguish the true causal QTL and false nearby QTL with a large sample size. This random allocation is explicit in an experimental cross such as a backcross or an intercross.[3] While this idea can be extended to natural populations, special attention must be paid to admixture, kinship, and other forms of relatedness.

Second, the explanation of causal inference among phenotypes requires the concept of conditional independence in DAGs composed of phenotypes and QTL nodes. In the next three paragraphs, we present some definitions and results that allow us to infer phenotype-to-phenotype causal relationships.

Here are definitions. In graph theory, a *path* is defined as any unbroken, non-intersecting sequence of edges in a graph, which may go along or against the direction of arrows. We say that a path p is *d-separated* [42, 43] by a set of nodes Z if and only if: (1) p contains a chain $i \rightarrow m \rightarrow j$ or a fork $i \leftarrow m \rightarrow j$ such that the middle node is in Z, or (2) p contains a collider $i \rightarrow m \leftarrow j$ such that the middle node m is not in Z and such that no descendant of m is in Z. We say that

---

[3] An experimental cross is generated by crossing inbred lines. An inbred line is obtained by repeated generations of inbreedings so that any genotype of the inbred line is homozygous, *AA*. Therefore, breeding within the inbred line produces genetically identical offspring to its parents. Both backcross and intercross first produce the first generation of population by mating two different inbred lines, *AA* and *BB*. The first generation is identical to each other with heterozygous genotypes, *AB*. The backcross population is produced by mating the first generation to one of its parental inbred lines such as *AA*. Then, the backcross population has genotypes either *AA* or *AB* in a ratio of 1 : 1. The intercross population is produced by mating the first generation itself so that it has genotypes *AA*, *AB*, or *BB* in a ratio of 1 : 2 : 1.

Z d-separates X from Y if and only if Z blocks every path from a node in X to a node in Y. The *skeleton* of a DAG is the undirected graph obtained by replacing its arrows by undirected edges. A *v-structure* is composed by two converging arrows whose tails are not connected by an arrow.

The equivalence concept plays a key role in learning the structure of networks from the data. Here we present three important equivalence relations for graphs or statistical models of graphs. Two graphs are *Markov equivalent* if they have the same set of d-separation relations [50]. Two structures m_1 and m_2 for Y are *distribution equivalent* with respect to the distribution family F if they represent the same joint distributions for Y, that is, for every θ_1, there exists a θ_2 such that $\mathbb{P}(Y \mid \theta_1, m_1) = \mathbb{P}(Y \mid \theta_2, m_2)$ [22]. In other words, m_1 and m_2 are distribution equivalent if the parameters θ_1 and θ_2 are simple reparametrizations of each other. If m_1 and m_2 are distribution equivalent, then the invariance principle of maximum likelihood estimates guarantees $\mathbb{P}(Y \mid \hat{\theta}_1, m_1) = \mathbb{P}(Y \mid \hat{\theta}_2, m_2)$, and m_1 and m_2 cannot be distinguished using the data. In this case we say that m_1 and m_2 are *likelihood equivalent*. In a Bayesian setting, we define likelihood equivalence using the prior predictive distribution, $\int \mathbb{P}(Y \mid \theta, m_1) \mathbb{P}(\theta \mid m_1) \, d\theta = \int \mathbb{P}(Y \mid \theta, m_2) \mathbb{P}(\theta \mid m_2) \, d\theta$. If models m_1 and m_2 are distribution equivalent and we adopt a proper prior $\mathbb{P}(\theta \mid m)$, it is often reasonable to expect $\mathbb{P}(Y \mid m_1) = \mathbb{P}(Y \mid m_2)$, so that we cannot distinguish m_1 and m_2 for any data set Y [22].

Now we state four important results regarding causal inference in systems genetics: (1) Two DAGs are Markov equivalent if and only if they have the same skeletons and the same set of v-structures [54]; (2) Distribution equivalence implies Markov equivalence, but the converse is not necessarily true [50]; (3) For a Gaussian regression model, Markov equivalence implies distribution equivalence [21]; (4) For the homogeneous conditional Gaussian regression model, Markov equivalence implies distribution equivalence [9].

Therefore, for the HCGR parametric family, two DAGs are distribution and likelihood equivalent if and only if they are Markov equivalent. This implies that we can simply check if any two DAGs have the same skeleton and the same set of v-structures in order to determine if they are likelihood equivalent and hence cannot be distinguished using the data.

Getting back to the idea of causal inference among phenotypes, let G_Y be a phenotype network represented by a standard Bayesian network of phenotypes, Y. Phenotype data alone can distinguish some network structures by its likelihood but may fail to distinguish some other network structures. For example, consider the three network structures in Table 7.1. Models G_Y^1 and G_Y^3 have the same skeleton ($Y_1 - Y_2 - Y_3$) and the same set of v-structures (no v-structure) and, thus, are distribution/likelihood equivalent. Model G_Y^2, on the other hand, has the same skeleton but a different set of v-structures and, hence, is not distribution/likelihood equivalent to models G_Y^1 and G_Y^3. Therefore, phenotype data alone can identify G_Y^2 but cannot distinguish G_Y^1 and G_Y^3.

Table 7.1 Models G_Y^1 and G_Y^3 are distribution/likelihood equivalent.

DAG structures	Skeletons	v-structures
$G_Y^1 = Y_1 \rightarrow Y_2 \rightarrow Y_3$	$Y_1 - Y_2 - Y_3$	$\varnothing$
$G_Y^2 = Y_1 \rightarrow Y_2 \leftarrow Y_3$	$Y_1 - Y_2 - Y_3$	$Y_1 \rightarrow Y_2 \leftarrow Y_3$
$G_Y^3 = Y_1 \leftarrow Y_2 \rightarrow Y_3$	$Y_1 - Y_2 - Y_3$	$\varnothing$

Table 7.2 Extended models G^1 and G^3 are no longer distribution/likelihood equivalent.

Extended DAG Structures	Skeletons	v-Structures
$G^1 = Q_1 \to Y_1 \to Y_2 \to Y_3$	$Q - Y_1 - Y_2 - Y_3$	$\emptyset$
$G^3 = Q_1 \to Y_1 \leftarrow Y_2 \to Y_3$	$Q - Y_1 - Y_2 - Y_3$	$Q \to Y_1 \leftarrow Y_2$

Adding causal QTL nodes to a phenotype network allows the inference of causal relationships between phenotypes that could not be distinguishable using phenotype data alone. For example, if we add a causal QTL Q_1 to Y_1 in phenotype networks G_Y^1 and G_Y^3 in the above example, then the corresponding extended network structures G^1 and G^3 have different v-structures as shown in Table 7.2.

7.2.4 QTL Mapping Conditional on Phenotype Network Structure

Now we examine the inference of QTLs conditional on a phenotype network. QTL mapping can be done in a conditional or unconditional fashion. In the unconditional mapping analysis, we measure the association of a trait Y_t and QTL Q using the LOD score (logarithm of odds):

$$ LOD(y_t, q) = \log_{10}\left(\frac{f(y_t \mid q)}{f(y_t)}\right), $$

where $f(y_t \mid q)$ represents the predictive density of a linear model with Q as an independent variable and $f(y_t)$ the predictive density of the baseline model. Here a predictive density is given by a maximized likelihood in a frequentist setting, or by the prior predictive density in a Bayesian setting. A high LOD score means that Y_t and Q are associated. Note that unconditional analysis can detect QTLs that directly affect the phenotype under investigation, as well as QTLs with indirect effects [9]. For example, if we consider the causal network of phenotypes and QTLs in Fig. 7.1, then the unconditional QTL mapping of Y_2 detects a direct QTL Q_2 as well as an indirect QTL Q_1 that affects Y_2 via Y_1. Fig. 7.2 shows the expected results of the unconditional analysis for each phenotype.

The conditional mapping analysis, on the other hand, incorporates other traits as covariates, and measures the association of Y_t and Q conditional on these covariates (say y_z) using the conditional LOD score:

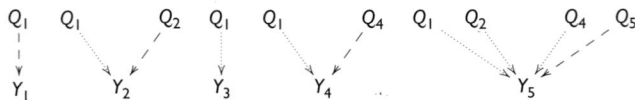

Fig 7.2 Output of the unconditional QTL mapping analysis for the phenotypes in Fig. 7.1. Dashed and pointed arrows represent direct and indirect QTL/phenotype causal relationships, respectively.

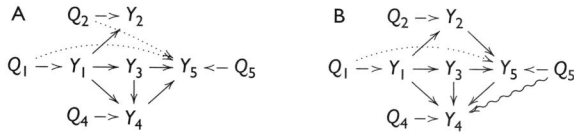

Fig 7.3 QTL mapping tailored to the network structure. Dashed, pointed, and wiggled arrows represent, respectively, direct, indirect, and incorrect QTL/phenotype causal relationships. (A) Mapping analysis of Y_5 conditional on Y_3 and Y_4 still detects Q_1 and Q_2 as QTLs for Y_5, since failing to condition on Y_2 leaves the paths $Q_1 \rightarrow Y_1 \rightarrow Y_2 \rightarrow Y_5$ and $Q_2 \rightarrow Y_2 \rightarrow Y_5$ in Fig. 7.1 open. In other words, (Y_3, Y_4) cannot d-separate (Q_1, Q_2) from Y_5 in the true causal graph. (B) Mapping analysis of Y_4 conditional on Y_1, Y_3, and Y_5 incorrectly detects Q_5 as a QTL for Y_4 because in the true network, the paths $Y_4 \rightarrow Y_5 \leftarrow Q_5$ and $Y_4 \leftarrow Y_3 \rightarrow Y_5 \leftarrow Q_5$ in Fig. 7.1 are open when we condition on Y_5.

$$LOD(y_t, q|y_z) = \log_{10}\left(\frac{f(y_t|q, y_z)}{f(y_t)}\right) - \log_{10}\left(\frac{f(y_t|y_z)}{f(y_t)}\right)$$
$$= LOD(y_t, q, y_z) - LOD(y_t, y_z).$$

Now, consider QTL mapping analysis tailored to a known phenotype network structure. In this situation, we can avoid detecting indirect QTLs by simply performing mapping analysis of the phenotypes conditional on their parents. For instance, in Fig. 7.1, if we perform QTL mapping of Y_5 conditional on Y_2, Y_3 and Y_4, we do not detect Q_1, Q_2, and Q_4 because of the following independence relations: $Y_5 \perp\!\!\!\perp Q_1 \mid Y_2, Y_3, Y_4$; $Y_5 \perp\!\!\!\perp Q_2 \mid Y_2, Y_3, Y_4$; and $Y_5 \perp\!\!\!\perp Q_4 \mid Y_2, Y_3, Y_4$. We only detect Q_5 due to the following relation: $Y_5 \not\perp\!\!\!\perp Q_5 \mid Y_2, Y_3, Y_4$.

In practice, however, the structure of the phenotype network is unknown, and performing QTL mapping conditional on a misspecified phenotype network structure can result in the inference of misspecified causal QTLs, as shown in Fig. 7.3. The mapping analysis of a phenotype conditional on downstream phenotypes in the true network, induces dependences between the phenotype and QTLs affecting downstream phenotypes. This leads to the erroneous inference that the phenotype includes downstream QTLs as its QTLs. For example, the mapping analysis of Y_4 conditioning on Y_1, Y_3 and a downstream phenotype Y_5 includes downstream Q_5 as its QTLs in Fig. 7.3B. However, a model with misspecified phenotype network and QTLs will generally have a lower marginal likelihood score than the model with the correct causal order for the phenotypes and correct QTLs. Since in practice QTLnet adopts a model selection procedure to traverse the space of network structures, it tends to prefer models closer to the true data-generating process. Simulation studies presented in [9] corroborate this point.

Note that, as pointed out in [9], the conditional LOD score can be adopted as a formal measure of independence between a phenotype and QTLs. Even though we restrict our attention to HCGR models, conditional LOD profiling is a general framework for the detection of conditional independences between continuous and discrete random variables. Contrary to partial correlations, the conditional LOD score does not require the assumption of multinormality of the data in order to formally test for independence, and it can handle QTL by covariate interactions.

7.2.5 Joint Inference of Phenotype Network and Causal QTLs

Subsection 7.2.4 describes the QTL mapping conditional on a phenotype network. In practice, the phenotype network is generally unknown and we cannot directly infer the correct causal QTLs. Therefore, we need to perform a joint inference of phenotype network and causal QTLs.

Recall that Y are phenotypes, X are genetic variations as defined in Subsection 7.2.2, and G is a Bayesian network structure of phenotypes and QTLs. Let G_Y represent a phenotype network and let $G_{Q \to Y}$ represent a graph from causal QTL nodes to phenotype nodes. Note that G_Y and $G_{Q \to Y}$ are subgraphs of the extended network structure G. Conforming to the HCGR model in equation (7.2), G_Y corresponds to the collection of causal relations from $pa(t)$ to trait t and $G_{Q \to Y}$ corresponds to the collection of causal relations from non-zero γ_{tk} to traits. Denote θ_G to be the parameter sets $(\beta_{tv}, \sigma_t^2, \mu_t, \theta_{tk})$. From equation (7.2), the likelihood of a Bayesian network of phenotypes and causal QTLs can be written as a product of normal densities:

$$\mathbb{P}(Y|G, X, \theta_G) = \mathbb{P}(Y|G_Y, G_{Q \to Y}, X, \theta_G)$$

$$= \prod_{t=1}^{T} \prod_{i=1}^{n} \mathcal{N}\left(\mu_{ti}^\star + \sum_{y_k \in pa(y_t)} \beta_{tk} y_{ki}, \sigma_t^2\right).$$

The marginal likelihood of phenotypes and causal QTLs $P(Y|G, X)$ is calculated by integrating parameters θ_G out in the Bayesian network:

$$\mathbb{P}(Y|G, X) = \int \mathbb{P}(Y|G, X, \theta_G)\, \mathbb{P}(\theta_G|G)\, d\theta_G.$$

The posterior probability of G conditional on the data is given by:

$$\mathbb{P}(G|Y, X) = \frac{\mathbb{P}(Y|G, X)\mathbb{P}(G)}{\sum_G \mathbb{P}(Y|G, X)\mathbb{P}(G)},$$

where $\mathbb{P}(G)$ represents the prior probability of the network structure G. In the next section, we devote our attention to the specification of $\mathbb{P}(G)$ using integrated biological knowledge.

Following [9], we adopt the QTLnet framework that jointly infers the phenotype network structure and causal QTLs. Most of the current literature in genetical network reconstruction has treated the problems of QTL inference and phenotype network reconstruction separately, generally performing QTL inference first, and then using QTLs to help determine the phenotype network structure [8, 62]. As indicated in Subsection 7.2.4, such a strategy can include QTLs with indirect effects into the network.

7.3 Causal Phenotype Network Incorporating Biological Knowledge

Besides the causal QTLs, biological knowledge is another useful and important information resource to enhance the construction of the phenotype network. Such knowledge can be integrated on top of the causal network to provide a more comprehensive picture of how genes are regulated. This integrated network could generate a new hypothesis on gene regulation, having an overall consistency with biological knowledge.

In this section, we propose a network inference method, QTLnet-prior, from phenotype data with genetic variations, integrating biological knowledge. The QTLnet-prior extends the framework of QTLnet referred to at the end of Subsection 7.2.5. It specifies the prior probability on phenotype network structures to integrate multiple sources of biological knowledge with flexible tuning parameters on confidence of knowledge [55]. The weighted integration of biological knowledge could produce a more predictive Bayesian network. The details of our extended framework, QTLnet-prior, are presented in Subsection 7.3.1. In Subsection 7.3.2, we sketch a Metropolis–Hastings MCMC scheme for QTLnet-prior implementation that integrates the sampling of network structures [19, 37], QTL mapping, and sampling of biological knowledge weights. In Subsection 7.3.3, we present how to encode biological knowledge into the prior distribution over phenotype network structures.

7.3.1 Model

EXTENDED MODEL

Denote by G a Bayesian network structure of phenotypes and QTLs. The graph G consists of a phenotype network (G_Y) and causal QTLs to phenotypes ($G_{Q \to Y}$). Let Y be phenotype data, X be genetic variations, and W represent weights set on various sources of biological knowledge B. The biological knowledge B is considered to be relations between phenotypes such as transcription factor binding, protein–protein interaction, and gene ontology annotation; that is, biological knowledge B can give a prior probability only for the phenotype network G_Y. The QTLnet framework presented in Section 7.2 assumes intrinsically a uniform prior over phenotype network structures. Additionally, we specify a prior distribution on the weights of biological knowledge in order to control the consistency between phenotype data and knowledge. Because the prior information can be inaccurate or incompatible with the phenotype data, it is important to quantify its uncertainty. We write the extended model as follows:

$$
\begin{aligned}
\mathbb{P}(G, W | Y, X, B) &\propto \mathbb{P}(Y | G, W, X, B)\mathbb{P}(G, W | X, B) \\
&= \mathbb{P}(Y | G, X)\mathbb{P}(G, W | X, B) \\
&= \mathbb{P}(Y | G, X)\mathbb{P}(G_Y, W | X, B)\mathbb{P}(G_{Q \to Y} | X, B) \\
&= \mathbb{P}(Y | G, X)\mathbb{P}(G_Y, W | B)\mathbb{P}(G_{Q \to Y} | X) \\
&= \mathbb{P}(Y | G, X)\mathbb{P}(G_Y | B, W)\mathbb{P}(W | B)\mathbb{P}(G_{Q \to Y} | X).
\end{aligned} \tag{7.3}
$$

In the first step, the posterior probability of a network G and weights W is calculated by multiplying the marginal likelihood $\mathbb{P}(Y | G, W, X, B)$ of the traits given the network G and the prior probability $\mathbb{P}(G, W | X, B)$ of a network and weights given genetic variations and biological knowledge. The marginal likelihood $\mathbb{P}(Y | G, W, X, B)$ can be simplified to be $\mathbb{P}(Y | G, X)$ as in the second step. In the third step, the prior probability $\mathbb{P}(G, W | X, B)$ can be decomposed into $\mathbb{P}(G_Y, W | X, B)$ and $\mathbb{P}(G_{Q \to Y} | X, B)$ by assuming the independence between a phenotype network G_Y along with the weights W and causal QTLs $G_{Q \to Y}$ given genetic variations X and biological knowledge B. The fourth step is provided by the fact that $\mathbb{P}(G_Y, W | X, B)$ is equal to $\mathbb{P}(G_Y, W | B)$ because the genetic variations are not included in the structure of the phenotype network G_Y, and $\mathbb{P}(G_{Q \to Y} | X, B)$ is equal to $\mathbb{P}(G_{Q \to Y} | X)$ because B affects G_Y but not $G_{Q \to Y}$. The extended model in equation (7.3) shows that prior distributions on phenotype network structure $\mathbb{P}(G_Y | B, W)$, biological knowledge weights $\mathbb{P}(W|B)$, and causal QTLs of traits $\mathbb{P}(G_{Q \rightarrow Y}|X)$ must be specified. We will describe how to set $\mathbb{P}(G_Y|B, W)$, $\mathbb{P}(W|B)$, and $\mathbb{P}(G_{Q \rightarrow Y}|X)$ in the following.

PRIOR ON PHENOTYPE NETWORK STRUCTURES $\mathbb{P}(G_Y|B, W)$

Incorporation of a priori biological knowledge into a prior on network structures can help to discriminate Bayesian networks having the same likelihood [55, 61]. If G^1 and G^2 have the same likelihood ($\mathbb{P}(Y|G^1) = \mathbb{P}(Y|G^2)$) but have different prior probabilities ($\mathbb{P}(G^1) \neq \mathbb{P}(G^2)$), the posterior probabilities would become different ($\mathbb{P}(G^1|Y) \neq \mathbb{P}(G^2|Y) \propto \mathbb{P}(Y|G^2)\mathbb{P}(G^2)$). For example, consider two graphs for nodes t and v: one is $t \rightarrow v$ and the other is $v \rightarrow t$. Their likelihoods are the same because they are Markov equivalent. If a prior indicates that one direction ($t \rightarrow v$) is more likely than the other direction ($t \leftarrow v$), then the posterior of one direction ($t \rightarrow v$) becomes higher than the other direction ($v \rightarrow t$). The biological knowledge B along with its weight W can therefore give different prior probabilities $\mathbb{P}(G_Y|B, W)$ for the phenotype network G_Y.

Various types of information can supplement the learning of a phenotype network. We can encode this supplementary information into unequal priors on network structures. A transcription factor binding location can be used to prefer the direction from a transcription factor to the target gene [4]. Pathway information can also guide to infer directions among phenotypes [55]. For example, consider a network with three nodes t, v, and u, where a path from t to v is known. Then, we can at least distinguish these two relations: $t \rightarrow v \leftarrow u$ and $t \rightarrow v \rightarrow u$. Regulation inference [41, 44, 57] from knockout data and protein–protein interaction [26] can be used as a prior for network structure. We will describe how to encode this information in Subsection 7.3.3. Since QTLnet is a Bayesian approach, we can flexibly incorporate various sources of biological knowledge by constructing meaningful priors for the network structures.

Now, it remains to set the prior distribution on phenotype network structure G_Y with respect to biological knowledge B. Since a Bayesian network distribution can be factored by its parent–child relations $\prod_t \mathbb{P}(Y_t|Y_{pa(t)})$, it is natural to assume the prior on DAG structures to be factored by its parent–child relations. Adapting the prior formulation over network structures in [55], we will show below that the prior satisfies the parent–child factorization.

Let us define the *energy* of a phenotype network G_Y relative to the biological knowledge B to be

$$\mathcal{E}(G_Y) = \sum_{i,j=1}^{T} |B(i,j) - G_Y(i,j)|, \qquad (7.4)$$

where B is an encoding meant to describe biological knowledge ranging from 0 to 1 and G_Y is represented by the adjacency matrix of a network structure. The adjacency matrix is a 0-1 matrix which assigns $G_Y(i,j)$ to be 1 if there is a directed edge from node i to j, and to be 0 otherwise. The energy $\mathcal{E}(G_Y)$ acts as a distance measure between biological knowledge and a network structure G_Y. For a fixed biological knowledge matrix B, network structures will have small energy if they agree with the biological knowledge, and will have large energy if they disagree with the knowledge.

The energy can be decomposed into the sum of local pseudo-energies defined by parent–child relations for each trait:

$$\mathcal{E}(G_Y) = \sum_{j=1}^{T} \left(\sum_{i \in pa(j)} (1 - B(i,j)) + \sum_{i \notin pa(j)} B(i,j) \right)$$

$$= \sum_{j=1}^{T} \left(\frac{|B|}{T} + \sum_{i \in pa(j)} (1 - 2B(i,j)) \right) = \sum_{j=1}^{T} \left(\frac{|B|}{T} + \mathcal{E}_{j,pa(j)}(G_Y) \right),$$

where $|B| = \sum_{i,j=1}^{T} B(i,j)$ and $\mathcal{E}_{j,pa(j)}(G_Y) = \sum_{i \in pa(j)}(1 - 2B(i,j))$, which is the local pseudo-energy defined by phenotype j and its parents. Therefore, the prior distribution on network structures can be constructed in terms of energy, and it is shown to be the Gibbs distribution factorized by parent–child relations:

$$\mathbb{P}(G_Y|B, W) = \frac{\exp(-W\mathcal{E}(G_Y))}{Z(W)} \tag{7.5}$$

$$= \frac{\prod_{j=1}^{T} \exp(-W\mathcal{E}_{j,pa(j)}(G_Y))}{Z'(W)}, \quad G_Y \in \text{DAG}$$

where $Z(W)$ is a normalizing constant given by $\sum_{G_Y \in \text{DAG}} \exp(-W\mathcal{E}(G_Y))$ and $Z'(W)$ is another normalizing constant given by $Z(W)/\exp(-W|B|)$. For a fixed W, network structures with small energy will have higher prior probabilities than network structures with large energy. The weight W of biological knowledge B is introduced to tune the confidence of biological information which sometimes can be inaccurate or incompatible with expression data. As W goes toward 0, the influence of a priori knowledge becomes negligible, and the prior distribution of network structure is assumed to be almost uniform. On the contrary, as W goes to the infinity, the prior on network structure peaks at the biological knowledge.

Multiple sources of biological knowledge can be integrated into a prior on network structures with different weights:

$$\mathbb{P}(G_Y|B, W) = \frac{\exp(-\sum_k W_k \mathcal{E}_k(G_Y))}{Z(W)}, \quad G_Y \in \text{DAG}$$

where B_k is an encoding matrix of biological knowledge from source k, B is the vector of biological knowledge matrices $(B_1, \ldots, B_k)$, W_k is the weight of B_k, W is the weight vector $(W_1, \ldots, W_k)$, and $Z(W)$ is the summation of the numerator over all DAGs.

PRIOR ON BIOLOGICAL KNOWLEDGE WEIGHTS $\mathbb{P}(W|B)$

The weight parameter is introduced to control the influence of biological knowledge on the phenotype network. A higher value of the weight would increase the influence of the biological knowledge on the posterior distribution of networks. Specifically, a large W puts significant prior probability on the phenotype network structures which consistently agree with biological knowledge B. Conversely, a small W puts fairly equal prior probabilities on all possible networks. If biological knowledge B is similar to the true network from which the expression data are generated, then the posterior probability will peak at high W. On the contrary, if biological knowledge deviates substantially from the true network, the posterior will peak at small W. This happens because a smaller W leads to a smaller ratio of prior probabilities of the deviated network and the true network. Consequently, the posterior of the true network can be larger than the posterior of the deviated network by the virtue of likelihood ratio overcoming the prior ratio at a small W.

For each biological knowledge B_k, we specify the prior probability distribution of the weight W_k to be an exponential distribution such that

$$\mathbb{P}(W_k|B_k) = \phi \exp(-\phi W_k), \tag{7.6}$$

with the rate parameter ϕ. Such an exponential prior for W_k has several advantages. First, it does not impose an upper bound on W_k. Second, it would not allow the weight to go to infinity too easily, since an infinite weight always results in a network closer to the biological knowledge regardless of expression data. Third, when biological knowledge is inaccurate or incompatible with expression data, the exponential distribution can control the contribution of negative biological knowledge more easily than a uniform distribution. The rate parameter ϕ is set to be 1 in our simulation because this rate balances the prior and likelihood well in the empirical study.

PRIOR ON CAUSAL QTLS $\mathbb{P}(G_{Q\to Y}|X)$

Without any specific information about the causal QTLs, we set the prior of causal QTLs to be a uniform distribution. Several alternative specifications can be found in Bayesian QTL mapping such as in [59] and in [58].

7.3.2 Sketch of MCMC

A main challenge in the reconstruction of networks is that the graph space grows superexponentially with the number of nodes. Exhaustive search over all network structures is impractical even for small networks. Hence, heuristic approaches are needed to efficiently traverse the graph space. We adopt a Metropolis-Hastings MCMC scheme that integrates the sampling of network structures [24, 37], QTL mapping, and the sampling of biological knowledge weights W. The MCMC scheme iterates between accepting a network structure G and accepting k weights $W_1, \cdots, W_k$ corresponding to k types of biological knowledge.

1. Sample a new phenotype network structure G_Y^{new} from a network structure proposal distribution $R(G_Y^{new}|G_Y^{old})$.

2. Given the phenotype network structure G_Y^{new}, sample a new set of causal QTLs $G_{Q\to Y}$ from a QTL proposal distribution $R(G_{Q\to Y}^{new}|G_{Q\to Y}^{old})$.

3. Accept the new extended network structure G^{new} composed of G_Y^{new} and $G_{Q\to Y}^{new}$ given the biological knowledge weights W with a probability

$$A_G = \min\left\{ 1, \frac{\mathbb{P}(Y|G^{new}, X)\mathbb{P}(G_Y^{new}|B, W)\mathbb{P}(G_{Q\to Y}^{new}|X)}{\mathbb{P}(Y|G^{old}, X)\mathbb{P}(G_Y^{old}|B, W)\mathbb{P}(G_{Q\to Y}^{old}|X)} \right.$$
$$\left. \times \frac{R(G_Y^{old}|G_Y^{new})R(G_{Q\to Y}^{old}|G_{Q\to Y}^{new})}{R(G_Y^{new}|G_Y^{old})R(G_{Q\to Y}^{new}|G_{Q\to Y}^{old})} \right\}.$$

4 For each biological knowledge k,

  (a) Sample a new weight W_k^{new} for biological knowledge B_k from a weight proposal distribution $R(W_k^{new}|W_k^{old})$.

  (b) Accept the new biological weight W_k^{new} given the phenotype network G_Y with a probability

$$A_{W_k} = \min\left\{1, \frac{\mathbb{P}(G_Y|W_k^{new}, W_{-k}^{old}, B)}{\mathbb{P}(G_Y|W^{old}, B)} \frac{\mathbb{P}(W_k^{new}|B)}{\mathbb{P}(W_k^{old}|B)} \frac{R(W_k^{old}|W_k^{new})}{R(W_k^{new}|W_k^{old})}\right\}.$$

5 Iterate steps 1–4 until the chain converges.

In step 1, a new phenotype network structure is proposed by a mixture of single edge operations (single edge addition, single edge deletion, single edge reversal) and edge reversal moves with orphaning [19]. The edge reversal move with orphaning consists of selecting an edge $i \rightarrow j$, removing the parents of each node on the selected edge, sampling new parents of node i (including node j), and sampling new parents of node j, as long as it does not make a cycle. It has been shown that edge reversal moves can significantly improve the convergence of an MCMC sampler [19]. The proposal distribution puts the same probability, summing to 1, to the graphs that can be reached by a corresponding edge move.

In step 2, causal QTLs can be sampled conditional on the phenotypes' parents. There are several ways to sample causal QTLs. One way is a Bayesian QTL mapping proposed in [59] for each phenotype. The prior distribution for the indicators of QTLs is $\prod w_k^{\gamma_k}(1-w_k)^{\gamma_k}$ where $w_k = \mathbb{P}(\gamma_k = 1)$ is the prior inclusion probability for the kth QTL. We can use this independent prior for the prior distribution and the proposal distribution for a causal QTL. Another way is the classical interval mapping of QTL for each phenotype conditional on its phenotypic parents. The classical interval mapping regresses a phenotype on a single QTL and picks every QTL over the significance threshold computed by permutations. Thus, this approach is deterministic as it chooses the same set of QTLs given the same set of parent phenotypes. It is a fast algorithm approximating the Bayesian mapping of QTL though it might fail to satisfy the irreducibility of the Markov Chain. We use the interval mapping for practical reasons.

In step 3, the computation of the ratio of marginal likelihoods, or Bayes factor, $\mathbb{P}(Y|G^{new}, X)/\mathbb{P}(Y|G^{old}, X)$, can be approximated by the difference of BIC scores [31] when the sample size is large:

$$\frac{\mathbb{P}(Y|G^{new}, X)}{\mathbb{P}(Y|G^{old}, X)} \approx \exp(-\frac{1}{2}(BIC_{G^{new}} - BIC_{G^{old}})).$$

The BIC score is defined to be $-2\log L + k\log n$, where L is the maximized value of the likelihood for the estimated model, k is the number of free parameters estimated, and n is the sample size.

In step 4, a new weight W_k^{new} can be sampled from a moving uniform distribution $U(W_k^{old} - 1, W_k^{old} + 1)$, and if the sampled W_k^{new} is less than 0, we take a negative of the new weight. This proposal distribution makes the ratio of proposal distributions, $R(W_k^{old}|W_k^{new})/R(W_k^{new}|W_k^{old})$, to be 1. In addition, we need to compute

$$\frac{\mathbb{P}(G_Y | W_k^{new}, W_{-k}^{old}, B)}{\mathbb{P}(G_Y | W^{old}, B)} = \frac{\frac{\exp(-W_k^{new}\mathcal{E}_k(G_Y) - \sum_{k' \neq k} W_{k'}^{old} \mathcal{E}_{k'}(G_Y))}{Z(W_k^{new}, W_{-k}^{old})}}{\frac{\exp(-\sum_k W_k^{old} \mathcal{E}_k(G_Y))}{Z(W^{old})}},$$

where $Z(W) = \sum_{G_Y \in \text{DAG}} \exp(-\sum_k W_k \mathcal{E}_k(G_Y))$ is a normalizing constant. Note that it is not feasible to compute the exact $Z(W)$ due to the exclusion of cyclic networks. We approximate the normalizing constant by the summation over directed graphs with restriction on the number of parents, e.g., 3 as adopted by [55].

After running an MCMC chain, we need to efficiently summarize the chain for the inference of a network structure. The choice by the highest posterior network structure might not produce a convincing model because the graph space grows rapidly with the number of phenotype nodes and the most probable network structure might still have a very low probability. Therefore, instead of selecting the network structure with the highest posterior probability, we perform Bayesian model averaging [8] over the causal links between phenotypes, to infer an averaged network. Explicitly, let Δ_{uv} represent a causal link from u to v, that is, $\Delta_{uv} = \{Y_u \rightarrow Y_v\}$. Then

$$\mathbb{P}(\Delta_{uv} \mid Y, X) = \sum_G \mathbb{P}(\Delta_{uv} \mid G, Y, X) \, \mathbb{P}(G \mid Y, X)$$

$$= \sum_G \mathbb{P}\{\Delta_{uv} \in G\} \, \mathbb{P}(G \mid Y, X).$$

The averaged network is represented by the causal links with maximum posterior probability or with posterior probability above a predetermined threshold, e.g., 0.5.

7.3.3 Summary of Encoding of Biological Knowledge

In equation (7.5), we have constructed a prior distribution on a network structure G_Y in terms of energy $\mathcal{E}(G_Y)$ relative to biological knowledge B. Now we describe how to encode a biological knowledge matrix B from several sources of biological information. Recall that B is an encoding meant to describe biological knowledge ranging from 0 to 1, and energy $\mathcal{E}(G_Y)$ is defined to be a distance measure between B and G_Y in equation (7.4). When there is no available biological knowledge, we would put every element in B as 1/2. Then all DAGs have the same energy, and therefore the probability of a network structure conditional on W is $1/K$, with K as the number of all DAGs. In contrast, when biological knowledge is available, we will look at several ways of encoding biological knowledge into B, such as transcription factor and DNA binding [4], protein–protein interaction [28], and gene ontology annotations [36].

TRANSCRIPTION FACTOR AND DNA BINDING

Chromatin immunoprecipitation with microarray experiments is used to investigate the interaction of proteins and DNA in vivo. This technology has been employed to generate putative lists of transcription factor/target gene interactions [33]. In [4], an approach was suggested to convert a p-value P_{ij}, quantifying the evidence that a transcription factor i binds to a putative target gene j, into a posterior probability for the presence and orientation of an edge in a Bayesian network. Following [4], we assume that the p-value P_{ij} follows a truncated exponential distribution with mean λ when the transcription factor i binds to a target gene j ($G_Y(i,j) = 1$) and a uniform distribution when the transcription factor does not bind to a target gene ($G_Y(i,j) = 0$).

$$\mathbb{P}_\lambda(P_{ij} = p | G_Y(i,j) = 1) = \frac{\lambda e^{-\lambda p}}{1 - e^{-\lambda}},$$

$$\mathbb{P}_\lambda(P_{ij} = p | G_Y(i,j) = 0) = 1.$$

The probability of the directed edge before observing any biological data is assumed to be $\mathbb{P}(G_Y(i,j) = 1) = 1/2$ so that without any biological data, the probability of presence of edge only depends on the expression data. By Bayes' rule, the probability of presence of an edge after observing a p-value is:

$$\mathbb{P}_\lambda(G_Y(i,j) = 1 | P_{ij} = p) = \frac{\lambda e^{-\lambda p}}{\lambda e^{-\lambda p} + (1 - e^{-\lambda})}.$$

Here λ is assumed to be uniformly distributed over the interval $[\lambda_L, \lambda_H]$ and the integration over λ is performed to obtain the probability of the presence of an edge:

$$\mathbb{P}(G_Y(i,j) = 1 | P_{ij} = p) = \frac{1}{\lambda_H - \lambda_L} \int_{\lambda_L}^{\lambda_H} \frac{\lambda e^{-\lambda p}}{\lambda e^{-\lambda p} + (1 - e^{-\lambda})} \, d\lambda.$$

This can be solved numerically, for instance, by choosing λ in the range $[0, 10000]$. We should thus obtain the following estimate: $B(i,j) = \mathbb{P}(G_Y(i,j) = 1 | P_{ij} = p)$.

PROTEIN–PROTEIN INTERACTION

Since protein–protein interaction is non-directional, we put the same probability on both directions. If we do not consider the diverse reliabilities of protein–protein interaction from several experiments, we set $B(i,j)$ and $B(j,i)$ to be $\delta > 1/2$ when we find any interaction on any experiment. If there are gold standards for positive and negative protein–protein interactions, and experiments have diverse reliabilities, then we can use the Bayes classifier proposed by [28] to combine heterogeneous data. Positive gold standards are well-known true protein–protein interactions, whereas negative gold standards are interactions which cannot happen, such as those between a pair of proteins in different subcellular compartments. An interaction experimental data set is a collection of observations over all pairs of proteins by binaries concerning whether the interaction is present or absent for each pair. Suppose there are L interaction experimental data sets with different false positive rates. We can calculate the posterior odds of an interaction from the binary observations $f_1, \ldots, f_L$ using the likelihood ratio LR:

$$O_{posterior} = \frac{\mathbb{P}(pos | f_1, \ldots, f_L)}{\mathbb{P}(neg | f_1, \ldots, f_L)} = O_{prior} \times LR$$

$$= \frac{\mathbb{P}(pos)}{\mathbb{P}(neg)} \times \frac{\mathbb{P}(f_1, \ldots, f_L | pos)}{\mathbb{P}(f_1, \ldots, f_L | neg)}.$$

In the positive gold standard interactions, we can find a set of interactions which have the observed values $f_1, \ldots, f_L$. The likelihood under the positive gold standard can be defined to be the proportion of the set with the values $f_1, \ldots, f_L$ in the positive gold standard. Similarly we define the likelihood $\mathbb{P}(f_1, \ldots, f_L | neg)$ under the negative gold standard. Then we can take the ratio of the two likelihoods to calculate the likelihood ratio LR. The prior odds O_{prior} can be defined by an expert. The encoding of B can be obtained by transforming the posterior odds into a posterior positive rate:

$$B(i, j) = B(j, i) = \frac{O_{posterior}}{1 + O_{posterior}}.$$

When the posterior odds is equal to 1, $B(i, j)$ and $B(j, i)$ are equal to 1/2. As the posterior odds increases, the values of $B(i, j)$ and $B(j, i)$ also increase.

GENE ONTOLOGY

The Gene Ontology (GO) [1] is a well-controlled vocabulary of terms describing the molecular functions, biological processes, and cellular components of a gene. A GO is structured as a DAG in which each node represents a GO term. The GO terms annotate a large fraction of genes. The distance between two genes can be defined in terms of their GO annotations. One well-defined distance is Lord's similarity [36]. This measure takes into account the hierarchy of GO and GO term occurrences in the myriad of genes. If two genes share a more specific GO term positioned in the lower part of the GO hierarchy, they are more likely to be similar. However, even if the shared GO terms lie in the same level of the hierarchy, the frequencies of the GO terms in the whole genes are different, which affects the similarity. Consider that two GO terms, c_1 and c_2, lie in the same level of the hierarchy. Suppose there are 100 genes annotated with term c_1 and there are 1000 genes annotated with term c_2. Then the chance of two genes sharing the term c_2 is higher than the chance of sharing the term c_1. This then implies that the term c_1 is more informative. The information content $IC(c)$ for a GO term c is defined to be the negative logarithm of the number of times the term or any of its descendant terms occurs in the myriad of genes, divided by the total number of occurrences of GO terms. The root of the hierarchy will have zero information content, whereas a leaf of the hierarchy will have high information content. Once the information content $IC(c)$ for each node in the GO is set up, we can define GO term similarity and gene similarity. The similarity between two GO terms is defined to be the maximum information content among the shared parents of the two terms, which is

$$\text{sim}(c_1, c_2) = \max_{c \in (pa(c_1) \cap pa(c_2))} IC(c).$$

Then, since a gene is annotated with a set of GO terms, the similarity between two genes g_1 and g_2 can be defined as the average of similarities calculated for all pairs of GO terms between two genes; that is,

$$\text{sim}(g_1, g_2) = \frac{\sum_{i=1}^{n} \sum_{j=1}^{m} \text{sim}(c_{1,i}, c_{2,j})}{nm}.$$

This Lord's measure can be used as an encoding of B if it is rescaled to be in the interval $[0, 1]$.

Table 7.3 Four methods for causal phenotype network inference, which differ in their use of genetic variation information and biological knowledge.

Method	Use of Genetic Variation Information	Use of Biological Knowledge
QTLnet-prior	YES	YES
QTLnet	YES	NO
WH-prior	NO	YES
Expression	NO	NO

7.4 Simulations

We performed a simulation study for comparing the proposed method (QTLnet-prior) with three other methods: QTLnet [9], WH-prior [55], and Expression. Table 7.3 provides a summary of these four methods in terms of using the genetic variation information and biological knowledge. QTLnet was implemented using R/QTLnet, QTLnet-prior was implemented with prior setting on R/QTLnet, and WH-prior was programmed as in [55], with a modification of approximating the marginal likelihood with the BIC score instead of using the BGe score [17].[4] Expression was programmed by modifying R/QTLnet to exclude QTL mapping.

We simulated expression data and a priori knowledge matrix according to the network structure in Fig. 7.1 and produced 100 simulated data sets. To generate expression data based on the network in Fig. 7.1, the genetic information was simulated first. The genetic map described five chromosomes of 100 cM with ten equally spaced markers in each chromosome, and the markers were simulated for 500 mice in an F2 population using R/qtl [7]. We assumed QTL Q_t was located in the middle of chromosome t. Then, each expression data set of the F2 population was generated with different genetic effects and partial regression coefficients between phenotypes. Genetic additive effects were sampled from a uniform distribution $U[0, 0.5]$, and dominance effects were sampled from $U[0, 0.25]$. The partial regression coefficients β_{ut} were sampled from $U[-0.5, 0.5]$. The residual phenotypic variance was 1. The biological knowledge matrix B was generated for several cases. The value $B(t, u)$ was generated from one of two $[0, 1]$-truncated normal distributions $\mathcal{N}_\pm(0.5 \pm \delta, 0.1)$ [16]. The distribution was truncated at 0 and 1 to guarantee that the value $B(t, u)$ ranged from 0 to 1. When no biological knowledge is available, the natural choice for $B(t, u)$ is 1/2. Consequently, the evidence for the presence (absence) of edge $t \to u$ is necessarily specified through a value of $B(t, u)$ greater (lower) than 1/2. In the simulation, the $B(t, u)$ value of true edge was generated from $\mathcal{N}_+$ and the $B(t, u)$ value of false edge was generated from $\mathcal{N}_-$.

---

[4] BGe stands for *Bayesian metric for Gaussian networks having score equivalence*. The BGe score was developed as a scoring metric for a Bayesian network of continuous variables under the assumption that the data are sampled from a multivariate Gaussian distribution. The BGe score is first derived for a complete Bayesian network where every pair of distinct nodes is connected by a direct edge. It assumes a prior on parameter to be a normal Wishart distribution so that one can obtain a closed-form marginal distribution. Under the assumption of parameter independence and modularity, the BGe score for an arbitrary Bayesian network is derived to be $\mathbb{P}(Y|G) = \prod_{t=1}^{T} \frac{\mathbb{P}(Y_t, Y_{pa(t)}|G_c)}{\mathbb{P}(Y_{pa(t)}|G_c)}$, where G_c is any complete DAG such that each node has the same parents as in G. It is known that Markov equivalent DAGs have the same BGe score. See [17] for details.

The parameter δ controls the accuracy of prior knowledge. We denote the generated biological knowledge to be positive knowledge, non-informative knowledge, or negative knowledge based on the sign of δ: +, 0, or –, respectively. We examined 11 cases of different accuracies of prior knowledge: $\delta \in \{\pm 0.1, \pm 0.08, \pm 0.06, \pm 0.04, \pm 0.02, 0\}$. In the extreme case when δ is equal to 0.5, the prior knowledge almost correctly reflects the network structure, whereas when δ is equal to –0.5, the prior knowledge incorrectly reflects the network structure almost in the opposite way. When δ is equal to 0, the information is generated with no distinction between true and false edges. For each simulated data set, we ran an MCMC for 30 300 iterations, discarded the first 300 iterations, sampled every ten iterations, and generated 3000 samples.

We assessed these four methods by using Receiver Operating Characteristic (ROC) curves of the proportion of recovered and spurious edges. Bigger areas under the ROC curve generally indicate better performance, as the area represents the probability that the classifier ranks true edges higher than false edges [13]. The ROC curves are obtained from the set of proportions of recovered edges and spurious edges for various posterior probability thresholds ranging from 0 to 1.

First, we evaluated the effect of incorporating genetic variation information. The effect of QTL mapping can be tested by comparing QTLnet-prior and WH-prior. QTLnet-prior is more effective in recovering the network structure than WH-prior in Fig. 7.4A, and we can conclude that QTL mapping increases the effectiveness. Better causal QTLs can be inferred by conditioning on the phenotype network. Similarly, a better phenotype network can be inferred by conditioning on causal QTLs. The gain is more apparent when the biological knowledge is negative.

Second, we evaluated the effect of incorporating biological knowledge. In Fig. 7.4A, when δ is positive, QTLnet-prior performs better than QTLnet and WH-prior performs better than

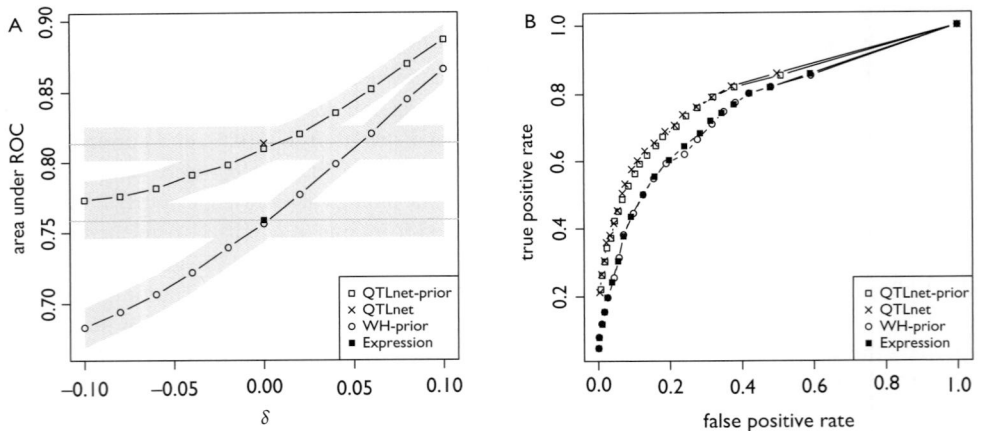

Fig 7.4 A comparison of the four methods of causal phenotype network inference by the area under the Receiver Operating Characteristic (ROC) curves with respect to the accuracy of biological knowledge. (A) The areas under ROC curves of QTLnet-prior, QTLnet, WH-prior, and Expression. The areas under ROC curves of QTLnet-prior and WH-prior are plotted against the accuracy of biological knowledge, δ. Since QTLnet and Expression do not incorporate biological knowledge, they are plotted in a single point each ($\times$, $\blacksquare$). The shaded area indicates the standard error of the area under ROC curve. (B) The ROC curves of QTLnet-prior and WH-prior are drawn when non-informative biological knowledge ($\delta = 0$) is incorporated. They are compared with the ROC curves of QTLnet and Expression which do not incorporate biological information.

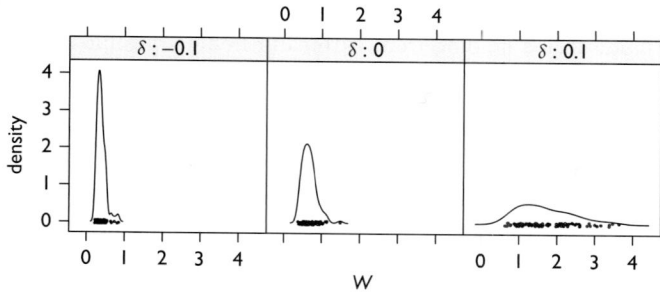

Fig 7.5 The distribution of median weight W of posterior sample by QTLnet-prior inference. Each panel shows the median W distribution when biological knowledge is defective ($\delta = -0.1$), non-informative ($\delta = 0$), and informative ($\delta = 0.1$).

Expression; whereas when δ is negative, QTLnet-prior performs worse than QTLnet and WH-prior performs worse than Expression. With a positive δ, as the accuracy of knowledge increases, QTLnet-prior and WH-prior benefit by prior knowledge incorporation. However, a negative δ, indicating that the knowledge disagrees with the true network structure, causes QTLnet-prior and WH-prior to be impaired by prior knowledge incorporation. The decreased performances in QTLnet-prior and WH-prior draw attention as to whether W can effectively control the influence of negative knowledge. Fig. 7.5 shows that the median of W in the posterior sample is close to 0 with negative knowledge. It implies that the weight W can effectively control the use of negative knowledge to some extent but not completely, based on the decreased recovery observed in comparison to the case of non-informative knowledge evidenced in Fig. 7.4A. In comparison with QTLnet and Expression, the reduced performance of QTLnet-prior and WH-prior can be explained by the remaining uncontrolled effect of prior probability incorporating negative knowledge. When non-informative knowledge is incorporated, there is no significant difference in the area under the ROC curve between QTLnet and QTLnet-prior (p-value = 0.82) and between Expression and WH-prior (p-value = 0.89), as shown in Fig. 7.4A and 7.4B.

7.5 Analysis of Yeast Cell-Cycle Genes

We used QTLnet-prior to reconstruct a network of 26 genes involved in the cell cycle in yeast (*Saccharomyces cerevisiae*), previously chosen by [4] for cell-cycle network analysis with time-dependent expression data and transcription factor binding information. Some genes express periodically according to cell-cycle phase (genome duplication phase, gap phase 2, cell division phase, and gap phase 1), and there are transcription factors that regulate these periodical genes [3]. The gene expression data and genetic variation information for this analysis were obtained from a backcross population of 112 segregates between BY4716 and RM11-1a [6]. In [6], the authors extracted gene expression data by constructing a backcross, isolating the RNA, and hybridizing the resulting complementary DNA (cDNA; see Chapter 1, Section 1.3) to microarrays. They also genotyped the population at 2957 genetic markers for genetic variations. In addition to gene expression data and genetic variations, we incorporated transcription factor binding information as biological knowledge for QTLnet-prior analysis. The *p*-value for evidence of transcription factor binding from chromatin immunoprecipitation with microarray experiments is available for 106 transcription factors from [33]. For the 26 genes in our analysis, 11 of them are transcription factors and the rest are known targets of one or more transcription factors. We transformed the *p*-values into the biological knowledge matrix B, as described in Subsection 7.3.3.

The construction of the causal network focused on the 26 phenotypes of 112 yeast segregates, incorporating genetic variation information at 2957 markers and biological knowledge of transcription factor binding. We ran an MCMC for 760 000 iterations, discarded the first 200 000 iterations, sampled every 100 iterations, and finally got 5 600 samples used for estimation. The computation took around 14 days of CPU time on a 2.66GHz Intel(R) Core(TM)2 Quad running Red Hat 4.1.2-50. To examine the mixing and convergence of the MCMC chain, we first computed the autocorrelation of Bayesian information criterion (BIC) scores and autocorrelation of W, respectively. As shown in Fig. 7.10 in Appendix 7.A, both autocorrelation values get close to 0. It indicates that the MCMC chain may not suffer from a slow mixing rate. Furthermore, we calculated Geweke's convergence diagnostic [18] to check the convergence of the Markov chain. The Geweke's diagnostic is asymptotically $\mathcal{N}(0, 1)$ when it is equal for the two means of the first 10% and the last 50% of the Markov chain. The Geweke's diagnostic for the BIC score is 0.34 and is −0.25 for W, suggesting the convergence of the chain. Fig. 7.6 shows the causal phenotype network reconstructed by QTLnet-prior. The full network of phenotypes and causal QTLs can be found in Fig. 7.9 in Appendix 7.A.

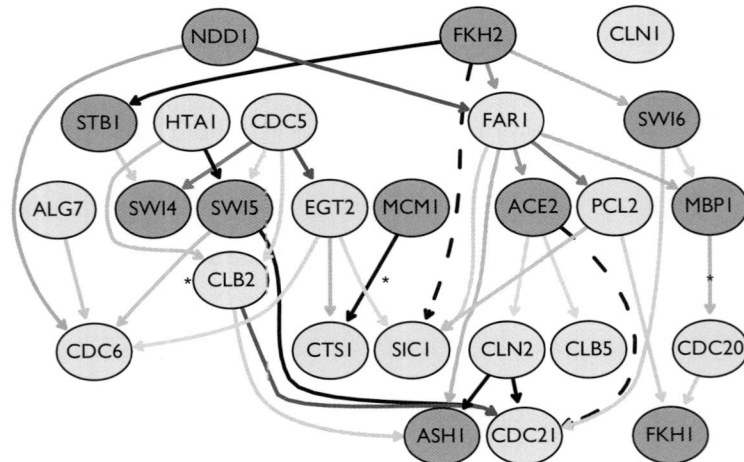

Fig 7.6 Yeast cell-cycle phenotype network by QTLnet-prior, integrating transcription factor binding information. A solid line represents an inferred edge with a posterior probability over 0.5, and the darkness of the edge is in proportion to the posterior probability. Dark nodes are transcription factors. The edge consistent with transcription factor binding information is marked with an asterisk. The transcription factor binding relation recovered by an indirect path in the inferred network is represented by a dashed line.

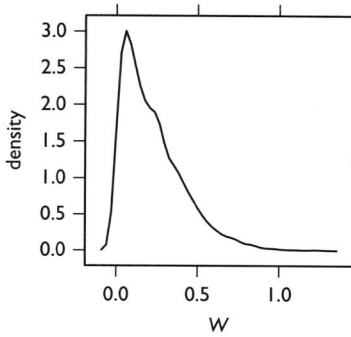

Fig 7.7 The posterior distribution of weight W of transcription factor information in reconstructing a yeast cell-cycle network by QTLnet-prior.

In the transcription factor biological knowledge matrix B, we defined a pair (i.e., an edge from node i to node j) to be *significant* if its $B(i, j)$ value is over 0.5. There are 44 significant transcription factor pairs in B. For the constructed network with 36 inferred edges in Fig. 7.6, we found three significant direct pairs (MBP1 $\rightarrow$ CDC20, SWI5 $\rightarrow$ CDC6, and MCM1 $\rightarrow$ CTS1) and two significant indirect pairs (ACE2 $\rightarrow$ CDC21 and FKH2 $\rightarrow$ SIC1). Interestingly, we did not find any reverse relations, that is, causal relations from target genes to transcription factors, in the inferred network. The remaining 39 causal relations in B were not inferred in the phenotype network.

Fig. 7.7 shows the posterior density distribution of the weight W of transcription factor information, which has a mode of approximately 0 with right skewness. To further examine the contribution of transcription factor information on phenotype network reconstruction, we applied QTLnet to construct the phenotype network without using transcription factor information. For the two networks inferred by QTLnet-prior and QTLnet, the posterior probability of every possible directed edge is very similar to each other, as shown in Fig. 7.8. Although the transcription factor knowledge did not improve the reconstruction of the cell-cycle network, it did not have a negative impact on reconstruction, either. The weight parameter was actually effective in protecting the network reconstruction against the inconsistent transcription factor information. The inconsistency between transcription factor information and expression data

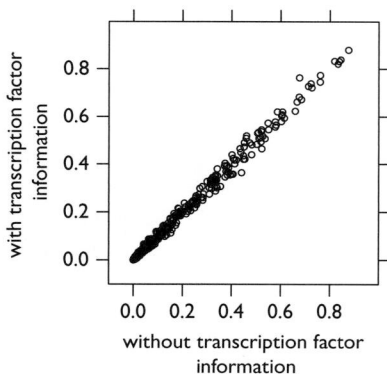

Fig 7.8 Comparison of the posterior probability of every possible directed edge between the network inferred by QTLnet-prior and the network inferred by QTLnet.

may be due to any of the following reasons: (1) inconsistency between physical regulation of transcription binding and transcriptional regulation level of expression changes; (2) necessary post-translational modification of transcription factor or construction of complexes with other proteins for regulation of target genes; (3) cell cycle phase or tissue-dependent transcription factor binding or false binding information; and (4) inability to capture the cyclicity of the cell-cycle network from static expression data relative to a single time point.

7.6 Conclusion

We have developed a phenotype network inference method (QTLnet-prior) to incorporate genetic variation information and biological knowledge. Genotypes are known to control phenotypes but not the other way and thereby can help to distinguish phenotype network structures. Biological knowledge can improve the clustering and directional inference between phenotypes. The simulation study shows that the proposed method can improve the reconstruction of the gene network by integrating genetic variation information and biological knowledge as long as knowledge agrees with data. When biological knowledge does not agree with data, the weight of knowledge controls the contribution of prior probability of biological knowledge on the likelihood of data, reducing (to some extent) the negative impact of the defective knowledge. We applied QTLnet-prior to estimate a yeast cell-cycle network of 26 genes with causal QTLs by integrating transcription factor binding information and compared its performance to QTLnet. The distribution of weight suggests that the transcription factor binding information was inconsistent with the expression data. Nonetheless, comparison with QTLnet's output showed a fairly similar result, suggesting the weight parameter of knowledge was effective in controlling the negative impact of inconsistent knowledge in this case.

When we interpret the inferred networks, we need to be cautious. Even though, in theory, the incorporation of causal QTLs allows us to distinguish network structures that would otherwise be likelihood equivalent, in practice some of the detected expression-to-expression causal relationships might be invalid. The possible explanation is that the inferred expression network represents a projection of real causal relationships that might take place outside the transcriptional regulation level. For instance, the true causal regulations could be due to transcription factor binding, direct protein–protein interaction, phosphorylation, methylation, etc. and might not be well reflected at the gene expression level. The incorporation of diffused biological knowledge, mined from different levels of biological regulation, could potentially improve the reconstruction of gene-expression regulatory networks. In any case, the inference of these networks can still play an important role in generating hypothetically possible causal relations.

There are several factors that could change the inference by QTLnet-prior. One is the prior distribution specification. We have used the Gibbs distribution as a prior distribution for network structures in equation (7.5) in terms of an absolute distance measure in equation (7.4) to incorporate biological knowledge. The exponential distribution is used for the weight of biological knowledge in equation (7.6) with the rate parameter (see Subsection 7.3.1). However, we could consider different choices of network structure distributions, measures to incorporate information, weight distributions, and hyperparameters. Another factor is the sample size of expression data. As the sample size increases, the contribution of biological knowledge will be generally reduced. This shows the limited contribution of biological knowledge on the reconstruction of networks, even though biological knowledge B can also be obtained from a number of experiments, as discussed in [55]. The third factor is the global control of biological knowledge on network reconstruction. Illustrated by the yeast cell-cycle network, every transcription factor/target regulation was controlled by the same weight parameter. It may have resulted in no contribution of any biological knowledge, even though five transcription factor/target regulations were inferred to be consistent with expression data. This suggests incorporation of biological knowledge by local control parameters when reconstructing a network. Finally, the encoding of biological knowledge plays an important role. We have proposed to use the encoding for a transcription factor and its targets from [4], protein–protein interactions from [28], and gene ontology annotations from [36]. These encodings are mainly about direct relationships in separate biological regulation levels. As discussed in the previous paragraph, this diffused biological knowledge can improve the Bayesian network reconstruction.

There are shortcomings to the QTLnet-prior framework inherited from QTLnet. One of the assumptions of QTLnet is an absence of latent variables. Latent variables can make it impossible to find the marginalized model in the class of DAGs as shown in [46] and can induce erroneous relations. Suppose there are three nodes y_1, y_2, and y_3, and y_1 and y_2 have a common parent c_1 while y_2 and y_3 have a common parent c_2. If the common parents c_1 and c_2 are not observed, we obtain the following independence relations: $y_1 \perp\!\!\!\perp y_3$ and $y_1 \not\!\perp\!\!\!\perp y_3 \mid y_2$. Then, we mistakenly infer that y_1 and y_3 are parents of y_2. To address this problem, one can consider the more general class of ancestral graphs, which takes care of latent variables. Ancestral graphs open up the possibility of latent variables, although they do not explicitly include the latent variables in the network structures [46].

A persistent challenge in Bayesian network analysis is how to cope with large networks, since the DAG space size grows superexponentially with the number of nodes. Approaches based on Markov blankets with and without restrictions on the number of parent nodes have been proposed [45, 47, 49]. In [27], the authors approximated the Bayesian network problem to a linear programming problem. In [51], the authors developed a parallel algorithm that infers subnetworks restricted on a Markov blanket and merges the subnetworks. Likewise, in phylogeny estimation, the supertree reconstruction from small trees has been studied [5]. We think the rigorous development of super Bayesian network methodology to integrate small subnetworks is a promising direction for the inference of a large network, since the inference of small subnetworks is computationally inexpensive and multiple subnetworks can be parallelized for computation. In this era of vast biological data and knowledge in various aspects, integrating them reasonably on a large scale can be an interesting topic for future research.

ACKNOWLEDGMENTS

The authors wish to thank NIH/NIDDK 58037 (JYM, BSY), and 66369 (JYM, BSY), NIH/NIGMS 74244 (JYM, BSY) and 69430 (JYM, BSY), NCI ICBP U54-CA149237 (ECN), and NIH R01MH090948 (ECN), for supporting this work. Alan D. Attie and Mark Keller motivated the work through our collaboration on causal models for diabetes and obesity.

APPENDIX 7.A INFERRED YEAST CELL CYCLE NETWORK WITH CAUSAL QTLS INTEGRATING TRANSCRIPTION FACTOR INFORMATION BY QTLNET-PRIOR

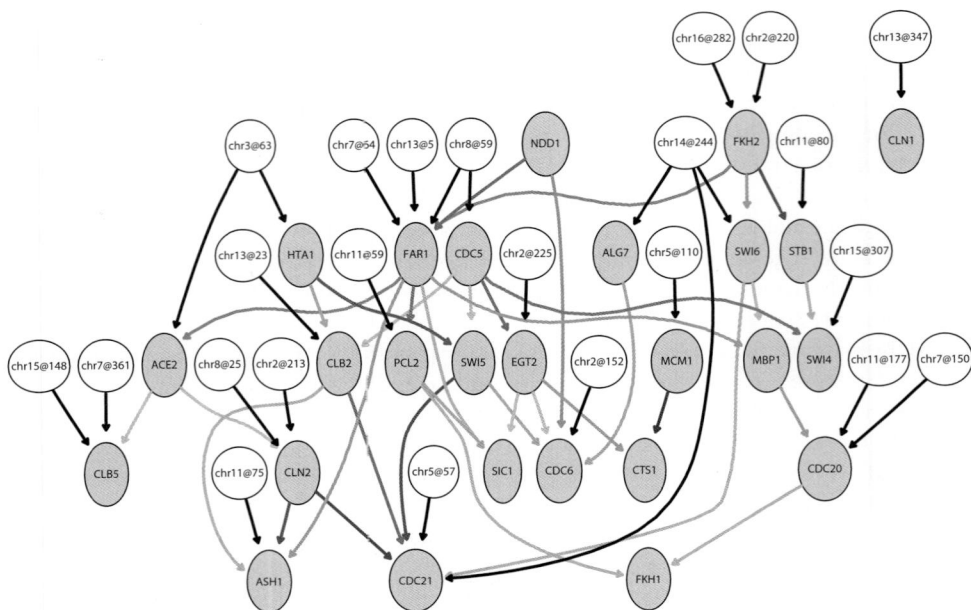

Fig 7.9 Yeast cell-cycle network integrating transcription factor binding information inferred by QTLnet-prior. The edge darkness is in proportion to the posterior probability.

APPENDIX 7.B CONVERGENCE DIAGNOSTICS OF YEAST CELL CYCLE NETWORK

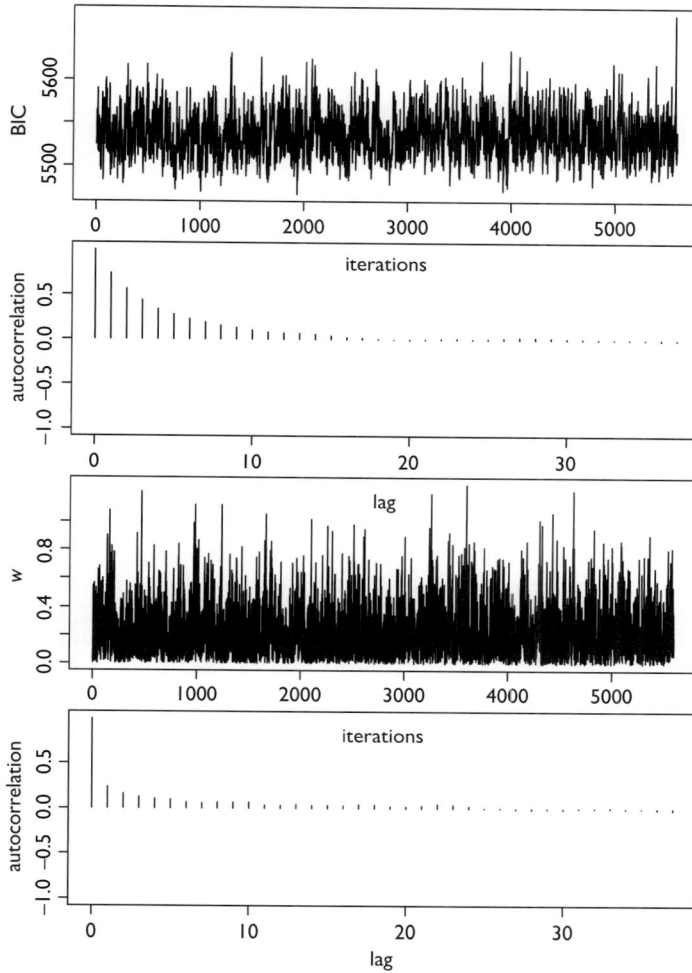

Fig 7.10 The top two figures are the trace plot and the autocorrelation plot of BIC scores for sampled causal networks. The bottom two figures are the trace and the autocorrelation plots of the sampled weights (*W*) on transcription factor binding information.

REFERENCES

[1] M. Ashburner, C. A. Ball, J. A. Blake, D. Botstein, H. Butler, J. M. Cherry, A. P. Davis, K. Dolinski, S. S. Dwight, J. T. Eppig, M. A. Harris, D. P. Hill, L. Issel-Tarver, A. Kasarskis, S. Lewis, J. C. Matese, J. E. Richardson, M. Ringwald, G. M. Rubin, G. Sherlock, and Consortium Gene Ontology. Gene Ontology: tool for the unification of biology. *Nature Genetics*, 25(1):25–29, 2000.

[2] J. E. Aten, T. F. Fuller, A. J. Lusis, and S. Horvath. Using genetic markers to orient the edges in quantitative trait networks: The NEO software. *BMC Systems Biology*, 2:34, 2008.

[3] J. Bähler. Cell-cycle control of gene expression in budding and fission yeast. *Annual Review of Genetics*, 39(1):69–94, 2005.

[4] A. Bernard and A. J. Hartemink. Informative structure priors: joint learning of dynamic regulatory networks from multiple types of data. *Pacific Symposium on Biocomputing 2005*, pages 459–470, 2005.

[5] O. R. P. Bininda-Emonds, J. L. Gittleman, and M. A. Steel. The (Super)tree of life: procedures, problems, and prospects. *Annual Review of Ecology and Systematics*, 33:265–289, 2002.

[6] R. B. Brem and L. Kruglyak. The landscape of genetic complexity across 5,700 gene expression traits in yeast. *Proceedings of the National Academy of Sciences of the United States of America*, 102(5):1572–1577, 2005.

[7] K. W. Broman, H. Wu, S. Sen, and G.A. Churchill. R/qtl: QTL mapping in experimental crosses. *Bioinformatics*, 19(7):889–890, 2003.

[8] E. Chaibub Neto, C. T. Ferrara, A. D. Attie, and B. S. Yandell. Inferring causal phenotype networks from segregating populations. *Genetics*, 179(2):1089–1100, 2008.

[9] E. Chaibub Neto, M. P. Keller, A. D. Attie, and B. S. Yandell. Causal graphical models in systems genetics: a unified framework for joint inference of causal network and genetic architecture for correlated phenotypes. *Annals of Applied Statistics*, 4(1):320–339, 2010.

[10] E. Chaibub Neto, M. P. Keller, A. T. Broman, A. D. Attie, and B. S. Yandell. Causal model selection tests in systems genetics. Technical Report 1157, Department of Statistics, University of Wisconsin-Madison, 2010.

[11] L. S. Chen, F. Emmert-Streib, and J. D. Storey. Harnessing naturally randomized transcription to infer regulatory relationships among genes. *Genome Biology*, 8(10):R219, 2007.

[12] S. Christley, Q. Nie, and X. Xie. Incorporating existing network information into gene network inference. *PLOS ONE*, 4(8):e6799, 2009.

[13] T. Fawcett. An introduction to ROC analysis. *Pattern Recognition Letters*, 27(8):861–874, 2006.

[14] N. Friedman, M. Linial, I. Nachman, and D. Pe'er. Using Bayesian networks to analyze expression data. *Journal of Computational Biology*, 7(3–4):601–620, 2000.

[15] T. S. Gardner, D. di Bernardo, D. Lorenz, and J. J. Collins. Inferring genetic networks and identifying compound mode of action via expression profiling. *Science*, 301(5629):102–105, 2003.

[16] F. Geier, J. Timmer, and C. Fleck. Reconstructing gene-regulatory networks from time series, knock-out data, and prior knowledge. *BMC Systems Biology*, 1:11, 2007.

[17] D. Geiger and D. Heckerman. Learning Gaussian networks. Technical Report MSR-TR-94-10, Microsoft Research, 1994.

[18] J. Geweke. Evaluating the accuracy of sampling-based approaches to the calculation of posterior moments. In J. M. Bernardo, J. O. Berger, A. P. Dawid, and A. F. Smith, editors, *Bayesian Statistics 4*, pages 169–193. Oxford University Press, 1992.

[19] M. Grzegorczyk and D. Husmeier. Improving the structure MCMC sampler for Bayesian networks by introducing a new edge reversal move. *Machine Learning*, 71:265–305, 2008.

[20] R. S. Hageman, M. S. Leduc, R. Korstanje, B. Paigen, and G. A. Churchill. A Bayesian framework for inference of the genotype-phenotype map for segregating populations. *Genetics*, 187:1163–1170, 2011.

[21] D. Heckerman and D. Geiger. Likelihoods and parameter priors for Bayesian networks. Technical Report MSR-TR-95-94, Microsoft Research, 1996.

[22] D. Heckerman, C. Meek, and G. Cooper. A Bayesian approach to causal discovery. In D. Holmes and L. Jain, editors, *Innovations in Machine Learning*, volume 194 of *Studies in Fuzziness and Soft Computing*, pages 1–28. Springer Berlin/Heidelberg, 2006.

[23] J. A. Hoeting, D. Madigan, A. E. Raftery, and C. T. Volinsky. Bayesian model averaging: a tutorial. *Statistical Science*, 14:382–417, 1999.

[24] D. Husmeier. Sensitivity and specificity of inferring genetic regulatory interactions from microarray experiments with dynamic Bayesian networks. *Bioinformatics*, 19(17):2271–2282, 2003.

[25] S. Imoto, T. Higuchi, T. Goto, K. Tashiro, S. Kuhara, and S. Miyano. Combining microarrays and biological knowledge for estimating gene networks via Bayesian networks. *Journal of Bioinformatics and Computational Biology*, 2(1):77–98, 2004.

[26] S. Imoto, S. Kim, T. Goto, S. Miyano, S. Aburatani, K. Tashiro, and S. Kuhara. Bayesian network and nonparametric heteroscedastic regression for nonlinear modeling of genetic network. *Journal of Bioinformatics and Computational Biology*, 1(2):231–52, 2003.

[27] T. Jaakkola, D. Sontag, A. Globerson, and M. Meila. Learning Bayesian network structure using LP relaxations. *Journal of Machine Learning Research*, 9:358–365, 2010.

[28] R. Jansen, H. Yu, D. Greenbaum, Y. Kluger, N. J. Krogan, S. Chung, A. Emili, M. Snyder, J. F. Greenblatt, and M. Gerstein. A Bayesian networks approach for predicting protein-protein interactions from genomic data. *Science*, 302(5644):449–453, 2003.

[29] R. C. Jansen and J.-P. Nap. Genetical genomics: the added value from segregation. *Trends in Genetics*, 17(7):388–391, 2001.

[30] C.-H. Kao and Z.-B. Zeng. Modeling epistasis of quantitative trait loci using Cockerham's model. *Genetics*, 160:1243–1261, 2002.

[31] R. E. Kass and A. E. Raftery. Bayes factors. *Journal of the American Statistical Association*, 90(430):773–795, 1995.

[32] D. Kulp and M. Jagalur. Causal inference of regulator-target pairs by gene mapping of expression phenotypes. *BMC Genomics*, 7(1):125, 2006.

[33] T. I. Lee, N. J. Rinaldi, F. Robert, D. T. Odom, Z. Bar-Joseph, G. K. Gerber, N. M. Hannett, C. T. Harbison, C. M. Thompson, I. Simon, J. Zeitlinger, E. G. Jennings, H. L. Murray, D. B. Gordon, B. Ren, J. J. Wyrick, J.-B. Tagne, T. L. Volkert, E. Fraenkel, D. K. Gifford, and R. A. Young. Transcriptional regulatory networks in *Saccharomyces cerevisiae*. *Science*, 298(5594):799–804, 2002.

[34] R. Li, S.-W. Tsaih, K. Shockley, I. M. Stylianou, J. Wergedal, B. Paigen, and G. A. Churchill. Structural model analysis of multiple quantitative traits. *PLOS Genetics*, 2(7):e114, 2006.

[35] B. Liu, A. de la Fuente, and I. Hoeschele. Gene network inference via structural equation modeling in genetical genomics experiments. *Genetics*, 178(3):1763–1776, 2008.

[36] P. W. Lord, R. D. Stevens, A. Brass, and C. A. Goble. Investigating semantic similarity measures across the Gene Ontology: the relationship between sequence and annotation. *Bioinformatics*, 19(10):1275–1283, 2003.

[37] D. Madigan and J. York. Bayesian graphical models for discrete data. *International Statistical Review*, 63:215–232, 1995.

[38] J. Millstein, B. Zhang, J. Zhu, and E. Schadt. Disentangling molecular relationships with a causal inference test. *BMC Genetics*, 10(1):23, 2009.

[39] J. H. Nadeau and A. M. Dudley. Systems genetics. *Science*, 331(6020):1015–1016, 2011.

[40] N. Nariai, S. Kim, S. Imoto, and S. Miyano. Using protein-protein interactions for refining gene networks estimated from microarray data by Bayesian networks. *Pacific Symposium on Biocomputing*, 336–47, 2004.

[41] O. Ourfali, T. Shlomi, T. Ideker, E. Ruppin, and R. Sharan. SPINE: a framework for signaling-regulatory pathway inference from cause-effect experiments. *Bioinformatics*, 23(13):i359–i366, 2007.

[42] J. Pearl. *Probabilistic Reasoning in Intelligent Systems:Networks of Plausible Inference.* Morgan Kaufmann Publishers, 1988.

[43] J. Pearl. *Causality: Models, Reasoning and Inference.* Cambridge University Press, 2000.

[44] T. Peleg, N. Yosef, E. Ruppin, and R. Sharan. Network-free inference of knockout effects in yeast. *PLOS Computational Biology,* 6(1):e1000635, 2010.

[45] E. Perrier, S. Imoto, and S. Miyano. Finding optimal Bayesian network given a super-structure. *Journal of Machine Learning Research,* 9:2251–2286, 2008.

[46] T. Richardson and P. Spirtes. Ancestral graph Markov models. *The Annals of Statistics,* 30(4):962–1030, 2002.

[47] C. Riggelsen. MCMC learning of Bayesian network models by Markov blanket decomposition. In J. Gama, R. Camacho, P. Brazdil, A. Jorge, and L. Torgo, editors, *Machine Learning: ECML 2005,* volume 3720 of *Lecture Notes in Computer Science,* pages 329–340. Springer Berlin/Heidelberg, 2005.

[48] E. E. Schadt, J. Lamb, X. Yang, J. Zhu, S. Edwards, D. Guhathakurta, S. K. Sieberts, S. Monks, M. Reitman, C. S. Zhang, P. Y. Lum, A. Leonardson, R. Thieringer, J. M. Metzger, L. M. Yang, J. Castle, H. Y. Zhu, S. F. Kash, T. A. Drake, A. Sachs, and A. J. Lusis. An integrative genomics approach to infer causal associations between gene expression and disease. *Nature Genetics,* 37(7):710–717, 2005.

[49] M. Schmidt, A. Niculescu-Mizil, and K. Murphy. Learning graphical model structure using L1-regularization paths. In *Proceedings of the 22nd National Conference on Artificial intelligence—Volume 2,* pages 1278–1283. AAAI Press, 2007.

[50] P. Spirtes, C. Glymour, and R. Scheines. *Causation, prediction, and search.* The MIT Press, 2nd edition, 2000.

[51] Y. Tamada, S. Imoto, and S. Miyano. Parallel algorithm for learning optimal Bayesian network structure. *Journal of Machine Learning Research,* 12:2437–2459, 2011.

[52] Y. Tamada, S. Kim, H. Bannai, S. Imoto, K. Tashiro, S. Kuhara, and S. Miyano. Estimating gene networks from gene expression data by combining Bayesian network model with promoter element detection. *Bioinformatics,* 19(suppl 2):ii227–ii236, 2003.

[53] B. D. Valente, G. J. M. Rosa, G. de los Campos, D. Gianola, and M. A. Silva. Searching for recursive causal structures in multivariate quantitative genetics mixed models. *Genetics,* 185:633–644, 2010.

[54] T. Verma and J. Pearl. Equivalence and synthesis of causal models. In G. Shafer and J. Pearl, editors, *Proceedings of the Sixth Conference on Uncertainty in Artificial Intelligence (UAI90),* pages 220–227. Morgan Kaufmann Publishers, 1990.

[55] A. V. Werhli and D. Husmeier. Reconstructing gene regulatory networks with Bayesian networks by combining expression data with multiple sources of prior knowledge. *Statistical Applications in Genetics and Molecular Biology,* 6:15, 2007.

[56] C. J. Winrow, D. L. Williams, A. Kasarskis, J. Millstein, A. D. Laposky, H. S. Yang, K. Mrazek, L. Zhou, J. R. Owens, D. Radzicki, F. Preuss, E. E. Schadt, K. Shimomura, M. H. Vitaterna, C. Zhang, K. S. Koblan, J. J. Renger, and F. W. Turek. Uncovering the genetic landscape for multiple sleep-wake traits. *PLOS ONE,* 4(4):e5161, 2009.

[57] C. H. Yeang, T. Ideker, and T. Jaakkola. Physical network models. *Journal of Computational Biology,* 11(2–3):243–262, 2004.

[58] N. J. Yi, D. Shriner, S. Banerjee, T. Mehta, D. Pomp, and B. S. Yandell. An efficient Bayesian model selection approach for interacting quantitative trait loci models with many effects. *Genetics,* 176:1865–1877, 2007.

[59] N. J. Yi, B. S. Yandell, G. A. Churchill, D. B. Allison, E. J. Eisen, and D. Pomp. Bayesian model selection for genome-wide epistatic quantitative trait loci analysis. *Genetics,* 170(3):1333–1344, 2005.

[60] J. Zhu, P. Y. Lum, J. Lamb, D. Guhathakurta, S. W. Edwards, R. Thieringer, J. P. Berger, M. S. Wu, J. Thompson, A. B. Sachs, and E. E. Schadt. An integrative genomics approach to the reconstruction of gene networks in segregating populations. *Cytogenetic Genome Research,* 105(2–4):363–374, 2004.

[61] J. Zhu, M. C. Wiener, C. Zhang, A. Fridman, E. Minch, P. Y. Lum, J. R. Sachs, and E. E. Schadt. Increasing the power to detect causal associations by combining genotypic and expression data in segregating populations. *PLOS Computational Biology*, 3(4):e69, 2007.

[62] J. Zhu, B. Zhang, E. N. Smith, B. Drees, R. B. Brem, L. Kruglyak, R. E. Bumgarner, and E. E. Schadt. Integrating large-scale functional genomic data to dissect the complexity of yeast regulatory networks. *Nature Genetics*, 40(7):854–861, 2008.

Structural Equation Models for Studying Causal Phenotype Networks in Quantitative Genetics

GUILHERME J. M. ROSA AND BRUNO D. VALENTE

P henotypic traits may exert causal effects between them. For example, high yield in agricultural species may increase the liability to certain diseases, and, conversely, the incidence of a disease may affect yield negatively. Likewise, the transcriptome may be a function of the reproductive status or developmental stage in plants and animals, which may depend on other physiological variables as well. Knowledge of phenotype networks describing such interrelationships can be used to predict the behavior of complex systems, e.g., biological pathways underlying complex traits such as diseases, growth, and reproduction. This chapter reviews the application of Structural Equation Models (SEMs) and related techniques to study causal relationships among phenotypic traits in quantitative genetics. It also discusses how genetic factors can confound the search for causal associations and how pedigree and genomic information can be used to control for such confounding effects and to aid causal inference.

8.1 Introduction

Phenotypic traits may exert causal effects between them. For example, high yield in agricultural species may increase the liability to certain diseases and, conversely, the incidence of a disease may affect yield negatively. Likewise, the transcriptome may be a function of the reproductive status or

Probabilistic Graphical Models for Genetics, Genomics, and Postgenomics. First Edition. Christine Sinoquet & Raphaël Mourad (Eds). © Oxford University Press 2014. Published in 2014 by Oxford University Press.

developmental stage in plants and animals, which may depend on other physiological variables as well. Knowledge of phenotype networks describing such interrelationships allow predicting the behavior of complex systems, e.g., biological pathways underlying complex traits related to diseases, growth, and reproduction.

Structural Equation Models (SEMs) can be used to study recursive and simultaneous relationships among phenotypes in multivariate systems such as genetical genomics, systems biology, and multiple trait models in quantitative genetics. Hence, SEMs can produce an interpretation of relationships among traits which differ from that obtained with traditional multiple trait models, in which all relationships are represented by symmetric linear associations among random variables, such as covariances and correlations.

This chapter reviews the application of SEMs and related techniques for the study of multiple phenotypes. In addition, it discusses how DNA polymorphism and pedigree information can be utilized to aid causal inference, by exploiting the concept of Mendelian randomization, and by accounting for confounding genetic effects. The chapter is organized as follows. In Section 8.2, the classical linear mixed-effects models commonly used in quantitative genetics for studying multiple traits are reviewed. Next, in Section 8.3, the mixed models are extended to accommodate functional relationships between phenotypic traits using structural equation models. In Section 8.4, a general data-driven approach to search for phenotypic causal relationships is presented, and in Section 8.5 it is discussed how genomic information can be exploited to aid causal inferences, and how pedigree or molecular marker data can be used to mitigate confounding effects related to pleiotropic polygenic effects. Lastly, a final section with concluding remarks is provided.

8.2 Classical Linear Mixed-effects Models in Quantitative Genetics

Mixed models provide a flexible tool for the analysis of data where responses are clustered around some average effects with random departures, such that there is a natural dependence between observations in the same cluster, e.g., family members in genetics studies. In quantitative genetics, and especially in animal and plant breeding applications, mixed models are commonly used for estimation of environmental effects, genetic parameters, and variance components associated with phenotypic traits measured on related individuals. Their popularity in this area stands from their flexibility to handle complex pedigrees, unequal family sizes, overlapping generations, sex-limited traits, assortative mating, and natural or artificial selection [15, 29]. In this section, a brief review of mixed models is provided, and later some of their applications in animal (and plant) breeding are discussed.

A linear mixed-effects model is defined as [18, 25]:

$$y = X\theta + Zb + \varepsilon, \tag{8.1}$$

where y is an $(n \times 1)$ vector of responses (observations), θ is a $(p \times 1)$ vector of fixed effects, b is a vector of random effects, X and Z are known incidence matrices relating y to the vectors θ and b, respectively, and ε is a vector of residual terms. Generally, it is assumed that b and ε are independent from each other and normally distributed with zero-mean vectors and variance–covariance matrices B and Σ, respectively.

In animal and plant breeding, a central goal of mixed-model applications refers to the prediction of random effects, such as the genetic merit (or breeding values) of selection candidates. In linear (Gaussian) models as in [29], such predictions are given by the conditional expectation of b given the data, i.e., $E[b|y]$. Given the model specifications above, the joint distribution of y and b is multivariate normal, expressed as:

$$\begin{bmatrix} y \\ b \end{bmatrix} \sim N \left(\begin{bmatrix} X\theta \\ 0 \end{bmatrix}, \begin{bmatrix} V & ZB \\ BZ' & B \end{bmatrix} \right),$$

where $V = ZBZ' + \Sigma$.

From the properties of multivariate normal distributions, $E[b|y]$ is given by:

$$E[b|y] = E[b] + \text{Cov}[b, y'] \text{Var}^{-1}[y](y - E[y]),$$

such that in this case:

$$\begin{aligned} E[b|y] &= BZ'V^{-1}(y - X\theta) \\ &= BZ'(ZBZ' + \Sigma)^{-1}(y - X\theta). \end{aligned}$$

This expression, however, depends on the fixed effects θ, which also need to be inferred from the data. The fixed effects are then typically replaced by their estimates ($\hat{\theta}$), such that predictions are made based on the following expression:

$$\hat{b} = BZ'V^{-1}(y - X\hat{\theta}).$$

To estimate the fixed effects θ, all random effects in model (8.1) can be combined into a single vector $e = Zb + \varepsilon$, such that the following fixed-effects model is obtained: $y = X\theta + e$, where $\text{Var}(e) = V = ZBZ' + \Sigma$. Under these settings, the distribution of y is multivariate normal with mean vector $X\theta$ and covariance matrix V, i.e., $(y) \sim N(X\theta, V)$, and the maximum likelihood estimator of θ can be shown to be:

$$\hat{\theta} = (X'V^{-1}X)^{-1}X'V^{-1}y,$$

which is distributed as $\hat{\theta} \sim N\left(\theta, (X'V^{-1}X)^{-1}\right)$. If the incidence (or design) matrix X is not full column rank, a generalized inverse[1] of $X'V^{-1}X$ must be used to obtain a solution $\theta^0 = (X'V^{-1}X)^- X'V^{-1}y$ of the system, from which estimable functions $\beta = L\theta$ are estimated as $\hat{\beta} = L\theta^0$.

The solutions $\hat{\theta}$ and $\hat{b}$ discussed above require V^{-1}. An alternative approach, which allows simultaneous estimation of θ and b without the need of computing V^{-1} directly, refers to the mixed model equations (MMEs) [14]. MMEs can be derived by maximizing (for θ and b) the joint density of y and b, yielding:

$$\begin{bmatrix} X'\Sigma^{-1}X & X'\Sigma^{-1}Z \\ Z'\Sigma^{-1}X & Z'\Sigma^{-1}Z + B^{-1} \end{bmatrix} \begin{bmatrix} \hat{\theta} \\ \hat{b} \end{bmatrix} = \begin{bmatrix} X'\Sigma^{-1}y \\ Z'\Sigma^{-1}y \end{bmatrix}.$$

---

[1] The generalized inverse of matrix M is denoted as M^-.

Using the second set of equations of the MME,

$$Z'\Sigma^{-1}X\hat{\theta} + (Z'\Sigma^{-1}Z + B^{-1})\hat{b} = Z'\Sigma^{-1}y,$$

we obtain:

$$\hat{b} = (Z'\Sigma^{-1}Z + B^{-1})^{-1}Z'\Sigma^{-1}(y - X\hat{\theta}).$$

It can be shown that this expression is equivalent to $\hat{b} = BZ'(ZBZ' + \Sigma)^{-1}(y - X\hat{\theta})$ and, more importantly, that $\hat{b}$ is the best linear unbiased predictor (BLUP) of b. Putting this result into the first set of equations of the MME, $X'\Sigma^{-1}X\hat{\theta} + X'\Sigma^{-1}Z\hat{b} = X'\Sigma^{-1}y$, we have:

$$X'\Sigma^{-1}X\hat{\theta} + X'\Sigma^{-1}Z(Z'\Sigma^{-1}Z + B^{-1})^{-1}Z'\Sigma^{-1}(y - X\hat{\theta}) = X'\Sigma^{-1}y,$$

such that:

$$\hat{\theta} = \{X'[\Sigma^{-1} - \Sigma^{-1}Z(Z'\Sigma^{-1}Z + B^{-1})^{-1}Z'\Sigma^{-1}]X\}^{-1}$$
$$X'[\Sigma^{-1} - \Sigma^{-1}Z(Z'\Sigma^{-1}Z + B^{-1})^{-1}Z'\Sigma^{-1}]y.$$

Similarly, this expression is equivalent to $\hat{\theta} = (X'V^{-1}X)^{-1}X'V^{-1}y$, which is the best linear unbiased estimator (BLUE) of θ.

It is important to note that $\hat{\theta}$ and $\hat{b}$ require knowledge of the variance–covariance matrices B and Σ. As these matrices are rarely known, the practical approach is to replace B and Σ by some sort of point estimates $\hat{B}$ and $\hat{\Sigma}$ into the MME.

Among many methods that have been proposed to estimate variance components in mixed-effects models, the maximum likelihood based approaches are currently the most popular (see, for example, [13]), especially the restricted (or residual) maximum likelihood estimation [30]. Additional literature on maximum likelihood methodology for inference in linear mixed models can be found, for example, in [27, 32, 41, 43, 49]. Alternatively, inferences can be based on a Bayesian approach, for which a prior distribution $\mathbb{P}(\theta, b, B, \Sigma)$ for all unknowns in the model need to be defined. The prior distribution is then updated with the likelihood to produce the joint posterior distribution $\mathbb{P}(\theta, b, B, \Sigma | y)$, given by:

$$\mathbb{P}(\theta, b, B, \Sigma | y) \propto \mathbb{P}(y | \theta, b, B, \Sigma)\, \mathbb{P}(\theta, b, B, \Sigma).$$

Markov chain Monte Carlo (MCMC) methods such as the Gibbs sampling can be used then to calculate features of the posterior distribution, such as marginal posterior means and credibility intervals (or Bayesian confidence intervals), e.g., highest posterior density (HPD) intervals. Additional discussion on specific prior distributions often adopted for the analysis of linear mixed models is provided later in this chapter. A more detailed discussion on Bayesian analysis and MCMC implementation in mixed models can be found, for example, in [10] and [43].

In the context of animal breeding and quantitative genetics, a standard linear mixed model widely used for the analysis of quantitative traits is the so-called "animal model." In its simplest form, with one phenotypic trait and a single observation (including missing values) per subject, the animal model can be represented as model (8.1), with the following specifications. As before, the vectors y and ε denote the observations (phenotypic scores) and residual effects. In most applications of animal models, however, residuals are assumed independent across animals, such that the residual covariance structure can be expressed as $\Sigma = I\sigma_\varepsilon^2$, where I is an identity matrix of appropriate order, and σ_ε^2 is the residual variance.

In the case of animal models, environmental factors affecting the phenotypes y, such as herd, year, and season of birth, are generally assumed as fixed effects, and so they are represented by the vector θ. In addition, the vector of random effects b has dimensions $(q \times 1)$ and represents breeding values. The vector b may actually include breeding values not only for the n animals with known phenotypes but also for the remaining animals in the pedigree, in which case q will be bigger than n.

Here, the matrix B describes the covariances among the breeding values, and it follows from standard results for the covariances between relatives, i.e., genetically related individuals. It is seen that the additive genetic covariance between two relatives i and i' is given by $2w_{ii'}\sigma_a^2$, where $w_{ii'}$ is the coefficient of coancestry between individuals i and i', and σ_a^2 is the additive genetic variance in the base population [50]. Hence, under the animal model, $B = A\sigma_a^2$, where A is the "additive genetic (or numerator) relationship matrix," having elements given by $a_{ii'} = 2w_{ii'}$.

As in the animal model $B^{-1} = A^{-1}\sigma_a^{-2}$ and $\Sigma^{-1} = I\sigma_\varepsilon^{-2}$, the MME are then reduced to:

$$\begin{bmatrix} X'X & X'Z \\ Z'X & Z'Z + \lambda A^{-1} \end{bmatrix} \begin{bmatrix} \hat{\theta} \\ \hat{b} \end{bmatrix} = \begin{bmatrix} X'y \\ Z'y \end{bmatrix},$$

where $\lambda = \frac{\sigma_\varepsilon^2}{\sigma_a^2} = \frac{1-h^2}{h^2}$, such that:

$$\begin{bmatrix} \hat{\theta} \\ \hat{b} \end{bmatrix} = \begin{bmatrix} X'X & X'Z \\ Z'X & Z'Z + \lambda A^{-1} \end{bmatrix}^{-1} \begin{bmatrix} X'y \\ Z'y \end{bmatrix}.$$

The quantity h^2 above refers to the heritability of the trait, which is defined as the proportion of the total phenotypic variance that is due to additive genetic effects, i.e., $h^2 = \frac{\sigma_a^2}{\sigma_a^2+\sigma_\varepsilon^2}$. On a related note, it is shown that A^{-1} can be obtained directly from the pedigree information, without the need to set up A [16, 33], which is computationally very convenient.

In Bayesian analysis using the animal model, the joint prior distribution of θ, b, σ_a^2, and σ_ε^2 is often defined as:

$$\mathbb{P}(\theta, b, \sigma_a^2, \sigma_\varepsilon^2) = \mathbb{P}(\theta)\,\mathbb{P}(b|\sigma_a^2)\,\mathbb{P}(\sigma_a^2)\,\mathbb{P}(\sigma_\varepsilon^2),$$

where $\mathbb{P}(\theta)$ is a (bounded) uniform distribution, $\mathbb{P}(b|\sigma_a^2) \sim N(0, A\sigma_a^2)$ as previously defined, and $\mathbb{P}(\sigma_a^2)$ and $\mathbb{P}(\sigma_\varepsilon^2)$ are scaled inverse chi-squared distributions with specific degrees of freedom and scale parameters. The joint posterior distribution of the unknowns in the model is then given by:

$$\mathbb{P}(\theta, b, \sigma_a^2, \sigma_\varepsilon^2|y) \propto \mathbb{P}(y|\theta, b, \sigma_\varepsilon^2)\mathbb{P}(\theta)\,\mathbb{P}(b|\sigma_a^2)\,\mathbb{P}(\sigma_a^2)\,\mathbb{P}(\sigma_\varepsilon^2).$$

The fully conditional distributions necessary for implementing the Gibbs sampling are shown to be either multivariate normal, for the case of $\mathbb{P}(\theta|y, b, \sigma_a^2, \sigma_\varepsilon^2)$ and $\mathbb{P}(b|y, \theta, \sigma_a^2, \sigma_\varepsilon^2)$, or scaled inverse chi-squared distributions, for $\mathbb{P}(\sigma_a^2|y, \theta, b, \sigma_\varepsilon^2)$ and $\mathbb{P}(\sigma_\varepsilon^2|y, \theta, b, \sigma_a^2)$, as a consequence of conjugacy [10, 43].

The animal model discussed above can be extended also to multiple (correlated) traits [19, 39]. For instance, consider as an example the analysis of k traits, in which the model for each trait is expressed as:

$$y_j = X_j\theta_j + Z_jb_j + \varepsilon_j,$$

where y_j, X_j, θ_j, Z_j, b_j, and ε_j are defined as before, but here have an additional index to indicate the trait ($j = 1, 2, \ldots, k$).

For a joint analysis of the k traits, the single trait models can be combined as:

$$y = X\theta + Zb + \varepsilon, \tag{8.2}$$

where $y = [y_1'\ y_2'\ \ldots\ y_k']'$, $\theta = [\theta_1'\ \theta_2'\ \ldots\ \theta_k']'$, $b = [b_1'\ b_2'\ \ldots\ b_k']'$, and $\varepsilon = [\varepsilon_1'\ \varepsilon_2'\ \ldots\ \varepsilon_k']'$, and the incidence matrices in this case are:

$$X = \begin{bmatrix} X_1 & 0 & \cdots & 0 \\ 0 & X_2 & \cdots & 0 \\ \vdots & \vdots & & \vdots \\ 0 & 0 & \cdots & X_k \end{bmatrix} \quad \text{and } Z = \begin{bmatrix} Z_1 & 0 & \cdots & 0 \\ 0 & Z_2 & \cdots & 0 \\ \vdots & \vdots & & \vdots \\ 0 & 0 & \cdots & Z_k \end{bmatrix}.$$

It is assumed that $\mathrm{Var}\begin{bmatrix} b \\ \varepsilon \end{bmatrix} = \begin{bmatrix} G_0 \otimes A & 0 \\ 0 & E \otimes I \end{bmatrix}$, where

$$G_0 = \begin{bmatrix} \sigma_{a_1}^2 & \sigma_{a_1 a_2} & \cdots & \sigma_{a_1 a_k} \\ \sigma_{a_1 a_2} & \sigma_{a_2}^2 & \cdots & \sigma_{a_2 a_k} \\ \vdots & \vdots & & \vdots \\ \sigma_{a_1 a_k} & \sigma_{a_2 a_k} & \cdots & \sigma_{a_k}^2 \end{bmatrix} \quad \text{and } E = \begin{bmatrix} \sigma_{\varepsilon_1}^2 & \sigma_{\varepsilon_1 \varepsilon_2} & \cdots & \sigma_{\varepsilon_1 \varepsilon_k} \\ \sigma_{\varepsilon_1 \varepsilon_2} & \sigma_{\varepsilon_2}^2 & \cdots & \sigma_{\varepsilon_2 \varepsilon_k} \\ \vdots & \vdots & & \vdots \\ \sigma_{\varepsilon_1 \varepsilon_k} & \sigma_{\varepsilon_2 \varepsilon_k} & \cdots & \sigma_{\varepsilon_k}^2 \end{bmatrix}$$

are the genetic and residual variance–covariance matrices, respectively, A and I are the numerator relationship matrix and an identity matrix, and $\otimes$ represents the direct (Kronecker) product. The multiple trait animal model (MTAM) described above (model (8.2)) can be represented graphically as in Fig. 8.1 (page 202), which depicts an example with three phenotypic traits.

The MME for multiple trait analyses are of the same form as before, i.e.,

$$\begin{bmatrix} X'(E^{-1} \otimes I)X & X'(E^{-1} \otimes I)Z \\ Z'(E^{-1} \otimes I)X & Z'(E^{-1} \otimes I)Z + G_0^{-1} \otimes A^{-1} \end{bmatrix} \begin{bmatrix} \hat{\theta} \\ \hat{b} \end{bmatrix} = \begin{bmatrix} X'(E^{-1} \otimes I)y \\ Z'(E^{-1} \otimes I)y \end{bmatrix},$$

from which the BLUE and the BLUP of θ and b can be obtained, respectively. In the case of Bayesian inference, the MTAM is very similar to the single trait model discussed previously, except that instead of scaled inverse chi-squared priors for the genetic and residual variance components, here inverted Wishart distributions are used for the variance–covariance matrices G_0 and E.

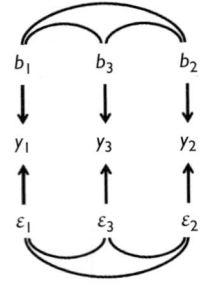

Fig 8.1 Graphical representation of a classical multiple trait animal model with three phenotypic traits (y_j, $j = 1, 2, 3$), in which b_j anc ε_j represent genetic and residual environmental effects affecting the traits. Arrows and arcs represent causal effects and unresolved linear associations, respectively.

There are many extensions and variations of the single- and multiple-trait animal models discussed above, for example, models with multiple random effects, as in the cases of repeated measurements of the same trait, or traits with maternal effects. For some discussion on such extensions and their applications refer, for example, to [17, 29, 35, 43].

In the animal model, the numerator relationship matrix A is derived from genetic theory, and it corresponds to the expected (given the pedigree) (co)variance structure among animals under an infinitesimal additive mode of gene action (e.g., [9, 50]). The animal model can be viewed as a special case of reproducing kernel Hilbert space regressions [5, 34], such that it can be extended also to other pedigree-based infinitesimal models (e.g., infinitesimal models for effects due to dominance or epistasis), or to models combining both molecular marker and pedigree information to arrive at some (co)variance structure (e.g., [8, 28]).

8.3 Mixed-effects Structural Equation Models

The MTAM discussed above (model (8.2)) can represent only symmetrical relationships (i.e., covariances and correlations) between phenotypic traits. Such phenotypic relationships are modeled with two components, which are described by the matrices G_0 and E, and refer respectively to genetic and environmental factors contributing simultaneously to pairs of traits (see Fig. 8.1). In this context, to accommodate functional relationships between traits, such as phenotypic causal effects, [43] proposed structural equation models (SEMs) within a mixed-effects framework, as discussed below.

Given a specific phenotypic causal structure (i.e., the information containing, for each trait, the subset of the remaining traits that directly affects it, which can be represented by a directed graph), an SEM embedded within an animal model context can be described as [12, 52]

$$y_i = \Lambda y_i + X_i \eta + d_i + \xi_i, \tag{8.3}$$

where y_i is a ($k \times 1$) vector of phenotypic records on subject i; Λ is a ($k \times k$) matrix with entries given by zeroes or structural coefficients according to the chosen causal structure; $X_i \eta$ represents the effects of exogenous covariates as linear regressions, in which the matrix X_i contains the covariates and η is a vector of fixed regression coefficients; d_i and ξ_i are ($k \times 1$) vectors of random additive genetic effects and model residuals, respectively, which are both associated with the ith subject. Furthermore, d_i and ξ_i are assumed to be distributed as $\begin{bmatrix} d_i \\ \xi_i \end{bmatrix} \sim N \left(\begin{bmatrix} 0 \\ 0 \end{bmatrix}, \begin{bmatrix} D_0 & 0 \\ 0 & \Psi_0 \end{bmatrix} \right)$,

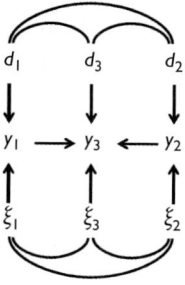

Fig 8.2 Graphical representation of a model with correlated genetic (d_j) and residual environmental (ξ_j) effects affecting three traits (y_j, $j = 1, 2, 3$), which also present some functional (causal) relationships among them. In the specific example depicted, traits 1 and 2 are parents of trait 3. Arrows and arcs represent causal effects and unresolved linear associations, respectively.

where D_0 and Ψ_0 are the additive genetic and residual covariance matrices, respectively. A graphical representation of a causal structure that could be applied to model (8.3) is provided in Fig. 8.2.

The model for n animals can be described as $y = (\Lambda \otimes I_n)y + X\eta + Zd + \xi$, with:

$$\begin{bmatrix} d \\ \xi \end{bmatrix} \sim N\left(\begin{bmatrix} 0 \\ 0 \end{bmatrix}, \begin{bmatrix} D_0 \otimes A & 0 \\ 0 & \Psi_0 \otimes I_n \end{bmatrix} \right),$$

where y, d, and ξ are, respectively, vectors of phenotypic records, additive genetic effects and model residuals sorted by trait and subject within trait, and X and Z are incidence matrices relating effects in η and d to y. This model may be rewritten as $\left[I_{kn} - (\Lambda \otimes I_n) \right] y = X\eta + Zd + \xi$, so that an equivalent reduced model can be obtained as in [12]:

$$y = \left[I_{kn} - (\Lambda \otimes I_n) \right]^{-1} X\eta + \left[I_{kn} - (\Lambda \otimes I_n) \right]^{-1} Zd + \left[I_{kn} - (\Lambda \otimes I_n) \right]^{-1} \xi.$$

The resulting sampling distribution of y, given all unknowns in the model, is normal with mean vector $\left[I_{kn} - (\Lambda \otimes I_n) \right]^{-1} (X\eta + Zd)$ and variance–covariance matrix $\left[I_{kn} - (\Lambda \otimes I_n) \right]^{-1} \Psi \left[I_{kn} - (\Lambda \otimes I_n) \right]^{'-1}$, where $\Psi = \Psi_0 \otimes I_n$.

By reducing the SEMs, the location and dispersion parameters are transformed into parameters of a standard MTAM [48, 52], as indicated below:

$$\begin{aligned} y_i &= (I_k - \Lambda)^{-1} X_i\eta + (I_k - \Lambda)^{-1} d_i + (I_k - \Lambda)^{-1} \xi_i \\ &= \mu_i^* + b_i^* + \varepsilon_i^*, \end{aligned}$$

where $\mu_i^* = (I_k - \Lambda)^{-1} X_i\eta$, $b_i^* = (I_k - \Lambda)^{-1} d_i$, and $\varepsilon_i^* = (I_k - \Lambda)^{-1} \xi_i$. In addition, the joint distribution of b_i^* and ε_i^* is

$$\begin{bmatrix} b_i^* \\ \varepsilon_i^* \end{bmatrix} \sim N\left(\begin{bmatrix} 0 \\ 0 \end{bmatrix}, \begin{bmatrix} G_0^* & 0 \\ 0 & R_0^* \end{bmatrix} \right),$$

with $G_0^* = (I_k - \Lambda)^{-1} D_0 (I_k - \Lambda)^{'-1}$ and $R_0^* = (I_k - \Lambda)^{-1} \Psi_0 (I_k - \Lambda)^{'-1}$.

Here, μ_i^*, b_i^*, ε_i^*, G_0^* and R_0^* are respectively the vectors of fixed effects, additive genetic effects, model residuals, and the genetic and residual covariance matrices of an MTAM.

However, an MTAM is just identified [52], such that changes in parametric values necessarily result in some change in the joint distribution of y. Conversely, SEMs carry extra parameters in Λ, resulting in an unidentifiable likelihood function. Nevertheless, it is possible to introduce constraints in SEMs to achieve parameter identifiability [52]. A constraint that is typically sufficient for acyclic models consists in coercing the residual covariance matrix Ψ_0 to be diagonal. After defining the causal structure and achieving parameter identifiability, one may apply standard statistical methods (e.g., [43]) to make inferences about model parameters. A Bayesian implementation of such models is described, for example, in [12] and [52].

SEMs have been used to study simultaneous and recursive relationships between phenotypes in various animal species, including goats [3], pigs [48], and cattle [4, 6, 7, 20, 21, 24, 51]. For a review and a discussion on such applications, the reader is referred to [36].

8.4 Data-driven Search for Phenotypic Causal Relationships

8.4.1 General Overview

As indicated by [12], [52], and [46], applications of mixed-effects SEMs in quantitative genetics are not as straightforward as fitting classic MTAM. Employing the former models implies coping with at least two additional challenges: (1) identifiability of model parameters, and (2) choice of causal structures.

As mentioned before, MTAMs are considered to be saturated models, which means that they have enough dispersion parameters to perfectly fit any joint distribution. One feature of a mixed-effects SEM is that it potentially presents every one of the dispersion parameters that also pertain to a MTAM, but more parameters are added in order to represent the magnitude of the causal association among phenotypes, which are extra sources of covariation. This feature results in model over-identification, such that depending on how loaded with parameters an SEM is, estimators for them may not be identifiable from the likelihood function. For that reason, model restrictions are necessary to fit an SEM. That could be attained in different ways, from parameter shrinkage using proper prior distributions in Bayesian analysis to assumptions of some conditional independences among the variables involved in the model [12, 48].

If there is any interest in the causal interpretation of the SEMs to be fitted, then the restriction applied must necessarily mirror prior causal knowledge/assumptions. As mentioned previously, the most common parameter restriction is considering the SEMs residual covariance matrix as diagonal, which is a sufficient restriction for fitting acyclic SEMs. Such restriction applied to the structure depicted in Fig. 8.2 is illustrated in Fig. 8.3. Fortunately, such restriction mirrors the causal assumptions adopted by some of the methods that tackle the second challenge in implementing SEMs in quantitative genetics: the search for causal structures. A summarized description of the problem and of the theoretical basis of the methods to perform this search is provided next.

As stated before, an SEM is presented conditionally on a causal structure, so that fitting a model that expresses causal associations among variables requires choosing a priori one of such structures. Such choice may be complicated since the space of structures is typically too large to allow for exhaustive comparison even when a few traits are studied. Furthermore, its magnitude grows

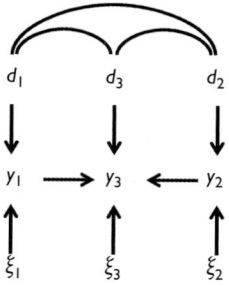

Fig 8.3 Graphical representation of a model with correlated genetic (d_j) and independent residual environmental (ξ_j) effects affecting three traits (y_j, $j = 1, 2, 3$), which also present some functional (causal) relationships among them. In the specific example depicted, traits 1 and 2 are parents of trait 3. Arrows and arcs represent causal effects and unresolved linear associations, respectively.

explosively as the number of traits under study increases [42]. In the applications of mixed-effects SEMs that followed [12], such choices were usually made based on prior knowledge/beliefs, which may be considered as a suboptimal exploration of the set of possible structures. Nevertheless, some algorithms have been developed to allow such exploration based on a multivariate sample and on a set of causal assumptions derived from a theory of causation [31, 44]. Here, we focus on the search for acyclic causal structures. Some terminology is defined next to support the presentation of involved concepts.

An acyclic recursive causal structure is represented by a directed acyclic graph (DAG), which consists of a set of variables connected by directed edges (arrows). If there is a direct causal relationship between a pair of variables, they are connected in the graph. A path in the causal structure is a sequence of connected variables, regardless of the direction of the arrows that connect them. Any path may allow flow of dependence between the pair of variables located in their extremities, unless such flow is blocked by a collider (variable with arrows pointing at them from opposite directions, like y_3 in $y_1 \rightarrow y_3 \leftarrow y_2$). In this structure, blocking the flow of dependence makes y_1 and y_2 independent. However, conditioning on a variable that is not in the extremities of the path switches its role in the aforementioned dependence flow: it blocks the flow of dependence if this variable is a non-collider (e.g., conditioning on y_3 in $y_1 \rightarrow y_3 \rightarrow y_2$, $y_1 \leftarrow y_3 \leftarrow y_2$ or $y_1 \leftarrow y_3 \rightarrow y_2$ makes y_1 and y_2 independent), or allows the flow of dependence if this variable is a collider. For any pair of variables y_1 and y_2 in a DAG, they are considered as d-separated conditionally on a subset S of variables if there are no paths that allow flows of dependence between y_1 and y_2 (i.e., no paths between y_1 and y_2 in a DAG such that all the colliders or their descendants are in S and no non-colliders are in S). On the basis of some assumptions (which will be discussed later), d-separations in the causal structure are mirrored as conditional independences in the joint distribution of the data. This feature can be explored to perform a data-driven search for a causal structure or a class of equivalent causal structures, which consists on causal structures that result in joint probability distributions with the same conditional independence relationships [31, 44].

To perform this search, it is necessary to guarantee the connection between joint distribution and a DAG. Given a set S of variables, such pair consisting of graph and joint distribution present Markov compatibility when the joint distribution may be factorized as $\mathbb{P}(S) = \prod_{y_j \in S} \mathbb{P}\left(y_j | parents(y_j)\right)$ without imposing additional constraints, where $parents(y_j)$ represents the set of parents of variable y_j according to the given causal structure. One criterion for this compatibility to hold is that every d-separation in the DAG must be mirrored by a conditional independence in $\mathbb{P}(S)$ [31].

Following Pearl [31], causal models such as SEMs with acyclic causal structures and independent residuals are referred to as Markovian models. The causal Markov condition states that Markovian causal models induce distributions that satisfy the Markov compatibility, which guarantees the connection between causal structures and joint distribution that allows for the data-driven search of the structure. Such condition implies assuming that there are no variables that causally affect two or more variables in the set of analyzed variables that are not already in the set, i.e., causal sufficiency assumption [44]. These concepts are necessary to guarantee that every d-separation in the causal structure is expected to be reflected as a stable conditional independence in the sampling distribution. The causal sufficiency assumption is generally deemed as a strong assumption, but this is equally assumed for SEMs fitted with independent residuals, which is a popular model restriction for SEM applications. Furthermore, in most of these applications known causal structures were assumed, which is not the case if search algorithms are used.

Other typical assumptions for the search algorithms are faithfulness (absence of unstable conditional independences due to improbable combination of parameter values), homogeneous causal structure throughout the population, and correct statistical decisions [44].

8.4.2 Search Algorithms

Based on the aforementioned theory, some algorithms search for causal structures that are compatible with the covariances obtained from a joint distribution by assuming that the conditional independences observed in a given sample are reflections from d-separations in the causal structure. Here we present the Inductive Causation (IC) algorithm, which consists in a set of queries concerning conditional independences, taking a correlation matrix as input and providing a partially oriented graph as output. Such graph represents a set of causal structures that are statistically equivalent, imposing the same set of conditional independences, which renders them undistinguishable from an observational sample. The edges that are left undirected by the algorithm may present one direction or the other in different structures within the class, such that no direction results in causal cycles or further unshielded colliders (substructures consisting of unlinked vertices with a common child, such as the trio $y_1 \rightarrow y_3 \leftarrow y_2$). The IC algorithm, when applied to a set S of variables (such as phenotypic traits), can be described as follows:

Step 1. For each pair of phenotypic traits y_j and $y_{j'}$ ($j \neq j' = 1, 2, \ldots, k$) in S, search for a set of traits $S_{jj'}$ such that y_j is independent of $y_{j'}$ given $S_{jj'}$. If y_j and $y_{j'}$ are dependent for every possible $S_{jj'}$, connect y_j and $y_{j'}$ with an undirected edge. This step returns an undirected graph U.

Step 2. For each pair of non-adjacent traits y_j and $y_{j'}$ with a common adjacent trait $y_{j''}$ in U (i.e., $y_j - y_{j''} - y_{j'}$), search for a set $S_{jj'}$ containing $y_{j''}$ such that y_j is independent of $y_{j'}$ conditional on $S_{jj'}$. If there is no such set, then add arrowheads pointing at $y_{j''}$ ($y_j \rightarrow y_{j''} \leftarrow y_{j'}$). Otherwise, continue.

Step 3. In the partially oriented graph returned by the previous step, orient as many undirected edges as possible in such a way that it does not result in new unshielded colliders or in cycles.

The network learning methods that construct classes of equivalent structures based on conditional independences observed on the joint distribution of variables are called *constraint-based algorithms*. The IC algorithm is one of such methods. However, other methods pertaining to the same group may be more computationally efficient and could be used instead. A second type of search algorithms uses an alternative approach, considering the task as a model selection problem and assigning scores for candidate structures. Because the search space is generally too large, these methods generally begin with a candidate structure and then proceed with a heuristic search, aiming to obtain a structure with maximum score. These are termed *score-based algorithms*. In this chapter, we focus on *constraint-based algorithms*, which are more related with learning structures accounting for its causal interpretation [23, 40].

8.5 Inferring Causal Structures in Genetics Applications

8.5.1 Genotypic information as Instrumental Variable

The IC algorithm and related methods discussed above are applicable in many different contexts to further our understanding regarding functional relationships and causal effects between variables. Especially in the context of genetics, such algorithms and methods can also benefit from the molecular biology knowledge that the flow of information (and as such, of causal effects) goes from genes to phenotypes. Moreover, as Thomas and Conti [45] pointed out, genetically randomized experimental populations that segregate naturally occurring allelic variants can provide a basis for the inference of networks of causal associations among variables such as genetic loci (or genes), physiological phenotypes, and disease states. In particular, the randomization of alleles that occurs during meiosis (generally referred to as Mendelian randomization) provides a setting that is akin to a randomized experimental design [36].

In this context, many authors have leveraged on such notion of Mendelian randomization to explore causal relationships among phenotypic traits. For example, [38] proposed an approach to infer the causal path involving a trio of variables: the expression of a particular gene, a genetic polymorphism on a specific locus, and a complex trait (e.g., a disease trait). Three possible structures were considered, which were referred to as causal, reactive, and independent models. A likelihood-based causal model selection was proposed, which uses conditional correlation measurements to determine which relationship among such a trio of variables is best supported by the data. Their procedure, however, is restricted to simple gene-phenotypes networks with three nodes, focusing on the identification of genes in the causal–reactive interval.

An extension of Schadt's approach [38] was proposed by Li et al. [26], who presented an SEM methodology to analyze multilocus, multitrait genetic data, and with different potential causal relationships among them. Their method comprises five steps, starting with a series of quantitative trait loci (QTL) genome scans run marginally for each individual phenotype, followed by conditional genome scans performed using one trait as a covariate in the analysis of another trait. The comparison between results from unconditioned and conditioned scans can give a first insight into the causal relationships among the phenotypes. A third step refers to the construction of an initial path model and its corresponding SEMs representation. After the path models are constructed, they are assessed in terms of goodness-of-fit by comparing the predicted and observed covariance matrices and by significance tests for individual path coefficients. Finally, an additional step is performed to refine the model, by proposing and assessing alternative models, which are generated by adding or removing edges in the initial model, or by reversing the causal direction of an edge.

Also using QTL information to orient edges connecting phenotypes, Chaibub Neto et al. [1] proposed a methodology comprising two main steps. First, an association network is constructed using either an undirected dependence graph [42] or a skeleton derived from the PC algorithm of Spirtes et al. [44]. Second, log-odds ratio (LOD) score tests are used to determine causal direction for every edge that connects a pair of phenotypes, conditional on QTL affecting the phenotypes. The authors assessed the performance of their methodology using simulations studies, showing that it can recover network edges and infer their causal direction correctly at a high rate.

The method proposed by Chaibub Neto et al. [1] depends on the availability of reliable information regarding QTL affecting the phenotypic traits of interest. However, as discussed by Chaibub Neto et al. [2], traditional QTL mapping approaches are based on single-trait analyses, in which the network structure among phenotypes is not taken into account. Such single-trait analyses may detect QTL that directly affect each phenotype, as well as QTL with indirect effects (relative to the set of traits analyzed), which directly affect phenotypes upstream of the specific phenotype being analyzed. Hence, traditional QTL mapping approaches that ignore the phenotype network result in poorly estimated genetic architecture of phenotypes, which may hamper correct inferences regarding causal relationships among phenotypes.

In view of this drawback of traditional QTL analyses and phenotype network reconstruction methods, Chaibub et al. [2] suggested a methodology that simultaneously infers a causal phenotype network and its associated genetic architecture. Their approach is based on jointly modeling phenotypes and QTL using homogeneous conditional Gaussian regression models and a graphical criterion for model equivalence. The concept of randomization of alleles during meiosis and the unidirectional relationship from genotype to phenotype are used to infer QTL with causal effects on phenotypes. Concomitantly, causal relationships among phenotypes are inferred exploring the QTL nodes, which might make it possible to distinguish among phenotype networks that would otherwise be distribution equivalent.

8.5.2 Accounting for Polygenic Confounding Effects

The phenotype network reconstruction approaches discussed in the previous section rely on information regarding QTL affecting the phenotypes, or on the availability of genetic marker information for joint inference regarding phenotype network and genetic architecture. The QTL are then used as parent nodes on putative networks, facilitating inferences on the remainder of the network, for example on the establishment of causal relationships among phenotypic traits.

However, even if genetic marker information is not available, SEM-related techniques can still be used in quantitative genetics for studying functional relationships between phenotypic traits. As discussed in Section 8.3, an adaptation of SEMs in mixed-effects model settings as typically applied to quantitative genetics has been described by [12], and since then applied by many researchers working with different species and phenotypic traits. In such applications, however, the causal structures were assumed known a priori (e.g., [7, 20]), or just a few putative structures selected using some prior knowledge were compared (e.g., [3, 4, 48, 51]). Nonetheless, it may be argued that even without information on QTL it may be possible to infer (at least partially) the causal relationships among phenotypic traits using data-driven algorithms that search for a causal structure, as presented in Section 8.4.

A difficulty in this respect, as pointed out by Valente et al. [46], is that in mixed-effects SEMs (as presented by Gianola and Sorensen [12]) associations between observed traits are explained not only by causal links between them, but also by genetic reasons, even if residuals are assumed independent. As a result, considering independent residuals is not sufficient to guarantee the connection between causal structure and the joint distribution of phenotypes. The unobserved correlated genetic effects considered in this context may confound the causal structure search if one tries to perform it based on the joint distribution of the phenotypes.

Nonetheless, as indicated by Valente et al. [46], genetic relationship information between individuals gives a mean of "controlling" for this confounder. Within this context, Valente et al. [46] proposed an approach to search for acyclic causal structures assuming that d-separations are reflected as conditional independences on the distribution of phenotypes after taking into account the additive genetic effects (i.e., the distribution of the phenotypes conditionally on the genetic effects). In this context, the connection between causal graph and distribution is reestablished. Given the model settings presented above, i.e., an SEM that accounts for additive genetic effects, the covariance matrix of the phenotypic vector y_i can be expressed as:

$$\mathrm{Var}(y_i) = (I_k - \Lambda)^{-1} G_0 (I_k - \Lambda)'^{-1} + (I_k - \Lambda)^{-1} \Psi_0 (I_k - \Lambda)'^{-1}.$$

Note that $(I_k - \Lambda)^{-1} G_0 (I_k - \Lambda)'^{-1}$ and $(I_k - \Lambda)^{-1} \Psi_0 (I_k - \Lambda)'^{-1}$ are the covariance matrices of additive genetic effects (G_0^*) and of residuals (R_0^*) obtained from a standard multiple trait mixed model that accounts for covariance between genetic effects and residuals from different traits, but not for causal relationships between phenotypes [12, 48]. The covariance matrix of y_i can be then rewritten as $\mathrm{Var}(y_i) = G_0^* + R_0^*$, and the covariance matrix between traits conditionally on the additive genetic effects can be represented as $\mathrm{Var}(y_i|u_i) = (I_k - \Lambda)^{-1} \Psi_0 (I_k - \Lambda)'^{-1} = R_0^*$. Therefore, estimates of R_0^* can be used to select a causal structure among phenotypes.

In Valente et al. [46], the (co)variance matrix R_0^* is inferred using Bayesian MCMC methods, in which samples are drawn from the posterior distribution of R_0^*. These samples are used then to obtain measures of uncertainty about this matrix, while accounting for uncertainty of all other parameters included in the reduced MTAM. In summary, the overall statistical approach proposed by [46] consists of three stages:

1. A Bayesian MTAM is fitted, and posterior samples of R_0^* are obtained.

2. The IC algorithm is applied to the posterior samples of R_0^* to make the required statistical decisions. Specifically, for each query about the statistical independence between variables y_j and $y_{j'}$ ($j \neq j' = 1, 2, \ldots, k$) given a set of variables $S_{jj'}$ and, implicitly, the genetic effects:

   a) Obtain the posterior distribution of residual partial correlation $\rho_{y_j, y_{j'}|S_{jj'}}$. These partial correlations are functions of R_0^*. Therefore their posterior distribution can be obtained by computing the correlation at each sample drawn from the posterior distribution of R_0^*.

   b) Compute the 95% HPD interval for the posterior distribution of $\rho_{y_j, y_{j'}|S_{jj'}}$.

   c) If the HPD interval contains 0, declare $\rho_{y_j, y_{j'}|S_{jj'}}$ as null. Otherwise, declare y_j and $y_{j'}$ as conditionally dependent.

3. Lastly, an SEM using the selected causal structure (or one member within the class of observationally equivalent structures retrieved by the IC algorithm) is fitted, as in [12], such that causal relationships (i.e., recursive effects) can be estimated.

Valente et al. [46] demonstrated that their method is able to retrieve the expected partially oriented graph among phenotypic traits by applying it to simulated data with different causal structures and sample sizes. That was followed by a first application of such method with real data,

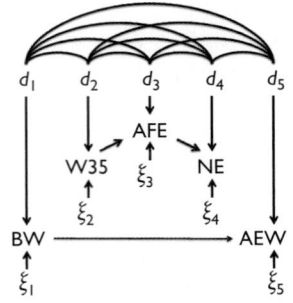

Fig 8.4 Graph selected for the example of analysis with European quail data, which was obtained by combining prior temporal information with the output of the IC algorithm with statistical decisions based on highest posterior intervals of 95% [46]. BW = birth weight, W35 = weight at 35 days, AFE = age at first egg, AEW = average egg weight from 77 to 110 days, and NE = number of eggs produced from 77 to 110 days.

in which Valente et al. [47] searched for causal networks involving five traits in European quail: birth weight, weight at 35 days of age, age at first egg, average egg weight from 77 to 110 days of age, and number of eggs laid in the same period. The data included phenotypes for all five traits recorded for 854 females and a pedigree file with a total of 10 680 birds. The posterior distributions of the partial correlations obtained are not very sharp, which makes the algorithm more prone to declare non-null partial correlations as null if decisions are based on 0.95 HPD intervals such as applied by [46]. Following Shipley [42], different HPD interval contents have been used for the statistical decisions, namely 0.7, 0.75, 0.8, 0.85, 0.9, 0.95 probabilities. Different results were obtained by changing the HPD interval, but the structures returned were completely undirected. In this application, prior knowledge regarding time sequence of the expression of each trait was applied to orient edges. After comparison of different structures and the just-identified MTAM, the model selected was a mixed-effects SEM with causal structure as depicted in Fig. 8.4.

8.6 Concluding Remarks

Although an SEM may be interpreted as a causal model, and therefore it may be used to express causal relationships among phenotypic traits, the causal structure of a fitted SEM may not reflect correctly the actual causal relationship among the modeled traits. Additionally, by studying observational data, inferring the causal structure is a much harder task than just describing the joint distribution of data by fitting a multivariate model. In this context, applying the IC algorithm and related techniques may be considered as a causal structure inference only if one is willing to accept the causal assumptions. If the assumptions are deemed as too strong, such algorithms could still be applied to explore the spaces of causal structures for SEMs constructed with diagonal residual covariance matrices.

Nonetheless, the latter applications may still produce interesting and useful results. Some causal learning does not require all of the aforementioned assumptions. Take as an example the structure depicted in Fig. 8.4 obtained by Valente et al. [47]. Even if one is not willing to assume causal sufficiency, the causal influence of age at first egg over number of eggs produced could be claimed based on the causal Markov condition and by acknowledging temporal sequence information regarding weight at 35 days, age at first egg, and number of eggs produced. Additionally, applications of such methods are useful as generator of causality hypotheses for subsequent research and investigation. Such hypotheses can then be supported or dismissed using additional data collected in other studies, or they might be tested experimentally through controlled interventions. Nevertheless, in a number of contexts, randomized experiments are un-feasible due to logistic or ethical constrains, such that observational data are the only available information regarding the system under study. In this context, SEMs and causal search tools such as the IC algorithm are handy.

Specifically in genetics and genomics studies, causal inference is aided by the concept of Mendelian randomization [45], in which allelic variants are randomized to zygotes during meiosis and eventually passed on from parents to offspring, analogously to a randomized experimental design. Applying SEM-related methodologies to QTL analysis and gene mapping with multiple traits not only improves inference regarding causal relationships among phenotypes by breaking statistical equivalence between different causal structures, but it also enhances detection power and precision of estimates, with the additional advantage of a distinction between direct and indirect genetic effects of QTL on each trait [2]. Furthermore, in genetics studies a putative causative mutation can be ultimately tested using, for example, gene knockout or knockdown technologies.

In addition to DNA polymorphism information and knowledge about genes or QTL that can be used as parent nodes in phenotype network reconstruction, the joint analysis of multilayer large-scale omics data such as the transcriptome, metabolome, and proteome can certainly provide added information and enhance the ability to infer causal phenotype relationships, although it also brings another level of statistical, computational, and data mining challenge [22]. Moreover, structural and functional data such as gene sequence, gene localization, transcription binding sites, gene ontology (GO), and metabolic pathway among others can also be used to verify and test putative gene and phenotype networks [22]. Such data can be used also as a priori information to aid network inference, the same way it has already been used in other omics applications such as microarray data [37].

SEMs have also been applied in contexts where pedigree information is available but not QTL or any genomic information [12]. As mentioned before, such a modeling approach allows a different interpretation of relationships among traits relative to standard multiple trait models traditionally used in animal breeding. In the classic models, all relationships are considered as symmetric linear associations among traits. In most applications of mixed-effects SEMs, prior belief was used as basis for specifying a causal structure, or for choosing a few structures that were compared using traditional model selection techniques. Alternatively, a broader data-driven search for recursive causal structures in the context of mixed models and quantitative genetics can be performed [46].

For the purpose of controlling for the genetic effects, which is necessary to search for acyclic causal structures in this context, [46, 47] applied classical infinitesimal additive genetic models involving a relationship matrix A constructed from pedigree information. The same could be attained more efficiently by using high-density molecular marker data (e.g., SNP genotypes) when such information is available. In that scenario, genetic merit prediction approaches can be employed such as Bayesian regression techniques [11] or kernel methods [5].

In conclusion, SEM applications provide additional insights and may present richer expression of relationships when compared with standard models for multiple traits. It allows specific characterization of pleiotropic and heterogeneous genetic effects of multiple loci on multiple traits, as well as causal relationships among phenotypes, which can be used to predict behavior of complex systems, e.g., biological pathways underlying disease traits. More specifically for livestock applications, knowledge regarding phenotype networks in the genetic analysis of quantitative traits may improve the prediction of external interventions, which may lead to more efficient breeding programs and aid in decisions involving farm management and veterinary practices.

REFERENCES

[1] E. Chaibub Neto, T.C. Ferrara, A.D. Attie, and B.S. Yandell. Inferring causal phenotype networks from segregating populations. *Genetics*, 179:1089–1100, 2008.

[2] E. Chaibub Neto, M.P. Keller, A.D. Attie, and B.S. Yandell. Causal graphical models in systems genetics: a unified framework for joint inference of causal network and genetic architecture for correlated phenotypes. *Annals of Applied Statistics*, 4:320–339, 2010.

[3] G. de los Campos, D. Gianola, P. Boettcher, and P. Moroni. A structural equation model for describing relationships between somatic cell score and milk yield in dairy goats. *Journal of Animal Science*, 84:2934–2941, 2006.

[4] G. de los Campos, D. Gianola, and B. Heringstad. A structural equation model for describing relationships between somatic cell score and milk yield in first-lactation dairy cows. *Journal of Dairy Science*, 89:4445–4455, 2006.

[5] G. de los Campos, D. Gianola, and G.J.M. Rosa. Reproducing kernel Hilbert spaces regression: a general framework for genetic evaluation. *Journal of Animal Science*, 87:1883–1887, 2009.

[6] E.L. de Maturana, G. de los Campos, X.-L. Wu, D. Gianola, K.A. Weigel, and G.J.M. Rosa. Modeling relationships between calving traits: a comparison between standard and recursive mixed models. *Genetics Selection Evolution*, 42:1, 2010.

[7] E.L. de Maturana, X.-L. Wu, D. Gianola, K.A. Weigel, and G.J.M. Rosa. Exploring biological relationships between calving traits in primiparous cattle with a bayesian recursive model. *Genetics*, 181:277–287, 2009.

[8] R.L. Fernando and M. Grossman. Marker-assisted selection using best linear unbiased prediction. *Genetics Selection Evolution*, 21:467–477, 1989.

[9] R.A. Fisher. The correlation between relatives on the supposition of Mendelian inheritance. *Transactions of the Royal Society of Edinburgh*, 52:399–433, 1918.

[10] A. Gelman, J.B. Carlin, H.S. Stern, and D.B. Rubin. *Bayesian Data Analysis*. Chapman & Hall, 2nd edition, 2004.

[11] D. Gianola, G. de los Campos, W.G. Hill, E. Manfredi, and R. Fernando. Additive genetic variability and the Bayesian alphabet. *Genetics*, 183:347–363, 2009.

[12] D. Gianola and D. Sorensen. Quantitative genetic models for describing simultaneous and recursive relationships between phenotypes. *Genetics*, 167:1407–1424, 2004.

[13] D.A. Harville. Maximum likelihood approaches to variance component estimation and to related problems. *Journal of the American Statistical Association*, 72:320–338, 1977.

[14] C.R. Henderson. Estimation of genetic parameters. *Annals of Mathematical Statistics*, 21:309, 1950.

[15] C.R. Henderson. Best linear unbiased estimation and prediction under a selection model. *Biometrics*, 31:423–447, 1975.

[16] C.R. Henderson. A simple method for computing the inverse of a numerator relationship matrix used in prediction of breeding values. *Biometrics*, 32:69–83, 1976.

[17] C.R. Henderson. *Applications of Linear Models in Animal Breeding*. University of Guelph, 1984.

[18] C.R. Henderson, O. Kempthorne, S.R. Searle, and C.N. Von Krosigk. Estimation of environmental and genetic trends from records subject to culling. *Biometrics*, 15:192–218, 1959.

[19] C.R. Henderson and R.L. Quaas. Multiple trait evaluation using relatives' records. *Journal of Animal Science*, 43:1188–1197, 1976.

[20] B. Heringstad, X.-L. Wu, and D. Gianola. Inferring relationships between health and fertility in Norwegian red cows using recursive models. *Journal of Dairy Science*, 92:1778–1784, 2009.

[21] J. Jamrozik, J. Bohmanova, and L.R. Schaeffer. Relationships between milk yield and somatic cell score in Canadian Holsteins from simultaneous and recursive random regression models. *Journal of Dairy Science*, 93:1216–1233, 2010.

[22] R.C. Jansen, B.M. Tesson, J. Fu, Y. Yang, and L.M. McIntyre. Defining gene and qtl networks. *Current Opinion in Plant Biology*, 12:241–246, 2009.

[23] D. Koller, N. Friedman *Probabilistic Graphical Models: Principles and Techniques*. The MIT Press, Adaptive Computation and Machine Learning, 1st edition, 2009.

[24] S. König, X.-L. Wu, D. Gianola, B. Heringstad, and H. Simianer. Exploration of relationships between claw disorders and milk yield in Holstein cows via recursive linear and threshold models. *Journal of Dairy Science*, 91:395–406, 2008.

[25] N.M. Laird and J.H. Ware. Random effects models for longitudinal data. *Biometrics*, 38:963–974, 1982.

[26] R. Li, S.W. Tsaih, K. Shockley, I.M. Stylianou, J. Wergedal, B. Paigen, and G.A. Churchill. Structural model analysis of multiple quantitative traits. *PLOS Genetics*, 2:e114, 2006.

[27] R. C. Littell, G. A. Miliken, W. W. Stroup, and R. Wolfinger. *SAS System for Mixed Models*. SAS Institute inc., 2nd edition, 2006.

[28] I. Misztal, A. Legarra, and I. Aguilar. Computing procedures for genetic evaluation including phenotypic, full pedigree, and genomic information. *Journal of Dairy Science*, 92:4648–4655, 2009.

[29] R. Mrode. *Linear Models for the Prediction of Animal Breeding Values*. CAB Int., 2nd edition, 2005.

[30] H.D. Patterson and R. Thompson. Recovery of inter-block information when block sizes are unequal. *Biometrika*, 58:545–554, 1971.

[31] J. Pearl. *Causality: Models, Reasoning and Inference*. Cambridge University Press, 2nd edition, 2009.

[32] J.C. Pinheiro and D.M. Bates. *Mixed-effects Models in S and S-Plus*. Springer, 2000.

[33] R.L. Quaas. Computing the diagonal elements of a large numerator relationship matrix. *Biometrics*, 32:949–953, 1976.

[34] G.K. Robinson. That blup is a good thing: the estimation of random effects. *Statistical Science*, 6:15–32, 1991.

[35] G.J.M. Rosa. Foundations of Animal Breeding. In R. A. Meyers, editor, *Encyclopedia of Sustainability Science and Technology*, pages 58–78. Springer, 2012.

[36] G.J.M. Rosa, B.D. Valente, G. de los Campos, S.-L. Wu, D. Gianola, and M.A. Silva. Inferring causal phenotype networks using structural equation models. *Genetics Selection Evolution*, 43:6, 2011.

[37] G.J.M. Rosa and A.I. Vazquez. Integrating biological information into the statistical analysis and design of microarray experiments. *Animal*, 4:165–172, 2010.

[38] E.E. Schadt, J. Lamb, X. Yang, J. Zhu, S. Edwards, D. Guhathakurta, S.K. Sieberts, S. Monks, M. Reitman, C. Zhang, P.Y. Lum, A. Leonardson, R. Thieringer, J.M. Metzger, L. Yang, J. Castle, H. Zhu, S.F. Kash, T.A. Drake, A. Sachs, and A.J. Lusis. An integrative genomics approach to infer causal associations between gene expression and disease. *Nature Genetics*, 37:710–717, 2005.

[39] L.R. Schaeffer. Sire and cow evaluation under multiple trait models. *Journal of Dairy Science*, 67:1567–1580, 1984.

[40] M. Scutari. Learning Bayesian networks with the bnlearn R package. *Journal of Statistical Software*, 35:3, 2010.

[41] S.R. Searle, G. Casella, and C.E. McCulloch. *Variance Components*. John Wiley & Sons, 1992.

[42] B. Shipley. *Cause and Correlation in Biology*. Cambridge University Press, 2002.

[43] D. Sorensen and D. Gianola. *Likelihood, Bayesian, and MCMC Methods in Quantitative Genetics*. Springer, 2002.

[44] P. Spirtes, C. Glymour, and R. Scheines. *Causation, Prediction and Search*. The MIT Press, 2nd edition, 2000.

[45] D.C. Thomas and D.V. Conti. Commentary: the concept of 'Mendelian randomization.' *International Journal of Epidemiology*, 33:21–25, 2004.

[46] B.D. Valente, G.J.M. Rosa, G. de los Campos, D. Gianola, and M.A. Silva. Searching for recursive causal structures in multivariate quantitative genetics mixed models. *Genetics*, 185:633–644, 2010.

[47] B.D. Valente, G.J.M. Rosa, R.B. Teixeira, and R.A. Torres. Searching for phenotypic causal networks involving complex traits: an application to European quails. *Genetics Selection Evolution*, 43:37, 2011.

[48] L. Varona, D. Sorensen, and R. Thompson. Analysis of litter size and average litter weight in pigs using recursive model. *Genetics*, 177:1791–1799, 2007.

[49] G. Verbeke and G. Molenberghs, editors. *Linear Mixed Models in Practice: a SAS-Oriented Approach*, volume 126. Lecture Notes in Statistics, Springer-Verlag, 1997.

[50] S. Wright. Systems of mating. i. the biometric relations between parents and offspring. *Genetics*, 6:111–123, 1921.

[51] X.-L. Wu, B. Heringstad, Y.M. Chang, G. de los Campos, and D. Gianola. Inferring relationships between somatic cell score and milk yield using simultaneous and recursive models. *Journal of Dairy Science*, 90:3508–3521, 2007.

[52] X.-L. Wu, B. Heringstad, and D. Gianola. Bayesian structural equation models for inferring relationships between phenotypes: a review of methodology, identifiability, and applications. *Journal of Animal Breeding and Genetics*, 127:3–15, 2010.

PART IV

Genetic Association Studies

Modeling Linkage Disequilibrium and Performing Association Studies through Probabilistic Graphical Models: a Visiting Tour of Recent Advances

CHRISTINE SINOQUET AND RAPHAËL MOURAD

This chapter offers an in-depth review of recent developments based on probabilistic graphical models (PGMs) and dedicated to two major concerns: the fundamental task of modeling dependences within genetic data, that is linkage disequilibrium (LD), and the downstream application to genome-wide association studies (GWASs). Throughout the whole chapter, the selected examples illustrate the use of Bayesian networks, as well as that of Markov random fields, including conditional and hidden Markov random fields. First, the chapter surveys PGM-based approaches dedicated to LD modeling. The next section is devoted to PGM-based GWASs and mainly focuses on multilocus approaches, where PGMs allow us to fully benefit from LD. This section also provides an illustration for the acknowledgment of confounding factors in GWASs. The next section is dedicated to the detection of epistastic relationships at the genome

Probabilistic Graphical Models for Genetics, Genomics, and Postgenomics. First Edition. Christine Sinoquet & Raphaël Mourad (Eds). © Oxford University Press 2014. Published in 2014 by Oxford University Press.

scale. A recapitulation and a discussion end the chapter. Finally, directions for future works are outlined.

9.1 Introduction

This chapter establishes a survey on the use of probabilistic graphical models (PGMs) designed for genetic purposes. PGMs have been widely used in bioinformatics, especially in fields such as DNA sequence analysis [14], functional genomics [49], and protein–protein interaction [31]. In the domain of genetics, gene expression analysis [13], linkage analysis [18], and molecular phylogenetics [23] have been conducive to the emergence of PGM-based methods. More recently, some successful approaches have been proposed to decipher the genetic factors of complex diseases. Intrinsically, to meet this purpose, these approaches often incorporate a modeling phase of the genetic data. In this context, some researchers have devoted efforts to model the complex structure of genetic data as an aim in itself. This chapter focuses on the PGM-based modeling of genetic data and on a major downstream application, genome-wide association studies.

Linkage disequilibrium (LD) is defined as the non-random association of alleles at different loci, in a population. LD mainly results from the transmission of entire stretches of DNA from parents to descendents. Recombination and mutation processes are two major factors that blur this transmission scheme. LD detection and the assessment of its intensity, extent, and distribution are fundamental steps in many genetic studies. For instance, in population genetics, LD patterns have been widely used to study the evolutionary and demographic processes involved in various animal and plant populations, such as admixtures, migration, and natural selection.

A hot topic of current genomics research is disease-gene association. Association studies consist in identifying which DNA variations are highly associated with a specific disease (i.e., the phenotype). In particular, single nucleotide polymorphisms (SNPs), which are a popular class of genetic markers, are widely used in disease-gene association studies. The rationale which grounds the detection of a genotype–phenotype association is the following: genetic variants that affect a phenotype of interest arose sometime in the past on a unique stretch of the genome; due to LD, this segment was then transmitted to subsequent generations together with flanking variants; in the case when the phenotype is a disease status (affected/unaffected), it is expected that the genotypes of the affected (cases) and unaffected (controls) be different around the causative mutation(s) because cases are carriers of the ancestral disease-bearing segment. As chromosomes from unrelated individuals have undergone more recombination events than can be found in any pedigree of realistic size, association studies provide a more accurate localization of disease susceptibility loci than allowed by linkage studies, which deal with related individuals (in pedigrees).

The advent of high-throughput genotyping technologies has brought the expectation of identifying genetic variations that underlie common diseases such as hypertension, cardiovascular diseases, diabetes, Alzheimer's disease, and various types of cancers [6]. Two main strategies of association studies have been developed: hypothesis-driven [27] and non-hypothesis-driven [10]. In this review chapter, we only focus on non-hypothesis-driven studies. Therein, no particular region (fine mapping study), neither gene nor set of genes (candidate gene study), is specifically suspected to be associated with the disease. Therefore, non-hypothesis-driven studies pose a formidable computational challenge, as they address the genome scale, by definition. Genome-wide association studies (GWASs) generally use brute-force methods to mine the whole genome in quest of putative associations: for example, testing each SNP in turn against the disease (single-SNP approach), based on the chi-square test or the Cochran–Armitage test [34].

Association studies may be subject to biases: population structure, family structure, and cryptic relatedness are well-known confounding factors that cause spurious associations. Addressing the presence of confounding factors at the genome scale is another challenge that has to be faced.

An additional challenge is the detection of epistasis at the genome scale. Epistasis is defined as the interaction between two or more genes, to affect a phenotype of interest: the impact on the phenotype cannot be predicted by merely combining the individual effects of the SNPs.

PGMs offer an attractive paradigm to account for dependences within high-dimensional data, in an uncertain framework. A probabilistic graphical model consists of two parts: a Markov graph that represents conditional independences between the variables observed for samples, and a collection of probability distributions that define the so-called parameters. Each node is associated with a probability distribution: the probability to observe a given value for the variable represented by the node is provided for each peculiar set of values that can be taken by the node's parent variables. In this chapter, we will keep to discrete or discretized variables. Due to conditional independence, the joint distribution of the variables breaks down into a product of simpler, lower-dimensional components.

Thanks to their well-understood mathematical formulation and their flexibility, PGMs are often the model of choice in various domains such as multimedia, astronomy, text and web mining, and bioinformatics applications. This framework allows us to perform computations on complex models in a tractable way. It has to be noted that the graphical models are not always made explicit.

In this review chapter, we provide a panorama of recent developments based on PGMs and dedicated to two major concerns: the fundamental task of LD modeling and the downstream application of GWASs. The rest of the chapter is organized as follows: PGM-based approaches dedicated to LD modeling are first presented in Section 9.2; then the following section is dedicated to PGM-based GWASs and mainly focuses on *multilocus* approaches, with which PGMs allow us to fully benefit from LD. This section also illustrates the incorporation of confounding factors. In Section 9.4, we review some PGM-based methods dedicated to epistasis detection at the genome scale. In the next section, we recapitulate and discuss the main advantages and drawbacks of the methods depicted. Finally, we indicate directions for future work.

9.2 Modeling Linkage Disequilibrium

Genetic data are remarkable in that they show high correlations. Such correlations define a dependence structure, called the linkage disequilibrium (LD). The pioneering works of [12, 17] described a block-based organization of the human genome, in which the blocks represent low haplotype diversity. The complex LD structure was then confirmed by the Hapmap project [50]. This project performed an extensive analysis of DNA variations across four reference populations, relying on Single Nucleotide Polymorphisms (SNPs). LD varies along the genome, showing regions of highly correlated SNPs that alternate with recombination hotspots. The latter are short regions characterized by very low correlations. In short, LD reflects the blurring of the ancestral genome, mainly, but not only, through recombination and mutation events.

However, this block-based organization of the genome is not so neat. First, within a population, each observed haplotype is a mosaic of a limited number of shared short haplotypes.

Nonetheless, the frontiers in the mosaic are not common to all subjects in the population: LD is fuzzy. Second, dependences between blocks are known to exist: various ranges of LD have been described [35, 41]. Typically, we will oppose the models able to handle long range LD—or cluster-based models—to the block-based models.

In most genome-wide association studies, data are obtained through the federation of efforts of several laboratories. Thus, it may happen that not all loci are genotyped over the whole population of individuals collected. Therefore, data imputation is often required. For this purpose, selecting a subset of SNPs that is sufficiently informative is the first step to implement the prediction of non-genotyped loci. The identification of such *tagging* SNPs is intrinsically linked to the availability of a good model for LD. Besides selection of tagging SNPs, such applications as evolution analysis, haplotype inference and gene mapping exploit an LD model. As will be explained later, ideally, association studies—including genome-wide association studies—should also exploit an LD model. Thus, linkage disequilibrium modeling is all the more a fundamental task as the quality of such aforecited downstream applications demands a faithful representation of LD.

However, LD modeling is a complex task. On the one hand, genetic data are known to exhibit both strong and weak dependences, together with dependences between close SNPs as well as multilocus dependences, the latter situation corresponding to LD varying from short-range to long-range. A consequence of this is the fuzzy nature of LD: the frontiers between the haplotype blocks are not accurately defined (see Fig. 9.1). On the other hand, efficient model learning is required to represent LD at genome scale.

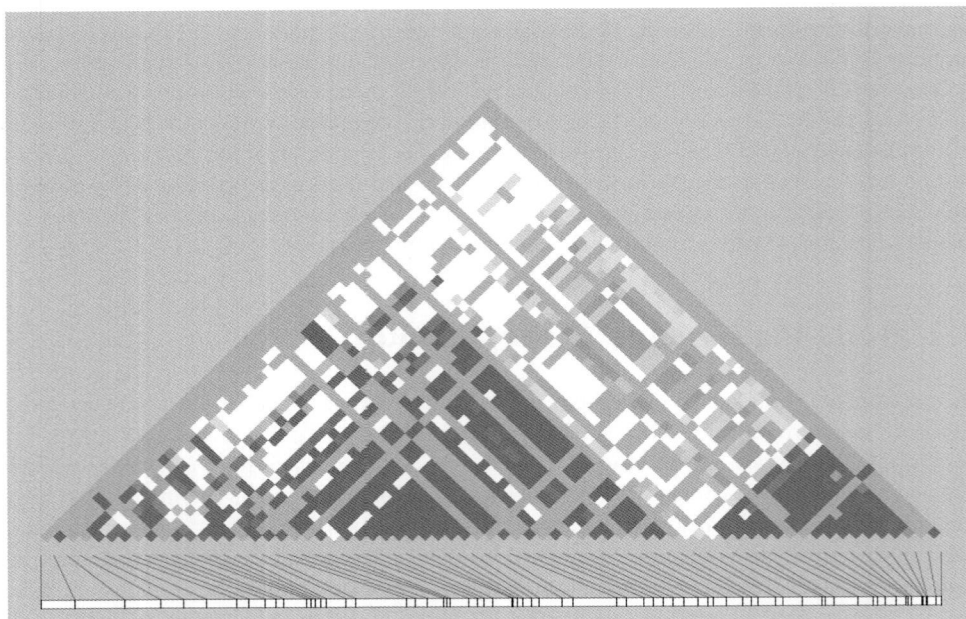

Fig 9.1 Linkage disequilibrium (LD) plot of a 500 kb SNP sequence. Human genome, chromosome 1, region [10 000 kb - 10 500 kb]. LD is revealed through the matrix of pairwise dependences between genetic markers. For a pair of SNPs, the color shade is all the darker as the correlation between the two SNPs is high. Reproduced in color in the color plate section.

Probabilistic graphical models offer an appealing framework to model complex dependences such as those defined by LD. Such specific cases of PGMs as k-order Markov models do not offer the flexibility required. Therefore, more sophisticated PGMs are preferred. We will present here the main approaches devoted to PGM-based modeling of linkage disequilibrium. Markov random fields (MRFs) and Bayesian networks (BNs) represent two prevailing categories within probabilistic graphical models. Both categories have been investigated to model linkage disequilibrium. We first provide a general panorama giving the main characteristics of the proposals. Then, a more detailed description of the models is given. The merits and caveats of these models are discussed on the fly, focusing on various criteria including presence of latent variables, block-based *versus* cluster-based modeling, and scalability. A final subsection recapitulates the approaches.

9.2.1 General Panorama

Thomas and Camp were the first to estimate discrete probabilistic graphical models to compute the joint distribution of allele frequencies [54]. This first approach relies on the general class of *decomposable* Markov random fields (DMRFs). As no operational characterization of the decomposability property was available, the mandatory try-and-test strategy seriously impeded the scalability. To circumvent this problem, Thomas's further works relied on navigating a restricted class of MRFs, known to be DMRFs: this restricted class is based on Interval Graphs (IG) [52, 53]. It is straightforward to design operations guaranteeing that the neighbor G' of the incumbent IG graph G be also an IG. However, this efficiency comes at a cost—the limitation of DMRFs to IG-based DMRFs. Therefore, recently, Abel and Thomas focused again on the class of DMRFs [1]. This time, scalability is obtained thanks to properties coming from graph theory.

In a decomposable Markov random field, the components of the factorized joint distribution of variables are cliques, whereas these components are the model vertices in a Bayesian network. Thus, Bayesian networks appear as more naturally flexible and interpretable tools than decomposable Markov random fields. Finally, the joint distribution of a Bayesian network is encoded *via* a *directed* acyclic graph (DAG), which suits any attempt of further interpretation related to causality.

Villanueva and Maciel have chosen to model LD through Bayesian networks because they were interested in learning the causality of the associations between loci [60]. As in this previous approach, Lee and Shatkay also model LD resorting to a Bayesian network [30].

Besides general Bayesian networks, all other BN-based proposals addressing LD modeling share the common use of latent variables (LVs). Zhang and Ji describe a two-layer Bayesian network which is a collection of (independent) latent class models [61]. Also a two-layer model, the embedded Bayesian network of Nefian is more complicated than the former model [40]. The shared drawback of the four BN-based methods abovementioned is the lack of scalability. In contrast, the forest of latent tree models (FLTM) of Mourad *et al.*, another BN-based proposal with latent variables, is a multilayer hierarchical model that meets the scalability criterion [37].

9.2.2 Decomposable Markov Random Fields

Together with various co-authors, Thomas has contributed much to the domain of LD modeling through Markov random fields. This led to the proposal of no less than four learning algorithms, two of which focused on a subclass of DMRFs that can be represented with an Interval Graph (IG).

ANTERIOR WORKS

Decomposable graphs allow the efficient computation of the likelihood of the structure, given the data. Thus, structure learning may be performed navigating the structure space while optimizing a log-likelihood-based score. In [54], to explore the DMRF space, operations based on connection or disconnection of randomly selected nodes were designed to build the neighborhood of an incumbent graph G. Whatever the method used for search space navigation (downhill search or simulated annealing), a severe issue alleviated the efficiency of the model learning algorithm: the operational characterization of general decomposable graphs was not available. Therefore, starting from the incumbent graph G, one had to check a posteriori whether a proposal G' in the neighborhood of G is decomposable. Consequently, the first model of Thomas and Camp was not scalable because in the learning algorithm, much time was wasted checking for graph decomposability and rejecting irrelevant structures.

To gain efficiency, Thomas next considered a subclass of DMRFs, that is the DMRFs which can be represented by Interval Graphs (IGs). The key to efficiency increase in model learning is the substitution of IG-representation sampling to the graph sampling. A graph may be represented by an interval graph if and only if its vertices can be made to correspond to intervals on the real line such that two vertices are connected provided that their corresponding intervals overlap. In the earliest IG-based proposal [52], the IG search space is navigated through simple operations. An example is the change in the length of an interval selected at random. To further decrease time complexity, a second IG-based version requires that any interval extends no more than a maximum value on each side [53].

BACK TO GENERAL DECOMPOSABLE MARKOV RANDOM FIELDS

The limitation of DMRFs to IG-based DMRFs was not satisfying with regard to modeling power. Recently, Abel and Thomas used rules to keep the graph decomposability property when sampling in the neighborhood of a general decomposable Markov random field [1]: in other words, it is now possible to enumerate all the DMRF neighbors of an incumbent DMRF. Central to this enumeration is the junction tree (JT) representation of a DMRF. A JT is a special kind of clique tree of the initial graph. We first recall that a clique tree is a tree-structured graph whose nodes are maximal cliques of the original graph. We also recall that the pairwise intersection of these maximal cliques define separator sets. A JT is then a clique tree such that for every pair of cliques (C_i, C_j) of the original graph, $C_i \cap C_j$ belongs to every separator set on the unique path from C_i to C_j (running intersection property).

In the IG strategy presented previously, sampling over the Interval Graph representations of DMRFs was substituted to sampling over the DMRFs with an IG representation. A similar principle governs the novel learning algorithm: sampling over the junction tree representations of DMRFs is the key point.

Exploiting former theoretical works [16, 19], Thomas and Green have characterized the pairs of vertices whose disconnection or connection maintains the decomposability property of a Markov graph [55, 56]. Updating a junction tree under this disconnection or connection scheme scales in $O(1)$. The novel algorithm of Abel and Thomas relies on this characterization [1]. Besides, this algorithm is specific in that a divide-and-conquer strategy is applied that alternates phase inference and model learning. This time, a first window moves along the genome, in which rounds of JT samplings alternate with haplotype imputation. A second window lagging a half-window behind the first one performs hill climbing optimization. This latest version, FitGMLD, can be found at http://balance.med.utah.edu/wiki/index.php/FitGMLD.

The largest data set tested with the seminal algorithm of Thomas and Camp describes 20 000 loci for 90 individuals. The revisited method based on decomposable graphs has been shown able to handle benchmarks describing 100 000 loci for 60 individuals and 500 loci for 12 500 individuals, respectively.

9.2.3 Bayesian Network-based Approaches without Latent Variables

SCORE- VERSUS CONSTRAINT-BASED ACKNOWLEDGMENT OF DATA DEPENDENCES

Villanueva and Maciel's structure learning algorithm relies on the standard K2 procedure, combined with a genetic algorithm [60]. Following a topological ordering of the vertices, the K2 procedure iteratively attempts to add up to u parents to each vertex in turn, where u is the maximum number of parents allowed for a vertex. To serve this purpose for, say, vertex X_i, each iteration considers the candidate in $\text{Prec}(X_i)$ whose addition best increases the score, and adds it to Pa_{X_i}, the set of parents of X_i. This iterative process terminates when u parents have been added or when no further addition can increase the score. The K2 score used in [60] is the BDe score. Prior knowledge obtained through conditional independence tests could be used to direct the choice of a specific topological ordering. Though, this line was not followed by the authors, who instead sample the space of topological orderings through a genetic algorithm. The so-called fitness of a topological ordering is the (best) score returned by the K2 heuristic. The method of Lee and Shatkay [30] implements hill climbing with random restarts. Therein, the search process is constrained by the greedy sparse candidate procedure, a standard used to accelerate structure learning in Bayesian networks by restricting the parents of each node to a small subset of candidates. The subset of parent candidates for node X_i is the set of variables most correlated with X_i. Thus, the method of Lee and Shatkay is far more direct than that of Villanueva and Maciel: hill climbing using Minimum Description Length (MDL) score, dependence tests between variables, and use of an upper bound on the number of parents allowed for each node are the main ingredients. Finally, the two previous strategies typically differ in the ways they account for data dependences. In [60], the scoring of the structures (likelihood-based K2 score) accounts for the goodness-of-fit to the dependences existing within the data. In contrast, the sparse candidate algorithm used in [30] implements an early exploitation of the knowledge on dependences.

MODEL POST-PROCESSING

On the other hand, though the above two approaches both potentially allow the most general BN topologies, comparing their running times would be unfair as in each case, a specific post-processing is performed. Villanueva and Maciel identify association blocks from equivalence classes of BNs. The process is the following: after convergence, further running the genetic algorithm yields equally K2-scored orderings corresponding to optimal BN structures; the latter structures are grouped into equivalent classes (all structures in the same equivalent class share the same skeleton, only differing in edge directions); a partially directed acyclic graph (PDAG) is built for each equivalent class; finally, in a PDAG, an association block is defined as a set with the largest number of consecutive SNPs that are pairwise marginally dependent. Lee and Shatkay's final objective is SNP tagging, that is the identification of sets of SNPs that best predict other SNPs. A greedy search is used to grow a set of candidate tagging SNPs. This search is guided by the conditional independence structure of the BN: only the parents of a vertex are allowed as predictors for this vertex.

Notwithstanding this difference in post-processings, in both cases, the heavy computational burden only allows the handling of small data sets. The program K2GA processes a few hundred haplotypes described by a few tens of SNPs in time ranging from a few seconds to less than ten minutes. Around eight hundred haplotypes described by about a hundred of SNPs are handled in nearly four hours by BNTagger, the software program of Lee and Shatkay.

9.2.4 Bayesian Network-based Approaches with Latent Variables

TWO-LAYER BN-BASED APPROACHES

The two-layer Bayesian network of Zhang and Ji [61] belongs to the category of models which exploits knowledge about haplotype block structure as follows: the haplotypes observed in a population tend to locally cluster into groups of similar haplotypes; thus, the number of clusters, and consequently, their sizes, vary along the genome; the principle of the modeling lies in assigning each locus to a cluster, each cluster being represented as a latent class model (see Fig. 9.2). The latent class model (LCM) is a BN where a *unique* latent variable is connected to each of its child variables, the child variables being independent of each other, conditional on their latent parent. In this category, Zhang and Ji's approach improves over the approach of Kimmel and Shamir, where the model is also a collection of LCMs but is block-based. In a block-based model, a cluster necessarily encompasses a physical stretch on the genome, whereas in the most flexible cluster-based model, SNPs from distant regions may be allowed in the same cluster. To learn Bayesian networks in the case of missing or latent variables, the standard strategy consists in running a structural expectation-maximization (SEM) procedure. SEM consists in successively optimizing the parameters conditional on the structure and then the structure conditional on the parameters.

In the simple model of Zhang and Ji, long-range LD and LD between close SNPs are represented on equal footing. In contrast, the embedded Bayesian network of Nefian [40] allows the representation of long-range LD as a higher-order dependence (see Fig. 9.3). Again a two-layer model, the model of Nefian is more complicated than Zhang and Ji's in that dependences are possibly allowed between the observed variables constituting the first layer (but only if the latter share the same parent—a latent variable). At the same time, dependences between the latent

Fig 9.2 The collection of independent latent class models of Zhang and Ji.

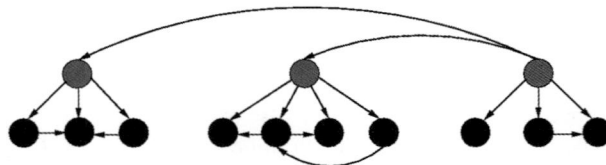

Fig 9.3 The embedded Bayesian networks of Nefian.

variables in the upper layer are allowed. The algorithm adapted to the construction of this embedded structure is naturally a SEM recursive procedure: independent runs of the standard SEM procedure are first executed to learn local structures as augmented LCMs (i.e., the child nodes may be connected within an LCM); once imputed, the latent variables rooting the local structures play the role of the observed variables for the global structure; then a final SEM step learns this global structure. In the approaches of Zhang and Ji on the one hand and of Nefian on the other hand, the structure learning step of SEM implements hill climbing. The moves implemented by Zhang and Ji are membership changes between LCMs (i.e., a SNP moves from one LCM to another one). Nefian uses standard moves (addition, removal, or reversal of an edge). The sophistication of Nefian's model comes at a cost to alleviate the computational burden, a strong constraint is imposed on the local structures: the genome is split into contiguous windows of fixed common size and as many LCMs are built from these blocks. The resulting fundamental difference between the models in [61] and [40] is that the first is cluster-based whereas the second is block-based. The lack of flexibility of the latter method is a severe drawback. The cluster-based approach of Zhang and Ji handles clusters of varying size. However, in spite of this improvement, the downside lies in that the number of clusters has to be specified.

SCALABLE HIERARCHICAL APPROACH

The hierarchical model of Mourad *et al.*, the FLTM (forest of latent tree models) [37], is a generalization of the model of Zhang and Ji, in which only one layer of latent variables was allowed (see Fig. 9.4). The latent tree model (LTM) can be described as a collection of embedded LCMs, thus defining layers of latent variables. Since pairwise dependences and multilocus LD can be simultaneously represented in the hierarchical structure, the FLTM allows a more faithful modeling of the genetic data than other methods. In particular, the hierarchical structure provides a flexible model, able to best reflect the fuzzy nature of LD. Moreover, this hierarchical structure is the key to meet the scalability requirement at genome scale.

The adapted ascending hierarchical clustering procedure used to learn an FLTM processes data as follows: at each agglomerative step, a partitioning method is used to identify cliques of pairwise-dependent variables; each such clique is intended to be subsumed into a latent variable, through an LCM. For each LCM, parameter learning is performed (through the standard expectation-maximization (EM) algorithm), which yields the marginal distribution of the latent variable and the conditional distributions of the child variables. Then, (linear) probabilistic inference is carried on, to impute the missing data corresponding to the latent variable. In the *forest* of LTMs, only the latent variables that are sufficiently informative about the subsumed variables must be kept. To recapitulate, the three main ingredients of the learning process are the clique partitioning, the LCM-based construction, and the LCM-based imputation.

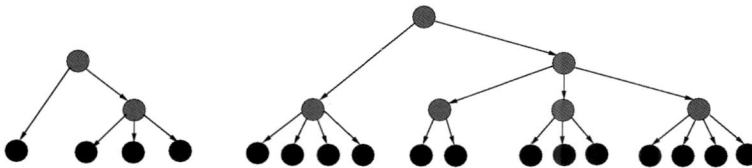

Fig 9.4 The forest of latent tree models of Mourad et al.

None of the two-layer BN models described above is scalable. For instance, not only does Nefian [40] use a divide-and-conquer strategy implemented via contiguous windows, she also resorts to parallelization. A window-based approach was also used in the first version of Mourad and co-workers' approach [37]. However, in contrast to Nefian's proposal, this first version reached scalability without resorting to parallelization. Later, this walking window process was more satisfactorily replaced with a sliding window approach: in the same vein as Thomas and co-workers' works [1, 53], SNPs too far apart on the genome are not allowed in the same clique. The first version could process data sets describing 10^5 SNPs for 2 000 individuals in around 15 hours, for an arbitrary window size of 100 SNPs [36]. When setting the sliding window size to 0.5 Mbp, a reasonable choice to capture LD, their novel algorithm runs in less than 12 hours [46, 47]. It has to be emphasized that as this novel version runs EM with 10 restarts, a significant improvement has been brought with respect to the initial version. Finally, the complexity roughly scales to $O(p)$ where p denotes the number of SNPs; the complexity is also a linear function of the sliding window size. The application software, CFHLC+ (Construction of Forests of Hierarchical Latent Class models), is available at http://sites.google.com/site/raphaelmourad/Home/programmes.

A compromise between the simplicity of Zhang and Ji's model and the sophistication of Nefian's, the FLTM model is the only approach that is scalable, due to its hierarchical structure. As highlighted in [39], imposing a hierarchical structure is the key to alleviate the computational complexity. In contrast with all PGM-based methods formerly described to model LD, local search is no more the rule; instead, a specific procedure is implemented, the greedy, iterative, bottom-up clustering of variables. This constraint on the BN structure is not a caveat to LD modeling accuracy since LD is hierarchical in essence.

9.2.5 Recapitulation

Table 9.1 recapitulates the main characteristics of the various models depicted to represent LD, and of their learning algorithms. The motivation of LD modeling is indicated for each approach: haplotype inference, identification of association blocks, and selection of tagging SNPs. Of the proposals, only three are scalable. In two cases, scalability is obtained through the use of specific data structures, the interval graph-based Markov random fields of Thomas *et al.* and the hierarchical Bayesian network of Mourad *et al.* Thus, specific efficient learning algorithms may be designed to build these constrained structures. However, a block-based modeling approach such as the interval graph-based one remains a rough approximation of LD. In contrast, the hierarchical FLTM model of Mourad et al. is a cluster-based approach. A recent work of Abel and Thomas also provided a scalable cluster-based approach thanks to the exploitation of properties derived from graph theory. A major difference between the two previous approaches lies in that Abel and Thomas's core model learning procedure handles phased data. Therefore, this core procedure is incorporated in a larger framework alternating model learning and phase inference. Thus, Abel and Thomas' approach also addresses haplotype inference. Haplotype inference was not an objective of the FLTM-based strategy. Finally, a common characteristic to both methods is the resort to maximum physical distances to control running times.

Table 9.1 Three Markov random fields and five Bayesian models dedicated to the modeling of linkage disequilibrium. In this context, haplotypes means phased genotypes.

Model	Purpose	Structure Learning Algorithm	Input Data	Scalability	Reference
Decomposable Markov random field	LD modeling	hill climbing on DMRFs (connection and disconnection of randomly selected pairs of vertices - 6 operations described), simulated annealing	haplotypes	medium	Thomas and Camp (2004) [54]
Interval graph-based Markov random field	haplotype inference	hill climbing on intervals (change in the length of an interval), interval tree [52], array [53]	haplotypes genotypes	yes yes	Thomas (2009) [52] Thomas (2009) [53]
Decomposable Markov random field	haplotype inference	hill climbing on junction trees	genotypes	yes	Abel and Thomas (2011) [1]
General Bayesian network	identification of association blocks	genetic algorithm in the space of topological orderings, combined with hill climbing in structures, conditional on a given topological ordering	haplotypes	no	Villanueva and Maciel (2010) [60]
General Bayesian network	selection of tagging SNPs	hill climbing with random restarts, with a constraint on the ordering of variables (sparse candidate algorithm)	haplotypes	no	Lee and Shatkay (2006) [30]
Two-layer Bayesian network Collection of latent class models	identification of association blocks	SEM, hill climbing (cluster membership change)	genotypes	no	Zhang and Ji (2009) [61]
Two-layer Bayesian embedded network Collection of latent class models augmented with dependences between siblings and dependences between latent variables	LD modeling	SEM, hill climbing (addition, removal and reversal of edges)	haplotypes	no	Nefian (2006) [40]
Forest of latent tree models	LD modeling	ascending hierarchical clustering of variables, combined with local structure learning for latent class models	genotypes	yes	Mourad et al. (2011) [37]

Note: DMRF: decomposable Markov random field, LD: linkage disequilibrium, SEM: structural expectation-maximization, SNP: single nucleotide polymorphism.

9.3 Single-SNP Approaches for Genome-wide Association Studies

The present section reviews probabilistic graphical models designed to achieve genome-wide association studies (GWASs) in the case when possibly several *independent* unknown causative genetic factors are suspected in the disease susceptibility: each causative SNP at each gene influencing the disease exerts a small additive effect on the disease phenotype. This situation is opposed to that of interacting factors, which we consider to be the definition of *epistasis* in the context of this chapter. In the most simple design, any SNP is tested alone against the disease, thus the designation of single-SNP approach. The most basic GWAS would consider each SNP in turn and test it against the phenotype studied. However, when confounding factors are suspected to interfere with the genotype to impact the phenotypes, models have to integrate these covariates. More sophisticated approaches will fully exploit LD knowledge, to exhibit an excess of haplotype sharing among cases, around a causative locus; these are called *multilocus* approaches.

Before we concentrate on the genome-wide scale, we briefly report some works dedicated to fine mapping and candidate-gene studies. In [43], a greedy search with random restarts is used to learn Bayesian networks in the presence of covariates such as ethnicity and cholesterol level. In [45], the K2 procedure employed uses a heuristic which relies on the following ordering of the variables: older SNPs are tested as children of more recent SNPs. The former SNPs are characterized by a more uniform distribution of their variants than that of recent SNPs in the population. Finally, another significant work is the extension of the aforementioned approach of Thomas to the discovery of SNP–phenotype associations [51].

Nevertheless, and not surprisingly, all the most recent works involving probabilistic graphical models fight on the ground of genome-scale tractability. Besides, probabilistic graphical models are an ideal framework to integrate various sources of data, to reinforce confidence in the results. In this section, we first describe an approach used to address the case of confounding factors. Then we depict three models developed to implement multilocus GWASs. This section will in particular exhibit two variations around Markov random fields: hidden Markov random field and conditional random field. Besides, a model closely connected to graphical models—the variable length Markov chain—will also be depicted.

9.3.1 Integration of Confounding Factors

The detection of the association between a given SNP and a studied phenotype may be hampered by confounding factors such as the existence of population structure, family structure, or cryptic relatedness in the population. The presence of confounding factors leads to spurious associations, whereas true associations may be missed. The standard methods used to run a GWAS in presence of covariates, such as linear mixed-effects models and principal component analysis, are subject to three main caveats: low power, presence of false positives, or intractability for a large number of individuals in the data set analyzed. Recently, Huang *et al.* have proposed a single-SNP approach based on a conditional random field [22].

In a Markov random field whose nodes describe variables X_is, the Markov property states that a given X_i is independent of any other X_j conditional on the neighbors of X_i. The concept of conditional random field (CRF) is defined as follows:

Definition 9.1 (Conditional random field)
Let $G = (V = X \cup Y,\ E)$ be a graph such that Y (output variables) is indexed by the vertices of G. (X, Y) is a conditional random field if, when conditioned on X (output variables), the random variables Y_is satisfy the Markov property: $\mathbb{P}(Y_i \mid X, Y_j, j \neq i) = \mathbb{P}(Y_i \mid X, Y_j, j \in \mathcal{N}_{Y_i})$, where $\mathcal{N}_{Y_i}$ denotes the neighborhood of Y_i in G.

The approach of Huang et al. focuses on the relation between *a given SNP* and *a given phenotype*, with the possible interference of covariates. To model these dependences, a conditional random field relates each individual's phenotype to each individual's SNP and other covariates. The nodes in Y describe the phenotypes of all individuals, and the undirected edges in E correspond to possible dependences between the phenotypes of pairs of individuals. X describes the covariates, including the SNP of interest. Each phenotype node Y_i is conditioned by the SNP and covariates observed for the corresponding individual i (together denoted X_i) (see Fig. 9.5).

In this framework, the probability of phenotype Y_i conditioned on the phenotypes in the neighborhood of Y_i, on the one hand, and on ith's individual SNP and covariates (X_i) on the other hand, is described as follows:

$$\mathbb{P}(Y_i \mid Y_{-i}, X, G, \theta, \beta) = \frac{\exp\left[-2\,Y_i \sum_{Y_j \in \mathcal{N}_{Y_i}} \theta_{ij}\,Y_j - 2\,Y_i\,\beta^T X_i\right]}{1 + \exp\left[-2\,Y_i \sum_{Y_j \in \mathcal{N}_{Y_i}} \theta_{ij}\,Y_j - 2\,Y_i\,\beta^T X_i\right]}, \tag{9.1}$$

where we recall that $\mathcal{N}_{Y_i}$ denotes Y_i's neighbors in the undirected graph connecting the phenotypes, and Y_{-i} represents the phenotypes deprived of Y_i. The influence of the genetic variations and covariate (X_i) is captured through the weight β. A large weight magnitude for a given covariate reflects the increased role of this covariate in determining the phenotype. The term θ_{ij} is a genetic similarity measure between the phenotypes of individuals i and j; in this respect, the matrix θ determines the neighborhood $\mathcal{N}_{Y_i}$ of phenotype Y_i in the graph G. Adding the contribution of phenotypes similar to Y_i serves the purpose of modeling confounding factors. The justification is as follows: the effects of confounding factors are all encoded in the set of SNPs carried by all the individuals; therefore, the similarities between individuals can be used to distinguish between spurious and true SNP-phenotype associations. Thus, equation (9.1) exhibits

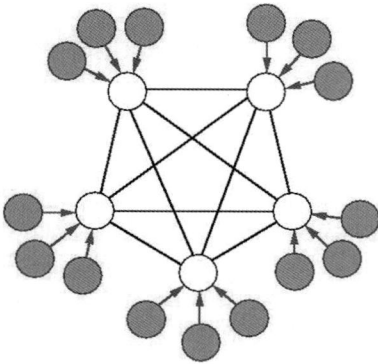

Fig 9.5 The conditional random field of Huang et al. This model specifies a joint probability over individuals' observed phenotypes, conditioned on each individual's covariates. The complete graph represents the dependences between the individuals' phenotypes. For each individual (white node), the three covariates represented here (in gray) include the studied SNP.

contributions for the phenotypes in $\mathcal{N}_{Y_i}$. Such contributions are weighted by the genetic similarities with phenotype Y_i. For this purpose, Huang et al. use a priori information and set pairwise similarities as Identity-by-State (IBS) values. The IBS value for two individuals is calculated as the percentage of SNP marker alleles shared by the individuals, across the whole set of SNPs analyzed in the GWAS.

Given individuals' phenotypes Y and covariates X and the matrix of genetic similarities θ, the objective is to assess the effect of the SNP of interest on the individuals' phenotypes by estimating the weight vector β. To this aim, a tractable procedure is performed, which minimizes the negative pseudo-likelihood criterion:

$$L(\beta) = -\sum_{Y_i \in Y} L_i(\beta)$$

$$= -\sum_{Y_i \in Y} [Y_i = 1] \log(p_i) + [Y_i = 0] \log(1 - p_i),$$

where 1 and 0 denote the two values for the disease status and p_i is the conditional probability of the ith individual's phenotype given the phenotypes over all other individuals Y_{-i}. This probability p_i writes as $p_i(\beta) = \mathbb{P}(Y_i = 1 \mid Y_{-i}, X, G, \beta)$.

Since all nodes have exponential distributions and all nodes are observed, the optimization is convex and thus yields the unique exact solution. Huang *et al.* use a gradient descent algorithm: the vector β is iteratively updated until convergence as $\beta \leftarrow \beta - \alpha\, g$, where $\alpha > 0$ is a learning rate parameter and g is the gradient of the negative pseudo-likelihood.

9.3.2 GWAS Multilocus Approach

In the introduction of Section 9.3 (page 228), we mentioned an alternative to the single-SNP approach: in a multilocus approach, the objective is to reinforce the assessment of an association between a putative *single* locus and a disease, through knowledge about linkage disequilibrium. Namely, association knowledge is expected to be reinforced from the multiple SNPs in LD with a given SNP. Linkage disequilibrium is likely to reveal an excess of haplotype sharing around a causative locus, among cases. In this subsection, we will first briefly return to the models of Thomas [54], on the one hand, and of Mourad et al. [37], on the other hand. We will show that it is straightforward to post-process these models for a GWAS purpose. Then, we will present the approach of Verzilli et al. [59], which is based on decomposable random Markov fields. A variant of the Markov chain model that grounds the gold standard method of Browning and Browning will be described [9]. On the other hand, hidden Markov random fields have also been investigated for a GWAS purpose: we will describe the method of Li et al. [32].

POST-PROCESSING MODELS REPRESENTING LINKAGE DISEQUILIBRIUM

In Subsection 9.2.2 (page 221), we have presented the works of Thomas and collaborators who pioneered the use of decomposable Markov random fields (DMRFs) to model linkage disequilibrium [1, 52–54]. When used to capture linkage disequilibrium, the DMRF model also allows testing for disease association. A simple way to achieve this objective is to include a vertex representing the phenotype (disease status indicator) in the DMRF. Then, the estimated model can be compared with the submodel obtained by removing any links between the phenotype and locus vertices. A standard chi-square likelihood-ratio test may be used for this purpose.

Mourad et al. indicate that their FLTM hierarchical model is relevant for a GWAS purpose: starting from nodes (i.e., latent variables) located at a high level in the hierarchy, a best-first search strategy would zoom through narrower and narrower regions of the genome, to pinpoint a region potentially associated with the disease. A severe drawback could be that the latent variables be not sufficiently informative about their child variables, especially at high levels in the hierarchy, which would entail a lack of power. To show that bottom up information decay in the hierarchy can be controlled to provide a reliable model, Mourad et al. performed intensive tests [47]: a single-SNP-phenotype association was simulated under 36 conditions. All latent variables were subject to an association test against the phenotype. All nodes that are ancestors of the causative SNP succeed in capturing an indirect association with the phenotype. In contrast, the other latent nodes globally show very weak associations. Thus the identification of such an ancestor node will allow pointing out potential causative markers as the leaf nodes of the tree rooted in this ancestor node.

ASSESSING AN ASSOCIATION THROUGH MODEL AVERAGING

In the same vein as Thomas and co-workers, Verzilli et al. rely on decomposable Markov random fields to implement GWASs [59]. This time, the cliques of the DMRF are explicitly interpreted as haplotype blocks; moreover, each clique is labeled with a boolean value (1/0) indicating the presence of at least one edge between some vertex in the clique and the phenotype vertex. A label set on 1 suggests the presence of a disease susceptibility locus in the region described by the clique. Verzilli et al. were able to implement searching in the space of general DMRFs: to escape the computationally demanding checking of the decomposability property, appropriate moves were tailored to preserve this property while browsing the search space of DMRFs. In contrast with the sophisticated graph theory rules used by Abel and Thomas [1], Verzilli et al. very naturally designed moves that involve changes in the set of cliques and separators. Three moves are described: merge step, split step, and switch-clique-label step. In the merge move, a randomly selected clique is merged with another one, provided that the size of the resulting clique does not exceed the maximum size allowed. The second clique is selected at random from the set of cliques that contain neighbors of the vertices of the former clique. The split move splits up a non-singleton clique selected at random. The switch-clique-label move changes the label of a clique C_j selected at random. If the clique was labeled $T_j = 1$, this meant that at least one vertex in the clique was linked with the phenotype vertex. In this case, the split move deletes all corresponding edges. If the clique was labeled $T_j = 0$, a set of separator markers is selected at random from the vertices in the clique. Edges connecting these vertices to the phenotype vertex are added to the graph.

A Markov chain Monte Carlo (MCMC) scheme is used to sample the DMRF search space. This MCMC relies on a Metropolis–Hastings procedure. The Bayesian model averaging paradigm is fully exploited for the GWAS purpose: from the posterior sample of graphs, the frequency with which a SNP vertex and the phenotype vertex are connected by an edge is an estimate of the posterior probability of association. For any SNP, in addition to the posterior probability of association, the Bayes factor to assess evidence in favor of association (H1) versus no association (H0) is given by the ratio of posterior odds $\frac{\mathbb{P}(H_1|D)}{\mathbb{P}(H_0|D)}$ to prior odds $\frac{\mathbb{P}(H_1)}{\mathbb{P}(H_0)}$. The prior probability distribution is modeled with a Poisson distribution.

GWAS THROUGH A VARIABLE LENGTH MARKOV CHAIN

In this subsection, we first present an approach based on a variant of the Markov chain, the variable length Markov chain. First, we recall some important notions as a prerequisite for further understanding.

A peculiar class of Bayesian networks is the class of Markov chains. A discrete-time Markov process with discrete state space is a sequence $X_0, X_1, X_2, X_3, \ldots$ of random variables taking their values in the state space. The value X_n is the state of the process at the instant n. In the case of Markov chains, the Markov property states that conditional on the present state of the process, its future and past are independent:

$$\mathbb{P}(X_{n+1} = x_{n+1} \mid X_n = x_n, X_{n-1} = x_{n-1}, \cdots, X_1 = x_1, X_0 = x_0)$$
$$= \mathbb{P}(X_{n+1} = x_{n+1} \mid X_n = x_n).$$

But often researchers are interested in delving into past state influences, introducing "memory" in the transitions through the use of a higher-order Markov model. A Markov chain of order k, where k $(k < n)$ is finite, is a process satisfying

$$\mathbb{P}(X_n = x_n \mid X_{n-1} = x_{n-1}, X_{n-2} = x_{n-2}, \cdots, X_1 = x_1, X_0 = x_0)$$
$$= \mathbb{P}(X_n = x_n \mid X_{n-1} = x_{n-1}, X_{n-2} = x_{n-2}, \cdots, X_{n-k} = x_{n-k}).$$

It is straightforward to transpose the concept of a time-varying random phenomenon to that of a random process where transitions concern contiguous positions in an ordered sequence. Typically, genomic data fall within this scope, and high-order Markov models are very popular in the field of bioinformatics. The class of first and higher-order Markov chains are among the simplest graphical models.

In high-order Markov models, each random variable depends on a fixed number of random variables. Nonetheless, some processes which are naturally fuzzy, such as linkage disequilibrium, would better be modeled through a more flexible model. The general graphical Markov model cannot allow a flexible memory. In contrast, in variable length Markov chains (VLMCs), the number of conditioning variables may vary based on the observations.

The VLMC-based approach of Browning and Browning [7, 8] falls in the category of multilocus-based methods. The aim is to identify local clusters of haplotypes in order to perform an association study between these clusters and the disease. Typically, in the ordered sequence of genetic markers, it may happen that when the variant 0 at location $t - 1$ is observed, the variant at location t only depends on that at location $t - 1$, whereas if the variant 1 is observed at location $t - 1$, then the variant at location $t - 2$ also determines the variant at location t.

In [7, 8], an efficient specific algorithm is used to fit the VLMC model, based on Ron et al.'s works [44]. The principle is the following: a tree is built from haplotype data, which will be transformed in a graph exhibiting *local* clusters of haplotypes. In both the initial tree and final graph, every SNP location is represented by a layer. In the initial tree, each path from the root to a leaf represents a distinct observed haplotype: in a given layer, each edge is labeled with a variant of the SNP at this location and with the number of individuals of the observed sample showing the haplotype ending in this variant. In a nutshell, the principle that governs the transformation of the initial tree into the final graph is the following: in the initial tree, two paths that encompass, say layers $\ell, \ell + 1 \ldots \ell + r$ and describe similar haplotypes of length $r + 1$ may be factorized into a unique path. The condition to factorize, that is similarity of haplotypes, means that, at each SNP, distributions (i.e., haplotype counts) are pairwise similar along the two paths. This principle readily extends to several paths as well as to subtrees (see Fig. 9.6). Thanks to a procedure meant to merge nodes which root *similar* subtrees, in the final graph, every edge

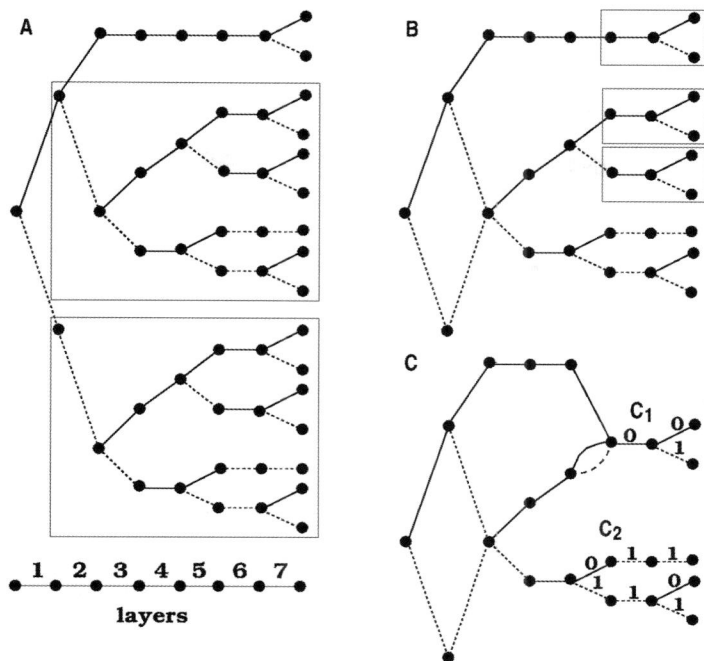

Fig 9.6 The variable length Markov chain of Browning and Browning. The layers correspond to the contiguous positions of the SNPs. The plain and dotted lines respectively correspond to the variants 0 and 1. The rectangles show the subtrees that can be factorized due to similar distributions along their edges in the graphs A) and B). C) The prediction of the 7th SNP only depends on the 6th SNP when the latter is equal to 0 (see subgraph C1). The prediction of the 7th SNP depends on the 5th and 6th SNPs when the latter SNP is equal to 1 (see subgraph C2).

represents a local cluster of haplotypes. However, not all edges are worth being tested for association: converging edges are tested for association, provided that edge counts are not too low (with respect to cases and controls). Considering case and control counts, an exact Fisher's test successively tests each edge against all other edges sharing the same endpoint vertex. Such edges reflect historical recombinations and thus represent clusters of haplotypes relevant for the association study.

In this method, the multiple testing issue is addressed through permutation tests. The same set of local clusters of haplotypes is kept throughout the permutations. The implementation for this approach is available as one of the multiple software programs provided by the package Beagle (http://faculty.washington.edu/browning/beagle/beagle.html). Due to fast model fitting, this method is renown as one of the few scalable multilocus approaches available for GWASs.

GWAS THROUGH A HIDDEN MARKOV RANDOM FIELD

The multilocus approach exploits the fact that if several SNPs are all in linkage disequilibrium with the true disease variants, such dependences are expected to increase the power to detect the SNPs associated with the disease. In [32], Li et al. propose to formally incorporate LD information in the Bayesian procedure designed to estimate the posterior probability that a given SNP is associated with the disease of interest. The key is that this posterior probability depends not only on the genotype data observed at this SNP, but also on the genotypes that are in strong LD with this particular SNP. The framework used by Li et al. is a hidden Markov random field.

We first recall that in a Markov random field whose nodes describe variables Z_is, the Markov property states that a given Z_i is independent of all other Z_j conditional on the neighbors of Z_i. A hidden Markov random field (HMRF) is a generalization of a hidden Markov model. Both models assume that the probabilistic nature of the observed X_is is determined by the unobservable H_is. In a hidden Markov model, the H_is are ruled by a Markov chain whereas in a hidden Markov random field, the H_is define a Markov random field.

The undirected weighted (w) graph of significantly dependent pairs of SNPs (the X_is) is constructed using a specified threshold τ (minimum r^2 correlation coefficient). The set of neighbors $\mathcal{N}_{X_i}$ of X_i then comprises all X_js satisfying the following condition: $w_{i,j} = r_{i,j}^2 \geq \tau$. Each X_i (i.e., each observed SNP i) is assigned a *latent* random variable H_i which is defined as the following indicator:

$$
H_i = \begin{cases} 1 \text{ if SNP } i \text{ is associated with the disease,} \\ 0 \text{ otherwise.} \end{cases}
$$

H, the vector of hidden states, follows a Markov random field. If an edge exists between SNPs i and j in the weighted graph, H_i and H_j are expected to be dependent, which can be stated in the conditional distribution:

$$
\mathbb{P}(H_i \mid H_{\mathcal{N}_{X_i}}, \theta) \propto \exp\left(\gamma H_i + \beta \sum_{j \in \mathcal{N}_{X_i}} w_{i,j} \, I(H_i = H_j) \right), \text{ with } \theta = (\gamma, \beta).
$$

Parameter β ($\beta \geq 0$) favors connected SNPs to have similar values for their latent variables. The genotype variable X_i is related to the model through the emission probabilities $\mathbb{P}(X_i \mid H_i = 0)$ and $\mathbb{P}(X_i \mid H_i = 1)$. The ultimate goal is to estimate the association of SNP i and the disease through sampling based on the conditional probability $\mathbb{P}(H_i \mid X, \hat{H}_{-i})$, where the last component denotes the vector of latent variables deprived of H_i.

An empirical Bayes model is used to derive the emission probabilities. It is based on the use of a Dirichlet prior (with parameter α), to model the genotype frequencies in both case and control populations. Thus, the derivations of $\mathbb{P}(X_i \mid H_i = 1, \alpha)$ and $\mathbb{P}(X_i \mid H_i = 0, \alpha)$ differ in that cases and controls may be distinguished in the first case, whereas that is not the case when SNP i is not associated with the disease ($H_i = 0$).

Finally all ingredients are in place to compute the posterior probability $\mathbb{P}(H_i \mid X, \hat{H}_{-i})$ used for sampling:

$$
\mathbb{P}(H_i \mid X, \hat{H}_{-i}) \propto \mathbb{P}(X_i \mid H_i, \hat{\alpha}) \, \mathbb{P}(H_i \mid \hat{H}_{\mathcal{N}_{X_i}}, \hat{\theta}).
$$

The model parameters ($\theta = (\gamma, \beta)$) in the MRF and the parameter α of the Dirichlet prior in the emission probabilities have to be estimated. An iterated conditional mode (ICM) procedure [5] is run for this purpose. This procedure successively updates the model parameters and the latent variables until convergence.

9.3.3 Strengths and Limitations

In this subsection, we first highlight the differences between the approaches, mainly focusing on the strategies implemented to find the putative associations and the procedures devoted to assess the significance of the association. Finally, we indicate through illustrations to which extent the methods are scalable.

FINDING ASSOCIATIONS

Four approaches have been presented, which address multilocus GWAS analysis. All four methods exploit a model of linkage disequilibrium. Alone in its category, the method of Huang and collaborators was developed to cope with confounding factors and focuses on similarities between individuals' phenotypes.

In the approaches of Thomas et al. and of Verzilli and co-workers, the cliques in the Markov random field model the regions where SNPs are highly correlated. Therefore, any edge between a SNP and the node representing the phenotype pinpoints the putative association of the disease with the region defined by this SNP and the SNPs that are in strong LD with it. In Thomas et al.'s method, post-processing the model is fast (likelihood ratio for the models with and without the edge having the phenotype as an endpoint). In Verzilli et al.'s approach, model averaging is performed: a posterior probability of association is estimated as the percentage of presence of the edge connecting a given SNP with the phenotype vertex, and a Bayes factor is calculated.

In the other graphical models described, the phenotype is not included as a node. In Mourad et al.'s forest of latent tree models (FLTM), the latent variables support the same interpretation as the cliques in the two above methods, that is regions whose SNPs are in strong linkage disequilibrium. The variable length Markov chain approach of Browning and Browning [7] resorts to localized haplotype clustering; such clusters of haplotypes may be seen as latent variables. Respectively in RMFs, FLTMs, and VLMCs, the organization in cliques, trees, and local clusters of haplotypes account for the fuzziness of linkage disequilibrium.

Moreover, the models of Mourad et al. and Browning and Browning were developed to reduce data dimensionality in GWASs (or any other downstream analyses): these models allow us to perform association tests on (a limited number of) latent variables and on local haplotypes respectively. Nonetheless, a major difference between the FLTM and VLMC models remains that the former is cluster-based whereas the latter is block-based. Finally, the FLTM model exhibits the highest degree of flexibility due to its hierarchical structure. This tree-based structure is the only one that can model simultaneously close- and long-range dependences.

In Li et al.'s approach, the LD pattern stochastically determines the pattern of dependences between the (latent) indicators of association (H_is) with the disease. Both Li et al.'s method and Verzilli et al.'s approach tackle the finding of associations using sampling processes. Verzilli et al.'s GWAS recipe relies on model averaging. The model post-processing operated by Li et al. samples the distribution of indicators H_is, to estimate the posterior probability of no association, for each SNP.

SIGNIFICANCE ASSESSMENT FOR SNPs

In the model-averaging framework, Verzilli and co-workers describe a straightforward procedure to test the statistical significance of the association of any SNP with the disease. It is completed with the calculation of a Bayes factor. Huang and co-workers rely on the asymptotical distribution of some statistic (a function of the β coefficient, see equation (9.1), page 229), to assess the statistical significance of each SNP. Permutation-based procedures correct for multiple testing in Browning and Browning's approach and Mourad and co-workers' method. In the latter case, a specific threshold is automatically derived for each layer of latent variables: such thresholds decrease from bottom to top, in the hierarchical structure.

To correct for multiple testing, Li et al. implement a false discovery controlling procedure. The false discovery rate (FDR) method aims at controlling the expected proportion of incorrectly rejected null hypotheses of "no association" ("false discoveries" or type I errors) [4]. The principle is to keep the number of type I errors low (but not null) among all significant results. This results in increasing the power, or stated equivalently, in diminishing the number of false negatives. Li et al. rely on a posterior probability-based definition of the FDR, shown to provide optimal control in the framework of hidden Markov Models [48]. After the convergence of the algorithm, the latent vector H is sampled through Gibbs sampling, based on the posterior probabilities $\mathbb{P}(H_i \mid X, \hat{H}_{-i})$. Based on these samples, the posterior probability $q_i = \mathbb{P}(H_i = 0 \mid X)$ can be estimated. Let $q(i)$ denote the value at ith position in the sorted vector of q_i values over all SNPs (increasing order). Let α a given threshold. If k is defined as $\max\{t \mid \frac{1}{t} \sum_{s=1}^{t} q(s) \leq \alpha\}$, then all null hypotheses of non-association must be rejected, from 1 to k.

EMPIRICAL COMPLEXITIES

In all above-cited approaches exploiting a model of linkage disequilibrium, the computational burden is alleviated by incorporating some constraint: Browning and Browning set conditions to control the size (width) of the variable length Markov chain. In the other three methods described, pairwise-dependent SNPs are allowed in the same clique, subject to a maximum physical distance (Thomas et al., Verzilli et al., Mourad et al.). In the recursive model of Mourad et al., latent variables are also subject to this constraint. This constraint reflects LD extent. In the VLMC approach, the conditions set on the width of the VLMC also indirectly account for LD extent, though the understanding is not as intuitive as for the other methods.

In addition to assessing the power of their method, for instance on a real data set describing 360 657 SNPs for 3400 individuals, Huang et al. analyzed to which extent their method is scalable. For comparison, a linear-mixed model ran in 33 hours on a benchmark of 7 579 SNPs and 37 830 individuals. In contrast, for the same data set, the execution time of the CRF-based algorithm amounted to 48 minutes. Importantly, as future studies tend to encompass tens of thousands of individuals, larger speedups are expected. The scalability of the method of Verzilli and co-workers has been checked on a benchmark describing 100 000 SNPs for 1018 individuals. In this case, only 4 minutes were required. Mourad et al.'s latest version was shown to run for 12 hours for 100 000 markers and 2000 individuals. The Beagle gold standard of Browning and Browning is known to handle large-scale genetic data sets with hundreds of thousands of markers genotyped on thousands of individuals (no running time reported). The reason for the linearity in number of markers of Beagle is that one does not have to look out all the way to the end of the chromosome when deciding whether to merge two nodes at some given level. Finally, Li and co-workers' approach is the only one for which scaling to genome-wide data is obtained through the use of contiguous windows. For instance, 30 216 SNPs describing 3075 cases and controls have been split into groups of 1000 SNPs (no running time reported).

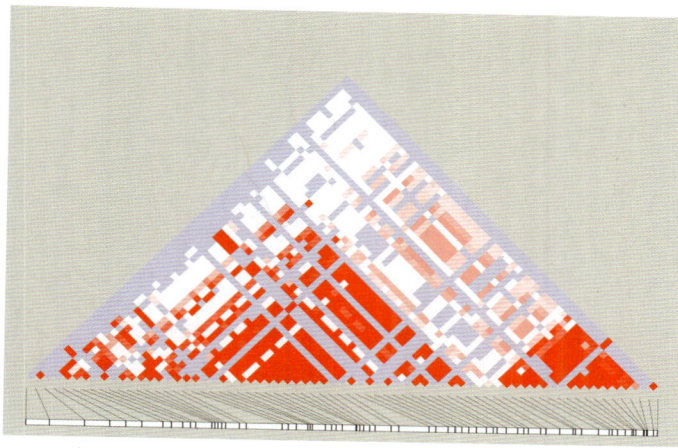

Fig 9.1 Linkage disequilibrium (LD) plot of a 500 kb SNP sequence. Human genome, chromosome 1, region [10 000 kb - 10 500 kb]. LD is revealed through the matrix of pairwise dependences between genetic markers. For a pair of SNPs, the color shade is all the darker as the correlation between the two SNPs is high.

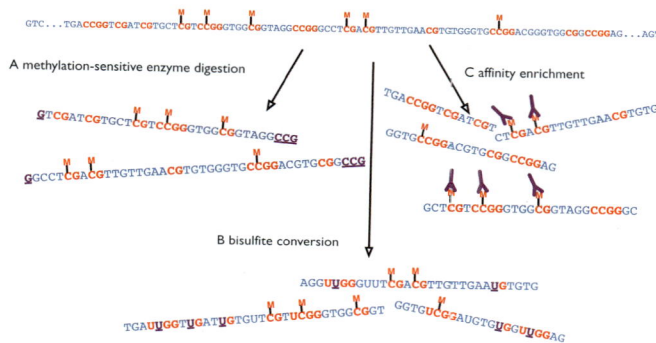

Fig 14.1 Three common techniques for genome-scale annotation of DNA methylation. (A) Enzyme digestion: the genomic DNA is digested with a methylation-sensitive restriction enzyme such as *HpaII*, which digests unmethylated CCGG sites. (B) Bisulfite conversion: converts cytosines that are not methylated to uracil. (C) Affinity enrichment: methylated cytosines in methylated regions are bound by antibodies or methyl-CpG binding proteins. M denotes methylation site.

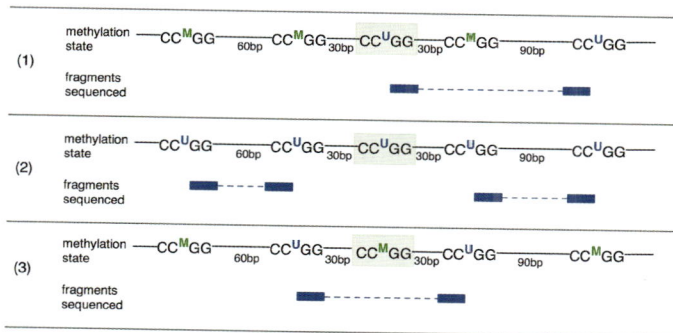

Fig 14.2 The methylation state of a site cannot always be determined from the number of fragments that originated at that site. In many cases, the methylation state of a site cannot be determined from the extent to which it was present at the end of sequenced fragments but can be determined by integrating sequencing data from its neighborhood. bp, base pair; M, methylated; U, unmethylated.

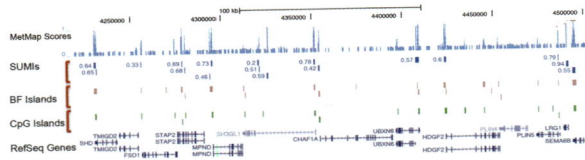

Fig 14.5 A section of the genome showing site-specific methylation scores (top panel) and unmethylated clusters (SUMIs, second panel) as inferred from one of human neutrophil samples. For the site-specific scores, a score of 0 determines a site as fully methylated. The third and fourth panels show BF islands as annotated by [5] and UCSC islands, respectively. While there is substantial overlap between SUMIs and the islands inferred by the sequence-based methods, a few novel SUMIs are seen in this figure, one of them at a transcription start site. RefSeq denotes genes annotated in the National Center for Biotechnology Information reference sequence database.

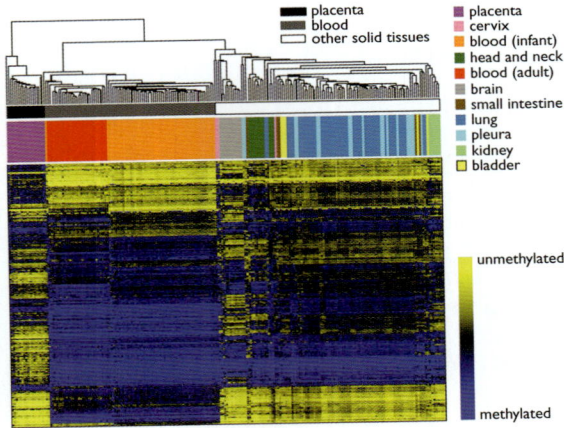

Fig 15.1 Clustering heat map showing DNA methylation patterns for 11 normal tissues [8]. Each cell represents an average beta value from the GoldenGate assay (Illumina). Rows represent one of 500 CpG dinucleotides, columns represent one of 211 individual samples.

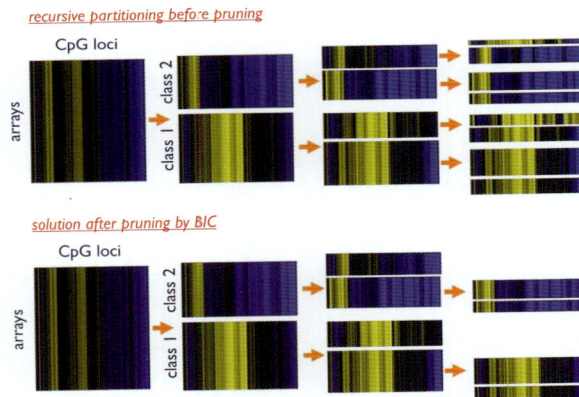

Fig 15.3 Schematic representation of the RPMM. Rows represent individual specimens or arrays, columns represent individual CpG loci. Initially, each array is assumed to be drawn from the same multivariate distribution consisting of a distinct distribution for each CpG (indicated by color). The data set is partitioned recursively into component data subsets using a two-part mixture model. Along the way, BIC is used to prune the tree, so that partitions that are likely to be unstable are never attempted.

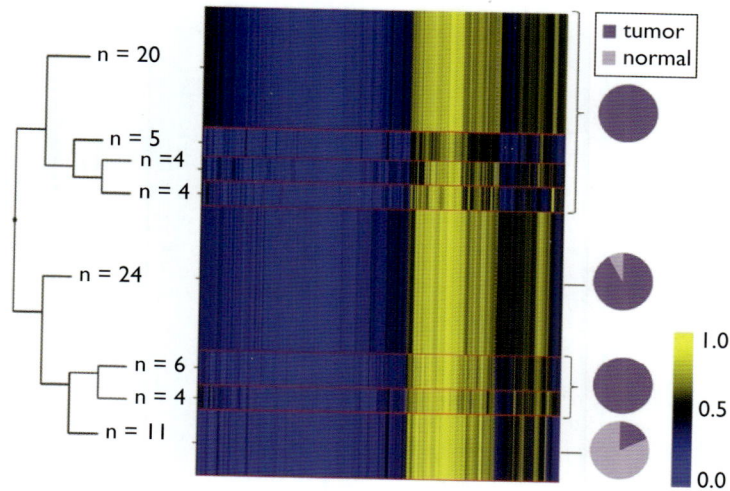

Fig 15.4 Recursive partitioning mixture model classification of normal and tumor head and neck tissues. The model was based on methylation values of 1413 autosomal loci measured using the GoldenGate assay produced by Illumina, and resulted in eight classes whose average methylation values are represented in the heat map. Distribution of normal and tumor samples within each class is depicted in pie charts on the right. Reproduced from [28], Figure 1B.

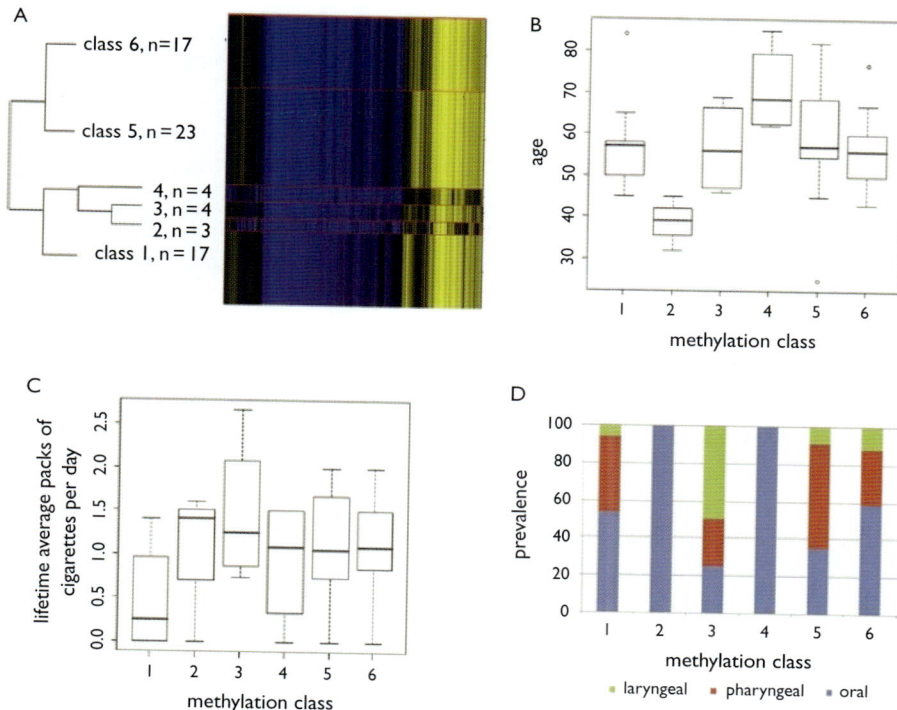

Fig 15.5 Recursive partitioning mixture model classification of HNSCCs. A) Six classes with average methylation values across loci depicted in the heat map. Associations of class membership with age, lifetime average packs of cigarettes smoked per day, and tumor location are shown in B, C, and D, respectively. Reproduced from [28], Figure 2A–D.

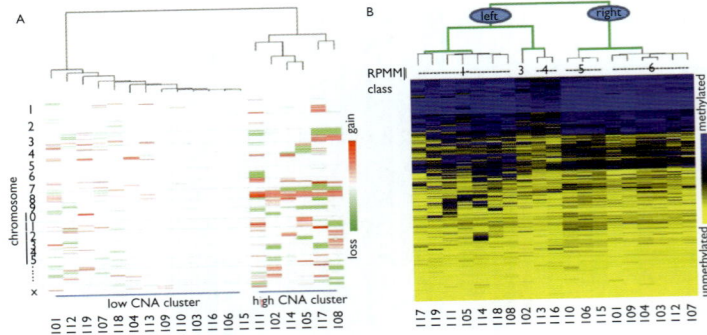

Fig 15.6 A) DNA copy number states are arranged by chromosome for 500 000 SNP loci. Copy number is red for amplified regions with three or more copies, white for two normal copies, and green for allele loss (no copies). Tumors are ordered by unsupervised hierarchical clustering and are dichotomized into low/high clusters of copy number alterations (CNAs). B) Methylation loci (more methylated = blue, less methylated = yellow) are grouped by Euclidean distance, and tumors samples are ordered first by RPMM class structure (green branches) then by simple hierarchical clustering (black branches). Tumor IDs are provided below each plot and "high CNA" samples are colored orange for reference. Reproduced from [33].

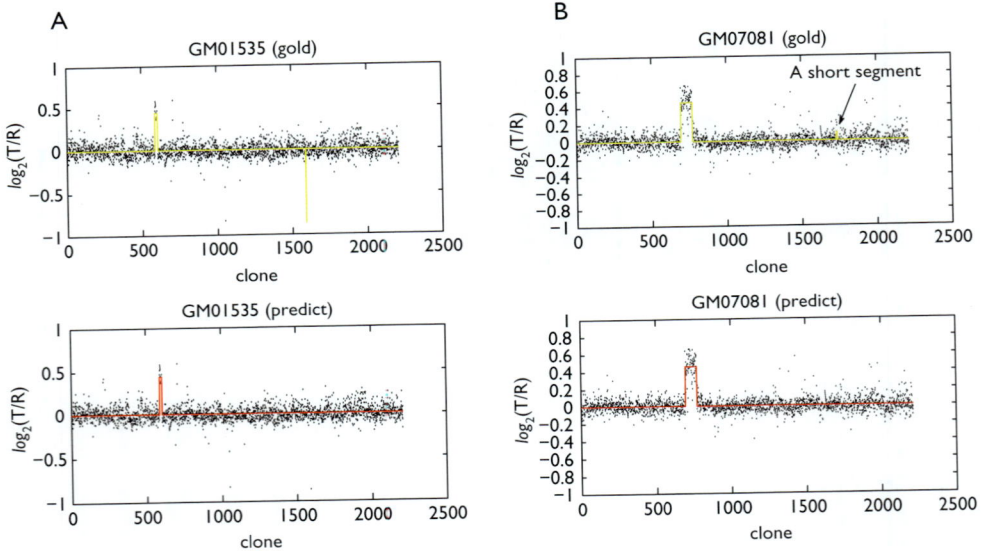

Fig 16.3 Predicted breakpoints by CRF-CNV (bottom) versus true breakpoints (top) on the two cell lines A) GM01535 and B) GM07081. R, gene expression level in the reference sample; T, gene expression level in the testing sample.

Table 9.2 Three Markov random field models and one variable length Markov chain model dedicated to single-SNP genome-wide association studies.

Model	Approach	Algorithm	Input Data	Scalability	Reference
Conditional random field	confounding factors	gradient-based exact optimization of pseudo-likelihood	genotypes	yes	Huang et al. (2011) [22]
Markov random field	multilocus	Markov chain Monte Carlo	genotypes	yes	Verzilli et al. (2006) [59]
Hidden Markov random field	multilocus	maximization of pseudo-likelihood, iterated conditional mode	genotypes	yes (contiguous windows)	Li et al. (2010) [32]
Variable length Markov chain	multilocus	specific heuristic, acyclic probabilistic finite automaton	haplotypes	yes	Browning and Browning (2007) [8] Browning (2006) [7]

RECAPITULATION

Table 9.2 recapitulates the PGM-based methods depicted in this section devoted to single SNP–phenotype association.

9.4 Identifying Epistasis at the Genome Scale

In the present context, the term epistasis is used to denote the departure from independence of the effects of different genetic loci on a studied phenotype: thus, the influence on phenotype cannot be directly predicted by combining the effects of the individual loci [11]. The individual loci involved in such genetic interactions often exert weak marginal influences, if not even negligible. Probabilistic graphical models naturally offer an appealing framework to detect interactions among the variables.

After reporting a general BN-based method proposed by Jiang and collaborators [26], the current section presents variations around the same simplified Bayesian model, by Han and Chen [20], on the one hand, and by Jiang et al. [25, 26], on the other hand. Then, the Markov blanket-based approach of Han et al. [21] is depicted.

9.4.1 Bayesian Network-based Approaches

The high combinatorics of epistasis compels us to consider a simplified Bayesian network. The adaptation to epistasis detection consists in adding a disease vertex in the directed acyclic graph (DAG) and allowing only incoming edges for this vertex (no edges are allowed between SNPs). The interpretation for such DAGs—called direct DAGs (DDAGs) according to [26]—is that the

SNPs are either direct or indirect causes of the disease. Counting all possible DAGs over n vertices may be performed thanks to the recurrence equation $a_n = \sum_{k=1}^{n} (-1)^{k-1} \binom{n}{k} 2^{k(n-k)} a_{n-k}$, where $a_0 = 1$ [42]. The space of DDAGs (2^n) is much smaller than the space of DAGs.

SEARCHING A SPACE OF SIMPLIFIED MODELS

Two algorithms have been proposed to sample the space of DDAGS. To speed up the search process, the learning algorithm of Han and Chen is a branch and bound strategy [20]. It is based on some score (Bayesian score or information theory score). To construct a neighbor for the incumbent graph, quite classically, the two moves allowed are the addition and the deletion of an edge. The search tree is traversed in a depth-first manner, which forbids the extension of a tree branch as soon as the score is shown to decrease at some tree vertex.

On the other hand, Jiang and collaborators have adapted a version of a BN learning algorithm called Greedy equivalent Search (GES). GES can learn the most concise DAG, assuming three properties: faithfulness condition, consistence of the scoring criterion, and composition property (see Appendix 9.A, page 243, for the corresponding definitions). The GES scheme is as follows: start with the empty DAG and iteratively add the edge that most increases the score until no edge fulfills this condition (forward search), then successively delete the edge whose deletion entails the highest increase until no edge checks this requirement (backward search). However, in the case of epistasis, searching over DDAGs with the GES algorithm would be prone to failure, in general: in the example of a two-way interaction, in most cases, neither SNP alone is associated with the disease. Therefore, the epistatic interaction does not satisfy the composition property required by the GES procedure. In the Multiple Beam Search (MBS) algorithm, Jiang et al. adapt the greedy search described above. They reduce the epistasis detection to the case where at least one pair of the interacting loci exerts a marginal effect. The adaptation is straightforward: instead of producing a unique model as above, the process iteratively generates models seeded with one of the SNPs and further extended through forward and backward searches. Thus, every pair of SNPs is investigated. The highest scoring models are output.

Interestingly, Jiang et al. propose an estimation procedure of the Bayesian network posterior probability (BNPP) for a k-locus epistatic model. A single-SNP approach needs only consider the null hypothesis of no association with the disease, H_0, and its alternative, H_1. However, the k-locus epistatic model has to be confronted with competing models (see an illustration with the 2-locus model, Fig. 9.7). The MBS algorithm only computes likelihoods for the models and does not consider their prior probabilities. Indeed, a large model would sometimes exhibit a high likelihood but not be very probable because of its small prior probability. Estimating the BNPP

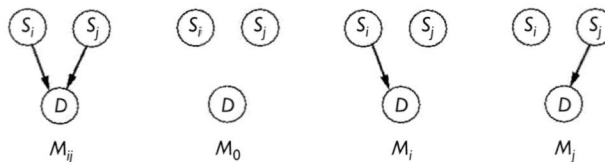

Fig 9.7 The Bayesian models competing with the 2-locus epistatic model. The 2-locus epistatic model, M_{ij}, is shown on the left.

accounts for this situation while it also considers multiple competing hypotheses. In the 2-locus example, the posterior probability of model M_{ij} is calculated as follows:

$$\mathbb{P}(M_{ij} \mid Data) = \frac{\mathbb{P}(Data \mid M_{ij}) \, \mathbb{P}(M_{ij})}{\mathbb{P}(Data)}$$

$$= \frac{\mathbb{P}(Data \mid M_{ij}) \, \mathbb{P}(M_{ij})}{\mathbb{P}(Data \mid M_{ij}) \, \mathbb{P}(M_{ij}) + \mathbb{P}(Data \mid M_0) \, \mathbb{P}(M_0) + \sum_k \mathbb{P}(Data \mid M_k) \, \mathbb{P}(M_k)}$$

where k sums over the two 1-locus models, and M_0 is the model with no dependence at all. To compute $\mathbb{P}(Data)$ without enumerating the competing models, Jiang and collaborators designed a recursive procedure [24].

SCALABILITY

The software program bNEAT of Han and Chen is reported to handle simulated data sets describing 100 SNPs for 2000 individuals. Han and Chen also describe the analysis of 97 327 SNPs for 146 individuals. However, in this case, it has to be noted that a two-stage procedure was used: indeed, the Bayesian network was built from the 200 top SNPs showing the highest marginal association. Three-way epistastic interactions have been evidenced by this method. To control the time complexity, Jiang and collaborators constrain the MBS algorithm to allow no more than a specified number of loci in the addition step. The time complexity is exponential in terms of the number of parents. This limitation forbids the detection of high-order interactions. Besides, if no model is returned by the algorithm, mixing combinatorial investigation of triplets with forward-backward selection is unfortunately intractable. The MBS method has been shown able to handle a data set describing 312 216 SNPs and 1411 individuals.

9.4.2 Markov Blanket-based Method

In the described above BN-based approaches of Jiang et al. and of Han and Chen, the DDAG structure is already a simplification of the initial Bayesian network. The simplified graph intends to focus only on the dependences between the SNPs and the disease status. Other works of Han and collaborators formulate the problem of epistasis detection as the identification of the Markov blanket of the phenotype vertex.

Definition 9.2 (Markov blanket)
In a Bayesian network (G, V), the Markov blanket $MB(T)$ of a target variable $T \in V$, is a minimal set verifying $X \perp\!\!\!\perp X_i \mid MB(T)$, for all $X \in V \setminus \{T\} \setminus MB(T)$. In other words, all other variables are probabilistically independent of the variable T, conditional on the Markov blanket of the variable T.

The Markov blanket completely shields the variable T from all other variables. In a Bayesian network, $MB(T)$ is composed of T's parents, its children, and its children's other parents (spouses) (see Fig. 9.8, page 240). Large-scale Markov blanket discovery requires heuristics. Several algorithms were designed to construct the Markov blanket of a variable: KS [29], GS [33], IAMB [57], MMMB [58], and HITON-MB [2].

Considering the disease status indicator as the target T, Han et al. propose to build an approximation of $MB(T)$ [21]. This set of SNPs is then expected to show a strong association with the disease while also containing few false positives. Starting from the empty set, the algorithm

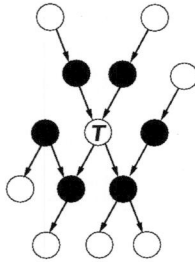

Fig 9.8 The Markov blanket of a target variable. The target variable is indicated as T. The nodes of the Markov blanket are represented in black.

DASSO-MB constructs $MB(T)$ relying on a forward phase and a backward phase. The forward phase consists in a loop that repeatedly identifies the SNP showing the maximal G^2 score with T, conditional to the current $MB(T)$. If this SNP is dependent of T, it is added to $MB(T)$. After this admission, false positives are possibly dismissed from $MB(T)$ through a (local) backward phase based on conditional independence tests. The "forward" loop stops when no further extension of $MB(T)$ is possible. Then, the second phase, a backward step, is applied. The standard backward phase would successively examine if any SNP X_i in $MB(T)$ may be "separated" from T conditional on $MB(T) \setminus \{X_i\}$. Han et al. adapted this previous test to take into account the potential joint effect of a set of SNPs S on the disease status: to dismiss X_I from the Markov blanket under construction, instead of testing for $X_i \perp\!\!\!\perp T \mid MB(T) \setminus \{X_i\}$, the test examines if there exists some non-empty subset of SNPs $S \subseteq MB(T) \setminus \{X_i\}$ such that $X_i \perp\!\!\!\perp T \mid S$.

9.4.3 Recapitulation

Table 9.3 recapitulates the characteristics of the three approaches depicted above.

Table 9.3 Three approaches based on Bayesian networks and dedicated to the detection of epistasis at the genome scale.

Model	Algorithm	Input Data	Scalability	Reference
Simplified Bayesian network (incoming edges for disease vertex)	branch and bound for depth-first strategy in search tree (addition and removal of incoming edges)	genotypes	no (two-stage analysis)	Han and Chen (2011) [20]
Same model as above	adapted forward and backward variable selection	genotypes	yes	Jiang et al. (2010) [26] Jiang et al. (2011) [25]
Markov blanket	adapted forward and backward selection including embedded backward procedure after each iteration of the forward procedure	genotypes	yes	Han et al. (2010) [21]

The approaches of Han and Chen, on the one hand, and the MBS approach of Jiang et al. on the other hand, rely on the same simplified Bayesian network model. To circumvent the complexity issue, Han and Chen preselect the SNPs, based on their marginal effects. However, some three-way interaction could be evidenced.

In contrast, to lower the complexity, Jiang et al. constrain the multiway interaction to contain at least a two-way epistatic interaction. In comparison with the two-stage strategy implemented by Han and Chen at the genome scale, in the approach of Jiang et al., the epistatic pattern need not be composed of SNPs with a marginal effect on the disease. Nonetheless, it is difficult to estimate which epistatic patterns are the more frequent: epistatic patterns containing a two-way epistatic interaction or epistatic patterns where all SNPs have an individual effect on the disease.

Finally, it has to be recalled that the MBS approach of Jiang et al. does not learn the whole Bayesian network, but a simplified DDAG network. Going a step further, the Markov Blanket-based method of Han et al. avoids learning the structure of the whole Bayesian network. Therefore, though necessarily complex at the genome scale, this method seems appealing to detect high-order epistatic patterns at this scale. Besides, parameter learning is short-circuited.

The same data sets could be analyzed in [20] and [21]. However, the Markov Blanket-based approach did not require the preselection of top-ranked SNPs.

9.5 Discussion

The approaches used to model linkage disequilibrium through probabilistic graphical models are empirical. Most often, they do not rely on population genetics concepts: PGM-based approaches estimate empirically the independences and dependences between loci caused by the recombination process, the mutation process, and population history without modeling any of these. The counterpart is that PGM-based models cannot be used to directly make inferences about how the population evolved or how it will evolve in the future.

Tractable models can be designed that represent good approximations of the linkage disequilibrium structure. Besides, the wide range of subsequent utilizations of PGM-based LD modeling include data dimension reduction, haplotype inference, SNP tagging, imputation of non-genotyped SNPs, and LD visualization. The latent variables in the forest of latent tree models (FLTMs) allow data dimension reduction (up to 80%) in [37]. In the field of haplotype imputation, FitGMLD, the software application of Abel and Thomas, provides results whose (high) accuracy is comparable to that obtained through the gold standard software suite Beagle of Browning and Browning.

In the domain of LD representation, a variety of coefficients has been proposed to quantify the intensity of the LD. Beyond standard pairwise LD measures such as the Lewontin measure D' and r^2 coefficient, multilocus LD coefficients, and standard heatmaps [3], PGM-based methods offer an appealing framework to display LD. Thanks to its hierarchical structure, the FLTM model has been shown to provide a compact view of both pairwise and multilocus linkage disequilibrium spatial structures [38].

Linkage disequilibrium arises when the recombination process along chromosomes has not had enough time to randomize the states of alleles at proximal genetic loci. Other forms of allelic associations can also arise due to selection, relatedness of individuals, and population admixture. PGM-based modeling is therefore potentially interesting in such situations. In this regard, the works of Huang and collaborators showed that a PGM-based approach is useful to correct for confounding factors while requiring significantly less running time than standard methods.

The works described in this chapter showed that in the field concerned, the PGM framework could provide scalable and accurate models. Therefore, such works also advocate the use of PGMs to build *generative* models. Though PGM-based LD models do not rely on population genetics concepts, they may be used to generate realistic genetic data. Moreover, an interesting application is assessing significance in association studies.

Another advantage over other models is that the PGM-based models offer biological insights. In FLTMs, a latent variable is likely to be interpreted as the shared ancestry of the multilocus haplotypes defined by the SNPs subsumed by this latent variable. Thus, each state of the latent variable may represent a (probabilistic) group of similar haplotypes. In the situation of limited ancestral recombination, similar haplotypes tend to share recent common ancestry. Low level latent variables subsume small genomic regions exhibiting strong LD and low recombination rates. Latent variables in the lower layers are therefore expected to bear such biological meaning. In VLMCs, many merges (where two edges direct into the same node) are expected at recombination hotspots. In regions of very low LD (due to recombination hotspots or to widely spaced markers), the number of nodes at the frontier between two contiguous layers may be reduced to one. Then the local haplotype clusters will simply correspond to the two SNP variants, with one edge corresponding to each variant, for each layer encompassed by the hotspot.

In the machine learning domain, Bayesian networks are a leading architecture for representing (direct and indirect) dependences as well as conditional independences, in an uncertain framework. PGMs allow us to build good approximations of the underlying reality. Therefore, they represent powerful tools for downstream analyses. Finally, although learning models from high-dimensional data is usually a concern, several PGM-based works have succeeded in designing scalable algorithms for the analysis of genome-scale data.

9.6 Perspectives

Advances in the use of probabilistic graphical models at the genome scale have been obtained recently. However, there is still room for progress.

With the ever decreasing cost of genotyping techniques, the ultimate aim is to apply the methods to substantial genomic regions with hundreds of thousands or millions of loci and substantial numbers of genotyped individuals (in the tens of thousands). Therefore, researchers will have to check that the scalability obtained so far still holds as more massive data are output or gathered through meta-analyses. It is likely that the design of highly scalable algorithms is required. In this line, structure learning procedures have to be tailored. It has been shown that specific procedures such as those based on graph theory or iterative hierarchical clustering of variables allow us to implement genome-scale methods. Exploiting the fact that linkage disequilibrium mostly concerns close SNPs opens the way to investigations around sparse graphical models. In this regard, such techniques as the LASSO-based method are promising [15].

Undoubtedly one of the most challenging issues due to its high combinatorial aspect, the detection of epistasis is still in its infancy. Methods based on the Markov Blanket concept short-circuit learning the whole Bayesian network. However, the complexity remains high. There is urgent need in this domain, as researchers now believe that epistasis may account for a significant portion of the dark matter of genetic risk for disease.

At the era of integration of omics data, the incorporation of biological priors is expected to both increase the accuracy of models and the efficiency of the algorithm designed to learn these models. For instance, when pairwise correlation between SNPs are explicitly considered, such as in the model of Li and collaborators, it would be worth integrating additional information about protein–protein interaction into the analysis. A multiscale approach would first consider the weighted LD graph for the SNPs within a given gene. Then, knowledge about gene network or protein–protein interaction network would provide a large combined SNP network. More generally, incorporating prior information on plausible gene regulatory networks relevant to the phenotype under study would certainly be useful to limit the space of possible dependences. Whatever the paradigm used, data integration is a promising solution. Probabilistic graphical models are known to be well suited for the purpose of integrating heterogeneous data: thus, genetics, gene expression, and proteomics may be integrated in an unified framework. Due to this flexibility, PGMs are expected to significantly help pave the way toward the improvement of prevention, diagnosis, and treatment of diseases.

APPENDIX 9.A THE THREE ASSUMPTIONS FOR THE GES ALGORITHM

Definition 9.3 (Faithfulness condition)

$(G, \mathbb{P})$ satisfies the faithfulness condition if graph G is a perfect map of probability distribution $\mathbb{P}$.

We recall that G is a perfect map of $\mathbb{P}$ if it is both an independence map and a dependence map of $\mathbb{P}$. G is an independent map of $\mathbb{P}$ if every conditional independence encoded through d-separation in G is also encoded in $\mathbb{P}$. Similarly, G is a dependent map of $\mathbb{P}$ if every conditional dependence encoded through d-connection in G is also encoded in $\mathbb{P}$.

Definition 9.4 (Inclusion in a DAG)

Probability distribution $\mathbb{P}$ is included in DAG G if $\mathbb{P}$ is the marginal of a probability distribution in a Bayesian network containing G.

Definition 9.5 (Consistence of a score)

Let s be a scoring criterion over some class of Bayesian networks and $\mathbb{P}$ the joint distribution determined by the data D. The score s is consistent for the class of Bayesian networks if it verifies the two following properties:

For sufficiently large data, if model M_1 includes $\mathbb{P}$ and M_2 does not, then $s(D, M_1) > s(D, M_2)$.

For sufficiently large data, if M_1 and M_2 both include $\mathbb{P}$ and M_1 has smaller dimension than M_2, then $s(D, M_1) > s(D, M_2)$.

Definition 9.6 (Composition property)

This property is verified when, given every variable $X_i \in X$ and every two subsets S, C, $\{X_i\} \not\perp S \mid C$ entails that there exists some $X_j \in S$ such that $X_i \not\perp X_j \mid C$.

. . . .

REFERENCES

[1] H.J. Abel and A. Thomas. Accuracy and computational efficiency of a graphical modeling approach to linkage disequilibrium estimation. *Statistical Applications in Genetics and Molecular Biology*, 10(1):5, 2011.

[2] C. Aliferis, I. Tsamardinos, and A. Statnikov. HITON, a novel Markov blanket algorithm for optimal variable selection. In *AMIA Annual Symposium Proceedings*, volume 2003, pages 21–25. American Medical Informatics Association 2003.

[3] J.C. Barrett, B. Fry, J. Maller, and M.J. Daly. Haploview: analysis and visualization of LD and haplotype maps. *Bioinformatics*, 21(2):263–265, 2005.

[4] Y. Benjamini and Y. Hochberg. Controlling the false discovery rate: a practical and powerful approach to multiple testing. *Journal of the Royal Statistical Society, Series B (Methodological)*, 57(1):289–300, 1995.

[5] J. Besag. On the statistical analysis of dirty pictures. *Journal of the Royal Statistical Society*, Series B, 48:259–302, 1986.

[6] T. Bishop and P. Sham. *Analysis of Multifactorial Diseases*. Academic Press, 1st edition, 2000.

[7] B.L. Browning and S.R. Browning. Efficient multilocus association testing for whole genome association studies using localized haplotype clustering. *Genetic Epidemiology*, 31:365–375, 2007.

[8] S.R. Browning. Multilocus association mapping using variable-length Markov chains. *The American Journal of Human Genetics*, 78(6):903–913, 2006.

[9] S.R. Browning and B.L. Browning. Rapid and accurate haplotype phasing and missing-data inference for whole-genome association studies by use of localized haplotype clustering. *The American Journal of Human Genetics*, 81(5):1084–1097, 2007.

[10] L. Cardon and J. Bell. Association study designs for complex diseases. *Nature Reviews Genetics*, 2:91–99, 2001.

[11] H.J. Cordell. Epistasis: what it means, what it doesn't mean, and statistical methods to detect it in humans. *Human Molecular Genetics*, 11(20):2463–2468, 2002.

[12] M.J. Daly, J.D. Rioux, S.F. Schaffner, T.J. Hudson, and E.S. Lander. High-resolution haplotype structure in the human genome. *Nature Genetics*, 29(2):229–232, 2001.

[13] A. Dobra, B. Jones, C. Hans, J. Nevins, and M. West. Sparse graphical models for exploring gene expression data. *Journal of Multivariate Analysis*, 90:196–212, 2004.

[14] Q. Morris, W. Zhang, N. Mohammad, B. Frey, and T. Hughes. Genrate: a generative model that finds and scores new genes and exons in genomic microarray data. *Pacific Symposium on Biocomputing*, 10:495–506, 2003.

[15] R. Tibshirani, J. Friedman, and T. Hastie. Sparse inverse covariance estimation with the graphical lasso. *Biostatistics*, 9:432–441, 2008.

[16] M. Frydenberg and S.L. Lauritzen. Decomposition of maximum likelihood in mixed interaction models. *Biometrika*, 76:539–555, 1976.

[17] S.B. Gabriel, S.F. Schaffner, H. Nguyen, J.M. Moore, J. Roy, B. Blumenstiel, J. Higgins, M. DeFelice, A. Lochner, M. Faggart, S.N. Liu-Cordero, C. Rotimi, A. Adeyemo, R. Cooper, R. Ward, E.S. Lander, M.J. Daly, and D. Altshuler. The structure of haplotype blocks in the human genome. *Science*, 296(5576):2225–2229, 2002.

[18] D. Geiger, C. Meek, and Y. Wexler. Speeding up HMM algorithms for genetic linkage analysis via chain reductions of the state space. *Bioinformatics*, 25(12):i196–i203, 2009.

[19] P. Giudici and P.J. Green. Decomposable graphical Gaussian model determination. *Biometrika*, 86:785–801, 1999.

[20] B. Han and X.-W. Chen. bNEAT: a Bayesian network method for detecting epistatic interactions in genome-wide association studies. *BMC Genomics*, 12(Suppl 2):S9, 2011.

[21] B. Han, M. Park, and X.-W. Chen. A Markov blanket-based method for detecting causal SNPs in GWAS. *BMC Bioinformatics*, 11(Suppl 3):S5, 2010.

[22] J.C. Huang, C. Meek, C. Kadie, and D. Heckerman. Conditional random fields for fast, large-scale genome-wide association studies. *PLOS ONE*, 6(7):e21591, 2011.

[23] D. Husmeier. Discriminating between rate heterogeneity and interspecific recombination in DNA sequence alignments with phylogenetic factorial hidden Markov models. *Bioinformatics*, 21:ii166–ii172, 2005.

[24] X. Jiang, M.M. Barmada, G.F. Cooper, and M.J. Becich. A Bayesian method for evaluating and discovering disease loci associations. *PLOS ONE*, 6(8):e22075, 2011.

[25] X. Jiang, R.E. Neapolitan, M.M. Barmada, and S. Visweswaran. Learning genetic epistasis using Bayesian network scoring criteria. *BMC Bioinformatics*, 12(89), 2011.

[26] X. Jiang, R.E. Neapolitan, M.M. Barmada, S. Visweswaran, and G.F. Cooper. A fast algorithm for learning epistatic genomic relationships. In *AMIA Annual Symposium Proceedings*, volume 2010, pages 341–345, 2010.

[27] T.J. Jorgensen, I. Ruczinski, B. Kessing, M.W. Smith, Y.Y. Shugart, and A.J. Alberg. Hypothesis-driven candidate gene association studies: practical design and analytical considerations. *American Journal of Epidemiology*, 170(8):986–993, 2009.

[28] G. Kimmel and R. Shamir. GERBIL: Genotype resolution and block identification using likelihood. *PNAS*, 102(1):158–162, 2005.

[29] D. Koller and M. Sahami. Toward optimal feature selection. In *ICML'96*, pages 284–292, 1996.

[30] P.H. Lee and H. Shatkay. BNTagger: improved tagging SNP selection using Bayesian networks. *Bioinformatics*, 22(14):211–219, 2006.

[31] S. Letovsky and S. Kasif. Predicting protein function from protein/protein interaction data: a probabilistic approach. *Bioinformatics*, 19(suppl 1):i197–i204, 2003.

[32] H. Li, Z. Wei, and J. Maris. A hidden Markov random field model for genome-wide association studies. *Biostatistics*, 11(1):139–150, 2010.

[33] D. Margaritis and S. Thrun. Bayesian network induction via local neighborhoods. In *Advances in Neural Information Processing Systems 12*, pages 505–511. The MIT Press, 1999.

[34] M.I. McCarthy, G.R. Abecasis, L.R. Cardon, D.B. Goldstein, J. Little, J.P. Ioannidis, and J.N. Hirschhorn. Genome-wide association studies for complex traits: consensus, uncertainty and challenges. *Nature Reviews Genetics*, 9(5):356–369, 2008.

[35] M.M. Miretti, E.C. Walsh, X. Ke, M. Delgado, M. Griffiths, S. Hunt, J. Morrison, P. Whittaker, E.S. Lander, L.R. Cardon, D.R. Bentley, J.D. Rioux, S. Beck, and P. Deloukas. A high-resolution linkage-disequilibrium map of the human major histocompatibility complex and first generation of tag single-nucleotide polymorphisms. *American Journal of Human Genetics*, 76(4):634–646, 2005.

[36] R. Mourad, C. Sinoquet, and P. Leray. Learning hierarchical Bayesian networks for genome-wide association studies. In Y. Lechevallier and G. Saporta, editors, *Nineteenth International Conference on Computational Statistics (COMPSTAT)*, pages 549–556, 2010.

[37] R. Mourad, C. Sinoquet, and P. Leray. A hierarchical Bayesian network approach for linkage disequilibrium modeling and data-dimensionality reduction prior to genome-wide association studies. *BMC Bioinformatics*, 12:16, 2011.

[38] R. Mourad, C. Sinoquet, and P. Leray. Visualization of pairwise and multilocus linkage disequilibrium structure using latent forests. *PLOS ONE*, 6(12):e27320, 2011.

[39] R. Mourad, C. Sinoquet, N.L. Zhang, T. Liu, and P. Leray. A survey on latent tree models and applications. *Journal of Artificial Intelligence Research*, 47:157–203, 2013.

[40] A.V. Nefian. Learning SNP dependencies using embedded Bayesian networks. In *IEEE Computational Systems, Bioinformatics Conference*, pages 1–6. 2006.

[41] T.J. Pemberton, C. Wang, J.Z. Li, and N.A. Rosenberg. Inference of unexpected genetic relatedness among individuals in HapMap Phase III. *American Journal of Human Genetics*, 87(4):457–464, 2010.

[42] R.W. Robinson. Counting unlabeled acyclic digraphs. In C. H. C. Little, editor, *Combinatorial Mathematics V*, volume 622 of *Lecture Notes in Mathematics*, pages 28–43, Springer, 1977.

[43] A. Rodin, T.H. Mosley, A.G. Clark, C.F. Sing, and E. Boerwinkle. Mining genetic epidemiology data with Bayesian networks application to APOE gene variation and plasma lipid levels. *Journal of Computational Biology*, 12(1):1–11, 2005.

[44] D. Ron, Y. Singer, and N. Tishby. On the learnability and usage of acyclic probabilistic finite automata. *Journal of Computer and System Sciences*, 56:133–152, 1998.

[45] P. Sebastiani, M.F. Ramoni, V. Nolan, C.T. Baldwin, and M.H. Steinberg. Genetic dissection and prognostic modeling of overt stroke in sickle cell anemia. *Nature Genetics*, 37(4):435–440, 2005.

[46] C. Sinoquet, R. Mourad, and P. Leray. Forests of latent tree models for the detection of genetic associations. In *Third International Conference on Bioinformatics Models, Methods and Algorithms (BIOINFORMATICS 2012)*, pages 1–10, 2012.

[47] C. Sinoquet, R. Mourad, and P. Leray. Forests of latent tree models to decipher genotype-phenotype associations In J. Gabriel, J. Schier, S. Van Huffel, E. Conchon, C. Correia, A. Fred, and H. Gamboa, editors, *Biomedical Engineering Systems and Technologies*, pages 113–134. Springer, 2013.

[48] W. Sun and T.T. Cai. Large scale multiple testing under dependency. *Journal of the Royal Statistical Society*, 71, Series B:393–424, 2009.

[49] D. Surmelia, O. Ratmannc, H.-W. Mewesa, and I.V. Tetkoa. Funcat functional inference with belief propagation and feature integration. *Computational Biology and Chemistry*, 32(5):375–377, 2008.

[50] The International HapMap Consortium. The international HapMap project. *Nature*, 426(6968):789–796, 2003.

[51] A. Thomas. Characterizing allelic associations from unphased diploid data by graphical modeling. *Genetic Epidemiology*, 29(1):23–35, 2005.

[52] A. Thomas. Estimation of graphical models whose conditional independence graph are interval graphs and its application to modelling linkage disequilibrium. *Computational Statistics & Data Analysis*, 53(5):1818–1828, 2009.

[53] A. Thomas. A method and program for estimating graphical models for linkage disequilibrium that scale linearly with the number of loci, and their application to gene drop simulation. *Bioinformatics*, 25(10):1287–1292, 2009.

[54] A. Thomas and N.J. Camp. Graphical modeling of the joint distribution of alleles at associated loci. *The American Journal of Human Genetics*, 74:1088–1101, 2004.

[55] A. Thomas and P. Green. Enumerating the decomposable neighbors of a decomposable graph under a simple perturbation scheme. *Computational Statistics & Data Analysis*, 53(4):1232–1238, 2009.

[56] A. Thomas and P.J. Green. Enumerating the junction trees of a decomposable graph. *Journal of Computational and Graphical Statistics*, 18(4):930–940, 2009.

[57] I. Tsamardinos, C. Aliferis, A. Statnikov, and E. Statnikov. Algorithms for large scale Markov blanket discovery. In *Sixteenth International FLAIRS Conference*, pages 376–380. AAAI Press, 2003.

[58] I. Tsamardinos, C.F. Aliferis, and A.R. Statnikov. Time and sample efficient discovery of Markov blankets and direct causal relations. In *Ninth ACM SIGKDD International Conference on Knowledge Discovery and Data Mining*, pages 673–678, 2003.

[59] C.J. Verzilli, N. Stallard, and J.C. Whittaker. Bayesian graphical models for genome-wide association studies. *The American Journal of Human Genetics*, 79:100–112, 2006.

[60] E. Villanueva and C.D. Maciel. Modeling associations between genetic markers using Bayesian networks. *Bioinformatics*, 26(18):i632–i637, 2010.

[61] Y. Zhang and L. Ji. Clustering of SNPs by a structural EM algorithm. In *International Joint Conference on Bioinformatics, Systems Biology and Intelligent Computing*, pages 147–150. IEEE, 2009.

CHAPTER 10

Modeling Linkage Disequilibrium with Decomposable Graphical Models

HALEY J. ABEL AND ALUN THOMAS

This chapter describes the use of decomposable graphical models (DGMs) to represent the dependences within genetic data, or linkage disequilibrium (LD), prior to various downstream applications. First, following the definition of DGMs, general model estimation algorithms are reviewed: schemes based on Markov chain Monte Carlo (MCMC) and related simulated annealing strategies are described. However, for tractable processing of high-dimensional data, it is shown that sampling the space of DGMs is efficiently replaced with the sampling of representations of DGMs—the junction trees. Then, a first application is considered, the phase imputation for diploid data, which consists in inferring the latent genetical phased haplotypes underlying the observed genetical unphased genotypes. Various strategies designed to allow model estimation on the genome-wide scale are also discussed. These strategies include limiting the length of the edges allowed in the graph, where length is either the physical distance in base pairs or the ordinal distance between the loci connected by the edge. In particular, it is shown that in the case of diploid data, decoupling the model estimation step from the phasing step allows scalability of the whole estimation process. The chapter ends with the illustration of the potentialities of DGMs through four applications. Again, the first application deals with phasing, used to impute the data when genotypes are missing, either due to random assay dropout, or to deliberate discarding of loci from the assay, for cost reasons. Besides, the incorporation of a genotyping error term in the probabilistic model allows us to detect and correct genotyping errors. The

Probabilistic Graphical Models for Genetics, Genomics, and Postgenomics. First Edition. Christine Sinoquet & Raphaël Mourad (Eds). © Oxford University Press 2014. Published in 2014 by Oxford University Press.

second useful application of graphical models for LD modeling is the ability to simulate from the joint distribution. Thus gene drop simulation—that is simulating the genotypes of related individuals in a pedigree—can be performed under LD. In a third application, the set of variables of the DGM is extended to admit variables representing the phenotypes, typically diseases, covariates, and population stratum information associated with any individual. In this line, a case study is discussed that focuses on the detection of causal genetic variants. Finally, DGMs can also serve the purpose of modeling admixture, which is the case arising when an individual's ancestors come from a mixture of populations. The adaptation of the model is described, and results are shown. The chapter concludes with some speculation on future work to model sequence data: there is an increasingly common use of high-throughput methods to sequence single DNA molecules to assess the genetic variants that an individual has; it is explained how to adapt a DGM to account for a bias inherent to the use of the alignment program involved in the process.

10.1 Introduction

The chromosomes making up the genomes that individuals receive from their parents consist of long strands of DNA. Much of these sequences of nucleotides is the same for all individuals; however, those positions at which variation is seen, called *polymorphisms*, are responsible for genetic differences between people. The various forms seen at a polymorphism are called *alleles*, and the particular combination of alleles along the polymorphisms of a chromosome is called a *haplotype*. Humans are *diploid*, that is, they carry two complete genomes and, hence, two complete sets of haplotypes, one inherited from each parent. When new alleles arise by mutation, they do so on a particular haplotype and will be associated with the alleles of that haplotype until shuffled in time by the process of genetic recombination. When mutations are sufficiently ancient or sufficiently far apart, this shuffling will be adequate to render the alleles observed at the polymorphisms independent; that is,

$$\mathbb{P}(x) = \mathbb{P}(X_1 = x_1, X_2 = x_2, \dots X_n = x_n) = \prod_{i=1}^{n} \mathbb{P}(X_i = x_i), \qquad (10.1)$$

where X_i is the allele at the ith polymorphism in the haplotype X ($X = \{X_1, \dots X_n\}$). In this case, it is said that the loci have reached linkage equilibrium (LE).

Until relatively recently, shortly after the turn of the millennium, the number of polymorphisms that could be readily assayed in an individual was small. A typical set of genetic markers would be around 1000 multiallelic short tandem repeats spread fairly evenly and sparsely throughout the genome. Under these circumstances LE was a fair assumption. In the last ten years, however, genotyping methods have been developed to assay, quickly and at low cost, on the order of 1 million markers, mostly biallelic single nucleotide polymorphisms (SNPs). Genotype assays are increasingly giving way to complete sequencing, which evaluates alleles at all of an individual's polymorphisms, SNPs as well as structural variants, including copy number variants (CNVs) and translocations. With these dense assays, it is no longer the case that the observed variants are ancient or well spaced, and it becomes necessary to consider *allelic association* or *linkage disequilibrium* (LD). We now have

$$\mathbb{P}(x) = \mathbb{P}(X_1 = x_1, \ X_2 = x_2, \ldots X_n = x_n) = \prod_j f_j(t_j) \qquad (10.2)$$

for some arbitrary collection of subsets $\{t_j\}$ of x and arbitrary factors $\{f_j(t_j)\}$ as functions on the variables in these subsets.

Since the shuffling of recombination is ongoing, we expect LD to be a local phenomenon, and hence expect the $\{t_j\}$ to represent relatively small sets of proximal loci. Under these circumstances, graphical models may offer a tractable way to model the complex joint haplotype distributions.

A graphical model consists of a conditional independence graph, or Markov graph, representing the conditional independences between the variables under consideration, and a set of potential functions that can be combined to form the joint distribution of these variables. The graph is defined, given equation (10.2), with a vertex representing each variable and an edge connecting any pair of variables that appears together in any of the subsets $\{t_j\}$. The potential functions may be arbitrary factors $\{f_j()\}$, or, as we describe below, elements of a canonical decomposition. In this chapter, we consider modeling LD by estimating decomposable graphical models (DGM). In the following section, we define DGMs and then review general methods for their estimation given a complete multinomial sample of observations. We then consider the particular problems encountered in the domain of LD modeling. First is the issue that haplotypes are not usually assayed. Instead, the unordered *genotypes*, the pairs of alleles an individual inherits from both parents at a polymorphism, are. Another slight, related complication is that, by nature of the molecular assay, genotypic observations are subject to dropout and error. The second major issue is the sheer scale of the number of variables involved, which necessitates some restriction on the model space. Finally, we consider some applications and extensions and conclude with an indication of future research directions.

We note that several existing methods for modeling LD involve, either implicitly or explicitly, types of graphical modeling. For example, the fastPHASE program [29] uses the graphical model described in [24]. In this model, variables are used to indicate what is termed an ancestral haplotype *cluster* from which an individual's haplotype is inherited subject to possible mutation. The graphical model is derived from the assumption that clusters along a chromosome are related in a first order Markov structure and that the mutation process occurs independent of the cluster of origin. The model underlying the BEAGLE program [6] is a special case of a split graphical model [16, 17]. A split graphical model is one in which the structure of the conditional independence graph can vary with states of the variables it contains. However, we consider here only estimating standard DGMs that specify a complete probability mass function for haplotypes and enable the full force of the general forward–backward machinery that makes graphical models such an attractive option.

10.2 Methods

10.2.1 Decomposable Graphical Models

A subgraph of a graph G is said to be *complete* if all pairs of vertices are connected in G. A complete subgraph is called a *clique* if it is maximal, that is, it is not contained in any other complete subgraph of G. A graphical model is *decomposable* if it satisfies the *running intersection property*;

that is, its cliques $\{C_1, C_2 \ldots C_m\}$ can be ordered and separators $\{S_i\}$ defined in such a way that

$$S_i = C_i \cup \bigcap_{k=i+1}^{m} C_k \subset C_j \quad \text{for some } j > i. \tag{10.3}$$

An equivalent condition is that G is a *triangulated* graph, in which any cycle of four or more vertices is shortcut by a chord: the term *chordal graph* is also used. If the conditional independence graph is decomposable, then there are canonical decompositions of the joint probability function into either clique conditional or clique marginal representations.

$$\mathbb{P}(X|G) = \prod_i \mathbb{P}(C_i|S_i) = \prod_i \frac{\mathbb{P}(C_i)}{\mathbb{P}(S_i)}. \tag{10.4}$$

We require the clique marginals to be *coherent* in the sense that if a separator S is contained in two cliques, C_1 and C_2, then it has the same marginal distribution in each clique. In this case, the model is specified by the clique marginals. Note that the requirement for coherence depends on the graph structure, and so cliques can, in principle, have different margins specified for different graphs. In practice, however, only cases where clique marginals are invariant to graph structure are considered, and we make this requirement in what follows.

It is important to recognize that the graph is only part of the specification of the graphical model: the form and the parameters of the clique marginals are also required. In equation (10.4), we have treated the clique and separator marginals as if they were fully specified; however, they may depend on parameter values that, potentially, have to be estimated. In a general Bayesian estimation framework, priors are assumed for these parameters, and a joint posterior distribution on parameter values and graph structure is obtained. However, as our interest is primarily in the graphical structure and as our aim is to address problems with large numbers of variables, we use simple maximum likelihood parameter estimates. For examples of the full Bayesian treatment, the reader is referred to the work of Jones et al. [19] or Giudici and Green [12].

The final important element is the *junction tree* representation of a decomposable graph. Consider an additional graph in which the cliques of G are vertices and each pair of such vertices is connected by an edge weighted by the cardinality of the intersection of the corresponding pair of cliques. Then any heaviest spanning tree J of this graph is a junction tree of G. Note that although defined in this way, there are far more efficient ways of finding J [30]. The junction tree has the *junction property* that the cliques containing any subset of vertices of G form a single connected subtree of J. A junction tree exists if and only if G is decomposable [13], and, in general, G will have many different equivalent junction trees that can be enumerated by the method of Thomas and Green [38]; that is, there is a one-to-many correspondence between decomposable graphs and junction trees. As we shall see below, the junction tree is a representation of G that allows for efficient manipulation.

Note that when G consists of several unconnected components, many authors define a corresponding *junction forest*. However, we follow Thomas and Green [37] in connecting the subtrees of a junction forest into a single junction tree by adding edges associated with the empty separator between the subtrees. This junction tree clearly has the junction property and avoids having to consider multiple special cases.

Finally, we should consider why a restriction to DGM is desirable. This is primarily because of tractability. Model estimation is far more efficient within this class [19], particularly as we will consider large numbers of variables. More importantly, the likely applications for the models we estimate will require the use of forward–backward methods which can only be carried out on DGM. For some applications, such as unconditional simulation from $\mathbb{P}(X)$, decomposability is not critical and forward-only algorithms in Bayesian networks can do this efficiently [5]. A good example of forward simulation in a Bayesian network is the *gene-drop* method commonly used in genetics to simulate genes in a pedigree or family tree [25]. Gene-dropping works by first allocating alleles to the founders of a pedigree according to the random distribution of alleles in a population. Then, as the name suggests, the alleles are dropped from generation to generation by mimicking simple Mendelian inheritance. This process is an efficient way to simulate the genotypes of relatives even for large and complex pedigrees. Recent versions also mimic the recombination process and so simulate alleles at multiple genetic loci [1], and we will consider extensions to this below. In contrast, efficient simulation of the values of all variables in a system, conditional on the values of a subset, requires the two-stage forward–backward approach. To use this, non-decomposable graphs have to be triangulated, that is, filled in with redundant edges to find an equivalent decomposable model: in effect, we multiply (10.2) by redundant factors of the form

$$f(x_i, x_j) = 1 \quad \forall x_i, x_j. \tag{10.5}$$

Furthermore, tractability of forward–backward methods is determined by the number of states in the clique marginals, which for multinomial models is exponential in the size of the clique; thus we need not only decomposability but also DGM with small cliques [22].

10.2.2 Estimating Decomposable Graphical Models

Bayesian graphical model estimation methods are based on the posterior distribution of the graph structure and the clique marginal parameters given the observed data. Let X be the observed data consisting of n complete multivariate observations of the v variables in the system. At this point, we do not allow for missing values or observational error, but this will be the focus of the next section. Let G be a particular graph structure, and let M be the collection of parameter estimates for all possible cliques. Thus, M contains all the parameters required for all possible graphs, not just the current graph G. Assuming independent priors for G and M, the posterior distribution that we wish to maximize or sample from is

$$\mathbb{P}(G, M|X) \propto \mathbb{P}(X|G, M)\mathbb{P}(G)\mathbb{P}(M). \tag{10.6}$$

This is a complex proposition and is usually addressed by MCMC methods. In order to simplify the problem, as noted above, we fix M to be $\hat{M}$, the maximum likelihood parameter estimates obtained from the data X. Because the variables we consider are discrete, the clique marginals will be simple multinomials, and maximum likelihood parameter estimates can be obtained from contingency tables. Although $\hat{M}$ is in principle the collection of parameter estimates for all possible cliques, in practice we only compute the estimates for the cliques we need for the values of G we encounter in the MCMC process. The probability distribution for G that we base our inference on is

$$\mathbb{P}(G|X, \hat{M}) \propto \mathbb{P}(X|G, \hat{M})\mathbb{P}(G). \tag{10.7}$$

In order to avoid overparameterization and favor tractability, we follow previous work [18] and choose a prior on G that penalizes complexity via a Gibbs measure on $df(G)$, the degrees of freedom, or number of free parameters, in the model

$$\mathbb{P}(G) \propto e^{-\alpha df(G)}. \tag{10.8}$$

The total degrees of freedom can be calculated from the clique and separator marginal degrees of freedom as follows:

$$df(G) = \sum_i df(C_i) - \sum_i df(S_i). \tag{10.9}$$

Under this choice of prior, the posterior distribution is

$$\mathbb{P}(G|X, \hat{M}) \propto \mathbb{P}(X|G, \hat{M})e^{-\alpha df(G)} \tag{10.10}$$

which also has a convenient decomposition as

$$\mathbb{P}(G|X, \hat{M}) \propto \prod_i \frac{\mathbb{P}(C_i|\hat{M})e^{-\alpha df(C_i)}}{\mathbb{P}(S_i|\hat{M})e^{-\alpha df(S_i)}}. \tag{10.11}$$

Graphs can be sampled from this posterior distribution using the methods of Metropolis [27] or Hastings [15]. Given an incumbent graph, G, a small perturbation such as adding or deleting an edge is applied to give a proposal G'. If the proposal scheme is symmetric, this is accepted as the new incumbent with probability

$$\max\left(1, \frac{\mathbb{P}(G'|X, \hat{M})}{\mathbb{P}(G|X, \hat{M})}\right) \tag{10.12}$$

according to Metropolis's method [27]. If the proposals are not symmetric, the method from [15] is used to specify the appropriate acceptance probability. These, and other MCMC methods, are, in effect, ways to integrate over the posterior distribution of graphical models and can be used to account for model uncertainty in model averaging approaches. On the other hand, if a single best model is required, maximization of the posterior distribution can be addressed by *simulated annealing* [20]. This can be considered as a variation of the method from [27] where the acceptance probability is

$$\max\left(1, \left[\frac{\mathbb{P}(G'|X, \hat{M})}{\mathbb{P}(G|X, \hat{M})}\right]^{\frac{1}{t}}\right). \tag{10.13}$$

The parameter, t, is usually referred to as the temperature and is varied during the course of the search. The temperature is initially set to a large value so that the acceptance probability is close to 1 and the search, in effect, performs a random walk over the space of decomposable graphs. It is then decreased gradually toward zero, which concentrates the sampling at the mode of the distribution. The goal of slowly decreasing t in this way is to avoid getting stuck in local optima. The schedule by which t is decreased is usually a pragmatic choice that depends on the specific context in which simulated annealing is used. The program package that the authors have developed includes an implementation of simulated annealing to fit a graphical model to multinomial data with a graphical user interface that allows the user to see the graph as it evolves and manually control the temperature. The package is called the Java Package for Statistical Genetics and Computational Statistics (JPSGCS) and is available from http://balance.med.utah.edu/wiki/index.php/JPSGCS.

Working within the subspace of decomposable graphs now involves some complexity, as arbitrary perturbations to a decomposable G do not necessarily result in decomposable G'. Moreover, checking for decomposability of G' directly is an $O(v + e)$ operation, where v is the number of vertices and e the number of edges of G'. Fortunately we have two simple and elegant results due to [10] and [12], which can be stated as follows:

- Removing the edge between adjacent vertices x and y in a decomposable graph G results in a decomposable graph G' if and only if x and y appear together in exactly one clique.
- Adding an edge between non-adjacent vertices x and y in a decomposable graph G results in a decomposable graph G' if and only if x and y appear in cliques that are connected in some junction tree J of G.

Given a junction tree J for G it is, therefore, straightforward to check that an edge removal is legal. Checking that a connection preserves decomposability is more involved, requiring inspection of J and possibly its replacement by a different junction tree of G. In the worst case this is an $O(v)$ process, but its average performance generally results in efficient checking. Additionally, it can be shown [12] that if G' is decomposable, a junction tree J' for G' can be found directly from J by changing some combination of four cliques and/or separators. In the case of disconnecting two vertices of G, determining the cliques and separators that need to be changed requires inspection of the junction tree in the neighborhood of the unique clique that contains the edge to be removed, and there are four separate cases that need to be considered. In the case of connecting two vertices of G, there are again four different possible cases which can be identified by inspection of the, possibly modified, junction tree. Each of the four updates corresponding to a disconnection is reversed by one of the four connection updates, so that the disconnection and connection types form four matching pairs of reversible updates. Green and Thomas [14] give a more complete description of these updates with figures to show the changes made to the junction tree.

As well as being an efficient way of operating on the junction trees, this also shows that the acceptance probability in (10.12) can be evaluated quickly in $O(1)$ time since, because of the clique-separator decomposition in equation (10.11), all but four of the terms in the ratio (10.12) cancel. The approach of Giudici and Green [12], which was also used by Jones et al. [19] in the decomposable case, therefore allows us to propose and evaluate Metropolis perturbations very efficiently. However, there remains the problem that for large G, the probability that a randomly chosen pair of vertices can be either connected or disconnected while preserving decomposability is typically very small; thus we spend much time rejecting non-decomposable proposals. This is addressed by a new junction tree sampler that restricts vertex connections to the vertices that are in cliques connected by a link in the current junction tree J [14]. This avoids the possible $O(v)$ operations to find a different junction tree for the same graph. Moreover, it allows the structure of J to restrict proposals to those that will result in decomposable G'. The result is a sampler with lower rejection rates that proposes and evaluates a proposal in $O(1)$ time. The proposal scheme is, however, asymmetric and requires use of Hastings's method [15].

10.2.3 Application to Diploid Data by Phase Imputation

So far, we have considered estimating DGM in the general multinomial case. The first application of graphical modeling to LD [35] followed this approach in special cases where haplotypes are directly observable, for instance, when genotyping male X-chromosomes. However, the problem of estimating LD models poses two interesting issues. The first of these is that of incomplete data and observational error. The molecular assays used in this field depend on the quality of DNA extraction, storage, and processing, and are subject to error and dropout. While perhaps not perfect, it is at least reasonable to assume that data are missing or mistaken at random. More specific to this particular problem is that, since humans are diploid, haplotypes are not typically observed. This is because the molecular genotyping assays involved operate not on individual molecules, but on populations of molecules derived from both of the alleles that an individual has inherited at a locus. Thus, for example we may find that at one SNP locus an individual has inherited the pair $\{A, G\}$ while a neighboring locus carries the pair $\{C, T\}$. It will not usually be known, however, whether the A and C alleles were inherited on one haplotype and the G and T on the other or if the A and T traveled together on one and the G and C on the other. This problem is referred to as not knowing the *phase* of the inherited haplotypes. Reconstructing this complete information from the unphased genotypes is called *phasing* the haplotypes. It is interesting to note that the most current sequencing methods do in fact assay single molecules and that molecular phasing information is available for loci that are very close together. In the final section, we speculate on how this information might be incorporated into the estimation framework.

Much of the appeal of graphical models is that they allow for partial and error-prone observation of data in an elegant way. Let X again be the notionally complete set of data, or the data that would be ideal but not necessarily possible to collect. Let D be the data that can in practice be collected: D will typically be an incomplete subset of X, or contain new variables whose values are derived from X by partial or erroneous observation. We will assume that the observational process is independent of the underlying graphical model, that is,

$$\mathbb{P}(D|X, G, M) = \mathbb{P}(D|X). \tag{10.14}$$

Hence, the extended system, including both the complete and observable variables, forms an augmented graphical model defined by

$$\mathbb{P}(G, M, X|D) \propto \mathbb{P}(D, X, G, M) = \mathbb{P}(D|X)\mathbb{P}(X|G, M)\mathbb{P}(G)\mathbb{P}(M). \tag{10.15}$$

This distribution can again be sampled from by MCMC methods. Given some starting reconstruction for X, we can estimate M by maximum likelihood and sample G from $\mathbb{P}(G|\hat{M}, X)$

by the Metropolis–Hastings methods described in Subsection 10.2.2. Given G, $\hat{M}$, and D, we can then make a new reconstruction, or imputation, of X by sampling its value from the distribution

$$\mathbb{P}(X|D, \hat{M}, G) \propto \mathbb{P}(D|X)\mathbb{P}(X|\hat{M}, G) \qquad (10.16)$$

using a standard forward–backward algorithm in the augmented graphical model. We then iterate around these conditional updates until convergence to the ergodic distribution. Note once more that, because of the computational burden, we fix M by maximum likelihood whereas a full Bayesian treatment would sample M from $\mathbb{P}(M|X)$.

If, instead of sampling from $\mathbb{P}(G, M, X|D)$, we wish to find its mode, the Metropolis–Hastings update of G can again be replaced with a random uphill search or simulated annealing, and a different forward–backward method can be used to find the value of X that maximizes $\mathbb{P}(X|D, M, G)$.

In the LD problem, X comprises two complete phase known haplotypes for each individual in the sample, while D is the list of genotypes. This can be represented in the graphical model shown in Fig. 10.1. Here, the haplotype LD model in the top of the picture has been applied in parallel to both of the individual's haplotypes. The alleles at the ith locus, $X_{i,1}$ and $X_{i,2}$ then determine the state of the ith genotype which, possibly subject to error, is observed as D_i. The conditional probabilities $\mathbb{P}(D_i|X_{i,1}, X_{i,2})$ will allocate most probability when D_i corresponds to the unordered

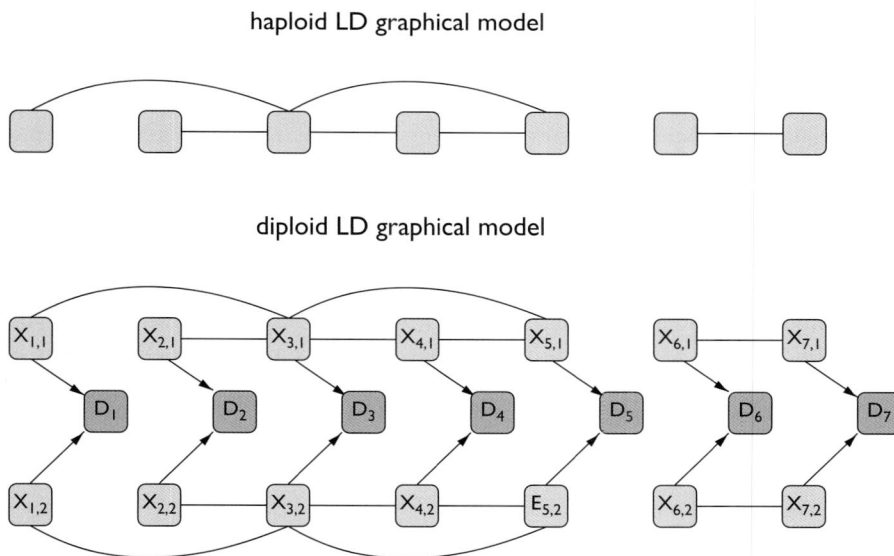

Fig 10.1 Conversion from haploid to diploid linkage disequilibrium (LD) graphical model. This figure shows how a graphical model for LD derived for haplotypes, shown above, can be duplicated to form a new graphical model representing the maternally and paternally inherited haplotypes. The augmented model, shown below, also contains variables representing the genotypes observed at the genetic loci. The paternal and maternal alleles at the ith locus are $X_{i,1}$ and $X_{i,2}$, respectively, and these variables are not directly observable. The observable genotype at the ith locus is labeled D_i. Some of the D_i may also be missing.

pair $\{X_{i,1}, X_{i,2}\}$, but may allow some probability at other values to allow for error. Reconstructing X from D for a particular individual then involves using a forward–backward algorithm to impute the haplotypes from the observed genotypes. Note that the diploid model shown will not typically be decomposable. This can be dealt with either by filling in edges, or by replacing each ordered pair $(X_{i,1}, X_{i,2})$ by a single variable encoding all the bivariate states. In this latter case, if the haplotype model is decomposable, so will be the derived diploid model.

10.2.4 Estimation on the Genome-Wide Scale

The second problem commonly encountered in linkage disequilibrium (LD) estimation is that of scale. In most applications of graphical model estimation, a system of 100 to 200 variables would be considered a moderate sized problem [19] and 1000 or more a large one; however, the LD problem is orders of magnitude larger than this. Current genotyping systems assay genotypes at over 1 000 000 specified genetic loci, mostly SNPs but also some CNVs, while sequencing methods can potentially assay all of an individual's variants, including ones at positions that were not previously known to be polymorphic. Recall that the human genome is approximately 3 000 000 000 bases long and that it is estimated that polymorphisms occur at a rate of about 1 per 1000 bases. Thus, a genome-wide analysis assayed by genomic sequencing might involve 3 000 000 variables. To some extent, this problem is mitigated if we consider the 22 autosomal chromosome pairs separately as they segregate independently; however, the larger chromosomes will still have several hundred thousand polymorphisms. To address this, we can exploit the local nature of LD. While other causes of allelic association, such as population admixture and natural selection, can form correlations between distant polymorphisms and even between those on different chromosomes, the shuffling induced by recombination makes LD a more localized effect. It makes sense, therefore, to consider restrictions on the structure of conditional independence graphs to represent this localization and to exploit the linear arrangement of polymorphisms along the chromosomes.

The most straightforward approach to this is to limit the length of the edges allowed in the graph, where length here is either the physical distance in base pairs or the ordinal distance between the loci connected by the edge. This approach, using ordinal distance, was taken by [2] in the FitGMLD program. As well as restricting the graph space, the limit on the edge length allows a *walking window* estimation algorithm. The two-stage MCMC estimation method described in Subsection 10.2.3 (page 254) is applied to a contiguous set, or window, of loci, where the width of the window is larger than the maximum edge length. The window is then shifted along the list, by an amount equal to half its length, and the estimation run on the new window. It was shown in [2] that a maximum edge length of 20 and a window width of 250 were optimal for the HapMap Yoruba chromosome 1 data set [31]. In this application, two windows were used. In the first, MCMC methods were used to sample from the posterior graph distribution. In the second, MCMC methods were used to maximize the probability under the posterior. Applying the schemes in these windows in this order is equivalent to an application of simulated annealing in which the temperature is initially set to 1 but switched to 0 instantaneously half-way through the process. The windows were both shifted along by half their width between estimation steps, with the maximizing window immediately following the sampling window. In this way, it is not necessary to simultaneously hold the data at all the loci for each of the individuals in the estimation data set. Only the data for the loci spanned by the two windows at each stage are required. Thus, even for large data sets of 1000 or more individuals, processor storage constraints do not limit the number of loci that can be considered in the same model. Note that to allow this, the data need to be arranged in a *locus-by-individual* array.

In addition to models and haplotype frequencies, programs such as fastPHASE and BEAGLE also output the phased haplotypes of the individuals in the training set. Phasing is also carried out by FitGMLD during the model fitting stage; however, this phasing uses only the information local to each window as it is processed. In contrast, although it is estimated in a piecewise fashion, the graphical model produced is a single structure representing the joint distribution of haplotypes over the complete range of the genetic loci in the data. FitGMLD, therefore, outputs only the graphical model, and phasing is deferred to another program, called Complete. Complete reads in the whole graphical model and a set of, possibly incomplete, unphased genotypes for a list of individuals, and outputs complete phased haplotypes. This phasing is done using the same approach as described in Subsection 10.2.3. Either a posterior maximum phasing for each individual is output, or a random imputation from the posterior. At this stage, although we store data for all the loci simultaneously, we need only do this for one individual at a time. Thus, space is again not a limiting factor; however, the data now needs to be read from an *individual-by-locus* array of input.

The effect of edge-length restrictions and decoupling of the model estimation step from the phasing step is that DGM estimation can be carried out on very large numbers of loci. Although fitting a model using FitGMLD requires considerable computation time, this was shown in [2] to grow linearly with both the number of loci and number of individuals. Importantly, once a DGM for LD has been estimated, the Complete program very quickly imputes or simulates complete phased haplotypes. Whereas alternative phasing methods have focused on obtaining single best effort imputations, it is now feasible to sample multiple imputations to better reflect and examine the effect of the uncertainty of the reconstruction. For very small numbers of loci, it is also feasible to list all possible haplotype reconstructions with their posterior probabilities; however, the number of possible reconstructions makes this infeasible for large numbers of loci. In some cases it may be possible to assess the effect of haplotype uncertainty by combining the graphical model for LD with the variables used in the subsequent analysis into a larger graphical model. Integration over the posterior haplotype distribution could then be performed exactly using the forward–backward algorithm. However, this would likely require considerable programming work and the derived graphical models might not be tractable. Assessing uncertainty of haplotype reconstruction by multiple random imputation is more straightforward: Complete can be used to produce multiple randomly imputed data sets that can then be analyzed using existing haplotype analysis programs. The number of imputations required will depend strongly on the frequency of the haplotypes being estimated and the specific use being made of them, so it is difficult to make general recommendations. However, as a rough guide, [34] showed that for estimating the cumulative distribution of the SNP streak statistics described in Subsection 10.3.2, 1000 simulations gave very consistent results.

Other model restrictions have also been considered, for instance, a restriction that the conditional independence graphs must be *interval graphs* [32]. An interval graph is one in which each vertex is associated with an interval of the real line and vertices are connected if and only if the intervals overlap by a certain amount. With the added requirement that each interval covers the location of its corresponding genetic locus, this becomes a heuristically appealing structure for LD as it represents both the location and extent of the LD around a locus. However, empirical studies showed that the edge length restriction dealt with in Subsection 10.2.4 gave more accurate imputations [2]. *Proper interval graphs* have also been considered [11]. A proper interval graph is an interval graph with the further requirement that the intervals are all of the same length. A consequence of this is that the graph has a junction tree representation with a simple linear structure. While less useful for general LD methods, this simplicity is proving useful for admixture mapping, as we discuss below.

10.3 Applications

We now consider applying graphical models for LD to some specific genetic problems.

10.3.1 Phasing

The most common use for LD modeling is to estimate phased haplotypes from unphased genotypes. This same process will also impute alleles when genotypes are missing, either due to random assay dropout, or at loci that were deliberately not assayed, for example, because of cost. Since we incorporate a genotyping error term into the graphical model, we are also able to detect and correct likely genotyping errors. In this subsection, we will illustrate the computational costs of phasing and imputation using the 223 110 loci on chromosome 1 genotyped by the HapMap project on 60 individuals from the Yoruba population of Nigeria [31]. Three programs were used for this: TransposeLinkage to turn standard LINKAGE [21] format files, which are individual-by-locus arrays, into locus-by-individual arrayed files; FitGMLD, to fit the graphical model; and Complete, for the phasing as described in Subsection 10.2.3. Full information about these programs is available from the JPSGCS wiki page. All the following programs were run on single Intel 2.80 GHz CPU.

Model estimation is by far the most computationally demanding part of the process. The computational time is a linear function of the number of loci [2]; however, estimating the full LD model for chromosome 1 still required an overnight run of 8 hours. Fig. 10.2 illustrates a typical example of the type of graph that FitGMLD outputs. This figure shows the conditional independence graph of the LD structure for the first 1000 SNP loci on chromosome 1 estimated from the 60 Yoruba individuals. The graph depicted is one that occurred about three-quarters of the way through the estimation process. The walking window has fit a model for the first 750 loci, shown in the lower left, but has not yet reached the final 250 loci, shown in the top right. The portion of the graph for the final 250 loci is still in the initial state, which is a second-order Markov structure.

In contrast to model fitting using FitGMLD, the phasing using Complete was very quick. Maximum posterior probability phases for all 60 individuals given the LD model and the observed genotypes were found in 172 seconds. This consisted of 36 seconds to read in the graphical model and set up the internal representation, and 2.27 seconds per individual to read the data, apply the LD graphical model to that data, and run the forward–backward algorithm to obtain a posterior most probable phasing.

For simulation of haplotypes from the posterior probability distribution, the computational economics are even better. Table 10.1 gives the times required for a range of numbers of simulations. These running times can be summarized in the following way: if n is the number of individuals and s is the number of simulations required, the total time required is

$$36 + n(1.66 + 0.229s) \text{ seconds.} \tag{10.17}$$

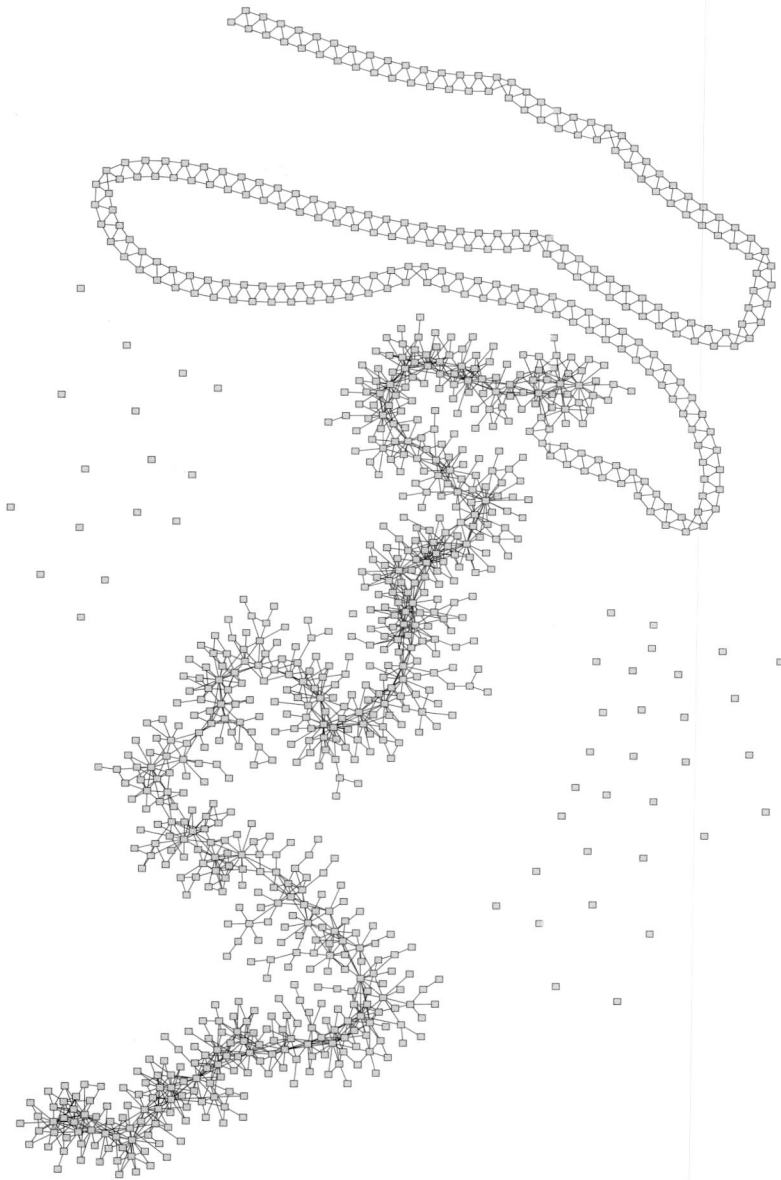

Fig 10.2 A typical output of software program FitGMLD. This figure shows a conditional independence graph for the first 1000 loci of chromosome 1 in the process of being estimated by the FitGMLD program. At this point in the process, the walking-window approach has estimated a model for about three-quarters of the loci, shown in the lower part of the graph, but not yet reached the final 250 loci. The graph connecting the unreached loci is still in its initial state which is a second-order Markov chain in which the order of the loci is determined by their physical order in the genome. The graph in the processed portion maintains a roughly linear global structure, but with considerable local deviation from strict linearity. The model was estimated from the HapMap data on 60 unrelated Yoruba individuals from Nigeria.

Number of Simulations	Time in Seconds
1	161
10	279
100	1 489
1000	13 862

So the marginal cost of a single random phasing for an individual is 0.229 seconds. Since the number of loci considered here is 223 110, this corresponds to approximately one second per simulation, per person, per million loci.

10.3.2 Unconditional Simulation

Another useful application of graphical models for LD is the ability to simulate from the joint distribution without any of the variables being fixed or weighted by observation. This approach was adopted [33] to extend the multilocus gene dropping method [1] to allow for LD. First, multilocus founder haplotypes were simulated from an LD model. The haplotypes of their children and other descendants were then assigned by mimicking the random segregation of chromosomes from parents to offspring, allowing for recombination, as with other gene dropping methods. Pedigree genotypes simulated in this way, under appropriate LD structure, can then be used to obtain empirical p-values for statistical tests that depend on observing alleles that relatives share. Such allele-sharing statistics have been proposed to detect regions associated with specific phenotypes using densely assayed genotype data in large, extended pedigrees. The conventional approach to this problem, called *linkage analysis*, is to compute log-likelihood ratio statistics. These statistics compare the probability of observing the marker data and phenotypes under the null hypothesis that the segregation of the alleles at the marker loci and the putative phenotype locus are independent with the probability under the alternative that the segregations are correlated because the loci are physically *linked*. The calculation of such statistics with multiple markers relies on the assumption of linkage equilibrium (LE) between the markers. While this is a very reasonable assumption for the highly informative, but sparse, microsatellite data that the methods were designed to exploit, it is untenable with dense marker loci. Alternative approaches have, therefore, been suggested. Thomas et al. [36] and Leibon et al. [23] proposed a simple SNP *streak* statistic that counts runs of consecutive loci at which relatives affected by the same disease share common SNP alleles. Long streaks of allele sharing should be rare under the null hypothesis of no linkage but more common around any gene affecting the disease that the individuals have been selected for. Long streaks are, therefore, indicative of an underlying shared genomic segment inherited from a common ancestor and, hence, a likely location for a gene influencing the disease. No analytic methods exist for assessing the statistical significance of such statistics, but the null distribution can be simulated with gene dropping. Thomas [33] showed by such simulation that, under the null hypothesis of no linkage, far longer random streaks of allele sharing were generated under LD than under LE, thus illustrating that an inappropriate assumption of LE can lead to false positive results when SNP streak statistics are used.

SNP streak statistics were first applied to a pedigree of eight men with prostate cancer, selected from a larger Utah pedigree. Previous conventional linkage analysis, using a sparse set of approximately 500 microsatellite markers, in the larger pedigree had shown evidence for linkage to a region of chromosome 1 [7]. When data on a relatively dense genome-wide set of 100 000 SNP markers became available for this pedigree, it was seen as an opportunity to test the SNP streak method. The longest SNP streak on chromosome 1 shared by all eight individuals was 64 loci long and occurred within the previously identified linkage region. The longest genomic segment shared by at least seven of the eight had length 495 and also fell within the previous linkage region. Sharing among seven of the eight was considered in order to allow for the possibility of genotyping error or the possibility that one case might be due to something other than shared genetics, a so-called *sporadic* case. An LD model estimated from 52 Utah controls was used to simulate the lengths of SNP streaks under the null hypothesis of no linkage. SNP streaks were also simulated using conventional gene dropping under LE. For the segment shared by at least seven of the eight affected men, this gave an empirical p-value of 0.01 under the LD model versus 0.012 under LE. However, for the case of all eight men sharing, the empirical p-value under LD was 0.034, compared to 0.0037 under the assumption of LE, a difference of an order of magnitude, thus showing again that properly accounting for LD between markers is needed to avoid inflating the significance of long SNP streaks.

10.3.3 Phenotypes and Covariates

Since DGM estimation is a general approach, it is not necessary to limit the variables involved to genetic loci. We can also introduce variables representing the phenotypes—typically diseases, covariates, and population stratum information associated with any individual. The values of these variables are duplicated and attached to each of the imputed complete multivariate observations representing the individual's haplotypes to give a set of observations on the extended system of variables. Such applications typically require extension of existing standard programs in order to build the system of variables, but the graphical modeling estimation machinery can be applied as before and, hence, give estimates of the structure of dependence between phenotypes, covariates, and genetic loci.

This approach was used to analyze the Genetics Analysis Workshop 17 data [3]. The GAW17 data set was a simulated exome sequence data set in which a disease trait was randomly derived under the assumption that it was common and affected by many, mostly rare, variants. Three quantitative genetically determined phenotypes, correlated with each other, were also simulated, as were the age, sex, ethnicity, and smoking status of the subjects. The quantitative phenotypes, age, sex, and smoking status were correlated with the disease. Two sets of data were created: one using closely related individuals in known pedigrees and the other using unrelated individuals randomly sampled from a large population. Abel and Thomas [3] analyzed the data on unrelated individuals.

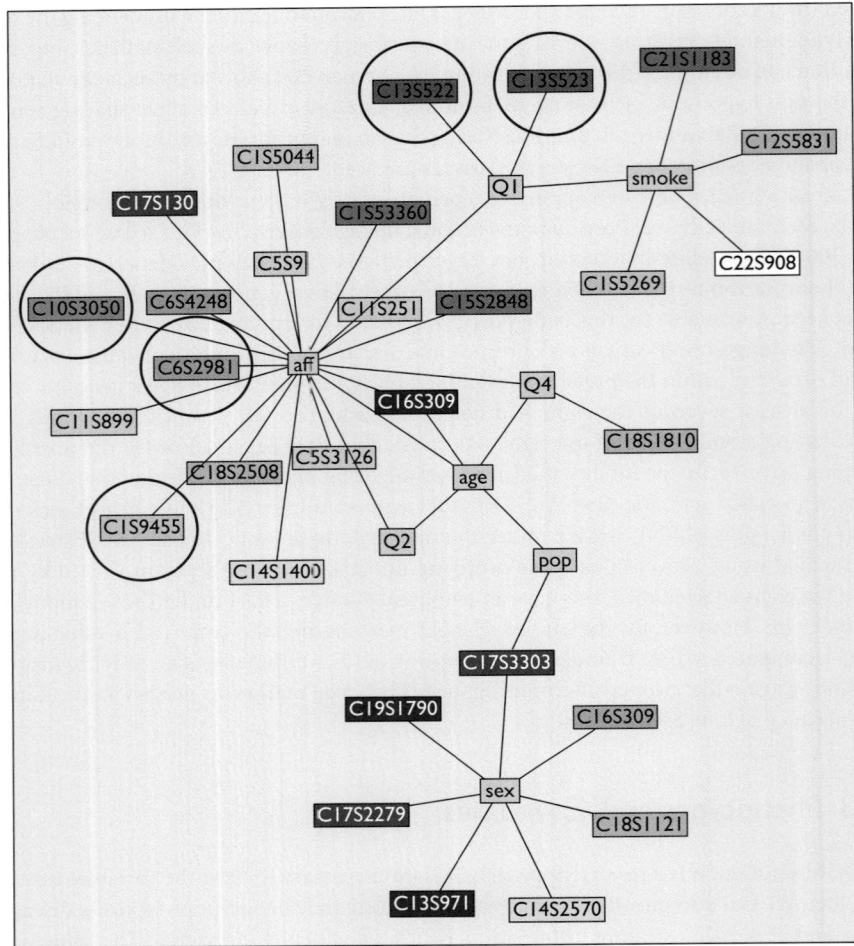

Fig 10.3 Recovery of relationships among outcomes and covariates. This is a graphical model estimated from the Genetics Analysis Workshop 17 mini-exome data. The disease phenotype variable is marked *aff*, other quantitative trait variables are marked Q1, Q2, and Q4, covariates are marked *age*, *sex*, and *smoke*, and the ethnicity of the subject is marked *pop*. The remaining nodes represent the alleles at genetic loci which the graphical model estimation process identified as first order neighbors of a trait or covariate. Genetic loci that are on the same chromosome are shown in the same color. Associations between the traits and the circled loci are true positives as these loci were used to influence the traits in the simulation that was used to construct the data. The associations with the other loci are false positives. The connections between the phenotypes and covariates model well the published simulation process.

Because of the complexity of the model and number of loci involved, we broke down the estimation process into two rounds. In the first round of model fitting, we estimated a graphical model for each chromosome separately. For each chromosome, the full joint distribution on the alleles at all loci on the chromosome, all phenotypes, the ethnicity, and all covariates was estimated. From these chromosome models, we removed all loci whose nodes in the conditional independence graph were more than three steps away from any of the nodes representing a phenotype, a covariate, or the ethnicity. Then, in the second round, a combined graphical model was estimated for all retained loci, the phenotypes, covariates, and ethnicity. All variants directly correlated, that is, connected by a single edge, with any of the non-locus variables at this final stage were deemed to be significantly associated. Fig. 10.3 shows a typical example. Here, five true causal variants, circled in red, were found to be significantly associated with either the affectation status (*aff*) or quantitative trait Q1. Furthermore, the model detected an appropriate, complex pattern of associations between the covariates and the outcomes. While, as is evident in this example, this method had a high false discovery rate (*FDR*), it correctly recovered the relationships among the outcomes and covariates, and had reasonable power to detect causal genetic variants. Among the causal variants detected were several common variants of relatively large effect. However, the approach also had some success in detecting both rare variants and those with only a small effect on the phenotype.

10.3.4 Admixture Mapping

While it is straightforward, as described above, to account for population stratification when the sampled individuals represent a single population, a more interesting problem is to model *admixture*, that is, when an individual's ancestors come from a mixture of populations. A first-generation admixed person will have one haplotype from each of the parental populations; however, in subsequent generations, the process of recombination will create haplotypes composed of segments mixed from the ancestral populations. The admixture problem is to map the ancestry of segments on each haplotype of an individual with mixed ethnicity. Such a mapping could be used to stratify a sample of study individuals in an adaptive fashion across the genome and avoid potential false results. When the admixed populations have differing incidences of a particular phenotype, it can also be used to localize genes influencing the trait. This approach has been used, for instance, to map loci influencing prostate cancer in a study of black Americans [9].

In recent work, we have used DGMs estimated from a set of ancestral populations to improve the statistical power, robustness, and scope of admixture inference [4]. To this end, we have developed a method to estimate a joint population-LD model for non-admixed individuals from two or more populations. While more general decomposable models are usually preferred, for reasons of tractability we restricted our search to conditional independence graphs in which the loci are connected in a proper interval graph. An additional variable is included to indicate the ethnicity of each sample member, and this variable is connected to all others in the conditional independence graph. A small example of the type of structure thus estimated is shown at the top of Fig. 10.4 (page 264).

This joint population-LD model is then extended to allow for admixture by replacing the single, global ethnicity variable by a series of local ethnicity variables, one for each clique of the proper interval graph. The ethnicity variables are connected in series by pairwise functions that give low probability to states where the ethnicity variables are discordant and, hence, penalize excessive switching between ethnicity indicators along the genome. Clique marginals for cliques that contain a single ethnicity variable are taken directly from the original global model. Clique marginals for cliques that contain two ethnicity variables are obtained by averaging the marginals from the global model over the margins given the values of the two ethnicity variables. The conditional independence graph for the graphical model thus obtained is shown at the bottom

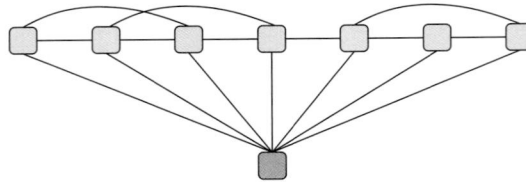

proper interval graph LD graphical model with a single ancestry variable

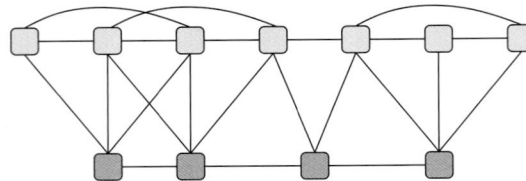

proper interval graph LD graphical model with an ancestry variable for each clique

Fig 10.4 Admixture inference. This figure illustrates how a graphical model for linkage disequilibrium (LD) and admixture that allows for a different population of origin for the chromosomal segments of an individual of mixed race can be constructed from a model estimated from a sample of non-admixed individuals from a selection of populations. The initial model, shown above, has variables representing alleles at genetic loci connected in a proper interval graph. The order of the loci from left to right is constrained by the physical order of the loci along the genome. Each of the allele variables, shown in light gray, is connected to the ancestry indicator, shown in dark gray. As the sampled individuals were not admixed, there is a single, global, ancestry indicator. In the model for admixture, shown below, the global ancestry indicator has been replaced by a sequence of local indicators, one for each clique in the proper interval graph. The local ancestry indicators are connected in a first-order Markov chain which has the effect of penalizing excessive, unrealistic flipping between population of origin states.

of Fig. 10.4. This extended model, once estimated for a set of reference populations, can then be used the estimate the local ancestry along the haplotypes of admixed individuals. Additionally, as for a typical graphical model for LD, it can be used for phasing and imputation in admixed populations. The HAPMIX software [28], which essentially extends the fastPHASE algorithm to allow for admixture, is in many ways comparable; however, it allows admixture of only two ancestral populations and is sensitive to genotyping error. DGMs for admixture mapping allow multiple ancestral populations and robust modeling of genotyping errors. They also give more accurate reconstructions of the admixture history, as shown in Table 10.2. The results shown are for a simulated admixed population of HapMap European and Yoruban ancestry. Unrelated European and Yoruban individuals were phased using BEAGLE [6] and divided into training and test sets. We used BEAGLE to create this phased data because, as it shared the same underlying models, using Complete might have favored our approach, while fastPHASE might have favored HAPMIX. Haplotypes for the test set were used as population founders and gene dropped onto a large, 20-generation pedigree, so that admixture occurred by random mating and recombination. The resulting genotypes were tested without genotype error, as well as with 1% or 2% of genotypes altered at random. The HAPMIX program and our DGM software were then used to estimate the number of African alleles at each locus, using the training haplotypes for LD model estimation. The table gives correlations between actual and estimated numbers of African alleles, averaged over all test individuals and all loci.

Table 10.2 A comparison of the accuracy of decomposable graphical models (DGMs) for admixture estimation with the existing HAPMIX program Correlations between actual and estimated number of African alleles at each locus, for a simulated admixed population.

Genotype Error Rate	HAPMIX Correlation	DGM Correlation
0	0.978	0.991
0.01	0.977	0.991
0.02	0.476	0.990

10.4 Application to Sequence Data

We conclude with some speculation on future work to model sequence data. Perhaps the most interesting recent development in genetic assays is the increasingly common use of high-throughput methods to sequence single DNA molecules to assess the genetic variants that an individual has. Typically, the sequences are aligned to a reference sequence, and variants at polymorphisms are determined by counting the evidence from individual reads. For example, if 30 sequences are found to cover an adenine (A)/cytosine (C) SNP and 16 of these show an A at the locus while 14 show a C, we would conclude that the individual was a heterozygote. Conversely 30 As or 30 Cs would clearly indicate a homozygote. Less clear cut is when, say, 28 As and 2 Cs are observed: is the individual a heterozygote, or are the Cs due to sequencing errors? There are several issues with sequence data that make this a challenging problem: the programs that call the sequence from the molecular readings make frequent errors, the number of sequences that cover any polymorphism is random, and determining the sequences that cover a polymorphism requires using alignment programs. Moreover, alignment programs preferentially find sequences that match the reference sequence perfectly and are more likely to misalign sequences that differ by one or more bases. There is, therefore, a bias toward underestimating the counts of sequences that differ from the reference. However, graphical modeling is a natural way to address such complex observational problems. For example, for SNP loci, if we assume that the total sequence coverage N at a particular locus is independent of the alleles, we might model the number of reads that match the reference as a binomial(N, p), where p depends on the genotype at the locus. For an individual homozygous for the reference, p would be $1 - \epsilon$, where ϵ allows for the possibility of erroneous sequences. Conversely, an individual homozygous for the non-reference allele would be characterized by $p = \epsilon$. For a heterozygote, we could use $p = \frac{1}{2} + \beta$, where β is a small value allowing for possible alignment bias. Choosing suitable values of ϵ and β will be assay dependent and requires further investigation. Whether values also need to be locus dependent or can be constant for sets of loci should also be determined.

Similar observational models can be derived for structural variants. For CNVs $(X_{i,1}, X_{i,2})$, specify the unknown number of copies on each gamete, and for the observed data, D_i, count the total number of sequence reads aligning to the copies. When there is low sequence coverage, the distribution of D_i can be modeled as a Poisson count with mean proportional to $X_{i,1} + X_{i,2}$; however, for high coverage assays, data appear to be overdispersed for a Poisson, and a mixture or a negative binomial may be more appropriate. Translocations may be more complex with $(X_{i,1}, X_{i,2})$ being indicators specifying the positions of the translocation in the reference and variant sequences. The data would consist of sequences that cover the junctions of the translocated element and its flanking sequence.

A more interesting exploitation of sequence data is that single molecule methods provide sporadic molecular phasing information. When a sequence is long enough to cover two polymorphisms, the phase between the alleles can be directly observed. As sequences are generally short, about 100 bases, this is not a common occurrence. More useful is that fact that sequences are usually assayed in pairs obtained from the ends of a longer molecule. This doubles the probability of coverage and elongates the potentially observable haplotype. There are also methods being developed that will sequence far longer DNA strands, perhaps up to 10 000 bases and longer, which will greatly enhance molecular phasing [8]. This type of information can again be incorporated into the graphical modeling framework. Suppose a sequence shows an A at SNP i and a guanine (G) at a nearby SNP j. We can represent this information with a conditional probability:

$$\mathbb{P}(\text{Sequence has A at locus } i \text{ and G at locus } j) =$$

$$\begin{cases} 1 - O(\epsilon) & \text{if } (X_{i,1}, X_{j,1}) = (A, G) \text{ and } (X_{i,2}, X_{j,2}) = (A, G), \\ \frac{1}{2} - O(\epsilon) & \text{if } (X_{i,1}, X_{j,1}) = (A, G) \text{ and } (X_{i,2}, X_{j,2}) \neq (A, G), \\ \frac{1}{2} - O(\epsilon) & \text{if } (X_{i,1}, X_{j,1}) \neq (A, G) \text{ and } (X_{i,2}, X_{j,2}) = (A, G), \\ O(\epsilon) & \text{if } (X_{i,1}, X_{j,1}) \neq (A, G) \text{ and } (X_{i,2}, X_{j,2}) \neq (A, G) \end{cases} \tag{10.18}$$

where $X_{i,1}$ and so on represent the alleles at these loci. We use $O(\epsilon)$ here as the error term because its appropriate form is something to be determined. In particular, the error probability may depend on the combination of alleles that appear in the reference sequence, which may again cause preferential alignment bias.

To date, these sequencing methods have mostly been used as straightforward genotyping assays. However, the established reliability and scalability of graphical models along with the flexibility to model complex modes of observation, error, and missing data may allow us to extend analysis of sequence data in a more robust and efficient manner and to exploit information not currently extracted. Finally, as can be seen from Fig. 10.2, the graphical models we estimate have a roughly linear structure, yet can accommodate significant departures from linearity at a local level. This is an excellent model for genetic variants because although polymorphic loci lie in a roughly linear order along the chromosomes, it is becoming increasingly apparent and interesting [26] that a considerable amount of the variation is in the structural arrangement itself.

ACKNOWLEDGMENTS

This work was supported in part by grant NIH R01GM081417 to Alun Thomas.

• •

REFERENCES

[1] G.R. Abecasis, S.S. Cherney, W.O. Cookson, and L.R. Cardon. Merlin—rapid analysis of dense genetic maps using sparse gene flow trees. *Nature Genetics*, 30:97–101, 2001.

[2] H.J. Abel and A. Thomas. Accuracy and computational efficiency of a graphical modeling approach to linkage disequilibrium estimation. *Statistical Applications in Genetics and Molecular Biology*, 10:5, 2011.

[3] H.J. Abel and A. Thomas. Case-control association testing by graphical modeling for the GAW 17 mini-exome sequence data. *BMC Proceedings*, 5:(Suppl 9):S62, 2011.

[4] H.J. Abel and A. Thomas. More accurate, robust and flexible admixture inference using graphical modeling. 2014. In preparation.

[5] L. Baum, T. Petrie, G. Soules, and N. Weiss. A maximization technique occurring in the statistical analysis of probabilitic functions of Markov chains. *Annals of Mathematical Statistics*, 41:164–171, 1970.

[6] S.R. Browning and B.L. Browning. Rapid and accurate haplotype phasing and missing data inference for whole genome association studies by use of localized haplotype clustering. *American Journal of Human Genetics*, 81:1084–1097, 2008.

[7] N.J. Camp, J. Swensen, B.D. Horne, J.M. Farnham, A. Thomas, L.A. Cannon-Albright, and S.V. Tavtigian. Characterization of linkage disequilibrium structure, mutation history, and tagging SNPs, and their use in association analyses: ELAC2 and familial early-onset prostate cancer. *Genetic Epidemiology*, 28:232–243, 2005.

[8] J. Eid, A. Fehr, J. Gray, K. Luong, J. Lyle, G. Otto, P. Peluso, D. Rank, P. Baybayan, B. Bettman, A. Bibillo, K. Bjornson, B. Chaudhuri, F. Christians, R. Cicero, S. Clark, R. Dalal, A. deWinter, J. Dixon, M. Foquet, A. Gaertner, P. Hardenbol, C. Heiner, K. Hester, D. Holden, G. Kearns, X. Kong, R. Kuse, Y. Lacroix, S. Lin, P. Lundquist, C. Ma, P. Marks, M. Maxham, D. Murphy, I. Park, T. Pham, M. Phillips, J. Roy, R. Sebra, G. Shen, J. Sorenson, A. Tomaney, K. Travers, M. Trulson, J. Vieceli, J. Wegener, D. Wu, A. Yang, D. Zaccarin, P. Zhao, F. Zhong, J. Korlach, and S. Turner. Real-time DNA sequencing from single polymerase molecules. *Science*, 323:133–138, 2009.

[9] M.L. Freedman, C.A. Haiman, N. Patterson, G.J. McDonald, A. Tandon, A. Waliszewska, K. Penney, R.G. Steen, K. Ardlie, E.M. John, I. Oakley-Girvan, A.S. Whittmore, K.A. Cooney, S.A. Ingles, D. Altshuler, B.E. Henderson, and D. Reich. Admixture mapping identifies 8q24 as a prostate cancer risk locus in African-American men. *Proceedings of the National Academy of Sciences of the United States of America*, 103:14068–14073, 2006.

[10] M. Frydenberg and S.L. Lauritzen. Decomposition of maximum likelihood in mixed interaction models. *Biometrika*, 76:539–555, 1989.

[11] F. Gardi. The Roberts characterization of proper and unit interval graphs. *Discrete Mathematics*, 307(22):2906–2908, 2007.

[12] P. Giudici and P.J. Green. Decomposable graphical Gaussian model determination. *Biometrika*, 86:785–801, 1999.

[13] M.C. Golumbic. *Algorithmic Graph Theory and Perfect Graphs*. Academic Press, 1980.

[14] P.J. Green and A. Thomas. Sampling decomposable graphs using a Markov chain on junction trees. *Biometrika*, 100(1):91–110, 2013.

[15] W.K. Hastings. Monte Carlo sampling methods using Markov chains and their applications. *Biometrika*, 57(1):97–109, 1970.

[16] S. Hojsgaard. *Split Models for Contingency Tables*. PhD thesis, Aarhus University, 1998.

[17] S. Hojsgaard. YGGDRASIL—a statistical package for learning split models. In C. Boutilier and M. Goldszmidt, editors, *Proceedings of the Sixteenth Conference on Uncertainty in Artificial Intelligence (UAI00)*, pages 274–281. Morgan Kaufman Publishers, 2000.

[18] S. Højsgaard and B. Thiesson. BIFROST—Block recursive models induced from relevant knowledge, observations, and statistical techniques. *Computational Statistics and Data Analysis*, 19:155–175, 1995.

[19] B. Jones, C. Carvalho, A. Dobra, C. Hans, C. Carter, and M. West. Experiments in stochastic computation for high-dimensional graphical models. *Statistical Science*, 20:388–400, 2005.

[20] S. Kirkpatrick, C.D. Gellatt, Jr., and M.P. Vecchi. Optimization by simulated annealing. Technical Report RC 9353, IBM, 1982.

[21] G.M. Lathrop, J.M. Lalouel, C. Julier, and J. Ott. Strategies for multilocus linkage analysis in humans. *Proceedings of the National Academy of Sciences of the United States of America*, 81:3443–3446, 1984.

[22] S.L. Lauritzen and D.J. Spiegelhalter. Local computations with probabilities on graphical structures and their applications to expert systems. *Journal of the Royal Statistical Society, Series B*, 50:157–224, 1988.

[23] G. Leibon, D.N. Rockmore, and M.R. Pollack. A SNP streak model for the identification of genetic regions identical-by-descent. *Statistical Applications in Genetics and Molecular Biology*, 7:16, 2008.

[24] N. Li and M. Stephens. Modeling linkage disequilibrium and identifying recombination hotspots using single-nucleotide polymorphism data. *Genetics*, 165:2213–2233, 2003.

[25] J.W. MacCluer, J.L. Vandeburg, B. Read, and O.A. Ryder. Pedigree analysis by computer simulation. *Zoo Biology*, 5:147–160, 1986.

[26] S.A. McCarroll and D.M. Altshuler. Copy-number variation and association studies of human disease. *Nature Genetics*, 39:S37–S42, 2007.

[27] N. Metropolis, A.W. Rosenbluth, M.N. Rosenbluth, and A.H. Teller. Equations of state calculations by fast computing machines. *Journal of Chemistry and Physics*, 21:1087–1091, 1953.

[28] A.L. Price, A. Tandon, N. Patterson, K.C. Barnes, N. Rafaels, I. Ruczinski, T.H. Beaty, R. Mathias, D. Reich, and S. Myers. Sensitive detection of chromosomal segments of distinct ancestry in admixed populations. *Public Library of Science, Genetics*, 5:e1000519, 2009.

[29] P. Scheet and M. Stephens. A fast and flexible statistical model for large-scale population genotype data: applications to inferring missing genotypes and haplotypic phase. *American Journal of Human Genetics*, 78:629–644, 2006.

[30] R.E. Tarjan. Decomposition by clique separators. *Discrete Mathematics*, 55:221–232, 1985.

[31] The International HapMap Consortium. A second generation human haplotype map of over 3.1 million SNPs. *Nature*, 449:851–861, 2007.

[32] A. Thomas. Estimation of graphical models whose conditional independence graphs are interval graphs and its application to modeling linkage disequilibrium. *Computational Statistics and Data Analysis*, 53:1818–1828, 2009.

[33] A. Thomas. A method and program for estimating graphical models for linkage disequilibrium that scale linearly with the number of loci, and their application to gene drop simulation. *Bioinformatics*, 25:1287–1292, 2009.

[34] A. Thomas. Assessment of SNP streak statistics using gene drop simulation with linkage disequilibrium. *Genetic Epidemiology*, 34:119–124, 2010.

[35] A. Thomas and N.J. Camp. Graphical modeling of the joint distribution of alleles at associated loci. *American Journal of Human Genetics*, 74:1088–1101, 2004.

[36] A. Thomas, N.J. Camp, J.M. Farnham, K. Allen-Brady, and L.A. Cannon-Albright. Shared genomic segment analysis. Mapping disease predisposition genes in extended pedigrees using SNP genotype assays. *Annals of Human Genetics*, 72:279–287, 2008.

[37] A. Thomas and P.J. Green. Enumerating the decomposable neighbours of a decomposable graph under a simple perturbation scheme. *Computational Statistics and Data Analysis*, 53:1232–1238, 2009.

[38] A. Thomas and P.J. Green. Enumerating the junction trees of a decomposable graph. *Journal of Computational and Graphical Statistics*, 18:930–940, 2009.

CHAPTER 11

Scoring, Searching and Evaluating Bayesian Network Models of Gene-phenotype Association

XIA JIANG, SHYAM VISWESWARAN, AND RICHARD E. NEAPOLITAN

The arrival of genome-wide association studies (GWASs) has opened the exciting possibility of identifying genetic variations (single nucleotide polymorphisms or SNPs) that underlie common diseases. However, our knowledge of the genetic architecture of common diseases remains limited. One likely reason for this is the complex interactions between genes, the environment and the studied disease. This chapter addresses three aspects which are expected to help make progress to reveal some of these complex interactions using GWAS data sets. First, results are shown that compare the performances of various Bayesian network scoring criteria. The purpose is to score candidate loci–disease models to identify which models are most noteworthy. In this comparison, simulated data sets are considered, in which two SNPs interact epistatically. Second, developing heuristic search algorithms for learning complex interactions from high-dimensional data is a recent vital area of research. The Multiple Beam Search (MBS) algorithm, a heuristic based on greedy search governed by stepwise search and elimination principles, is then presented. This procedure is meant to learn a Bayesian network best describing the links between the genetic variants and the disease. This heuristic can detect epistatic 2-SNP

Probabilistic Graphical Models for Genetics, Genomics, and Postgenomics. First Edition. Christine Sinoquet & Raphaël Mourad (Eds). © Oxford University Press 2014. Published in 2014 by Oxford University Press.

interactions. The limitations of this heuristic are discussed. Third, the hypothesis testing involved in genome-wide epistasis detection is substantially different from that involved in a standard GWAS analysis, where only a null hypothesis and an alternative are considered. As there are different alternative models for an interaction model (2-SNP, 3-SNP, 4-SNP . . .), the goal is to compute the posterior probability of such competing models, given the data. This third aspect is addressed by developing the BN posterior probability (BNPP). The second aspect was evaluated using simulated data sets. Real data sets were used to evaluate the third aspect.

11.1 Introduction

The advent of high-throughput genotyping technology has brought the promise of identifying genetic variations that underlie common diseases such as hypertension, diabetes mellitus, cancer, and Alzheimer's disease. However, our knowledge of the genetic architecture of common diseases remains limited; this is in part due to the complex relationship between the genotype and the phenotype. One likely reason for this complex relationship arises from gene–gene and gene–environment interactions. So an important challenge in the analysis of high-throughput genetic data is the development of computational and statistical methods to identify gene–gene interactions. In this chapter, we apply Bayesian network scoring criteria to identifying gene–gene interactions from GWAS data.

The rest of the chapter has the following sections. Section 11.2 gives background related to epistasis and GWASs; Section 11.3 describes a Bayesian network model that represents epistasis; Section 11.4 describes several BN scoring criteria and evaluates and compares them using synthetic genotype data; Section 11.5 describes a heuristic search strategy for identifying genetic interactions in real GWAS data; Section 11.6 describes how to determine if an identified genetic interaction is sufficiently noteworthy for further investigation.

11.2 Background

11.2.1 Epistasis

In Mendelian diseases, a genetic variant at a single locus may give rise to the disease [2]. However, in many common diseases, it is likely that manifestation of the disease is due to genetic variants at multiple loci, with each locus conferring a modest risk of developing the disease. For example, there is evidence that gene–gene interactions may play an important role in the genetic basis of hypertension [24], sporadic breast cancer [30], and other common diseases [25]. The interaction between two or more genes to affect a phenotype such as disease susceptibility is called *epistasis*. Biologically, epistasis likely arises from physical interactions occurring at the molecular level. Statistically, epistasis refers to an interaction between multiple loci such that the net effect on phenotype cannot be predicted by simply combining the effects of the individual loci. Often, the individual loci exhibit weak marginal effects; sometimes they may exhibit none.

The ability to identify epistasis from genomic data is important in understanding the inheritance of many common diseases. For example, studying genetic interactions in cancer is essential to further our understanding of cancer mechanisms at the genetic level. It is known that cancerous cells often develop due to mutations at multiple loci, whose joint biological effects lead to uncontrolled growth. But many cancer-associated mutations and interactions among the mutated loci remain unknown. For example, highly penetrant cancer susceptibility genes, such as BRCA1 and BRCA2, are linked to breast cancer [1]. However, only about five to ten percent of breast cancers can be explained by germ-line mutations in these single genes. "Most women with a family history of breast cancer do not carry germ-line mutations in the single highly penetrant cancer susceptibility genes, yet familial clusters continue to appear with each new generation" [16]. This kind of phenomenon is not yet well understood, and undiscovered mutations or undiscovered interactions among mutations are likely responsible.

The most common genetic variation is the Single Nucleotide Polymorphism (SNP) which results when a nucleotide that is typically present at a specific location on the genomic sequence is replaced by another nucleotide. In most cases, a SNP is biallelic; that is, it has only two possible values among adenine (A), guanine (G), cytosine (C), and thymine (T), the four DNA nucleotide bases. The less frequent (rare) allele must be present in 1% or more of the population for a site to qualify as a SNP [4]. The human genome is estimated to contain 15–20 million SNPs. In what follows, we will refer to SNPs as the loci investigated when searching for a correlation of some loci with a phenotype such as disease susceptibility.

When trying to learn epistatic interactions from genomic data, in some way we must score candidate SNP models to determine which models are most noteworthy. Briefly, a SNP model is a Bayesian network model that describes the relationship between SNPs and disease status; Bayesian network models are described in detail in Section 11.3. Standard techniques such as multiple linear regression may not be appropriate because both the predictors and the target are discrete. A well-known technique, which was designed to address the problem of handling discrete variables, is *Multifactor Dimensionality Reduction (MDR)* [13]. MDR combines two or more variables into a single variable (hence leading to dimensionality reduction); this changes the representation space of the data and facilitates the detection of non-linear interactions among the variables. MDR has been successfully applied to detect epistatic interactions in diseases such as sporadic breast cancer [30] and type II diabetes [6]. However, the determination of the best way to score candidate SNP models so as to identify epistasis remains an open question. MDR is often called a *combinatorial approach* because the researchers who developed it investigated all combinations of the SNPs in the analysis.

11.2.2 Genome-wide association studies

The advent of high-throughput technologies has enabled genome-wide association studies (GWAS). A GWAS involves genotyping about 500 000 representative SNPs in individuals sampled from a population. A data set in which each record has such a large number of attributes is called *high-dimensional*. In case-control GWAS, we identify the disease status along with the values of the SNPs. Such studies provide researchers unprecedented opportunities to investigate the complex genetic basis of diseases. That is, we can investigate the association of each SNP by itself with the disease and also investigate combinations of SNPs looking for epistasis. Such an investigation is called an *agnostic study* because we have no special prior belief concerning any particular locus. By looking at single-locus associations, researchers have identified over 150 risk loci associated with 60 common diseases and traits [8, 15, 20, 22, 28, 33].

However, single-SNP investigations could not detect complex epistatic interactions in which each locus by itself exhibits little or no marginal effect. To fully exploit these data and possibly reveal a great deal of the dark matter of genetic risk, it is critical that we analyze such data using multilocus methods. However, we suffer from combinatorial explosion with such data. For example, if we only investigated all 1-, 2-, 3-, and 4-SNP combinations when there are 500 000 SNPs, we would need to investigate 2.604×10^{21} combinations. So researchers have recently endeavored to develop heuristic approaches for guiding the search when investigating epistatic interactions using a GWAS data set [3, 19, 23, 39]. However, the successful analysis of epistasis using high-dimensional data sets remains an open and vital problem.

These opportunities for learning potential disease risk from high-dimensional data sets present us with another challenge: namely, how do we analyze and interpret our results when there are possibly billions of hypotheses? The hypothesis testing involved here is substantially different from that involved in a typical analysis, where we might analyze the effect of a new drug. In this latter case, we are analyzing only one hypothesis, and the drug has a fairly high prior probability of being effective, otherwise the study would not have been considered. In discovery studies involving many hypotheses, each hypothesis has a very low prior probability.

Historically, the most common strategy for handling this multiple hypothesis testing problem has been to control type I error (false discovery) by using the Bonferroni correction to constrain the familywise error rate. For example, the results in [28] were reported as being significant with Bonferroni correction. However, as discussed in [17] and at the beginning of Section 11.6, the Bonferroni correction encounters difficulties, particularly in this domain. Corrected results often fail to duplicate across studies [12]. One reason might be owing to these difficulties. So another challenge in learning from GWAS data is determining a way to decide which SNP models to flag as sufficiently noteworthy for further investigation.

In summary, we have identified the following three difficulties when learning from high-dimensional GWAS data sets:

1 Determining the best way to score candidate SNP models to identify which mode s are most noteworthy.

2 Developing heuristic search algorithms for learning epistatic interactions from high-dimensional data.

3 Determining a way to decide which discovered SNP models to flag as *sufficiently* noteworthy for further investigation.

Sections 11.4, 11.5, and 11.6 present progress we have made addressing each of these difficulties. They all use the same Bayesian network model, which we describe first.

11.3 A Bayesian Network Model

It is assumed that the reader is familiar with *Bayesian networks* (*BNs*). If not, the reader should consult a text such as [26] for an introduction. Briefly, a BN consists of a *directed acyclic graph* (*DAG*) whose nodes are random variables, and whose edges represent relationships among the random variables that are often causal; the prior probability distribution of every root variable in the DAG; and the conditional probability distribution of every non-root variable given each set of values of its parents. If G is the DAG in the BN and $\mathbb{P}$ is the joint probability distribution, we denote the BN by $(G, \mathbb{P})$. If we have a DAG whose nodes consist of random variables but no

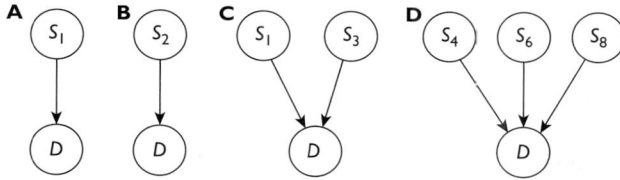

Fig 11.1 Four DDAG models.

joint probability distribution, then the DAG is called a *DAG model*. Such a model only represents possible relationships among the random variables.

In theory we could develop a DAG model that represents many factors that affect phenotype, including inheritable allele variation, somatic mutations in alleles, environmental factors, and epigenetic phenomena such as DNA methylation. A subnetwork of that model contains only variables that represent inheritable allele variation (the variables whose values are obtained in a GWAS). If, for example, we have a 5-way interaction, two 3-way interactions and two 2-way interactions, there are 15 SNPs in this subnetwork, all which have edges to the disease node D. We have neither the data nor the computational time to score such networks. So we must settle for trying to learning pieces of the network, such as particular interactions, separately. These small subnetworks are the focus of this paper, and we call them *direct DAG (DDAG) models*. Examples of such models appear in Fig. 11.1. The first two represent when a single SNP by itself is associated with the disease, the third one represents when two SNPs together are associated with the disease, and the fourth one represents when three SNPs together are associated with the disease. It is important to recognize that, for example, the model in Fig. 11.1C does not entail that S_1 and S_3 are interacting to affect D. Each could be affecting it separately. Without making specialized mathematical assumptions, it is difficult to distinguish these two situations from data alone. However, in Section 11.6 we provide one way to do this.

11.4 Scoring Candidate Models

This section describes an evaluation of various scoring criteria for DDAG models. It is based on results in [18]. First we review these criteria.

11.4.1 Bayesian Network Scoring Criteria

The *Bayesian score* for BNs is the probability of the observed data given a DAG model G [9]. In the case of a DDAG model G and data *Data* concerning the SNPs and disease in the model, the Bayesian score is given by:

$$\mathbb{P}(Data|G) = \prod_{j=1}^{q} \frac{\Gamma(a_{j1} + a_{j2})}{\Gamma(a_{j1} + a_{j2} + s_{j1} + s_{j2})} \frac{\Gamma(a_{j1} + s_{j1})}{\Gamma(a_{j1})} \frac{\Gamma(a_{j2} + s_{j2})}{\Gamma(a_{j2})}, \tag{11.1}$$

where q is the number of different states of the parents of the disease node D, a_{j1} is the ascertained prior belief concerning the number of individuals having the disease and having the parents in their jth state, a_{j2} is the ascertained prior belief concerning the number of individuals not having the disease and having the parents in their jth state, s_{j1} is the number of individuals in the data set having the disease and having the parent nodes in their jth state, and s_{j2} is the number of individuals not having the disease and having the parent nodes in their jth state. The parameters a_{ji} are known as hyperparameters. Hyperparameters are used to penalize complex DAGs; typically higher values will penalize complex DAGs less than lower values. Different Bayesian scores differ in the values used for the hyperparameters as shown below.

When both a_{j1} and a_{j2} are equal to 1 for all j, the score is called the K2 score [9]. When

$$a_{j1} = a_{j2} = \frac{\alpha}{2q}$$

for a parameter α, the score is called the *Bayesian Dirichlet equivalent uniform (BDeu)* score, and α is called the prior equivalent sample size [14].

The Minimum Description Length (MDL) principle is an information-theoretic principle [29] which states that the best model is the one that minimizes the sum of the encoding lengths of the data and the model itself. To apply this principle to scoring DAG models, we must determine the number of bits needed to encode a DAG and the number of bits needed to encode the data given the DAG. Suzuki [34] developed the following MDL scoring criterion for an arbitrary DAG model:

$$\frac{d}{2} \log_2 m - m \sum_{j=1}^{q} \sum_{k=0}^{1} \mathbb{P}(D = k, pa_j) \log_2 \frac{\mathbb{P}(D = k, pa_j)}{\mathbb{P}(D = k)\mathbb{P}(pa_j)}, \tag{11.2}$$

where d is the number of parameters needed to represent the conditional probability distributions of the disease node D, m is the number of observations, q is the number of different values the parents of D can jointly assume, pa_j is the jth value of the parents of D, and the probabilities are estimated from the data. In equation (11.2), the first sum is the DAG penalty, which is the number of bits sufficient to encode the DAG model, and the second term is the number of bits sufficient to encode the data given the model.

Other MDL scores assign different DAG penalties and therefore differ in the first term in equation (11.2) but encode the data in the same way. For example, the *Akaike information criterion (AIC)* score is an MDL scoring criterion that uses d as the DAG penalty. So, the AIC score is as follows:

$$d - m \sum_{j=1}^{q} \sum_{k=0}^{1} \mathbb{P}(D = k, pa_j) \log_2 \frac{\mathbb{P}(D = k, pa_j)}{\mathbb{P}(D = k)\mathbb{P}(pa_j)}.$$

The Bayesian score does not explicitly include a DAG penalty for network complexity. However, a DAG penalty is implicitly determined by the hyperparameters. Silander et al. [32] show that if we use the BDeu score, then the DAG penalty decreases as α increases. The K2 score uses hyperparameters in a way that can be related to a prior equivalent sample size. When a node is modeled as having more parents, the K2 score effectively assigns a higher prior equivalent sample size to that node, which in turn decreases its DAG penalty.

11.4.2 Experiments

METHOD

We evaluated the performance of MDR, the K2 score, the BDeu score with α taking its value in $\{3, 6, 9, 12, 15, 18, 21, 24, 30, 36, 42, 54, 162\}$, Suzuki's MDL score, and the AIC score. We evaluated the scoring criteria using simulated data sets developed in [36]. Each simulated data set was developed from one of 70 epistasis models. Each model represents a probabilistic relationship in which two SNPs together are correlated with the disease but neither SNP is individually predictive of the disease. The relationships represent various degrees of penetrance, heritability, and minor allele frequency[1]. Data sets were generated with a case-control ratio (ratio of individuals with the disease to those without the disease) of 1:1. To create one data set, [36] fixed the genetic model by specifying the penetrance, heritability, and minor allele frequency. Based on the model, genotype data for the two epistatic SNPs were generated to which were added genotype data for 18 additional SNPs that were not predictive of disease. For each of the 70 models, 100 data sets were generated providing a total of 7000 data sets. This procedure was followed for data set sizes equal to 200, 400, 800, and 1600.

For each of the simulated data sets, we scored all 1-SNP, 2-SNP, 3-SNP, and 4-SNP DDAGs. The total number of DDAGs scored for each data set was therefore 6195. Since in a real setting we would not know the number of SNPs in the model generating the data, all models were treated equally in the learning process.

RESULTS

We say that a method correctly learns the model generating the data if it scores the DDAG representing the generating model highest out of all 6195 models. Table 11.1 shows the number of times out of 7000 data sets that each BN scoring criterion correctly learned the generating model for each sample size. In this table, the scoring criteria are listed in descending order of correctness. Table 11.1 shows a number of interesting results. First, the AIC score performed reasonably well on small sample sizes, but its performance degraded at larger sample sizes. Unlike the other BN scores, the DAG penalty in the AIC score does not increase with the sample size. Second, MDR performed well overall but substantially worse than the best performing scores. Third, the best results were obtained with the BDeu score at moderate values of α. However, the results were very poor for large values of α, which assign very small DAG penalties.

BDeu scores with values of α in the range 12–18 performed substantially better than all other scores. If our goal is only to find a score that most often scores the correct model highest on low-dimensional simulated data sets like the ones analyzed here, then our results support the use of these BDeu scores. However, in contrast, we are interested in the discovery of promising SNP-disease associations that may be investigated for biological plausibility. So perhaps more relevant

---

[1] Penetrance is the probability that an individual will have the disease given that the individual has a genotype that is associated with the disease. Heritability is the proportion of the disease variation due to genetic factors. The minor allele frequency is the relative frequency of the less frequent allele for a given locus.

Table 11.1 The number of times out of 7000 data sets that each scoring criterion identified the correct model for sample sizes of 200, 400, 800, and 1600. The last column gives the total accuracy over all sample sizes. The scoring criteria are listed in descending order of total accuracy.

	Scoring Criterion	200	400	800	1600	Total
1	$score_{\alpha=15}$	4379	5426	6105	6614	22524
2	$score_{\alpha=12}$	4438	5421	6070	6590	22519
3	$score_{\alpha=18}$	4227	5389	6095	6625	22336
4	$score_{\alpha=9}$	4419	5349	5996	6546	22313
5	$score_{\alpha=21}$	3989	5286	6060	6602	21934
6	$score_{\alpha=6}$	4220	5165	5874	6442	21701
7	$score_{\alpha=24}$	3749	5156	5991	6562	21448
8	$score_{MDR}$	4112	4954	5555	5982	20603
9	$score_{\alpha=3}$	3839	4814	5629	6277	20559
10	$score_{\alpha=30}$	3285	4779	5755	6415	20234
11	$score_{Suz}$	3489	4580	5521	6215	19805
12	$score_{\alpha=36}$	2810	4393	5464	6150	18817
13	$score_{\alpha=42}$	2310	4052	5158	5895	17415
14	$score_{K2}$	1850	3475	5095	6116	16536
15	$score_{\alpha=54}$	1651	3297	4492	5329	14769
16	$score_{AIC}$	2497	1967	1462	1126	7052
17	$score_{\alpha=162}$	26	476	1300	2046	3848

than whether the correct model scores the highest is the recall of the correct model relative to the highest scoring model. The recall is given by

$$recall(S, T) = \frac{\#(S \cap T)}{\#(S)},$$

where S is the set of SNPs in the correct model, T is the set of SNPs in the highest scoring model, and # returns the number of SNPs in a set. The value of the recall is 0 if and only if the two sets do not intersect, while it is 1 if and only if all the SNPs in the correct model are in the highest scoring model. Therefore, recall is a measure of how well the SNPs in the correct model are discovered. Recall does not measure, however, the extent to which the highest scoring model has additional SNPs that are not in the correct model (i.e., false positives).

Table 11.2 The sum of the recall for each scoring criterion over 7000 data sets for sample sizes of 200, 400, 800, and 1600 The last column gives the total recall over all sample sizes. The scoring criteria are listed in descending order of total recall.

	Scoring Criterion	200	400	800	1600	Total
1	$score_{\alpha=162}$	5259	6043	6566	6890	24758
2	$score_{AIC}$	5186	5960	6481	6830	24457
3	$score_{\alpha=54}$	5223	5941	6473	6813	24450
4	$score_{K2}$	5303	5962	6371	6747	24383
5	$score_{\alpha=42}$	5203	5902	6425	6794	24324
6	$score_{\alpha=36}$	5181	5866	6395	6768	24210
7	$score_{\alpha=30}$	5147	5816	6352	6754	24069
8	$score_{\alpha=24}$	5080	5767	6300	6725	23872
9	$score_{\alpha=21}$	5031	5733	6265	6704	23733
10	$score_{MDR}$	4870	5710	6324	6748	23652
11	$score_{\alpha=18}$	4973	5681	6230	6681	23565
12	$score_{\alpha=15}$	4902	5622	6183	6647	23354
13	$score_{\alpha=12}$	4786	5531	6119	6605	23041
14	$score_{\alpha=9}$	4649	5416	6026	6547	22638
15	$score_{\alpha=6}$	4383	5219	5901	6453	21956
16	$score_{\alpha=3}$	3953	4862	5652	6285	20752
17	$score_{Suz}$	3618	4696	5595	6251	17760

Table 11.2 shows the recall for the various scoring criteria. The criteria are listed in descending order of total recall. Overall, these results are the reverse of those in Table 11.1. The BDeu scores with large values for α and the AIC score appear at the top of the list. Part of the explanation for this is that these BDeu scores and the AIC score incorporate small DAG penalties, which results in larger models often scoring higher. A larger model has a greater chance of containing the two interacting SNPs. MDR again performed well but substantially worse than the best performing scores.

Perhaps the smaller DAG penalty is not the only reason that the BDeu scores with larger values for α performed best. It is possible that the BDeu scores with larger values for α can better detect the interacting SNPs than the BDeu scores with smaller values but that the scores with larger values do poorly at scoring the correct model (the one with only the two interacting SNPs) highest because they too often pick a larger model containing those SNPs. To investigate this possibility,

Table 11.3 The number of times out of 500 that each scoring criterion correctly learned the correct model in the case of the most difficult models (55–59) for sample sizes of 200, 400, 800, and 1600 The last column gives the total accuracy over all sample sizes. The scoring criteria are listed in descending order of accuracy.

	Scoring Criterion	200	400	800	1600	Total
1	$score_{\alpha=54}$	14	48	167	352	681
2	$score_{\alpha=162}$	1	21	146	355	563
3	$score_{\alpha=36}$	13	46	155	318	532
4	$score_{\alpha=21}$	12	43	106	289	450
5	$score_{\alpha=18}$	11	37	91	274	413
6	$score_{MDR}$	3	25	79	245	352
7	$score_{\alpha=12}$	7	25	65	215	312
8	$score_{\alpha=9}$	5	20	48	186	259
9	$score_{\alpha=3}$	3	6	13	86	108
10	$score_{Suz}$	0	1	2	41	44

we investigated how well the scores discovered models 55–59 (see Supplementary Table 1 in [36]). These models have the weakest heritability (0.01) and a minor allele frequency of 0.2, and are therefore the most difficult to detect.

Table 11.3 shows the number of times the correct hard-to-detect model scored highest for a representative set of the scores. The BDeu score with large values for α performed substantially better than all other scores.

Among all the scores examined, the BDeu scores with large α values were the best at identifying the difficult models; however, overall these scores performed poorly when all models were considered. An explanation for this phenomenon is that these scores can indeed find interacting SNPs better than scores with smaller values for α. When the interacting SNPs are fairly easy to identify, the larger DAG penalties of the BDeu scores with large α values make it harder for them to identify the correct model relative to other scores. On the other hand, when the interacting SNPs are hard to detect, their better detection capability more than compensates for their increased DAG penalty.

11.5 Searching over the Space of Models

As mentioned in the introduction, besides the difficulty of choosing an appropriate scoring criterion, a second challenge is designing an efficient search algorithm. In particular, there is combinatorial explosion when learning epistatic interactions in high-dimensional GWAS data.

Therefore, researchers have endeavored to develop heuristic search algorithms. This section describes one such algorithm that was developed in [19].

A quadratic-time BN learning algorithm called Greedy Equivalent Search (GES) [5] will learn the most concise DAG representing a probability distribution under the assumptions that the scoring criterion is consistent (a consistent scoring criterion assigns the same score to equivalent DAGs that represent the same conditional independences), and that the probability distribution admits a faithful DAG representation and satisfies the composition property. The reader is referred to [26] for a discussion of these assumptions. Briefly, the algorithm starts with the empty DAG and in sequence greedily adds the edge to the DAG that most increases the score until no edge increases the score. Then, in sequence, the algorithm greedily deletes the edge from the DAG for which the deletion increases the score the most until no edge decreases the score.

An initial strategy for learning an epistatic interaction might be to try to learn the interacting SNPs by using the GES algorithm to search all DDAGs. However, a more thorough examination reveals that this could not work. Suppose we have a true epistatic relationship between two SNPs and disease variable D and that all other SNPs are independent of D. Then, each SNP by itself is independent of D. Suppose further that we have a data set so large that the generative distribution is represented exactly in the data set. In this case the GES algorithm will learn the correct DAG if the assumptions in the algorithm are met. However, in the first step of the algorithm, all SNPs will score the same because they are all independent of D, none of them will increase the score, and the algorithm will halt.

Epistatic relationships do not satisfy the composition property, which is necessary to the success of the GES algorithm. The composition property entails that if the disease is dependent on SNPs S_1 and S_2 together, then it must be dependent on at least one of them by itself. Clearly, if S_1 and S_2 interact epistatically to affect the disease, then the disease is independent of each of them by itself. We can make progress toward breaking through this barrier to epistasis learning by expanding each of the SNPs using greedy search rather than expanding only the single SNP that initially increases the score the most. In this way, we will definitely investigate every 2-SNP combination. If epistasis is occurring, we will thereby score two of the SNPs involved in the epistatic relationship. Once we score two of them, often we should also find possible 3rd and 4th and so on SNPs involved in the relationship. The algorithm follows. In this algorithm, $score(A_i)$ denotes the score of the model that has edges from the SNPs in A_i to D:

---

for each SNP SNP_i
    $A_i \leftarrow \{SNP_i\}$;
    do
        if adding any SNP to A_i increases $score(A_i)$
            add the SNP to A_i that increases $score(A_i)$ the most;
        endif
    while adding some SNP to A_i increases $score(A_i)$;
    do
        if deleting any SNP from A_i increases $score(A_i)$
            delete the SNP from A_i that increases $score(A_i)$ the most;
        endif
    while deleting some SNP from A_i increases $score(A_i)$;
endfor;
report k highest scoring sets A_i.

We call this algorithm *Multiple Beam Search* (MBS). It clearly requires $O(n^3)$ time in the worst case where n is the number of SNPs. However, in practice, if the data set was large, we would add at most m SNPs in the first step, where m is a parameter. So the time complexity would be $O(mn^2)$.

This technique does not work if there is a probabilistic dependence between k SNPs and D but every proper subset of the k SNPs independent of D. This case is called *pure, strict epistasis*. The MBS algorithm is effective for handling epistatic relationships in which we have k SNPs interacting, each of which is individually probabilistically independent of the disease, and there is some probabilistic dependence between the disease and at least one pair of the interacting SNPs.

11.5.1 Experiments

METHOD

Using the synthetic Velez data sets discussed at the beginning of Subsection 11.4.2, we compared the following two methods: (1) scoring using a BN score and looking at all 1-SNP, 2-SNP, 3-SNP, and 4-SNP DDAGs (this method is called *BayCom*); and (2) scoring using the same BN score and searching using MBS with a maximum of $m = 4$ SNPs added in the first step.

RESULTS

Table 11.4 shows the number of times the correct model scored highest over all 7000 data sets. Table 11.5 shows the running times. MBS performed as well as BayCom at identifying the correct model and was up to 28 times as fast.

11.6 Determining Whether a Model is Sufficiently Noteworthy

As mentioned in the introduction, the exciting possibility of learning potential disease risk from high-dimensional data sets gives rise to new challenges concerning the analysis and interpretation of the results. The hypothesis testing involved here is substantially different than that involved in a typical analysis in which we might analyze the effect of a new drug. In this latter case, we are analyzing only one hypothesis, and that hypothesis has a fairly high prior probability (we would not have done the study if we had not thought there was a good chance the drug was effective). In

Table 11.4 Number of times the correct model scored highest for MBS and BayCom algorithms.

Data Set Size	MBS	BayCom
200	4049	4049
400	5111	5111
800	5881	5881
1600	6463	6463

Table 11.5 Average running times over all 7000 data sets for MBS and BayCom algorithms.

Data set Size	MBS	BayCom
200	0.108 s	2.000 s
400	0.191 s	5.150 s
800	0.361 s	9.610 s
1600	0.629 s	18.000 s

these discovery studies involving many models, each model has a very low prior probability. In [38], two important endeavors in analyzing GWAS data sets and similar data sets are identified:

1 Ranking a list of SNPs to carry forward to the next stage of study, when the size of the list has already been decided upon.
2 Calibrating inference to allow estimation of the number of false discoveries, or false non-discoveries, or posterior probability of the null hypothesis given the data.

As also mentioned in the introduction, the must typical strategy for addressing the second endeavor has been to control type I error (false discovery) by using the Bonferroni correction to control the familywise error rate, and that correction has been used to analyze results obtained in a number of GWASs. We briefly describe the corrections and the problem with using it.

When testing multiple hypotheses as in a GWAS concerning many SNPs, one of the hypotheses is likely to have a significant p-value by chance. As a result, researchers often use the Bonferroni correction to control the familywise error rate by multiplying the p-value by the number of hypotheses n. For example, if $p = 2.1 \times 10^{-6}$ and $n = 500\,000$, then the corrected p-value is $np = 500\,000 \times 2.1 \times 10^{-7} = 0.105$, the result would not be deemed significant by most standards, and the null hypothesis would not be rejected. A related correction is the Šidák correction, which is $1 - (1 - p)^n$.

In [38], it is noted that in the case of a GWAS, the Bonferroni correction will often be an overly conservative procedure, since at least in the early stages of such studies, we are more concerned with avoiding missed associations; and making some false discoveries is not too high a cost to pay to find real associations. Neapolitan [27] raises a more fundamental problem with the Bonferroni correction. He argues that it is a misguided practice, and that the significance we attach to a result concerning a particular hypothesis cannot depend on the number of hypotheses we happen to test along with that hypothesis. Briefly, the argument is as follows. Suppose on a particular day a statistician tested the effectiveness of 1000 different drugs using randomized controlled experiments. The statistician should treat each of those experiments no differently than if it was the only experiment done that day, and no correction should be applied. The reason is that each alternative hypothesis has a reasonably high prior probability of being true (as mentioned earlier), and performing the other tests does not affect this. In most cases involving multiple hypotheses (e.g., in a GWAS) each hypothesis has a very low prior probability, and the corrections serve as surrogates for these low prior probabilities. However, they are poor surrogates, because the prior probability depends on the number of hypotheses we happen to test in the study.

Regardless of one's stance on this matter, there are clear difficulties in applying the Bonferroni correction in GWASs. Suppose that one study investigates 100,000 SNPs while another investigates 500 000 SNPs. Suppose further that the data concerning a particular SNP and the disease are identical in the two studies. Owing to the different corrections, that SNP could be reported as significant in one study but not the other. Yet the data concerning the SNP are identical in the two studies. As noted earlier, in GWASs, results are often not duplicated across studies. Part of the explanation would be the practice of using different corrections in different studies. Initially GWAS data were analyzed by investigating only 1-SNP correlations. So, if there were 100 000 SNPs, there would be 100 000 hypotheses. Based on these studies, quite a few results have been reported as significant with correction [1, 3, 17–19]. It is becoming increasingly popular to also investigate 2-SNP models in the effort to identify epistasis [8–10]. If there are 100 000 SNPs, there are about 5×10^9 2-SNP models. If the researchers who previously reported significant results had also investigated the 2-SNP models, the corrections would have been based on many more models, and the results probably would have all been reported as insignificant.

Realizing that there are difficulties in using the Bonferroni correction in the analysis of GWAS data sets, researchers have developed alternative techniques, which are summarized in [17]. As noted in [17], these techniques all suffer from the following two difficulties:

1 Only two possibilities are considered. So, in our SNP-disease application, these techniques could only consider two values of a SNP. However, we may want to model three possible values of a SNP.

2 They can only consider a null hypothesis H_0 and an alternative H_1. Only considering two hypotheses is a severe limitation; if we model 2-SNP, 3-SNP models, etc., there are several different alternative models for the one whose probability we are computing, and they all have different likelihoods.

To overcome these difficulties, Jiang et al. [17] developed the Bayesian network posterior probability, which we discuss next.

11.6.1 The Bayesian Network Posterior Probability (BNPP)

Each of the DDAGs in Fig. 11.1 represents a model or hypothesis that the SNPs in the DDAG are associated with the disease. Our goal is to compute the posterior probability of such models M given the *Data*. We can do that using Bayes' theorem as follows:

$$\mathbb{P}(M|Data) = \frac{\mathbb{P}(Data|M)\mathbb{P}(M)}{\mathbb{P}(Data)}. \tag{11.3}$$

The value of $\mathbb{P}(Data|M)$ can be computed using the BDeu score with a particular choice of α. The value of $\mathbb{P}(M)$ is the prior probability of M, which we discuss shortly. We call the posterior probability in equation (11.3) the *Bayesian Network Posterior Probability (BNPP)*. First we show how to compute the BNPP; then we discuss assessing the prior probabilities.

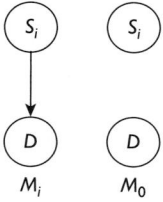

Fig 11.2 The 1-SNP model and its competing model. The 1-SNP model where S_i has a causal influence on D all by itself is on the left and the model where it does not is on the right.

Consider first a 1-SNP model. Let M_i be the model that S_i *all by itself* has an effect on D and M_0 be the model that it does not (See Fig. 11.2). Then, the posterior probability of M_i is given by

$$\mathbb{P}(M_i|Data) = \frac{\mathbb{P}(Data|M_i)\mathbb{P}(M_i)}{\mathbb{P}(Data|M_i)\mathbb{P}(M_i) + \mathbb{P}(Data|M_0)\mathbb{P}(M_0)}. \tag{11.4}$$

The BDeu score is decomposable in the sense that it is the product of terms, where each term is computed locally for an individual node. So the terms involving each S_i are the same in model M_i and M_0, which means that they cancel out in equation (11.4), and we need only compute the values of terms involving D. This is true regardless of the number of SNPs in the model. Note that the model in Fig. 11.2 is not that S_i is merely associated with the disease, but rather that it is associated all by itself. For example, if S_i was involved in an epistatic interaction with no marginal effects, the model would be false. Note further that S_i can have any number of discrete values in the model. We are not restricted to only two values. So we can represent all three values of a SNP, or if we are representing an environmental feature with many values, we can represent all of them. If the environmental feature is continuous, we can discretize it. So we overcome the first difficulty mentioned above.

Fig. 11.3 shows the model M_{ij} where S_i and S_j both have an effect on D (without needing other interacting SNPs). Note that this model includes the possibility that there is epistasis with no marginal effects and that each SNP by itself has an effect with no epistasis. The three competing models are on the right. Note that the model denoted as M_i is not the same as the model M_i in Fig. 11.2. The current model entails that S_j does not have a causal influence on D, whereas the model in Fig. 11.2 says nothing about S_j.

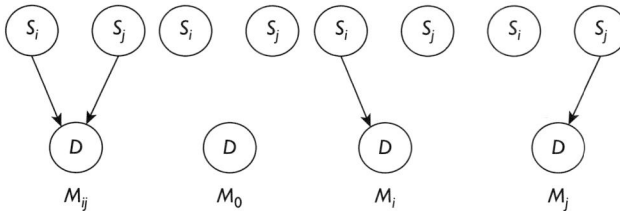

Fig 11.3 The 2-SNP model and its three competing models. The 2-SNP model that both S_i and S_j have causal influences on D is on left; the three competing models are on the right.

No other known method (discussed in [17]) considers these multiple competing hypotheses. These methods would only consider the null hypothesis M_0 in which no association with D would hold. However, if either model M_i or M_j were correct, we would observe an association of the two SNPs together with D even though M_{ij} is incorrect.

The posterior probability is as follows:

$$\mathbb{P}(M_{ij}|Data)$$
$$= \frac{\mathbb{P}(Data|M_{ij})\mathbb{P}(M_{ij})}{\mathbb{P}(Data|M_{ij})\mathbb{P}(M_{ij}) + \mathbb{P}(Data|M_0)\mathbb{P}(M_0) + \sum_k \mathbb{P}(Data|M_k)\mathbb{P}(M_k)},$$

where k sums over the two 1-SNP models. Note that there are several competing models (hypotheses) rather than just a single null hypothesis, and all have different likelihoods. So we overcome the second difficulty mentioned above (see page 282). A method that only considers a null hypothesis and an alternative hypothesis does not have the same representational power.

Fig. 11.4 shows a 3-SNP model and the competing models. The number and complexity of the competing models increases dramatically with the size of the model. However, we need not identify all the competing models because we have developed the following novel recursive algorithm to compute $P(Data)$ for an arbitrary number of SNPs (Algorithm 1). $P(Data)$ is the denominator in the formula for the posterior probability of a model.

There are n SNPs in the model being analyzed. The algorithm proceeds by calling procedure *ComputeLikelihood* for every $m \leq n$. For each value of m, this routine then computes the contribution of all m SNP models to the likelihood by recursively visiting all such models. Since every subset of the n SNPs determines a competing model, the likelihoods for 2^n models are computed. However, since ordinarily there are at most five SNPs in a model, this computation is feasible.

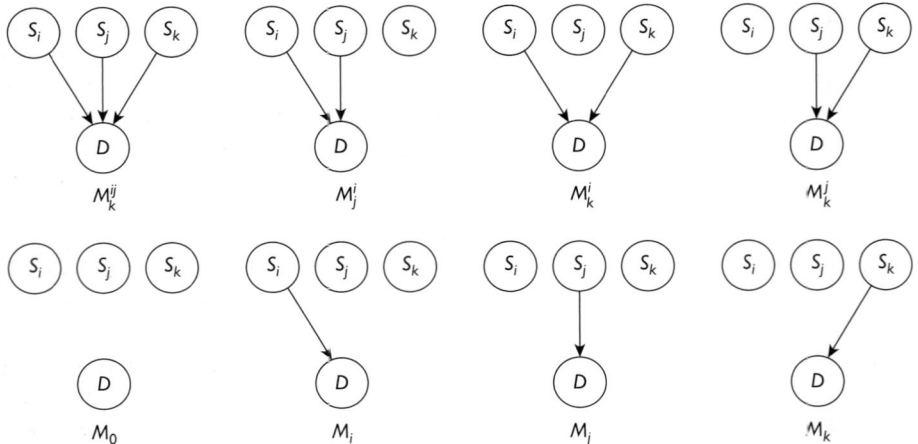

Fig 11.4 The 3-SNP model and its seven competing models.

Algorithm 1 Compute $\mathbb{P}(Data)$

The SNPs in the model evaluated are $S[1], S[2], \ldots, S[n]$.
prior$[m]$ is the prior probability of an m-SNP model.

1: $\mathbb{P}(Data) \leftarrow 0$;
2: **for** $m \leftarrow 0$ **to** n
3:     *likelihood* $\leftarrow 0$;
4:     $M \leftarrow \varnothing$;
5:     *ComputeLikelihood*$(1, m)$;
6:     $\mathbb{P}(Data) \leftarrow \mathbb{P}(Data) + likelihood \times prior[m]$;
7: **endfor**

Algorithm 2 *ComputeLikelihood*(k, m)
---
m is the size of the models.
1: **if** $m = 0$
2:     *likelihood* $\leftarrow$ *likelihood* $+ \mathbb{P}(Data|M)$;
3: **else**
4:     **for** $i \leftarrow k$ **to** $n - m + 1$
5:         add $S[i]$ to M;
6:         *ComputeLikelihood*$(i + 1, m - 1)$;
7:         remove $S[i]$ from M;
8:     **endfor**
9: **endif**

11.6.2 Prior Probabilities

The greatest difficulty in most Bayesian analyses is arguably the assessment of prior probabilities. The Bayesian paradigm was mostly abandoned years ago largely owing to the perceived arbitrary nature of these assessments. For example, in 1921 R.A. Fisher [11] stated, "The Bayesian approach depends upon an arbitrary assumption, so the whole method has been widely discredited." However, multiple hypothesis testing reveals that priors are sometimes needed. Wakefield [38] points out that "as more genome-wide association studies are carried out lower bounds on $\pi_1 = 1 - \pi_0$ will be obtained from the confirmed 'hits'—it is a lower bound since clearly many non-null SNPs for which we have low power of detection will be missed." By π_1 he means the prior probability that the model is correct. We agree that, in time, results will help us to determine priors. In the meantime, the following is an initial effort to assess priors.

Researchers in the Wellcome Trust Case Control Consortium [7] assessed that there are 1 000 000 regions of correlated SNPs in the genome, with an expectation of 10 regions having an effect on phenotype. They therefore assign a prior probability of 0.00001 that a SNP is associated with the disease; it notes that "other plausible estimates may vary from this (0.00001) by an order of magnitude or so in either direction." Based on similar assumptions, Wacholder et al. [37] arrive at a prior probability between 0.0001 and 0.00001 that a randomly selected non-synonymous variant is associated with a complex disease. However, they note that the prior probability that a variant of a gene with functional data is suggestive of association with a disease could possibly be in the range 0.01 to 0.001.

Based on these analyses, in an agnostic search we assume that each SNP has between a 0.0001 and a 0.00001 prior probability of being associated with the disease. The following discussion assumes a value of 0.00001. The calculations can be repeated using a value of 0.0001. For a given SNP, we assume that the prior probability of 0.00001 is equally distributed among the possibilities of it exhibiting an association all by itself, of it requiring one other SNP to exhibit an association, of it requiring two other SNPs to exhibit an association, and so on up to 10 possible SNPs. So we assume that the prior probability of each of these events is $0.1 \times 0.00001 = 1.0 \times 10^{-6}$. When we say that the SNP requires one other SNP to exhibit an association, we mean that the SNP shows no association by itself, and there is at least one other region with which it does show an association.

We then have the following:

1 Prior probability of a 1-SNP model M_i:[2]

$$\mathbb{P}(M_i) = 1.0 \times 10^{-6}.$$

2 Prior probability of a 2-SNP model M_{ij}:

$$\mathbb{P}(M_{ij}) = \mathbb{P}(M_i)\mathbb{P}(M_j)$$
$$+\mathbb{P}(M_{ij}|S_j \text{ requires two SNPs})\mathbb{P}(S_j \text{ requires two SNPs})$$
$$= \left(1.0 \times 10^{-6}\right)\left(1.0 \times 10^{-6}\right) + \frac{1}{999999} \times \frac{1}{9}\left(\sum_{i=1}^{9} i\right)\left(1.0 \times 10^{-6}\right)$$
$$= 6.0 \times 10^{-12}.$$

When we say "S_j requires two SNPs," we mean one other SNP besides itself. The far right term is obtained as follows: S_j comes from one of the 1 000 000 regions. There are 999 999 remaining regions. We assume that nine of these regions are associated with the phenotype, and that it is equally likely that S_j shows a 2-SNP effect with only one of them, with two of them, etc., up to all nine of them. The probability of a single region containing S_i is 1/999999, the probability of two regions containing S_i is 2/999999, and so on. So, $\mathbb{P}(M_{ij}|S_j \text{ requires two SNPs})$, the probability of S_j showing an effect with S_i (given S_j does show a 2-SNP effect) is then as follows:

$$\frac{1}{999999} \times \frac{1}{9} + \frac{2}{999999} \times \frac{1}{9} + \cdots + \frac{9}{999999} \times \frac{1}{9} = \frac{1}{999999}\frac{1}{9}\left(\sum_{i=1}^{9} i\right).$$

---

[2] Note that this applies to the models in Figs. 11.2 and 11.3 that only have one edge because the prior probability of a SNP not having an edge is almost 1.

3 Prior probability of a 3-SNP model M_{ijk}:

$$\mathbb{P}(M_{ijk})$$
$$= \mathbb{P}(M_i)\mathbb{P}(M_j)\mathbb{P}(M_k) + 3\mathbb{P}(M_{ij})\mathbb{P}(M_k)$$
$$\quad + \mathbb{P}(M_{ijk}|S_k \text{ requires three SNPs})\mathbb{P}(S_k \text{ requires three SNPs})$$
$$= \left(1.0 \times 10^{-6}\right)^3 + 3 \times 6.0 \times 10^{-12} \times 1.0 \times 10^{-6}$$
$$\quad + \frac{1}{\binom{999999}{2}} \frac{1}{\binom{9}{2}} \left(\sum_{i=1}^{\binom{9}{2}} i\right)\left(1.0 \times 10^{-6}\right)$$
$$= 5.6 \times 10^{-17}.$$

4 Prior probability of a 4-SNP model M_{ijkl}:

$$\mathbb{P}(M_{ijkl})$$
$$= \mathbb{P}(M_i)\mathbb{P}(M_j)\mathbb{P}(M_k)\mathbb{P}(M_l) + 6\mathbb{P}(M_{ij})\mathbb{P}(M_{kl}) + 4\mathbb{P}(M_{ijk})\mathbb{P}(M_l)$$
$$\quad + \mathbb{P}(M_{ijkl}|S_l \text{ requires 4 SNPs})\mathbb{P}(S_l \text{ requires four SNPs})$$
$$= \left(1.0 \times 10^{-6}\right)^4 + 6 \times \left(6.0 \times 10^{-12}\right)^2$$
$$\quad + 4 \times 5.6 \times 10^{-17} \times 1.0 \times 10^{-6} + \frac{1}{\binom{999999}{3}} \frac{1}{\binom{9}{3}} \left(\sum_{i=1}^{\binom{9}{3}} i\right)\left(1.0 \times 10^{-6}\right)$$
$$= 6.96 \times 10^{-22}.$$

We can repeat these calculations using 0.0001 as the prior probability of a SNP being associated with a disease. In this way we obtain upper and lower priors.

11.6.3 Experiments

Reiman et al. [28] analyzed a GWAS late onset Alzheimer's disease (LOAD) data set concerning 312 317 SNPs from an Affymetrix 500K chip, and a locus in the APOE gene, which is known to be predictive of LOAD, was examined. The APOE ε_4 genotype is known to be associated with LOAD. The data set consists of three cohorts containing a total of 1411 participants. Of the 1411 participants, 861 had LOAD and 550 did not. In addition, 644 participants were APOE ε_4 carriers, who carry at least one copy of the APOE ε_4 genotype, and 767 were APOE ε_4 non-carriers. Reiman et al. [28] found that the APOE gene is significantly associated with LOAD, the GAB2 gene is not significantly associated with LOAD, the GAB2 gene is significantly associated with LOAD in APOE ε_4 carriers, and the GAB2 gene is not significantly associated with LOAD in the APOE ε_4 non-carriers. These results indicate that APOE and GAB2 may interact epistatically to affect LOAD.

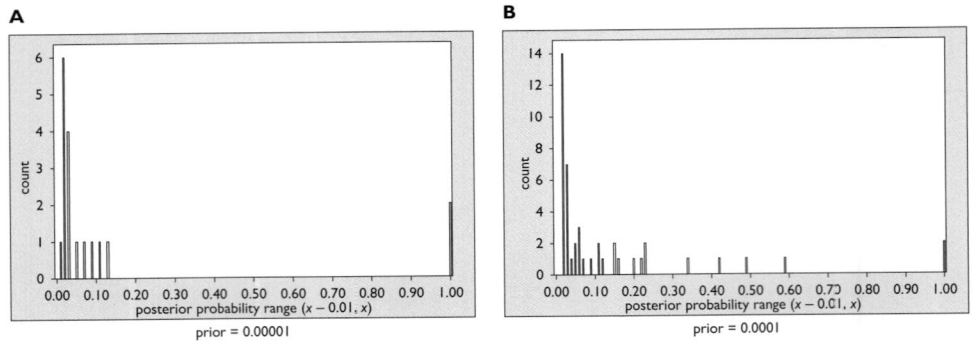

Fig 11.5 Bar charts showing the number of SNPs in each posterior probability range (based on low and high priors). The posterior probabilities used to create this chart are those of each SNP being associated with LOAD.

METHOD

Since the APOE gene is included in the study and it is not a SNP, we refer to the models as n-locus models in what follows. Using the same data as Reiman et al. [28], we computed the posterior probability of each locus being associated with LOAD (1-locus models), and the posterior probability of each locus together with APOE being associated with LOAD (2-locus models).

RESULTS

The average posterior probability of all one-locus models is 2.85×10^{-5} when the individual SNP prior is set equal to 0.0001, and 9.18×10^{-6} when that prior is set equal to 0.00001. Furthermore, the numbers of models (loci) with posterior probabilities less than 0.01 were respectively 312 301 and 312 273 for the two priors. Fig. 11.5 shows bar charts depicting the results concerning the remaining loci.

Table 11.6 shows the loci in the ten most probable models. APOE has a posterior probability of about 1, regardless of the prior, as does SNP rs41377151. SNP rs41377151 is on the APOC1 gene, which is in strong linkage disequilibrium (LD) with APOE and for which previous studies have indicated that the two genes predict LOAD equally well [34]. The 3rd most probable locus is rs1082430, which is located on the PRKG1 gene. There are a number of previous studies associating this gene with LOAD [10], [21]. Of the seven remaining probable loci, there is some previous evidence linking four of them to LOAD [31].

As mentioned in Subsection 11.6.2, as more GWASs are carried out, we will better be able to assess appropriate priors. These results indicate that 0.00001 may be more appropriate than 0.0001, since the latter prior resulted in fairly high posterior probabilities for three SNPs that have no known previous association with LOAD; nonetheless, these might be valid predictors of LOAD that have not been appreciated previously.

The average posterior probability of all 2-locus models, in which one of the loci was APOE, is 1.05×10^{-4} when the individual SNP prior is set equal to 0.0001 and 1.41×10^{-5} when that prior is set equal to 0.00001. Furthermore, the numbers of models with posterior probabilities less than 0.01 were respectively 312 267 and 312 028 for the two priors. Fig. 11.6 shows bar charts depicting the results concerning the remaining models. Table 11.7 shows the loci in the ten most probable models. Eight of those loci are SNPs located on the GAB2 gene. The prior probability

Table 11.6 The SNPs most probably associated with LOAD.

SNP	$\mathbb{P}$ range (based on low and high priors)	Previous LOAD Association
APOE	(1, 1)	yes
rs41377151	(1, 1)	yes
rs10824310	(0.124, 0.586)	yes
rs4356530	(0.086, 0.485)	no
rs17330779	(0.066, 0.416)	yes
rs6784615	(0.048, 0.335)	yes
rs10115381	(0.027, 0, 222)	no
rs12162084	(0.024, 0.217)	yes

of a two-SNP model is 6×10^{-10} when the individual SNP prior is 0.0001 and 6×10^{-12} when that prior is 0.00001. We see from Table 11.7 that the posterior probabilities of 2-locus models containing APOE and a GAB2 SNP are much greater than these prior probabilities. On the other hand, the 1-locus models containing GAB2 SNPs have posterior probabilities about equal to their prior probabilities. These results together indicate that GAB2 by itself does not affect LOAD, but that GAB2 interacts with APOE to affect LOAD.

The two loci in the top ten two-locus models that are not located on GAB2, namely SNPs rs6784615 and rs12162084, are among the 10 most probable one-locus models (see Table 11.6). These results together indicate that each of these SNPs may affect LOAD independently of APOE. As indicated in Table 11.6, previous studies have associated these SNPs with LOAD.

Another interesting result is that APOE and rs41377151 (the two loci with posterior probabilities about equal to 1 in Table 11.6), when considered together, have posterior probabilities of

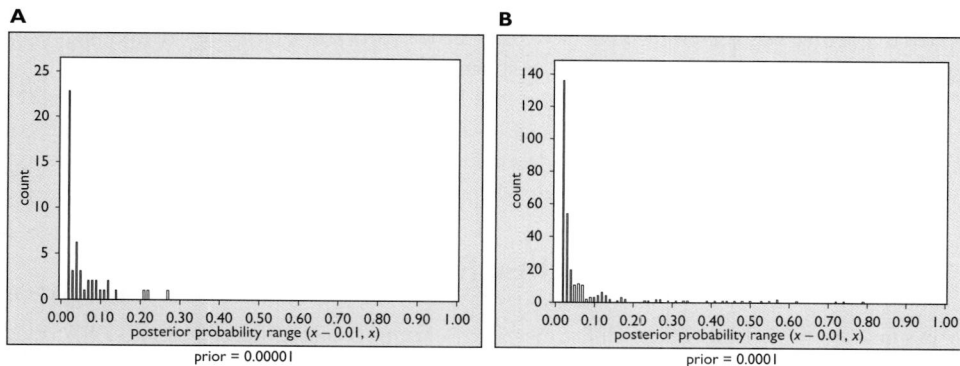

Fig 11.6 Bar charts showing the number of SNPs in each posterior probability range (based on low and high priors). The posterior probabilities used to create this chart are those of each SNP together with APOE being associated with LOAD.

Table 11.7 The SNPs, which together with APOE, are most probably associated with LOAD, based on two-SNP models.

SNP	$\mathbb{P}$ range (Based on Low and High Priors)	GAB2 ?
rs1007837	(0.265, 0.784)	yes
rs7101429	(0.214, 0.731)	yes
rs901104	(0.201, 0.715)	yes
rs4291702	(0.139, 0.617)	yes
rs4945261	(0.144, 0.564)	yes
rs12162084	(0.144, 0.563)	no
rs7115850	(0.103, 0.547)	yes
rs10793294	(0.099, 0.523)	yes
rs2450130	(0.083, 0.491)	yes

1.25×10^{-4} and 1.25×10^{-5} for the individual SNP priors of 0.0001 and 0.00001 respectively. This result indicates that the model containing both loci is incorrect. As mentioned above, SNP rs41377151 is located on the APOC1 gene, and previous investigations have shown that APOE and APOC1 are in linkage disequilibrium and each of them predicts LOAD as well as the other [35]. However, we know of no previous study substantiating that the two loci identify the same single causal mechanism of LOAD. This result could not have been obtained with a method that only considers the null hypothesis that the two loci together are not associated with LOAD, and the alternative hypothesis that they are. For example, using Pearson's chi-square test, we obtained p-values all close to 0 for APOE alone, rs41377151 alone, and APOE and rs41377151 together (the two-locus model). The BNPP determined that the 2-locus model is improbable because it also evaluated the competing hypotheses that only one locus is directly causative of LOAD. To learn that the two-locus model is not significantly better than the one-locus model using commonly applied frequentist statistics, we would need to perform an analysis such as stepwise regression or regression on the two loci followed by an investigation of the coefficients.

The three interesting results just discussed (the first concerning GAB2, the second rs6784615 and rs12162084, and the third rs41377151) follow from computing the posterior probabilities of all one-locus models and all two-locus models containing APOE. It was not necessary to suspect any of them ahead of time or perform a focused analysis.

11.7 Discussion and Further Research

We put forward the following three difficulties when learning from high-dimensional GWAS data sets:

1 determining the best way to score candidate SNP models to identify which models are most noteworthy,

2 developing heuristic search algorithms for learning epistatic interactions from high-dimensional data,

3 determining a way to decide which discovered SNP models to flag as *sufficiently* noteworthy for further investigation.

We provided three efforts at attacking the difficulties. As to the first difficulty, we showed results comparing the performance of BN scoring criteria using simulated data sets in which two SNPs interact epistatically. We found that the BDeu score with fairly large values of α performed best. These studies need to be repeated using data sets with three, four, and perhaps five interacting SNPs to determine if this result extends to higher-order interactions.

Concerning the second difficulty, we showed that the heuristic search algorithm MBS can do just as well as an exhaustive search. However, even this algorithm does not scale up to high-dimensional data sets. If we have, for example, 500 000 SNPs, we could not start a beam from each SNP and complete in a reasonable amount of time. Furthermore, MBS would not detect a pure, strict epistatic 3-SNP interaction.

Finally, we addressed the third difficulty by developing the BNPP. The BNPP requires the assessment of prior probabilities of DDAG models. We provided such an assessment in Subsection 11.6.2. These priors are only an initial effort. As noted in that section, as more genome-wide association studies are carried out, we will be able to refine these priors.

· ·

REFERENCES

[1] B.M. Armes, A.J. Egan, M.C. Southey, G.S. Dite, M.R. McCredie, G.G. Giles, J.L. Hopper, and D.J. Venter. The histologic phenotypes of breast carcinoma occurring before age 40 years in women with and without BRCA1 or BRCA2 germline mutations. *Cancer*, 83:2335–2345, 2000.

[2] W. Bateson. *Mendel's Principles of Heredity*. Cambridge University Press, 1909.

[3] D. Brinza, J. He, and A. Zelkovsky. Optimization methods for genotype data analysis in epidemiological studies. In I. Mandoiu and A. Zelikousky, editors, *Bioinformatics Algorithms: Techniques and Applications*, pages 395–415, Wiley, 2008.

[4] A.J. Brookes. The essence of SNPs. *Gene*, 234:177–186, 1999.

[5] D. Chickering and C. Meek. Finding optimal Bayesian networks. In A. Derwiche and N. Friedman, editors, *Proceedings of the Eighteenth Conference on Uncertainty in Artificial Intelligence*, Morgan Kaufmann Publishers, 2002.

[6] Y.M. Cho, M.D. Ritchie, J.H. Moore, J.Y. Park, K.-U. Lee, H.D. Shin, H.K. Lee, and K.S. Park. Multifactor dimensionality reduction reveals a two-locus interaction associated with type 2 diabetes mellitus. *Diabetologia*, 47:549–554, 2004.

[7] The Wellcome Trust Case Control Consortium. Genome-wide association study of 14,000 cases of seven common diseases and 3,000 shared controls. *Nature*, 447:661–678, 2007.

[8] K.D. Coon, A.J. Myers, D.W. Craig, J.A. Webster, J.V. Pearson, D.H. Lince, V.L. Zismann, T.G. Beach, D. Leung, L. Bryden, R.F. Halperin, L. Marlowe, M. Kaleem, D.G. Walker, R. Ravid, C.B. Heward, J. Rogers, A. Papassotiropoulos, E.M. Reiman, J. Hardy, and D.A. Stephan. A high-density whole-genome association study reveals that APOE is the major susceptibility gene for sporadic late-onset Alzheimer's disease. *Journal of Clinical Psychiatry*, 68:613–618, 2007.

[9] G.F. Cooper and E. Herskovits. A Bayesian method for the induction of probabilistic networks from data. *Machine Learning*, 9:309–347, 1992.

[10] M.D. Fallin, M. Szymanski, R. Wang, A. Gherman, S.S. Bassett, and D. Avramopoulos. Fine mapping of the chromosome 10q11-q21 linkage region in Alzheimer's disease cases and controls. *Neurogenetics*, 11(3):335–348, 2010.

[11] R.A. Fisher. The arrangement of field experiments. *Journal of the Ministry of Agriculture of Great Britain*, 33:503–513, 1926.

[12] A. Galvin, J.P.A. Ioannidis, and T.A. Dragani. Beyond genome-wide association studies: genetic heterogeneity and individual predisposition to cancer. *Trends in Genetics*, 3(26):132–141, 2010.

[13] L.W. Hahn, M.D. Ritchie, and J.H. Moore. Multifactor dimensionality reduction software for detecting gene-gene and gene-environment interactions. *Bioinformatics*, 19:376–382, 2003.

[14] D. Heckerman, D. Geiger, and D. Chickering. Learning bayesian networks: the combination of knowledge and statistical data. Technical Report MSR-TR-94-09, Microsoft Research, 1994.

[15] A. Herbert, N.P. Gerry, and M.B. McQueen. A common genetic variant is associated with adult and childhood obesity. *Journal of Computational Biology*, 312:279–384, 2006.

[16] National Cancer Institute: http://www.cancer.gov/cancertopics/understandingcancer/cancergenomics/AllPages.

[17] X. Jiang, M.M. Barmada, G.F. Cooper, and M.J. Becich. A Bayesian method for evaluating and discovering disease loci associations. *PLOS ONE*, 6(8):e22075, 2011.

[18] X. Jiang, R.E. Neapolitan, M.M. Barmada, and S. Visweswaran. Learning genetic epistasis using Bayesian network scoring criteria. *BMC Bioinformatics*, 12:89, 2011.

[19] X. Jiang, R.E. Neapolitan, M.M. Barmada, S. Visweswaran, and G.F. Cooper. A fast algorithm for learning epistatic genomics relationships. In *AMIA Annual Symposium Proceedings*, pages 341–345. American Medical Informatics Association, 2010.

[20] J.-C. Lambert, S. Heath, G. Even, D. Campion, K. Sleegers, M. Hiltunen, O. Combarros, D. Zelenika, M.J. Bullido, and B. et al. Tavernier. Genome-wide association study identifies variants at clu and cr1 associated with Alzheimer's disease. *Nature Genetics*, 41:1094–1099, 2009.

[21] X. Liang, M. Slifer, E.R. Martin, N. Schnetz-Boutaud, J. Bartlett, B. Anderson, S. Züchner, H. Gwirtsman, J.R. Gilbert, M.A. Pericak-Vance, and J.L. Haines. Genomic convergence to identify candidate genes for Alzheimer disease on chromosome 10. *Human Mutation*, 30(3):463–471, 2009.

[22] T.A. Manolio and F.S. Collins. The HapMap and genome-wide association studies in diagnosis and therapy. *Annual Review of Medicine*, 60:443–456, 2009.

[23] D.J. Miller, Y. Zhang, G. Yu, Y. Liu, L. Chen, C.D. Langefeld, D. Herrington, and Y. Wang. An algorithm for learning maximum entropy probability models of disease risk that efficiently searches and sparingly encodes multilocus genomic interactions. *Bioinformatics*, 25(19):2478–2485, 2009.

[24] J.H. Moore and S.M. Williams. An algorithm for learning maximum entropy probability models of disease risk that efficiently searches and sparingly encodes multilocus genomic interactions. *Annals of Medicine*, 34:88–95, 2002.

[25] R.I. Nagel. Epistasis and the genetics of human diseases. *C R Biologies*, 328:606–615, 2005.

[26] R.E. Neapolitan. *Learning Bayesian Networks*. Prentice Hall, 2004.

[27] R.E. Neapolitan. A polemic for Bayesian statistics. In D. Holmes and L. Jain, editors, *Innovations in Bayesian Networks, Studies in Computational Intelligence*, volume 8, pages 7–32 Springer Verlag, 2008.

[28] E.M. Reiman, J.A. Webster, A.J. Myers, J. Hardy, T. Dunckley, V.L. Zismann, K.D. Joshipura, J.V. Pearson, D. Hu-Lince, and M.J. Huentelman. Gab2 alleles modify Alzheimer's risk in APOE carriers. *Neuron*, 54:713–720, 2007.

[29] J. Rissanen. Modeling by shortest data description. *Automatica*, 14:465–471, 1978.

[30] M.D. Ritchie, L.W. Hahn, N. Roodi, R. Bailey, W.D. Dupont, F.F. Parl, and J.H. Moore. Multifactor-dimensionality reduction reveals high-order interactions among estrogen-metabolism genes in sporadic breast cancer. *American Journal of Human Genetics*, 69:138–147, 2001.

[31] H. Shi, C. Medway, J. Bullock, K. Brown, N. Kalsheker, and K. Morgan. Analysis of genome-wide association study (GWAS) data looking for replicating signals in Alzheimer's disease (AD). *International Journal of Molecular Epidemiology and Genetics*, 1(1):53–66, 2010.

[32] T. Silander, P. Kontkanen, and P. Myllymäki. On sensitivity of the MAP Bayesian network structure to the equivalent sample size parameter. In R. Parr, L. C. van der Gaag, editors, *Proceedings of the Twenty-third Conference on Uncertainty in Artificial Intelligence (UAI07)*, pages 360–367. AUAI Press, 2007.

[33] M. Spinola, P. Meyer, S. Kammerer, F.S. Falvella, M.B. Boettger, C.R. Hoyal, C. Pignatiello, R. Fischer, R.B. Roth, U. Pastorino, K. Haeussinger, M.R. Nelson, R. Dierkesmann, T.A. Dragani, and A. Braun. Association of the PDCD5 locus with lung cancer risk and prognosis in smokers. *American Journal of Human Genetics*, 55:27–46, 2001.

[34] J. Suzuki. Learning Bayesian belief networks based on the minimum description length principle: basic properties. *IEICE Transactions on Fundamentals*, E82-A, 10:2237–2245, 1999.

[35] B. Tycko, J.H. Lee, A. Ciappa, A. Saxena, C.M. Li, L. Feng, A. Arriaga, Y. Stern, R. Lantigua, N. Shachter, and R. Mayeux. APOE and APOC1 promoter polymorphisms and the risk of Alzheimer disease in African American and Caribbean Hispanic Individuals. *Archives of Neurology*, 61(9):1434–1439, 2004.

[36] D.R. Velez, B.C. White, A.A. Motsinger, W.S. Bush, M.D. Ritchie, S.M. Williams, and J.H. Moore. A balanced accuracy function for epistasis modeling in imbalanced data sets using multifactor dimensionality reduction. *Genetic Epidemiology*, 31:306–315, 2007.

[37] S. Wacholder, S. Chanock, M. Garcia-Closas, L. El Ghormli, and N. Rothman. Assessing the probability that a positive report is false; an approach for molecular epidemiology studies. *Journal of the National Cancer Institute*, 96:434–432, 2004.

[38] J. Wakefield. Reporting and interpreting in genome-wide association studies. *International Journal of Epidemiology*, 37(3):641–653, 2008.

[39] J. Wu, B. Devlin, S. Ringguist, M. Trucco, and K. Roeder. Screen and clean: a tool for identifying interactions in genome-wide association studies. *Genetic Epidemiology*, 34:275–285, 2010.

CHAPTER 12

Graphical Modeling of Biological Pathways in Genome-wide Association Studies

MIN CHEN, JUDY CHO, AND HONGYU ZHAO

Genome-wide association studies (GWASs) are widely used to identify genetic variants associated with complex diseases. In GWASs, it is desired to prioritize genes, i.e., to identify good candidates of disease-associated genes that are of interest for further follow-up studies. On the other hand, much knowledge has been accumulated in the literature on biological pathways and interactions. It is conceivable that appropriate incorporation of such prior knowledge may improve the likelihood of making genuine discoveries in GWASs. A number of methods have been developed to incorporate prior biological knowledge when prioritizing genes. However, most methods treat genes in a specific pathway as an exchangeable set without considering the topological structure of the pathway. How genes are functionally related to each other in a pathway may be very informative for GWAS analysis, and such information can be utilized to increase the power of detecting real associations. When associations have been firmly established for some genes either through GWAS or prior candidate gene-based studies, one can take advantage of this knowledge to examine other genes related to these known genes through the same pathways they all participate in. Based on results obtained from a standard association study on a Crohn's disease cohort, it is first verified that neighboring genes in a pathway are more likely to share the same disease status. Then, a Markov random field (MRF) model is proposed, to

Probabilistic Graphical Models for Genetics, Genomics, and Postgenomics. First Edition. Christine Sinoquet & Raphaël Mourad (Eds). © Oxford University Press 2014. Published in 2014 by Oxford University Press.

incorporate pathway topology for association analysis. We show that the conditional distribution of our MRF model takes on a simple logistic regression form. Finally, we evaluate our model on real data.

12.1 Introduction

Genome-wide association studies (GWASs) are widely used to identify genetic variants associated with complex diseases. A case-control sample design is often used to select and genotype individuals at a large number of single nucleotide polymorphisms (SNPs) and other variants, e.g., copy number variations (CNVs). Researchers then examine these markers to identify their associations with disease, or to prioritize markers for follow-up studies. In most published studies, the search is limited to single markers. However, this approach may lack adequate statistical power for true discoveries. Frequently the relative risks, known as the effect size, of individual markers are small, and the sample size is not sufficiently large, especially when the minor allele frequency is low. As a result, single marker analysis may be less informative than considering multiple markers and multiple genes together because genes may interact with each other and may jointly affect disease risks. Besides, a large amount of knowledge about biological pathways and gene–gene interactions has been accumulated from past biological and bioinformatics studies. It is likely that appropriate incorporation of prior knowledge on genes and pathways may improve the chance of making genuine discoveries. To boost the statistical power, many approaches have been developed, including (1) aggregation of multiple markers located in the same gene region or in the same haplotype block, and (2) incorporation of information from other sources into the GWAS analysis. It has been reported that the gene level analysis can identify new associations in addition to the ones identified using individual SNPs [23, 28]. Gene-based analyses include those using the most significant SNP within and near a gene [28], combinatorial statistics (Fisher, Sidat, and Simes) from all individual markers [23], principal component analysis regressions [2] and the sparse partial least squares regressions [9].

One information-rich resource is the existing databases containing known gene pathways and protein-protein interactions, like BioCarta (http://www.biocarta.com/genes/index.asp), GeneMAPP [25], KEGG (Kyoto Encyclopedia of Genes and Genomes) [16] and the HPRD (the Human Protein Reference Database, http://www.hprd.org/). Since genes interact with each other in biological processes, it is possible that they may jointly affect the risk of a complex disease. As will be shown later, one can take advantage of graphical modeling to explicitly model the relationships of genes in a pathway. In GWAS it is desired to prioritize genes, i.e., to identify good candidates of disease-associated genes that are of interest for further follow-up studies. A number of methods have been developed to incorporate prior biological knowledge when prioritizing genes. Some examples include Prioritizer [11], Endeavour [1], CGI [20], CANDID [15], Gene-Wanderer [18], CIPHER [31], GIN [24], and the pathway-based gene set enrichment approach [28]. These studies have shown that it is useful to incorporate prior biological information in GWASs. However, functional relationships among genes are not considered because these approaches in general take a list of genes, instead of the complete network structure, as their input data. In this way, genes in the list are treated as exchangeable, and the regulatory relationships and interactions revealed by a pathway are not taken into account. As a result, information about the pathway topology and interactions among genes is usually ignored. However, how genes are functionally related to each other in a pathway may be very informative for GWAS analysis, and such information can be utilized to increase the power of detecting real associations. When associations have been firmly established for some genes either through GWASs or prior candidate gene-based studies, one can take advantage of this knowledge to examine other genes related to these known genes through the same pathways they all participate in.

In this chapter we introduce a Markov Random Field (MRF) model, one of the graphical modeling methods, to incorporate biological pathway information in GWASs. This model was discussed in [8], and here we provide more elaborations and mathematical derivations that were omitted from the previous publication. We note that several papers have considered MRFs to combine data from different sources in genomics studies, e.g., a spatial normal mixture model [29] for gene expression and ChIP-chip data, a Gamma-Gamma model and MRF for messenger RNA microarray data [30] and prioritizing genes by combining gene expression and protein interaction data [20]. Li et al. [19] proposed a hidden MRF for GWASs in the context of jointly analyzing markers in linkage disequilibrium was proposed.

12.2 MRF Modeling of Gene Pathways

We start by considering a graphical model in which a biological pathway is represented by an undirected graph. An example of such a graph is shown in Fig. 12.1, in which each node represents a gene and each edge denotes the interaction between a pair of genes. We define a graph G to be (V, E) where V is the set of n genes (nodes) in the graph and E denotes the set of all edges:

$$V = \{1, \ldots, n\},$$
$$E = \{< i, j >: i \text{ and } j \text{ are directly connected}\}.$$

For the ith gene in V, let N_i denote the set of its neighbors and d_i be the number of its neighbors:

$$N_i = \{j :< i, j >\in E\},$$
$$d_i = |N_i|.$$

Define S_i as the true association status:

$$S_i = +1 \quad \text{if gene } i \text{ is associated with the disease,}$$
$$S_i = -1 \quad \text{if gene } i \text{ is NOT associated with the disease.}$$

The values ± 1 are referred to as labels of nodes hereafter. Let $S = (S_1, \ldots, S_n)$, the labeling of V. Thus, S is a spatial random vector whose elements may be correlated with each other. Note that each node can be labeled either -1 or $+1$, and so there are a total of 2^n unique labelings, also called configurations, of the graph. The ultimate goal is to infer the value of S_i based on the graph topology determined by the underlying biological pathway and the observed association data.

Next we need to assign a probability measure to S. First we present a motivating example from a GWAS of Crohn's disease [10]. As will be shown next, the result clearly suggests that genes in the same neighborhood within a pathway tend to show similar association statuses. This Crohn's disease cohort includes 401 cases and 433 controls, and Illumina HumanHap300 Bead-Chips were used for genotyping. To analyze these data, we first mapped SNPs to genes and then

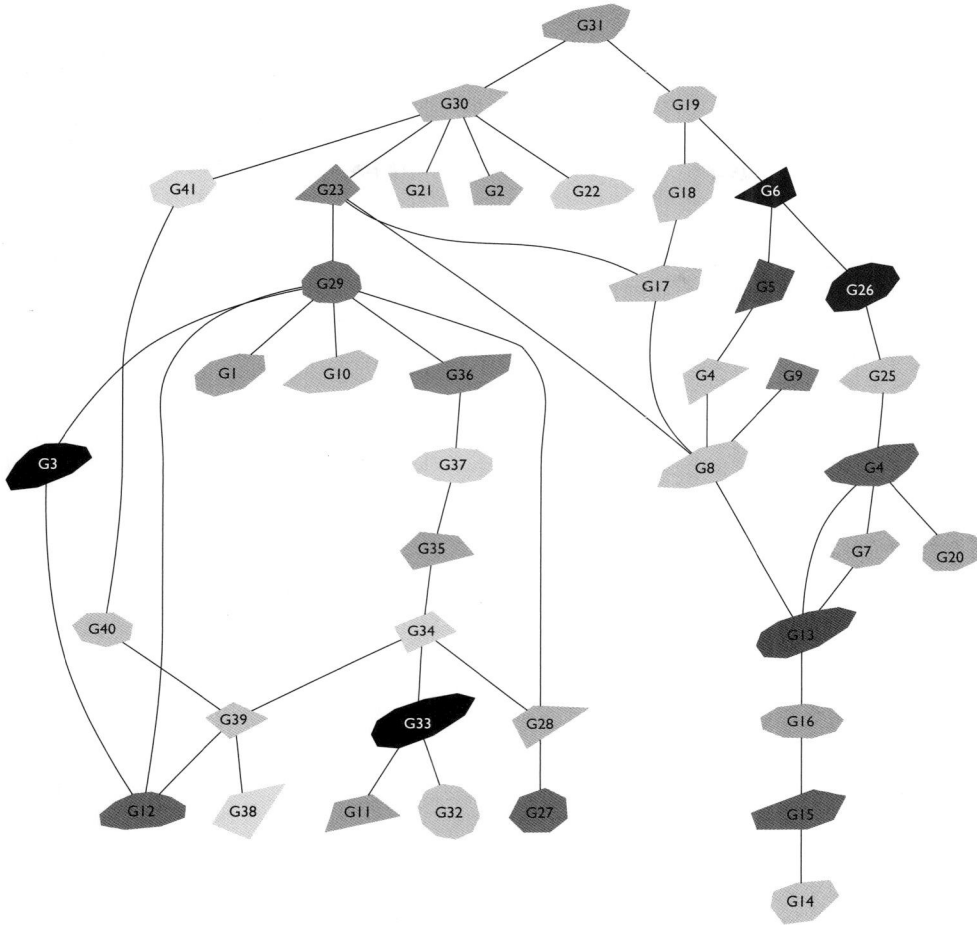

Fig 12.1 A gene pathway modeled as an undirected graph.

applied principal component analysis regressions to obtain gene-level p-values of the association tests with Crohn's disease status [2]. More details about this data set are given later in the real data example. We then obtained pathway information from BioCarta, GeneMAPP, and KEGG. We considered a total of 3735 genes in over 350 pathways. To avoid effects caused by possible linkage disequilibrium (LD), genes were excluded if they were on the same chromosome and were within 1 million base pairs. To see whether genes tend to show similar evidence for association when connected with each other in the same pathway, we used a cut-off value of 0.15. Genes were considered interesting and were labeled with +1 if their p-values were below this cut-off value. Note that we used a relatively loose threshold so that a sufficiently large number of genes are called "interesting", and this loose cut-off also reflects our belief that many genes have weak effects and only show moderate evidence of association. In a pathway k, we consider the number of edges connecting a pair of "interesting" genes, which depends on the labels of all genes. We denote this number as D_k. A large value of D_k would suggest that "interesting" genes are more likely to be neighboring genes. To assess the statistical evidence for the tendency to observe large D_k values, we employ a permutation procedure as follows. The null hypothesis is that there is no tendency for neighboring genes to have similar disease association status, i.e., "interesting" or not. In each permutation, we randomly permute the "interesting" labels of all genes and derive a permuted statistic. These permuted statistics are used to arrive at an empirical distribution of D_k under the null hypothesis. We then compare the observed D_k statistic with the empirical distribution. Finally the p-value of the observed D_k in this empirical distribution is calculated. A p-value close to 0 indicates that "interesting" genes tend to be neighbors. This procedure is repeated for all pathways, and the histogram of p-values of D_k for all pathways is plotted in Fig. 12.2. It is evident that this distribution is highly skewed to the left, which suggests associated genes tend to be neighbors in a given pathway.

To formalize the idea that neighboring genes tend to have similar association status, we need a probability measure so that nodes connected with each other tend to have the same labels. Here we consider a nearest neighbor Gibbs measure [17] that has the following form:

$$\mathbb{P}(\mathbf{S}|\theta_0) = \frac{1}{z(\theta_0)} \exp\left\{ h \sum_{i \in V} I_1(S_i) + \tau_0 \sum_{<i,j> \in E} (w_i + w_j)I_{-1}(S_i)I_{-1}(S_j) \right.$$

$$\left. + \tau_1 \sum_{<i,j> \in E} (w_i + w_j)I_1(S_i)I_1(S_j) \right\}, \tag{12.1}$$

where $\theta_0 = (h, \tau_1, \tau_0)$, $I_1(\cdot)$ and $I_{-1}(\cdot)$ are indicator functions, $w_i = d_i^{1/2}$, and $z(\theta_0)$ is a normalizing function that is the sum over all 2^n possible configurations:

$$z(\theta_0) = \sum_{\mathbf{S}} \exp\left\{ h \sum_{i \in V} I_1(S_i) + \tau_0 \sum_{<i,j> \in E} (w_i + w_j)I_{-1}(S_i)I_{-1}(S_j) \right.$$

$$\left. + \tau_1 \sum_{<i,j> \in E} (w_i + w_j)I_1(S_i)I_1(S_j) \right\}. \tag{12.2}$$

Note that it is prohibitive to evaluate $z(\theta_0)$ when n is large. Here τ_0 and τ_1 assign weights to edges connecting two non-associated nodes and two associated nodes, respectively. The function w_i will be later elaborated in more details in the context of the conditional probability.

Fig 12.2 Histogram of p-values of D_k, the number of edges connecting a pair of "interesting" genes in a pathway k, which depends on the labels of all genes. A large value of D_k would suggest that "interesting" genes are more likely to be neighbors. A permutation procedure is used to derive an empirical distribution of D_k under the null hypothesis. The p-value of an observed D_k is calculated with respect to this empirical distribution. See Section 12.1 for more details.

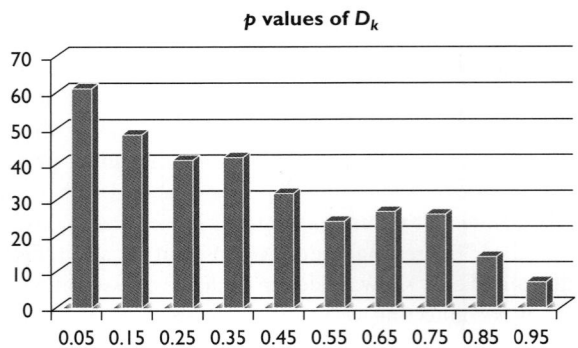

p values of D_k

In (12.1), the second sum is taken over all edges connecting immediate neighbors in which both end nodes are labeled -1, and the third sum is taken over all edges in which both end nodes are labeled 1. Positive τ_0 and τ_1 will put more weights on configurations in which directly linked nodes have the same labels, which is desirable in our context. The parameter h determines the marginal probability of S_i when $\tau_0 = \tau_1 = 0$, i.e., if all nodes are treated as singletons that are independent:

$$\mathbb{P}(S_i = 1 | h, \tau_0 = \tau_1 = 0) = \frac{\exp(h)}{\exp(h) + 1}.$$

The simple form Gibbs measure in (12.1) has a local Markov property that makes it attractive to model a biological pathway, in which directly linked genes interact with each other. Next we show that it defines an MRF on S, which by definition is $\mathbb{P}(S_i | S_{V-i}) = \mathbb{P}(S_i | S_{N_i})$, where $V - i$ denotes all nodes but i, and N_i is the set of all immediate neighbors of node i. Also, the conditional distribution of the MRF has a logistic regression form as shown below. Recall that the logistic regression uses the logit, defined as follows, to model the probability p of a dichotomous outcome variable:

$$\text{logit}(p) = \log\left(\frac{p}{1-p}\right).$$

The quantity $p/(1 - p)$ is also known as the odds, and its logarithm is assumed to be a linear function of predictors in logistic regression.

Proposition 12.1
The spatial random vector S, under the Gibbs measure in (12.1), is Markovian and thus defines an MRF satisfying

$$\mathbb{P}(S_i | S_{V-i}, \theta_0) = \mathbb{P}(S_i | S_{N_i}, \theta_0).$$

Moreover, the conditional distribution has a logistic regression form:

$$\text{logit}(\mathbb{P}(S_i | S_{N_i}, \theta_0)) = h + \tau_1 \left(w_i J_i^{(1)} + \sum_{j \in N_i} w_j I_1(S_j) \right)$$
$$- \tau_0 \left(w_i J_i^{(-1)} + \sum_{j \in N_i} w_j I_{-1}(S_j) \right), \quad i = 1, \ldots, n, \tag{12.3}$$

where $J_i^{(l)} = \sum_{k \in N_i} I_l(S_k)$, $l = \pm 1$. *Equivalently, (12.3) can be rewritten as a system of linear equations:*

$$\text{logit}(\mathbb{P}(S_i | S_{N_i}, \theta_0)) = \beta_{i0} + \beta_{i1} S_1 + \cdots + \beta_{in} S_n, \quad i = 1, \ldots, n, \tag{12.4}$$

where

$$\beta_{i0} = h,$$
$$\beta_{ij} = \begin{cases} 0 & \text{if } i = j \text{ or } <i, j> \notin E \\ (w_i + w_j)\{\tau_1 I_1(S_j) - \tau_0 I_{-1}(S_j)\} & \text{if } <i, j> \in E. \end{cases}$$

For the proof of this proposition, see Appendix 12.A (page 313).

The Markov property implies that the conditional distribution of S_i, given all other node labels in the network, is equivalent to the conditional distribution of S_i given all its immediate neighbors. It follows immediately from (12.4) that if S_i and S_j are not neighbors, then they are conditionally independent.

Now we give an interpretation of w_i. From (12.4), it is clear that the conditional distribution of S_i depends on the weighted sum of labels of its neighbors, with the weight being $(w_i + w_j)\tau_1$ if $S_j = 1$ and $-(w_i + w_j)\tau_0$ if $S_j = -1$. Here $(w_i + w_j)$ is the sum of weights on both ends of a linking edge. We set w_i to be the square root of d_i, which is the degree of gene i. As a result, a gene that interacts with many other genes in the pathway has a large weight because it may play a central role in a biological process, and thus it is likely to have a large influence.

The Markovian property of (12.1) can be derived directly from a more general result [17], which states that a nearest neighbor Gibbs measure determines an MRF. Our proof is specific to (12.1) and is needed to derive the logistic model in (12.3). We note that under the setting of a rectangular lattice, a special graph in the form of rectangular grids, [5, 6] presented a general logistic model called the autologistic model.

12.3 A Bayesian Framework

12.3.1 Prior Specification and Likelihood Function

Here we describe a Bayesian framework that utilizes the graphical model in the previous section as a prior distribution for a biological pathway. The following is a brief overview of the Bayesian framework. First we specify a model for the observed data, denoted by $y = (y_1, \ldots, y_n)$, in the form of a likelihood function $f(y|S, \mu, \sigma^2)$, where S is the true but unknown association status, and μ and σ^2 are unknown distribution parameters (more details on the likelihood function will be given later in this section). In Bayes' approach, we suppose that the parameters (S, μ, σ^2) are random quantities that have a prior distribution $\mathbb{P}(S, \mu, \sigma^2)$. Assuming a priori independence of S and (μ, σ^2), the prior distribution can be written as $\mathbb{P}(S, \mu, \sigma^2) = \mathbb{P}(S|\theta_0)\mathbb{P}(\mu, \sigma^2|\theta_1)$, where θ_0 and θ_1 are hyperparameters. We are interested in making inferences about S based on its posterior distribution $\mathbb{P}(S|y, \theta_0, \theta_1)$. Note that since we are not interested in (μ, σ^2), we seek to integrate them out from the joint distribution function:

$$ f(y|S, \theta_1) = \int \int f(y|S, \mu, \sigma^2)\mathbb{P}(\mu, \sigma^2|\theta_1) \, d\mu \, d(\sigma^2). $$

Therefore, the posterior distribution is

$$ \mathbb{P}(S|y, \theta_0, \theta_1) = \frac{f(y|S, \theta_1) \, \mathbb{P}(S|\theta_0)}{\sum_S f(y|S, \theta_1) \, \mathbb{P}(S|\theta_0)}. $$

We use the MRF model introduced in the previous section as the prior distribution $\mathbb{P}(S|\theta_0)$. It can explicitly model the interactions of genes in a neighborhood defined by the topology of a pathway. In the Bayesian setting, the parameters in θ_0 are known as hyperparameters for the prior distribution (12.1). As mentioned before, one desirable feature of the prior is that positive τ_0 and

τ_1 will put more weights on configurations in which directly linked nodes have the same labels. One advantage of the graphical prior specification is that, as will be shown later, the posterior distribution of the association status also defines an MRF whose conditional distribution has a logistic form, which provides great interpretability and facilitates the MCMC sampling of the posterior distribution.

Now we discuss the likelihood function of the observed data. Recall that in GWASs, case and control samples are compared to search for genetic variants that are associated with the disease. As mentioned in Section 12.1, there are many methods to summarize or aggregate multiple SNPs to gene-level scores. Usually a statistical test is employed to test the association. We consider the situation where the association is examined at the gene level and the observed evidence of association is summarized by a p-value. Further, for gene i we define the response variable y_i to be the normal score corresponding to the p-value:

$$y_i = \Phi^{-1}(1 - p_i),$$

where p_i is the p-value and $\Phi(\cdot)$ is the cumulative distribution function of $\mathcal{N}(0, 1)$. Let $y = (y_1, \ldots, y_i, \ldots, y_n)$. Here the p-values, and hence y, are assumed to be conditionally independent given the true association status S. Under the null hypothesis that there is no association, each p-value has a uniform distribution on $(0,1)$. Therefore, under the null case where $S_i = -1$, the density of y_i is $f_0(y_i) \sim \mathcal{N}(0, 1)$. However, if there is an association between the gene and the disease, i.e., $S_i = +1$, the distribution of y_i is usually unknown. For simplicity, we assume that it is from $\mathcal{N}(\mu_i, \sigma_i^2)$, where μ_i is the location parameter and σ_i is the scale parameter that usually depends on the true effect size, allele frequencies, and the sample size. Note that μ_i is positive because when $S_i = +1$, one will observe many small p-values that correspond to large y_i values. As a result, the distribution of y_i tends to shift to the right. To account for the uncertainty about the parameters, we can put prior distributions on μ_i and σ_i^2, and marginalize over them to obtain the marginal distribution of y_i. Among many choices for the prior distributions, for the sake of computational convenience, we consider conjugate priors (e.g., see [13]), a well-known and widely used family of distributions. They are conjugate to the likelihood function, which means the posterior distribution belongs to the same family as the prior. Here we consider conjugate priors $\mu_i | \sigma_i^2 \sim \mathcal{N}(\bar{\mu}, \sigma_i^2/a)$ and $\sigma_i^2 \sim$ Inverse Gamma$(\nu/2, \nu d/2)$, or equivalently $\sigma_i^{-2} \sim$ Gamma$(\nu/2, \nu d/2)$.[1] The latter form is useful for sampling and for the derivation of the marginal distribution of y_i. We define $\theta_1 = (\bar{\mu}, a, \nu, d)$. The prior mean of μ_i is $\bar{\mu}$, and its prior variance is σ_i^2/a. The prior mean of σ_i^{-2} is $1/d$, and the prior variance is $\text{Var}[\sigma_i^{-2}] = 2/(\nu d^2)$. This normal-inverse-gamma prior specification for $\mu_i | \sigma_i^2$ and σ_i^2 is of conjugate form so that the integration over μ_i and σ_i^2 is analytically tractable, as will be shown next. We note that the hyperparameters $(\bar{\mu}, a, \nu, d)$ can be estimated from the observed data via an empirical Bayes method [8]. Under this prior setting, analytical simplification yields the marginal density of y_i:

$$f_1(y_i | S_i = 1, \theta_1) = \int \frac{1}{\sqrt{2\pi\sigma_i^2}} \exp\left[\frac{-(y - \mu_i)^2}{2\sigma_i^2}\right] \cdot \frac{\sqrt{a}}{\sqrt{2\pi\sigma_i^2}} \exp\left[\frac{-a(y - \bar{\mu})^2}{2\sigma_i^2}\right]$$

$$\times \frac{(\nu d/2)^{\nu/2}}{\Gamma(\nu/2)} (\sigma_i^{-2})^{\nu/2-1} \exp\left[-(\sigma_i^{-2})\frac{\nu d}{2}\right] d\mu_i \, d(\sigma_i^{-2})$$

---

[1] The density function of the gamma distribution is $\frac{(\nu d/2)^{\nu/2}}{\Gamma(\nu/2)}(\sigma_i^{-2})^{\nu/2-1} \exp\left[-(\sigma_i^{-2})\frac{\nu d}{2}\right]$.

$$= \int \frac{1}{\sqrt{2\pi \frac{a+1}{a} \sigma_i^2}} \exp\left[\frac{-a(y-\bar{\mu})^2}{2(a+1)\sigma_i^2}\right] \frac{(vd/2)^{\frac{v}{2}}}{\Gamma(v/2)} (\sigma_i^{-2})^{\frac{v}{2}-1} \exp\left[\frac{-vd\sigma_i^{-2}}{2}\right] d(\sigma_i^{-2})$$

$$= \frac{1}{\sqrt{2\pi}} \frac{\sqrt{a}}{\sqrt{a+1}} \frac{(vd/2)^{\frac{v}{2}}}{\Gamma(v/2)} \int (\sigma_i^{-2})^{\frac{v+1}{2}-1} \exp\left[-\sigma_i^{-2}\left\{\frac{vd}{2} + \frac{a(y-\bar{\mu})^2}{2(a+1)}\right\}\right] d(\sigma_i^{-2})$$

$$= \frac{1}{\sqrt{2\pi}} \frac{\sqrt{a}}{\sqrt{a+1}} \frac{(vd/2)^{v/2}}{\Gamma(v/2)} \Gamma\left(\frac{v+1}{2}\right) \cdot \left\{\frac{vd}{2} + \frac{1}{2}\frac{a}{a+1}(y-\bar{\mu})^2\right\}^{-(v+1)/2}$$

$$= \pi^{-1/2}(vd)^{v/2} \frac{\sqrt{a}}{\sqrt{a+1}} \frac{\Gamma((v+1)/2)}{\Gamma(v/2)} \left(\frac{a}{a+1}(y_i-\bar{\mu})^2 + vd\right)^{-(1+v)/2}. \tag{12.5}$$

This is equivalent to $(y_i - \bar{\mu})/\sqrt{(a+1)\,d/a} \sim t(v)$, Student's t-distribution with v degrees of freedom,[2] when $v = 1, 2$, and others.

12.3.2 Posterior Distribution

Now we present the posterior distribution of association status S after combining the structure of the gene pathway (V, E) and the evidence provided by the observed association statistics y. The joint density of y is

$$f(y|S, \theta_1) = \prod_{\{j:S_j=-1\}} f_0(y_j) \times \prod_{\{j:S_j=+1\}} f_1(y_j|S_j = 1, \theta_1).$$

Thus, the posterior distribution of S given the observed data y is

$$\mathbb{P}(S|y, \theta_0, \theta_1) \propto f(y|S, \theta_1)\,\mathbb{P}(S|\theta_0). \tag{12.6}$$

Similar to the MRF interpretation of the prior distribution (12.1), the posterior also has a nice conditional distribution and actually defines an MRF, as will be shown next.

In Section 12.2, it is shown that the simple form Gibbs measure in (12.1) has the local Markov property

$$\mathbb{P}(S_i|S_{V-i}, \theta_0) = \mathbb{P}(S_i|S_{N_i}, \theta_0).$$

Consequently, it defines an MRF on S. To see that the posterior distribution also defines an MRF, note that for node i,

$$\mathbb{P}(S_i = +1|y, S_{V-i}, \theta_0, \theta_1) \propto f_1(y_i|\theta_1)\,\mathbb{P}(S_i = +1|S_{V-i}, \theta_0)$$
$$= f_1(y_i|\theta_1)\,\mathbb{P}(S_i = +1|S_{N_i}, \theta_0).$$

---

[2] The density function of $t(v)$ is $f(x; v) = \frac{\Gamma((v+1)/2)}{\Gamma(v/2)}(\pi v)^{-1/2}(1 + x^2/v)^{-(1+v)/2}$. A transformation $x = (y_i - \bar{\mu})/\sqrt{(a+1)\,d/a}$ yields the density function in (12.5).

Thus, the conditional posterior distribution of S_i given all other nodes only depends on its neighbors, which means the posterior distribution leads to an MRF. The conditional posterior log odds of S_i is

$$h + \log LR(y_i; \theta_1) + \tau_1 \left(w_i J_i^{(1)} + \sum_{j \in N_i} w_j I_1(S_j) \right)$$

$$- \tau_0 \left(w_i J_i^{(-1)} + \sum_{j \in N_i} w_j I_{-1}(S_j) \right), \tag{12.7}$$

where

$$LR(y_i; \theta_1) = \frac{f_1(y_i | \theta_1)}{f_0(y_i)}$$

is the marginal likelihood ratio. Therefore, (12.7) is the product of the marginal likelihood ratio, reflecting the evidence from the data for association with the disease, and of the conditional prior odds, reflecting the effect from interactions among neighboring genes from the biological pathway.

To make it clear, we can rewrite (12.7) in the form of a system of autologistic regression equations:

$$\text{logit}(\mathbb{P}(S_i | y, S_{V-i}, \theta_0, \theta_1)) = \beta'_{i0} + \beta_{i1} S_1 + \cdots + \beta_{in} S_n, \quad i = 1, \ldots, n, \tag{12.8}$$

where

$$\beta'_{i0} = h + \log(LR(y_i); \theta_1),$$

$$\beta_{ij} = \begin{cases} 0 & \text{if } i = j \text{ or } (i, j) \notin E \\ (w_i + w_j)\{\tau_1 I_1(S_j) - \tau_0 I_{-1}(S_j)\} & \text{if } (i, j) \in E. \end{cases}$$

Here we make a few observations. First, it is easy to see that the posterior conditional logit form in (12.7) is the same as the prior conditional logit in (12.4) except its intercept is $h + \log(LR(y_i); \theta_1)$. Thus, the observed log-likelihood ratio provides a fixed additive effect to the prior logit. Second, the coefficient matrix is symmetric, i.e., $\beta_{ij} = \beta_{ji}$. Similar to equation (12.4), if gene i and j are not neighbors, then $\beta_{ij} = \beta_{ji} = 0$ and they are conditionally independent. On the other hand, if i and j are neighbors, then the impacts between each other are equal. Third, genes i and j are in general correlated in their joint posterior distribution, even if they are not neighbors and are conditionally independent. Moreover, the more common neighbors they share with each other, the stronger the correlation is between the two genes.

When n is large, since it is prohibitive to evaluate posterior probabilities on the entire space of configurations, we implement a Markov chain Monte Carlo (MCMC) method to sample from the posterior distribution. As shown before, due to the MRF property, the posterior has a nice closed-form conditional distribution that can greatly facilitate the MCMC. To sample from the posterior distribution, here we implement a Gibbs sampler that is well suited for an MRF. The algorithm is described as follows. First we set an initial value for S, say $s^{(0)}$, where $s = (s_1, \ldots, s_n)$. Note that we use the upper case S to denote a random vector and use the lower case s to denote a realization of the random vector, i.e., a vector consisting of +1s and −1s. Also we write $\mathbb{P}(S = s)$, the probability function evaluated at s, as $\mathbb{P}(s)$. Then in step k, we update the labels sequentially for $i = 1, \ldots, n$ according to (12.8):

$$\text{logit}(\mathbb{P}(s_i^{(k)}|y, s_1^{(k)}, \ldots, s_{i-1}^{(k)}, s_{i+1}^{(k-1)}, \ldots, s_n^{(k-1)}, \theta_0, \theta_1))$$
$$= \beta_{i0}' + \beta_{i1} s_1^{(k)} + \cdots + \beta_{i,i-1} s_{i-1}^{(k)} + \beta_{i,i+1} s_{i+1}^{(k-1)} + \cdots + \beta_{in} s_n^{(k-1)},$$

to obtain $s^{(k)}$ from $s^{(k-1)}$. In each cycle, we may want to randomize the order in which the nodes are updated.

12.3.3 Making Inference Based on the Posterior Distribution

Most GWASs lead to a set of candidate genes/SNPs that will need to be validated in follow-up studies. Therefore, it is important to include as many truly associated genes as possible among the top ranked genes. Our proposed method allows us to rank genes as detailed below.

There are several ways of inferring the labels according to the posterior distribution of S. The first one is to use maximum a posteriori (MAP) estimate, which is a point estimate for S with the largest posterior probability. Let us denote it by $\hat{s}^A = (\hat{s}_1^A, \ldots, \hat{s}_n^A)$. The MAP is the maximizer of the joint posterior distribution:

$$\hat{s}^A = \arg\max_s f(y|s, \theta_1)\mathbb{P}(s|\theta_0).$$

A Gibbs sampler outlined above can be applied to stochastically search for the solution to the above optimization problem. Multiple restarts with different initial configurations are recommended. An alternative approach is to base the estimate on the posterior conditional probability of S_i, $\mathbb{P}(s_i|y, s_{V-i}, \theta_0, \theta_1)$, given the observed data and all the other nodes s_{V-i}. We can estimate s_i by maximizing this conditional probability:

$$\hat{s}_i^C = \arg\max_{s_i} \mathbb{P}(s_i|y, s_{V-i}, \theta_0, \theta_1)$$
$$= \arg\max_{s_i} f(y_i|s_i, \theta_1)\mathbb{P}(s_i|s_{N_i}, \theta_0). \qquad (12.9)$$

The advantage of this approach is that the above problem is trivial to solve. The second factor in formula (12.9) can be evaluated in closed form using equation (12.4). In [7], an algorithm known as iterated conditional modes (ICM) was proposed that iteratively updates s_i. Note that the convergence of ICM is assured. To see this, we can write $\mathbb{P}(S = s|y, \theta_0, \theta_1)$ in equation (12.6), the posterior probability when $S = s$, as the following:

$$\mathbb{P}(s|y, \theta_0, \theta_1) = \mathbb{P}(s_i|y, s_{V-i}, \theta_0, \theta_1)\mathbb{P}(s_{V-i}|y, \theta_0, \theta_1).$$

The first factor is the objective function of the optimization in (12.9) and thus is non-decreasing at each iteration. The second factor is a constant because at each iteration, we only maximize over the variable s_i without altering the values of s_{V-i}. As a result, the value of the posterior distribution $\mathbb{P}(s|y, \theta_0, \theta_1)$ never decreases from iteration to iteration. So it is easy to see ICM will converge to a local maximum in the posterior distribution. Since ICM runs fast and usually converges in several iterations, multiple restarts with different initial configurations are recommended. Finally, the resulting configurations can be compared by evaluating $f(y|\hat{s}^C, \theta_1)\mathbb{P}(\hat{s}^C|\theta_0)$ up to a normalizing constant to pick the largest one.

The inference can also be based on the marginal posterior probability. Let $m_i = \mathbb{P}(S_i = 1|y)$. We consider a decision rule in the form $\delta(m_i) = I(m_i \geq m^*)$, where $I(\cdot)$ is an indicator function and m^* is the sought decision threshold. If $\delta(m_i) = 1$, the decision is positive (also referred to as discovery), and gene i is considered to be associated with the disease. Likewise if $\delta(m_i) = 0$, the decision is negative. To address the problem of multiple comparisons, we consider loss functions associated with making wrong decisions (false discoveries and false negatives), and solve the decision problem by minimizing the expectation of the loss functions under the posterior distribution. Here we consider two loss functions. First, if we are interested in the 0–1 loss function $L_1(S, \delta) = \sum_{i=1}^n |I_1(S_i) - \delta(m_i)|$, we may want to minimize the expected loss

$$m_1^* = \arg\min_{m^*} \mathbb{E}\{L_1(S, \delta)|y, \theta_0, \theta_1\}$$

$$= \arg\min_{m^*} \sum_S \left\{ \sum_{i=1}^n |I_1(S_i) - \delta(m_i)| \right\} \cdot \mathbb{P}(S|y, \theta_0, \theta_1), \tag{12.10}$$

under the posterior distribution of S. The solution is $m_1^* = 0.5$. Note that L_1 assigns equal loss to the false positive and false negative errors. This is to minimize the expected frequency of making wrong calls for the association status. Note that the performance of the decision rule δ is based on the frequentist operating characteristic in the Bayesian framework, which is common in medical decision makings [21]. The second loss function we consider is the false discovery rate (FDR):

$$FDR = L_2(S, \delta) = \frac{\sum_i \delta(m_i)I_{-1}(S_i)}{\sum_i \delta(m_i)}. \tag{12.11}$$

Suppose the goal is to control the expected FDR, under the posterior distribution, such that it is no more than α, i.e., $\mathbb{E}\{L_2(S, \delta)|y, \theta_0, \theta_1\} \leq \alpha$. If we rank all genes by their posterior means from the largest to the smallest, and let $m_{(i)}$ denote the ith order statistics, then the solution is to choose a cut-off value $m_2^* = m_{(j)}$ where j is the largest integer that makes $j^{-1} \sum_{i=1}^j m_{(i)} \geq (1 - \alpha)$. We should mention that more complicated loss functions can be considered under the framework of our model (see [21] for other examples).

12.3.4 Numerical Studies

SIMULATIONS TO STUDY PRIOR EFFECTS

The first numerical study is focused on the effects of prior settings. The simulation study is based on the network shown in Fig. 12.3 (page 306). This network was adapted from the BioCarta "Human Rho cell motility signaling pathway," and we deleted a few genes either absent from our Crohn's disease data or in isolated small groups ("gene islands") that are disconnected from the other genes. We assume three different sets of truly associated genes, plotted in triangles, rectangles, and pentagons, each of which contains three, five, and seven nodes, respectively. To simulate different levels in the power of the association tests, for each gene with $S_i = +1$, the p-value is computed from a two-sided z test for which the z scores are randomly drawn from

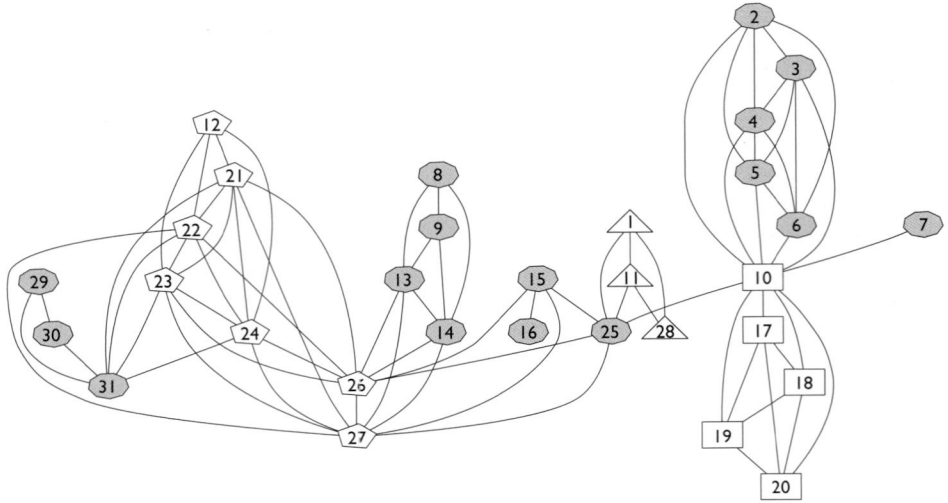

Fig 12.3 A 31-node network adapted from the BioCarta "Human Rho cell motility signaling pathway." Triangles, rectangles, and pentagons denote three different sets of truly associated genes, each of which contains three, five, and seven nodes, respectively.

$\mathcal{N}(1,1)$, $\mathcal{N}(1.5,1)$, and $\mathcal{N}(2,1)$, respectively, corresponding to the power 0.16 (low), 0.32 (median), and 0.51 (high) in the association tests. The p-values for $S_i = -1$ are generated randomly from Uniform(0, 1).

Table 12.1 Eight priors from four main groups indexed by numbers 1 through 4, and two subgroups indexed by letters a and b. Groups 1 through 4 favor a small, median small, median large, and large number of nodes labeled with +1, respectively. Nodes with identical labels are a priori more likely to be next to each other in subgroup b than subgroup a.

Group	Subgroup	Hyperparameters			Estimates		
		h	τ_1	τ_0	$\mathbb{E}[\mathbb{P}(S_i = 1)]$	$\mathbb{E}[\mathbb{P}(S_i = S_j = 1)]$	$\mathbb{E}[\mathbb{P}(S_i = S_j = -1)]$
1	a	−3.00	0.10	0.01	0.044	0.003	0.917
	b	−3.00	0.25	0.10	0.049	0.043	0.923
2	a	−2.00	0.10	0.01	0.156	0.047	0.710
	b	−2.50	0.20	0.05	0.141	0.119	0.776
3	a	−1.25	0.05	0.01	0.250	0.081	0.563
	b	−3.00	0.25	0.05	0.254	0.264	0.602
4	a	−1.50	0.10	0.01	0.355	0.227	0.402
	b	−2.00	0.25	0.10	0.405	0.412	0.466

To examine the effects of hyperparameters on the network, we consider eight priors, listed in Table 12.1, that roughly form four main groups indexed by numbers 1 through 4, and two subgroups indexed by letters a and b. For each set of hyperparameters, a Gibbs sampler is run to draw samples from the corresponding prior distribution, and we can estimate $\mathbb{P}(S_i = 1)$, the prior mean, and $\mathbb{P}(S_i = S_j = 1)$ and $\mathbb{P}(S_i = S_j = -1)$ where $(i, j) \in E$, the probabilities of edges (i, j) linking two nodes with identical labels. The averages of the estimated probabilities are listed in the last three columns of Table 12.1. The hyperparameters (h, τ_1, τ_0) jointly determine the prior knowledge or beliefs of the network, e.g., the prior probabilities of $\mathbb{P}(S_i = 1)$, $\mathbb{P}(S_i = S_j = 1 | (i, j) \in E)$ and $\mathbb{P}(S_i = S_j = -1 | (i, j) \in E)$. We choose values for (h, τ_1, τ_0) in a way such that the average prior means of all nodes are approximately 0.05, 0.15, 0.25, and 0.4, respectively for the four main groups. Furthermore, values of (τ_0, τ_1) in subgroup b are larger than those in subgroup a, meaning that nodes with identical labels are a priori more likely to be next to each other in subgroup b than subgroup a, as can be seen from the last two columns in Table 12.1. Because the posteriors are found to be insensitive to the hyperparameters $(\bar{\mu}, a, \nu, d)$ when ν is large, they are set to $(3, 1, 10, 1)$. Recall that ν is the hyperparameter in the inverse gamma distribution that we assign as a prior to σ_i^2. A large value of ν puts a large prior variance on σ_i^2, which allows a wide range of values for both μ_i and σ_i^2.

We simulate 200 data sets for each combination of the three power settings (low, median, and high) and three truly associated sets (three, five, and seven nodes). For each data set, we run eight Gibbs samplers using eight different hyperparameters described above. Each Gibbs sampler is run with 100 restarts, and each start contains 100 steps. Genes can be ranked by either p-values or the marginal posterior means $\mathbb{P}(S = 1 | y)$ using our method. For each ranking of genes and for an arbitrary cut-off value, we can calculate the true positive rate (TPR) and the false positive rate (FPR):

$$\text{TPR} = \frac{\text{number of true positives}}{\text{number of true positives} + \text{number of false negatives}},$$

$$\text{FPR} = \frac{\text{number of false positives}}{\text{number of true negatives} + \text{number of false positives}}.$$

Then by varying the cut-off threshold we can plot true positive rates versus false positive rates to obtain a Receiver Operating Characteristic (ROC) curve. Finally, the area under the ROC curve (AUC) can be computed. We compared the average AUC of 200 simulated data sets using p-values and marginal posterior means and plotted the results in Fig. 12.4 (page 308). In general, the AUC of the Bayesian method is larger than that obtained when using p-values alone. The Bayesian method provides good AUC values if the prior mean is close to the truth, especially when the power is low. For example, in the middle column panels where there are five truly associated genes, prior settings 2 and 3, favoring median number of truly associated nodes, outperform prior settings 1 and 4. Similarly, in the right panel where the true model contains seven genes, prior settings 3 and 4, which are in favor of large models, perform better than the other prior settings. Furthermore, priors in subgroup b are higher than in subgroup a in general. It is not surprising because the priors in subgroup b encourages nodes labeled with +1 to group together, which agrees with the simulation setting.

SIMULATIONS OF A LARGE PATHWAY WITH A FEW ASSOCIATED GENES

To evaluate the control of the false positive rates and false discovery rates of the Bayesian method in relatively large pathways with only a few associated genes, we conducted another simulation

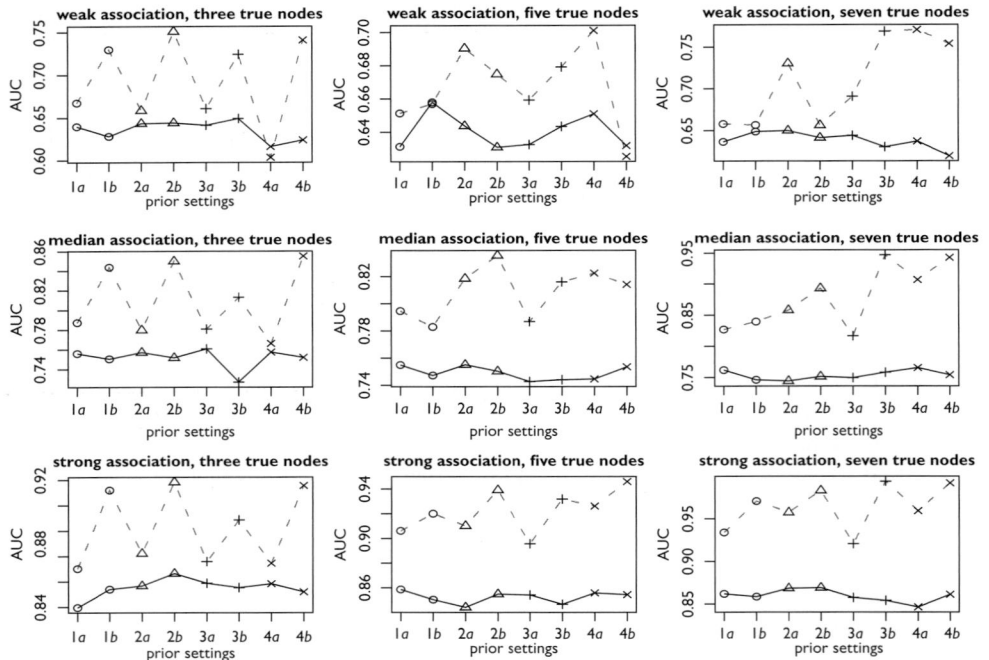

Fig 12.4 Comparison under different priors. The three rows of panels correspond to weak (top panel), median (middle panel), and strong (bottom panel) association signals. The three columns of panels correspond to three sets of truly associated genes, i.e., three (left), five (middle), and seven (right) nodes, respectively. The dotted lines link AUC values of the proposed method and the solid lines connect AUC values using p-values. Circles, triangles, plus signs, and crosses denote prior hyperparameter groups 1, 2, 3, and 4, respectively.

study based on a simulated network shown in Fig. 12.5 that contains 60 nodes. We considered three truly associated gene sets, namely (2, 11, 19), (2, 11, 19, 41), and (2, 11, 19, 20, 41), and labeled them as models 1, 2, and 3 in Table 12.2. Similar to the previous study, we simulated p-values from Z-scores randomly drawn from $\mathcal{N}(1,1)$, $\mathcal{N}(1.5,1)$, and $\mathcal{N}(2,1)$, corresponding to weak, median, and strong associations, respectively. Three prior settings were considered for (h, τ_1, τ_0), namely $(-1.5, 0.15, 0.02)$, $(-1.50, 0.10, 0.01)$ and $(-2, 0.2, 0.01)$, whose average prior probability $\mathbb{P}(S_i = 1)$ was approximately 0.2, and average prior probabilities $\mathbb{P}(S_i = S_j = 1)$ for $(i, j) \in E$ were roughly 0.13, 0.11, and 0.08, respectively. For the Bayesian method, we considered three decision rules. The first one (PM1) used the posterior mean with a cut-off value $m_1^* = 0.5$ as in (12.10), the second one was maximum conditional probability (MCP) as in (12.9), and the third one (PM2) was to control the *FDR* at 0.1, as in (12.11). Then we compared the three methods with the p-value method (P-value) with a cut-off value set at 0.05, as well as the *FDR* correction method (BH) in [4]. For each scenario, we simulated 100 data sets and ran a Gibbs sampler with 100 restarts where each start contained 100 iterations. For each simulated data set, we calculated the false positive rate (FPR), true positive rate (TPR, also known as sensitivity or power), *FDR*, and AUC as before. Table 12.2 lists the average values of the 100 simulation runs. In general, PM1 and maximum conditional probability control the false positive rate below the 0.05 level and have lower *FDRs* than p-value while achieving better or similar sensitivity as the p-value method. In

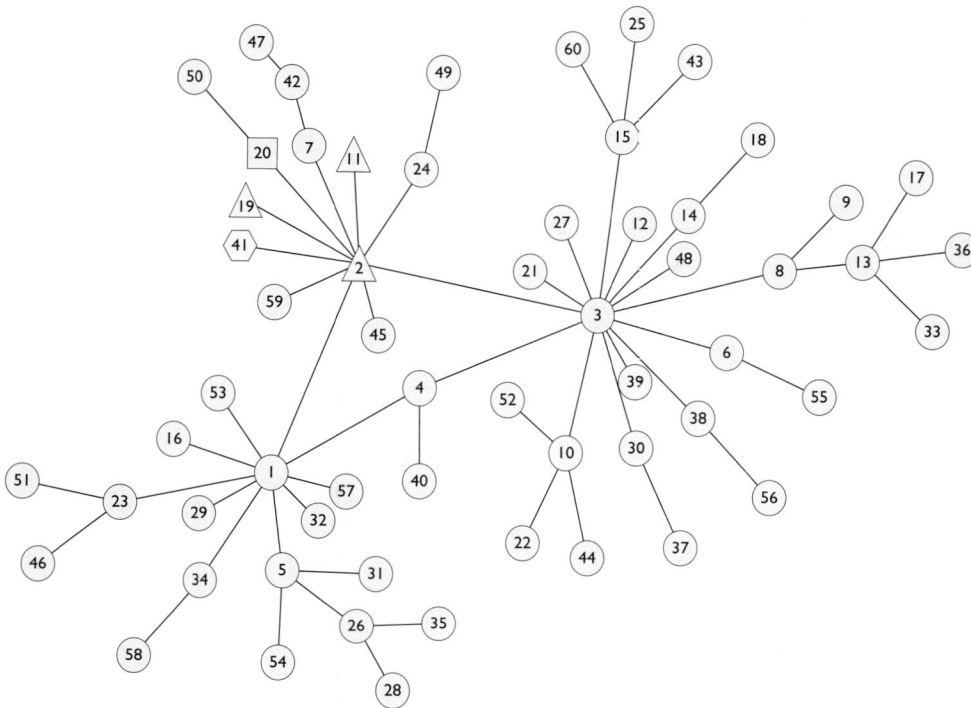

Fig 12.5 A simulated 60-node network.

terms of controlling *FDR*, PM2 controls the *FDR* around 0.1, and it has smaller false positive rates or better power than the BH method in most cases when it achieves similar or better *FDR*.

12.3.5 Real Data Example—Crohn's Disease Data

The Crohn's disease [10] data set was used to evaluate the performance of the Bayesian model. Crohn's disease is a type of inflammatory bowel disease characterized by chronic inflammation of discontinuous segments of the intestine. The disease is found to be related to the interaction of several factors including genetic susceptibility, the intestinal microbial flora of the patient, the patient's immune response to these microbiota, and environmental triggers [26]. It has been well established that Crohn's disease has a strong genetic component [22].

The cohort used in the analysis includes 401 cases and 433 controls. SNPs with a call rate greater than 0.9, minor allele frequency greater than 0.01, and Hardy–Weinberg equilibrium[3] *p*-value greater than 0.001 were kept, while subjects with a call rate less than 0.95 were removed from the analysis. Finally, 397 cases and 431 controls remained in the analysis. SNPs were considered

---

[3] Significant departure from the Hardy–Weinberg equilibrium is mostly likely due to genotyping errors, i.e., systematic mistyping of heterozygotes as homozygotes or vice versa [14].

Table 12.2 Average false positive rate (FPR), true positive rate (TPR), false discovery rate (FDR), and area under the receiver operating characteristic curve (AUC) in the 60-node simulated network.

Model	Method		Weak Association				Median Association				Strong Association			
			FPR	TPR	FDR	AUC	FPR	TPR	FDR	AUC	FPR	TPR	FDR	AUC
1		P-value	0.0470	0.157	0.817	0.629	0.0516	0.297	0.745	0.739	0.0523	0.533	0.621	0.863
		BH	0.0021	0.037	0.085		0.0014	0.043	0.070		0.0032	0.187	0.098	
	Prior 1	PM1	0.0412	0.167	0.790	0.657	0.0468	0.297	0.695	0.776	0.0496	0.567	0.591	0.899
		MCP	0.0370	0.150	0.768		0.0418	0.253	0.696		0.0454	0.523	0.593	
		PM2	0.0018	0.030	0.087		0.0014	0.050	0.075		0.0026	0.187	0.120	
	Prior 2	PM1	0.0404	0.163	0.792	0.653	0.0437	0.290	0.689	0.768	0.0472	0.543	0.585	0.890
		MCP	0.0375	0.150	0.779		0.0416	0.277	0.689		0.0449	0.523	0.583	
		PM2	0.0016	0.030	0.085		0.0021	0.090	0.110		0.0025	0.177	0.115	
	Prior 3	PM1	0.0300	0.143	0.688	0.690	0.0326	0.293	0.592	0.795	0.0377	0.577	0.483	0.907
		MCP	0.0253	0.140	0.648		0.0270	0.247	0.594		0.0337	0.500	0.480	
		PM2	0.0023	0.040	0.107		0.0012	0.053	0.065		0.0019	0.187	0.100	
2		P-value	0.0457	0.173	0.730	0.629	0.0514	0.330	0.668	0.738	0.0505	0.465	0.565	0.840
		BH	0.0018	0.018	0.090		0.0027	0.035	0.100		0.0038	0.178	0.094	
	Prior 1	PM1	0.0389	0.168	0.694	0.659	0.0450	0.340	0.621	0.788	0.0446	0.508	0.523	0.879
		MCP	0.0370	0.145	0.701		0.0416	0.318	0.618		0.0411	0.490	0.515	
		PM2	0.0020	0.018	0.110		0.0016	0.050	0.085		0.0016	0.178	0.075	

Continued

Table 12.2 (Continued).

Prior 2	PM1	0.0379	0.170	0.692	0.653	0.0439	0.323	0.624	0.775	0.0430	0.490	0.522	0.869
	MCP	0.0370	0.153	0.694		0.0413	0.305	0.626		0.0413	0.468	0.530	
	PM2	0.0020	0.018	0.110		0.0016	0.048	0.085		0.0020	0.158	0.095	
Prior 3	PM1	0.0289	0.143	0.639	0.683	0.0364	0.320	0.583	0.803	0.0352	0.485	0.469	0.888
	MCP	0.0252	0.125	0.610		0.0321	0.288	0.577		0.0293	0.455	0.454	
	PM2	0.0014	0.018	0.080		0.0013	0.048	0.065		0.0027	0.205	0.108	
3	P-value	0.0478	0.148	0.756	0.648	0.0476	0.326	0.591	0.744	0.0458	0.524	0.468	0.856
	BH	0.0038	0.028	0.129		0.0015	0.050	0.051		0.0029	0.166	0.052	
Prior 1	PM1	0.0402	0.136	0.744	0.675	0.0425	0.336	0.544	0.794	0.0409	0.584	0.422	0.897
	MCP	0.0364	0.124	0.737		0.0387	0.320	0.545		0.0373	0.568	0.403	
	PM2	0.0027	0.022	0.150		0.0011	0.056	0.053		0.0020	0.230	0.060	
Prior 2	PM1	0.0404	0.134	0.738	0.669	0.0411	0.318	0.559	0.779	0.0398	0.556	0.427	0.886
	MCP	0.0380	0.124	0.749		0.0393	0.302	0.559		0.0375	0.546	0.414	
	PM2	0.0025	0.026	0.140		0.0009	0.052	0.045		0.0018	0.206	0.062	
Prior 3	PM1	0.0282	0.120	0.654	0.700	0.0333	0.310	0.486	0.806	0.0315	0.556	0.381	0.904
	MCP	0.0247	0.100	0.620		0.0280	0.270	0.479		0.0276	0.540	0.358	
	PM2	0.0020	0.014	0.110		0.0020	0.046	0.055		0.0015	0.216	0.052	

Note: BH, FDR correction method; MCP, maximum conditional probability; PM1, posterior mean with $m_1^* = 0.5$; PM2, FDR controlled at 0.1 p-value, p-valve cut-off at 0.05.

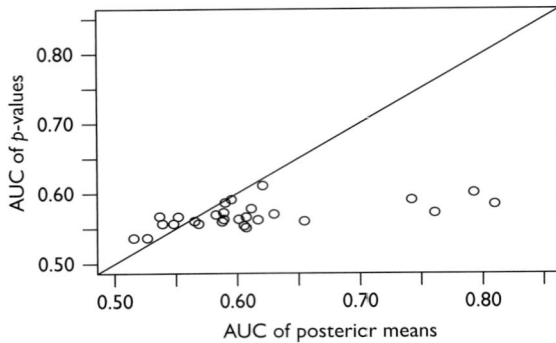

Fig 12.6 AUC comparison of rankings by *p*-values and posterior means for Crohn's disease data. AUC values of the rankings by *p*-values are on the *y* axis and that of the posterior means are on the *x* axis for pathways containing three or more target genes.

mapped to a gene if their physical locations were within the range between 10 kb upstream of the transcription start site and 10 kb downstream of the transcription end site of the gene as given by the RefSeq annotation at the National Center for Biotechnology Information website. Then, principal component analysis was performed on all SNPs mapped to the gene. Finally, the gene-level *p*-value was obtained by regressing disease status on top principal components that accounted for at least 85% of the variation in those SNPs [2, 12, 27]. The R code and an example of a pathway as well as the gene-level *p*-values can be found at http://bioinformatics.med.yale.edu/group.

We ran our algorithm on 289 pathways that had at least 20 genes with non-missing *p*-values. These pathways contained a total of over 3700 unique genes. The hyperparameters (h, τ_1, τ_0) were chosen such that the average prior mean was roughly between 0.2 and 0.4 based on the simulation findings. To evaluate the performance, we considered 32 target genes that were confirmed to be related to Crohn's disease [3]. Of these genes, ten could be mapped to 66 pathways. In Fig. 12.6, we plot the AUC values of the rankings by *p*-values on the *y* axis and posterior means on the *x* axis for pathways containing three or more target genes. A majority of AUC values are improved if genes are ranked by the posterior mean. The average AUC based on *p*-values is 0.568, while it is 0.613 when based on posterior means.

12.4 Discussion

In this chapter, we described an MRF model and a Bayesian framework to incorporate prior knowledge of biological pathways into GWASs. One limitation of the MRF model is that the Gibbs sampler tends to move around local maxima for a long time and thus can be slow in convergence to the posterior distribution. We recommend running the MCMC with multiple random restarts and examining the sampling distribution of network statistics such as the number of genes labeled with +1 and the proportion of edges linking genes with identical labels. In our studies, we found that a Markov chain initially moves very rapidly from its starting state, usually within the first 10 to 20 steps, before it reaches some steady state and stabilizes for a long period thereafter. We suggest running 100 Gibbs steps for each random starting state, and conducting the simulation with 100 or more restarts. The computing time of this scheme is about one minute for the 60-gene network in Subsection 12.3.4 on a PC with a 2.5 GHz Intel Core 2 Duo CPU and 4 GB memory. It takes from less than a minute to a few minutes on the same PC to run pathways for Crohn's disease data. Another limitation of pathway-based analysis is that not all the genes can be associated with pathways. It is likely that with knowledge accumulation, more genes will be mapped to pathways.

ACKNOWLEDGMENTS

This work was supported in part by NIH grants GM 59507, U01 DK062422, 1R01DK072373, and UL1 RR024139 and NSF grant DMS-0714817. We thank Dr. David Ballard for providing gene level p-values of the Crohn's disease data set.

APPENDIX 12.A PROOF OF PROPOSITION 12.1

Proof 12.1 *The probability function in (12.1) can be expressed as*

$$\mathbb{P}(S_1, \ldots, S_n | \theta_0) = \frac{1}{z(\theta_0)} \prod_{i \in V} \exp(U_i),$$

where

$$U_i = h\, I_1(S_i) + \frac{\tau_0}{2} I_{-1}(S_i) \sum_{j \in N_i} (w_i + w_j)\, I_{-1}(S_j) + \frac{\tau_1}{2} I_1(S_i) \sum_{j \in N_i} (w_i + w_j)\, I_1(S_j).$$

The conditional probability of S_i given all other nodes in S_{V-i} is

$$\mathbb{P}(S_i | S_{V-i}) = \frac{\mathbb{P}(S_1, \ldots, S_n)}{\sum_{S_i \in \{-1,1\}} \mathbb{P}(S_1, \ldots, S_i, \ldots, S_n)}$$

$$= \frac{\exp(U_i) \cdot \prod_{k \in N_i} \exp(U_k) \cdot \prod_{l \in V-i-N_i} \exp(U_l)}{\prod_{l \in V-i-N_i} \exp(U_l) \cdot \sum_{S_i \in \{-1,1\}} \left\{ \exp(U_i) \cdot \prod_{k \in N_i} \exp(U_k) \right\}}$$

$$= \frac{\exp(U_i) \cdot \prod_{k \in N_i} \exp(U_k)}{\sum_{S_i \in \{-1,1\}} \left\{ \exp(U_i) \cdot \prod_{k \in N_i} \exp(U_k) \right\}}. \tag{12.12}$$

The last step is true because for any $l \in V - i - N_i$, since i and l are not neighbors, U_l is independent of S_i. Note that for any $k \in N_i$, U_k is a function of S_i:

$$U_k = h\, I_1(S_k) + \frac{\tau_0}{2}\, I_{-1}(S_k) \sum_{j \in N_k} (w_k + w_j)\, I_{-1}(S_j)$$

$$+ \frac{\tau_1}{2}\, I_1(S_k) \sum_{j \in N_k} (w_k + w_j)\, I_1(S_j)$$

$$= h\, I_1(S_k) + \frac{\tau_0}{2}(w_k + w_i)\, I_{-1}(S_k)\, I_{-1}(S_i) + \frac{\tau_0}{2}\, I_{-1}(S_k) \sum_{j \in N_k-i} (w_k + w_j)\, I_{-1}(S_j)$$

$$+ \frac{\tau_1}{2}(w_k + w_i) I_1(S_k) I_1(S_i) + \frac{\tau_1}{2} I_1(S_k) \sum_{j \in N_k-i} (w_k + w_j)\, I_1(S_j)$$

$$= V_k(S_i) + C_k,$$

where

$$V_k(S_i) = \frac{\tau_0}{2}(w_k + w_i)\, I_{-1}(S_k)\, I_{-1}(S_i) + \frac{\tau_1}{2}(w_k + w_i)\, I_1(S_k)\, I_1(S_i)$$

and

$$C_k = h\, I_1(S_k) + \frac{\tau_0}{2}\, I_{-1}(S_k) \sum_{j \in N_k-i} (w_k + w_j)\, I_{-1}(S_j) + \frac{\tau_1}{2}\, I_1(S_k) \sum_{j \in N_k-i} (w_k + w_j)\, I_1(S_j).$$

Here C_k does not depend on S_i. Thus, the numerator of (12.12) can be written as

$$\exp(U_i) \cdot \prod_{k \in N_i} \exp(U_k) = \exp\left\{ h\, I_1(S_i) + \sum_{j \in N_i} V_j(S_i) \right\} \prod_{k \in N_i} \exp\{V_k(S_i)\} \prod_{k \in N_i} \exp(C_k)$$

$$= \exp\left\{ h\, I_1(S_i) + 2 \sum_{j \in N_i} V_j(S_i) \right\} \prod_{k \in N_i} \exp(C_k).$$

Finally, (12.12) becomes

$$\mathbb{P}(S_i | S_{V-i})$$

$$= \frac{\exp\left\{ h\, I_1(S_i) + 2 \sum_{j \in N_i} V_j(S_i) \right\} \prod_{k \in N_i} \exp(C_k)}{\sum_{S_i \in \{-1,1\}} \left[\exp\left\{ h\, I_1(S_i) + 2 \sum_{j \in N_i} V_j(S_i) \right\} \prod_{k \in N_i} \exp(C_k) \right]}$$

$$= \frac{\exp\left\{ h\, I_1(S_i) + 2 \sum_{j \in N_i} V_j(S_i) \right\}}{\sum_{S_i \in \{-1,1\}} \left[\exp\left\{ h\, I_1(S_i) + 2 \sum_{j \in N_i} V_j(S_i) \right\} \right]}$$

$$= \frac{\exp\left\{ h\, I_1(S_i) + \tau_0\, I_{-1}(S_i) \sum_{j \in N_i} (w_i + w_j)\, I_{-1}(S_j) + \tau_1\, I_1(S_i) \sum_{j \in N_i} (w_i + w_j)\, I_1(S_j) \right\}}{\exp\left\{ h + \tau_1 \sum_{j \in N_i} (w_i + w_j)\, I_1(S_j) \right\} + \exp\left\{ \tau_0 \sum_{j \in N_i} (w_i + w_j)\, I_{-1}(S_j) \right\}}$$

$$= \mathbb{P}(S_i | S_{N_i}).$$

because it only depends on S_{N_i}. Therefore, the odds is

$$\frac{\mathbb{P}(S_i = +1 | S_{N_i})}{\mathbb{P}(S_i = -1 | S_{N_i})}$$

$$= \frac{\exp\left\{h + \tau_1 \sum_{j \in N_i} (w_i + w_j) I_1(S_j)\right\}}{\exp\left\{\tau_0 \sum_{j \in N_i} (w_i + w_j) I_{-1}(S_j)\right\}}$$

$$= \exp\left\{h + \tau_1 \sum_{j \in N_i} (w_i + w_j) I_1(S_j) - \tau_0 \sum_{j \in N_i} (w_i + w_j) I_{-1}(S_j)\right\},$$

which is a logistic regression model that can be rewritten as

$$logit(\mathbb{P}(S_i | S_{N_i})) = h + \tau_1 \left(w_i J_i^{(1)} + \sum_{j \in N_i} w_j I_1(S_j) \right)$$

$$- \tau_0 \left(w_i J_i^{(-1)} + \sum_{j \in N_i} w_j I_{-1}(S_j) \right).$$

REFERENCES

[1] S. Aerts, D. Lambrechts, S. Maity, P. Van Loo, B. Coessens, F. De Smet, L.C. Tranchevent, B. De Moor, P. Marynen, B. Hassan, P. Carmeliet, and Y. Moreau. Gene prioritization through genomic data fusion. *Nature Biotechnology*, 24:537–544, 2006.

[2] D.H. Ballard, J. Cho, and H. Zhao. Comparisons of multi-marker association methods to detect association between a candidate region and disease. *Genetic Epidemiology*, 34:201–212, 2010.

[3] J.C. Barrett, S. Hansoul, D.L. Nicolae, J.H. Cho, R.H. Duerr, J.D. Rioux, S.R. Brant, M.S. Silverberg, K.D. Taylor, M.M. Barmada, A. Bitton, T. Dassopoulos, L.W. Datta, T. Green, A.M. Griffiths, E.O. Kistner, M.T. Murtha, M.D. Regueiro, J.I. Rotter, L.P. Schumm, A.H. Steinhart, S.R. Targan, R.J. Xavier; NIDDK IBD Genetics Consortium, C. Libioulle, C. Sandor, M. Lathrop, J. Belaiche, O. Dewit, I. Gut, S. Heath, D. Laukens, M. Mni, P. Rutgeerts, A. Van Gossum, D. Zelenika, D. Franchimont, J.P. Hugot, M. de Vos, S. Vermeire, E. Louis; Belgian-French IBD Consortium; Wellcome Trust Case Control Consortium, L.R. Cardon, C.A. Anderson, H. Drummond, E. Nimmo, T. Ahmad, N.J. Prescott, C.M. Onnie, S.A. Fisher, J. Marchini, J. Ghori, S. Bumpstead, R. Gwilliam, M. Tremelling, P. Deloukas, J. Mansfield, D. Jewell, J. Satsangi, C.G. Mathew, M. Parkes, M. Georges, and M.J. Daly. Genome-wide association defines more than 30 distinct susceptibility loci for Crohn's disease. *Nature Genetics*, 40:955–962, 2008.

[4] Y. Benjamini and Y. Hochberg. Controlling the false discovery rate: a practical and powerful approach to multiple testing. *Journal of the Royal Statistical Society, Series B (Methodological)*, 57:289–300, 1995.

[5] J. Besag. Nearest-neighbour systems and the auto-logistic model for binary data. *Journal of the Royal Statistical Society, Series B (Methodological)*, 34:75–83, 1972.

[6] J. Besag. Spatial interaction and the statistical analysis of lattice systems. *Journal of the Royal Statistical Society, Series B (Methodological)*, 36:192–236, 1974.

[7] J. Besag. On the statistical analysis of dirty pictures. *Journal of the Royal Statistical Society, Series B (Methodological)*, 48:259–302, 1986.

[8] M. Chen, J. Cho, and H. Zhao. Incorporating biological pathways via a Markov random field model in genome-wide association studies. *PLOS Genetics*, 7:e1001353, 2011.

[9] H. Chun, D.H. Ballard, J. Cho, and H. Zhao. Identification of association between disease and multiple markers via sparse partial least-squares regression. *Genetic Epidemiology*, 35:479–486, 2011.

[10] R.H. Duerr, K.D. Taylor, S.R. Brant, J.D. Rioux, M.S. Silverberg, M.J. Daly, A.H. Steinhart, C. Abraham, M. Regueiro, A. Griffiths, T. Dassopoulos, A. Bitton, H. Yang, S. Targan, L.W. Datta, E.O. Kistner, L.P. Schumm, A.T. Lee, P.K. Gregersen, M.M. Barmada, J.I. Rotter, D.L. Nicolae, and J.H. Cho. A genome-wide association study identifies IL23R as an inflammatory bowel disease gene. *Science*, 314:1461–1463, 2006.

[11] L. Franke, H. van Bakel, L. Fokkens, E.D. de Jong, M. Egmont-Petersen, and C. Wijmenga. Reconstruction of a functional human gene network, with an application for prioritizing positional candidate genes. *The American Journal of Human Genetics*, 78:1011–1025, 2006.

[12] W.J. Gauderman, C. Murcray, F. Gilliland, and D.V. Conti. Testing association between disease and multiple SNPs in a candidate gene. *Genetic Epidemiology*, 31:383–395, 2007.

[13] A. Gelman, J.B. Carlin, H.S. Stern, and D.B. Rubin. *Bayesian Data Analysis*. CRC press, 2004.

[14] I. Gomes, A. Collins, C. Lonjou, N.S. Thomas, J. Wilkinson, M. Watson, and N. Morton. Hardy-Weinberg quality control. *Annals of Human Genetics*, 63:535–538, 1999.

[15] J.E. Hutz, A.T. Kraja, H.L. McLeod, and M.A. Province. Candid: a flexible method for prioritizing candidate genes for complex human traits. *Genetic Epidemiology*, 32:779–790, 2008.

[16] M. Kanehisa and S. Goto. KEGG: Kyoto encyclopedia of genes and genomes. *Nucleic Acids Research*, 28:27–30, 2000.

[17] R. Kindermann and J.L. Snell. *Markov Random Fields and their Applications*. American Mathematical Society, 1980.

[18] S. Köhler, S. Bauer, D. Horn, and P.N. Robinson. Walking the interactome for prioritization of candidate disease genes. *The American Journal of Human Genetics*, 82:949–958, 2008.

[19] H. Li, Z. Wei, and J.M. Maris. A hidden Markov random field model for genome-wide association studies. *Biostatistics*, 11:139–150, 2010.

[20] X. Ma, H. Lee, L. Wang, and F. Sun. CGI: a new approach for prioritizing genes by combining gene expression and protein-protein interaction data. *Bioinformatics*, 23:215–221, 2007.

[21] P. Müller, G. Parmigiani, C. Robert, and J. Rousseau. Optimal sample size for multiple testing: the case of gene expression microarrays. *Journal of the American Statistical Association*, 99:990–1001, 2004.

[22] M. Peeters, H. Nevens, F. Baert, M. Hiele, A.M. de Meyer, R. Vlietinck, and P. Rutgeerts. Familial aggregation in Crohn's disease: increased age-adjusted risk and concordance in clinical characteristics. *Gastroenterology*, 111:597–603, 1996.

[23] G. Peng, L. Luo, H. Siu, Y. Zhu, P. Hu, S. Hong, J. Zhao, X. Zhou, J.D. Reveille, L. Jin, C.I. Amos, and M. Xiong. Gene and pathway-based second-wave analysis of genome-wide association studies. *European Journal of Human Genetics*, 18:111–117, 2010.

[24] S.F. Saccone, N.L. Saccone, G.E. Swan, P.A.F. Madden, A.M. Goate, J.P. Rice, and L.J. Bierut. Systematic biological prioritization after a genome-wide association study: an application to nicotine dependence. *Bioinformatics*, 24:1805–1811, 2008.

[25] N. Salomonis, K. Hanspers, A.C. Zambon, K. Vranizan, S.C. Lawlor, K.D. Dahlquist, S.W. Doniger, J. Stuart, B.R. Conklin, and A.R. Pico. GenMAPP 2: new features and resources for pathway analysis. *BMC Bioinformatics*, 8:217, 2007.

[26] R.B. Sartor. Mechanisms of disease: pathogenesis of Crohn's disease and ulcerative colitis. *Nature Clinical Practice, Gastroenterology & Hepatology*, 3:390–407, 2006.

[27] K. Wang and D. Abbott. A principal components regression approach to multilocus genetic association studies. *Genetic Epidemiology*, 32:108–118, 2008.

[28] K. Wang, M. Li, and M. Bucan. Pathway-based approaches for analysis of genomewide association studies. *The American Journal of Human Genetics*, 81:1278–1283, 2007.

[29] P. Wei and W. Pan. Incorporating gene networks into statistical tests for genomic data via a spatially correlated mixture model. *Bioinformatics*, 24:404–411, 2008.

[30] Z. Wei and H. Li. A Markov random field model for network-based analysis of genomic data. *Bioinformatics*, 23:1537–1544, 2007.

[31] X. Wu, R. Jiang, M.Q. Zhang, and S. Li. Network-based global inference of human disease genes. *Molecular Systems Biology*, 4:189, 2008.

CHAPTER 13

Bayesian, Systems-based, Multilevel Analysis of Associations for Complex Phenotypes: from Interpretation to Decision

PÉTER ANTAL, ANDRÁS MILLINGHOFFER, GÁBOR HULLÁM, GERGELY HAJÓS, PÉTER SÁRKÖZY, ANDRÁS GÉZSI, CSABA SZALAI, AND ANDRÁS FALUS

The relative scarcity of the results reported by genetic association studies (GASs) has prompted many research directions, such as the use of univariate Bayesian analysis and the use of multivariate, complex, or integrated models. Despite the centrality of the concept of association in GASs, refined concepts of association are missing, meanwhile various feature subset selection methods became de facto standards for defining multivariate relevance. On the other hand, probabilistic graphical models, including Bayesian networks (BNs), are more and more popular, as they can learn non-transitive, multivariate, non-linear relations between complex phenotypic descriptors and heterogeneous explanatory variables. To integrate the advantages of Bayesian statistics and BNs, the BN-based Bayesian multilevel analysis of relevance (BN-BMLA) was proposed. This approach allows the processing of multiple target variables, while ensuring scalability and providing a multilevel view of the results of multivariate analysis. This chapter

Probabilistic Graphical Models for Genetics, Genomics, and Postgenomics. First Edition. Christine Sinoquet & Raphaël Mourad (Eds). © Oxford University Press 2014. Published in 2014 by Oxford University Press.

discusses the use of Bayesian BN-based analysis of relevance in exploratory data analysis, optimal decision and study design, and knowledge fusion, in the context of GASs. First, various BN-based concepts of association and relevance are overviewed. In particular, the chapter analyzes the connections between BNs and strong/weak relevance and Markov blankets/boundaries or related sets; relevance relations are defined. Then the advantages of the Bayesian statistical approach for sufficiently characterizing and exploring weakly significant results are shown. For this purpose, the focus is set on the posteriors that are defined over the relevance relations abovementioned. The next section discusses Bayes optimal decisions on multivariance relevance in GAS results. In the last section, it is shown that the Bayesian BN-based approach provides a framework for the fusion of the results obtained through various genetic data analyses. This last section describes a procedure dedicated to estimating the posteriors of complex features, such as those involved in the hierarchical, interrelated hypotheses of the BN-BMLA framework.

13.1 Introduction

The relative scarcity of results reported by genetic association studies has led to several approaches such as univariate Bayesian analysis [5, 52] and the use of multivariate, complex or integrated models [61, 65]. Probabilistic graphical models, including Bayesian networks (BNs), are more and more popular as they can learn non-transitive, multivariate, non-linear relations between complex phenotypic descriptors (also known as target variables, dependent variables, or dependent outcomes) and heterogeneous, mostly genetic, explanatory variables (input variables, factors or predictors, also known as attributes or features in statistics). We discuss the limitations of conventional statistical association here and show how they can be circumvented using the notion of the so-called relevance of input variables to one or more target variables. We will show that relevance is a useful extension of association in Subsection 13.2.1. Recently we proposed the Bayesian network-based Bayesian multilevel analysis of relevance (BN-BMLA), which integrates the advantages of Bayesian statistics and BNs [4]. Moreover, this approach allows investigation of multiple target variables and provides scalable intermediate levels between univariate strong relevance and full multivariate relevance to interpret the results at partial multivariate levels. Thus, BN-BMLA provides a multilevel view on multivariate analysis. We discuss the use of the BN-BMLA relevance analysis in data exploration, optimal decision, and knowledge fusion.

First, we overview some of the structural properties of BNs, with special emphasis on systems-based analysis of relevance in Section 13.2. Section 13.3 shows the advantages of the Bayesian statistical approach in the characterization and exploration of weakly significant results. In Section 13.4, we discuss the application of Bayesian decision theory to GAS. In Section 13.5, we deal with the Bayesian interpretation and fusion of results from data analysis. In this chapter, we also consider the practical and computational aspects of the Bayesian inference: we apply the methods described in the domain of asthma.

Input variables in the current application typically correspond to single nucleotide polymorphisms (SNPs), whereas target variables represent asthmatic and possibly other phenotypic aspects. The notation is as follows: input variables (typically SNPs) are denoted with X ($X = \{X_1, \ldots, X_n\}$), target variables (typically asthma and possibly other phenotypic variables) are denoted with Y ($Y = \{Y_1, \ldots, Y_m\}$) and V ($V = X \cup Y$) denotes all the variables. The data set with N samples is denoted D_N.

13.2 Bayesian network-based Concepts of Association and Relevance

Refined concepts of association are missing in genetic association studies despite the centrality of the concept. Therefore, various feature subset selection (FSS) methods became de facto standards for quantifying the joint relevance of multiple variables and their interactions, which will be referred to as *multivariate relevance* (for an overview of FSS the reader is referred to [48]). In this section, we discuss the use of Bayesian network structural properties to define such refined concepts.

13.2.1 Association and Strong Relevance

The extension of standard statistical pairwise association (univariate case) toward many-to-one or many-to-many relations (multivariate case) is a challenging task, because several goals can be formulated, such as assessing the predictive performance and interdependence of multiple predictors.

In the *predictive approach* to the identification of relevant variables, relevance to the target variable Y is defined in the wrapper framework below.[1] The inherent limitation of a wrapper approach is that it is biased by the class of predictive model used, by the optimization algorithm, by the data set, and by the loss function[2] quantifying false and missed discovery [32]. A typical example setting in case-control studies is the use of logistic regression fitted by a gradient descent method to minimize misclassification error, optionally with complexity regularization to minimize overfitting for a given data set. The standard conditional probabilistic version of relevance types is free of model class, optimization, data set, or loss function, and is defined as follows:

Definition 13.1
A feature (input variable) X_i is strongly relevant to Y, if there exists an $X_i = x_i, Y = y$, and $s_i = x_1, \ldots, x_{i-1}, x_{i+1}, \ldots, x_n$, $S_i = \{X_1, \ldots, X_{i-1}, X_{i+1}, \ldots, X_n\}$ such that $\mathbb{P}(x_i, s_i) > 0$ and $\mathbb{P}(y|x_i, s_i) \neq \mathbb{P}(y|s_i)$. A feature X_i is weakly relevant if it is not strongly relevant and there exists a subset of features S'_i of S_i for which there exists some x_i, y, and s'_i such that $\mathbb{P}(x_i, s'_i) > 0$ and $\mathbb{P}(y|x_i, s'_i) \neq \mathbb{P}(y|s'_i)$. A feature is relevant if it is either weakly or strongly relevant; otherwise it is irrelevant [32].

Strong relevance formalizes the necessary and sufficient set of predictors, whereas weak relevance does not require necessity. For example, direct associations correspond to strong relevance, whereas indirect associations, mediated by other variables in linkage, correspond to weak relevance (for direct, indirect, and confounded associations, see [15]).

Besides the predictive aspects, relevance can also be defined using BNs to represent complex dependence patterns between phenotypic descriptors and multiple predictors, and to express causal relationships between variables. In general, a BN represents the joint probability distribution $\mathbb{P}(V)$ with two main components. The first main component is a directed acyclic graph (DAG) G, whose nodes are random variables from the set V. We often refer to G as the structure of the BN. A directed edge pointing from a so-called parent variable X_p to its child variable X_c implies not only that X_c depends on X_p but also that X_p is a direct cause of X_c in the sense that

---

[1] For a brief definition of the wrapper framework, see Appendix 13.A (page 354) or [32].
[2] For a brief definition of loss functions in the feature subset selection context, see Appendix 13.A (page 354).

the BN does not contain any other variable X_i that mediates the effect of X_p on X_c. The strength of such dependence of X_c on X_p is given by the extent to which the probability distribution of X_c shifts in response to a certain change in the value of X_p. This gives rise to a separate set of parameters defining the conditional distribution for each variable, given each possible value of its parent variables; these parameter sets altogether form θ, the second main component of the BN.

The BN representation is unique in offering three integrated levels to formulate these aspects [42]:

1 *Predictive performance:* Multivariate relevance can be based on the concepts of predictivity, cardinality and redundancy of the predictors.

2 *Global patterns of independences:* The global characterization of relevance relations based on the overall independence map[3] of the application domain is of fundamental importance to clarify direct, indirect and confounded associations [15].

3 *Causal aspects:* A causal model[4] also helps to dissect and estimate the effect size of genetic mechanisms related to complex phenotypes described by multiple variables [42].

The *independence map-based approach* provides the definition of joint relevance of multiple input variables (therefore, this approach is more naturally suited to multivariate analysis) [41]:

Definition 13.2
A set of variables $X' \subseteq V$ is called a Markov blanket set (MBS) of Y with respect to the distribution $\mathbb{P}(V)$, if $(Y \perp\!\!\!\perp V \setminus X'|X')_p$, where $\perp\!\!\!\perp$ denotes conditional independence. A minimal Markov blanket is called a Markov boundary.

The probabilistic definitions of Markov blanket and Markov boundary became fundamental concepts in the modern multivariate conceptualization of multivariate associations and provide foundations and motivations for many feature subset selection methods [1, 34, 56]. It follows from these basic characteristics of BNs that they offer a graphical representation of relevance derived both from the predictive approach (cf. strong and weak relevance) and the independence map-based one (cf. Markov blankets and boundaries). Thus, BNs provide a unifying framework for both approaches. Additionally, this common framework allows the extension of the concept of relevance relations with causality. The connections between BNs, strong/weak relevance, and Markov blankets/boundaries are provided by the following theorems:

Theorem 13.1
For a distribution $\mathbb{P}(V)$ defined by the BN (G, θ) the variables $bd(Y, G)$ form a (not necessarily unique or minimal) Markov blanket of Y, where $bd(Y, G)$ denotes the set of parents, children, and the children's other parents for Y [41].

Theorem 13.2
If $\mathbb{P}(V)$ is a stable distribution[5] and DAG G is Markov compatible with $\mathbb{P}(V)$,[6] then $bd(Y, G)$ is the unique, minimal Markov blanket (called a Markov boundary). In stable distributions, $bd(Y, G)$ also identifies the strongly relevant variables [56].

---

[3] For the definition of independence map, see Appendix 13.A (page 354) or [41].
[4] For the definition of a causal model, see Appendix 13.A (page 354) or [26, 42].
[5] For the definition of stable distribution, see Appendix 13.A (page 354).
[6] $\mathbb{P}$ is Markov compatible with respect to G if $\mathbb{P}$ can be factorized by G [42].

A BN structure explicitly represents both aspects of the Markov boundary: on the one hand, $bd(Y,G)$ is *necessary* because it contains only strongly relevant variables; on the other hand, $bd(Y,G)$ is *sufficient* because none of the remaining variables in V (i.e., the remaining nodes in G) is strongly relevant.

Definition 13.3

The induced pairwise relation $MBM(Y,X_i,G)$ with respect to G between Y and X_i is called Markov blanket membership (MBM):

$$MBM(Y,X_i,G) \Leftrightarrow X_i \in bd(Y,G). \tag{13.1}$$

As Theorem 13.2 shows, the boundary graph $bd(Y,G)$ exactly represents the (probabilistically defined) Markov boundary, and the MBMs $MBM(Y,X_i,G)$ exactly represent the (probabilistically defined) strongly relevant variables X_i for a stable distribution $\mathbb{P}(V)$ and a Markov compatible DAG G. The following Subsection 13.2.2 describes the mathematical background of the soundness of representing relevance by $bd(Y,G)$. Readers interested only in the application of the concepts can continue at Subsection 13.2.3 (page 323).

13.2.2 Stable Distributions, Markov Blankets and Markov Boundaries

The limitation of the assumption about the stability of the distribution is a subtle issue. The main reason lies in the independences, which are only parametrically and not structurally defined [41, 42]. Regardless of whether our hypotheses are the BN structures (DAGs G) or the *observational equivalence* classes of BNs[7] ($G^\sim$), then for a given G or $G^\sim$ the structurally implied conditional independences through d-separation[8] hold in any Markov compatible distribution $\mathbb{P}(V)$ [42]. This means that $bd(Y,G)$ represents a Markov blanket in any Markov compatible distribution and $MBM(Y,X_i,G)$ represents relevant variables. Note that the graphical definition of MBM satisfies the probabilistic definition of Markov blanket in Definition 13.2 according to Theorem 13.1 assuming a Markov compatible distribution with G, because $bd(Y,G)$ is always a Markov blanket. Furthermore, the definition of relevance based on $bd(Y,G)$ avoids the problem of the multiplicity of Markov blankets and defines mutually exclusive and exhaustive events (a sample space), which will be necessary to define a probability distribution over the sets of strongly relevant variables (see Section 13.3, page 328).

However, according to Theorem 13.2, in case of a stable distribution, the implied results are sharper: $bd(Y,G)$ represents a Markov boundary and $MBM(Y,X_i,G)$ represents strongly relevant variables. Whereas only Markov blanket and relevance can be used generally, the non-stable distributions are rare in a technical sense. This property formally appears in the Bayesian approach we are using in this chapter as follows. Using a continuous parameter distribution $\mathbb{P}(\theta|G)$, the specified distribution $\mathbb{P}(V)$ is stable with probability 1 [38]. Thus, in that Bayesian context, $bd(Y,G)$ and $MBM(Y,X_i,G)$ almost surely represent Markov boundaries and strong relevance and therefore we can use the stricter Markov boundary concept, despite the more widespread use of the term of Markov blanket.

For simplicity, we also refer to $bd(Y,G)$ as the Markov boundary set of Y using the notation $MBS(Y,G)$, assuming that the hypothesized distribution(s) is (are) stable with respect to G. Under

---

7 For a brief definition of observational equivalence classes of BNs, see Appendix 13.A (page 354).
8 For the definition of d-separation, see Appendix 13.A (page 354) or [42].

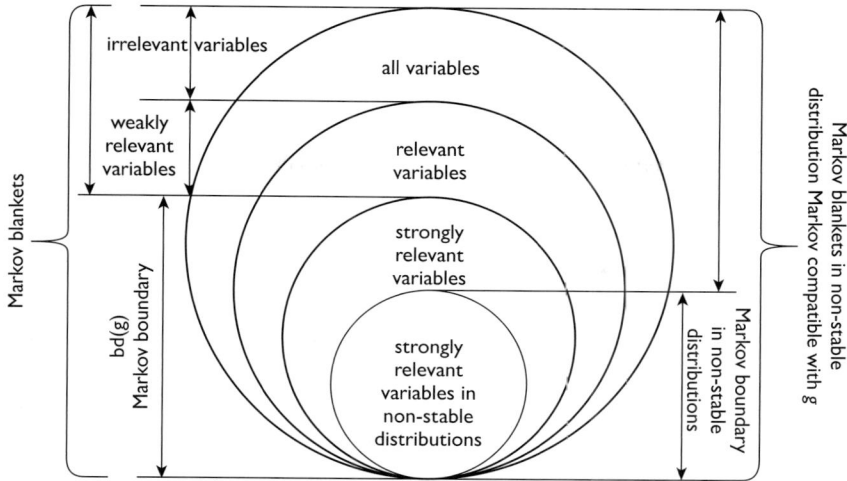

Fig 13.1 Possible relevance status of the variables in distributions Markov compatible with G. The left and right sides indicate the cases of stable and non-stable distributions respectively (in both cases Markov compatible with G). In non-stable distributions, the set of strongly relevant variables and a Markov blanket for Y can be smaller than $bd(Y, G)$, but $bd(Y, G)$ is always a Markov blanket for Y.

the same assumption, we also refer to $MBM(Y, X_i, G)$ as a strong relevance relation. According to Theorem 13.2, this graph-based definition of the MBM relation is equivalent to the probabilistic definition of the strong relevance relation assuming a Markov compatible stable distribution. Fig. 13.1 shows the possible relevance status of the variables, in cases of stability and non-stability of distributions Markov compatible with G.

13.2.3 Further relevance types

The concept of multivariate relevance can be further enriched using the common conceptual framework of BNs to represent all the relevant interactions between the strongly relevant variables as follows:

Definition 13.4
A subgraph of BN structure G is called the Markov blanket graph or mechanism boundary graph (MBG) MBG(Y, G) of variable Y if it includes the nodes in the Markov blanket defined by bd(Y, G) and the incoming edges into Y and its children [3].

An MBG represents all the necessary and sufficient interactions with respect to the Markov boundary of a given variable; thus it has a fundamental role both in the prediction of a target variable and in the interpretation of the interactions and mechanisms for a target variable.

Throughout the chapter, we use the term feature in a broad sense to denote a function $F(V', G) = f$ with arguments V' ($V' \subseteq V$) and DAG G, e.g., $MBG(Y, G) = mbg$, $MBS(Y, G) = mbs$ and $MBM(Y, X_i, G)$; that is, we concentrate only on the structural aspects of DAGs for a given subset of variables V'. However, the concept of features over DAGs is also used to denote a quantitative property of a given DAG, such as maximal in-degrees, out-degrees, clique sizes, or length

Table 13.1 Graphical model based definition of relevance types.

Relation	Notation	Graphical
Direct causal relevance	$DCR(X, Y)$	There is an edge between X and Y.
Indirect causal relevance	$ICR(X, Y)$	There is directed path between X and Y.
Causal relevance	$CR(X, Y)$	$DCR(X, Y)$ or $ICR(X, Y)$.
Confounded relevance	$ConfR(X, Y)$	X and Y have a common ancestor.
(Pairwise) Association	$A(X, Y)$	$DCR(X, Y)$ or $ICR(X, Y)$ or $ConfR(X, Y)$.
Interactionist relevance	$IR(X, Y)$	X and Y have a common child.
Strong relevance	$SR(X, Y)$	$IR(X, Y)$ or $DCR(X, Y)$ (also $MBM(X, Y)$).
Mixed conditional relevance	$MCondR(X, Y)$	There is an undirected path between X and Y with a node with converging edges.
Conditional relevance	$CondR(X, Y)$	$IR(X, Y)$ or $MCondR(X, Y)$.
Weak relevance	$WR(X, Y)$	$MCondR(X, Y)$ or $ICR(X, Y)$ or $ConfR(X, Y)$

of directed paths. Additionally, the term "feature" is also used to denote variables, e.g., in the case of feature subset selection (FSS) problems.

The Markov blanket-based extensions of the relevance concepts express local aspects, i.e., neighboring variables. The advantages of BNs for representing global dependence maps and relevance relations are well known, but their direct application in FSS is hindered by their high computational and sample complexity.[9] Now we consider the use of the global representational properties of BNs to characterize relevance relations, e.g., the type of association between non-neighboring, distant variables. The subtypes of relevance are listed in Table 13.1 along with the notation assigned to them.

For their causal interpretation under the causal Markov assumption, the reader is referred to [26, 42].[10] It needs to be emphasized that these relations represent different aspects of relevance and there are subtle differences with respect to their use in genetics, because of the possibility of multiple target variables and the possibility of non-genetic predictors. Following the usual genetic terminology [15], direct relevance (DR) formalizes the concept of direct association, although it covers direct consequences as well. Indirect causal relevance (ICR) and confounded relevance (CR) separate and express the concepts of indirect association and confounded association. Pairwise association (A) represents the usual association, which is the union of direct, indirect, and confounded associations. Interactionist relevance (IR) deviates from the purely epistatic relation [16], because this latter is the consequence of vanishing marginal effects of the individual variables, which can be modeled by contextual dependences (see Definition 13.5, page 325). To summarize, the most notable difference between the standard concept of pairwise association and

---

9   For a brief definition of sample complexity, see Appendix 13.A (page 354).

10   Note that the present definition of confounded relevance covers only a partial aspect of confounding [8].

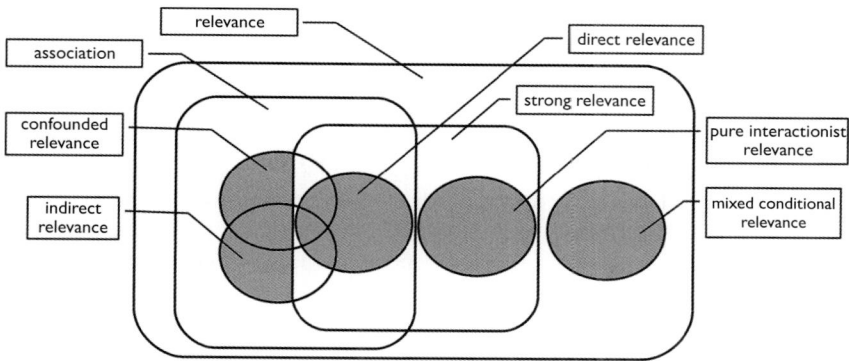

Fig 13.2 Overview of types of pairwise dependences. Weak relevance is constituted by mixed conditional relevance, indirect causal relevance, and confounded relevance (see Table 13.1 and Fig. 13.1).

strong relevance is that association includes certain forms of weak relevance (i.e., confounded relevance (ConfR) and transitive causal relevance (TCR)) and that it does not include the case represented by interactionist relevance. A direct consequence is that predictors exclusively in interactionist relevance will be filtered out by pairwise methods typical in high-dimensional studies, as they are not associated by definition. Also note that most of these relations are not mutually exclusive, e.g., a predictor may have direct and indirect effects at the same time, which can also be confounded. An overview of these relations is shown in Fig. 13.2. The applications of these relevance types are described in Subsection 13.3.6.

Hitherto, the definitions of relevance were based on the general concept of conditional independence, but conditional independence can be made more specific by introducing *contextual independence* when independence is present only in a given context (for its use in the field of BNs, the reader is referred to e.g., [6]). In genetics, this representation can express a related, seemingly complementary phenomenon if a variation has no effect with respect to a given target except in the presence of other variations [14, 45].

Definition 13.5

A variable X_i, which is strongly relevant to Y, is also contextually relevant to Y, if there exists some $X_i = x_{i_1}, x_{i_2}$ and context $s_i = x_1, \ldots, x_{i-1}, x_{i+1}, \ldots, x_n$ for which $\mathbb{P}(x_{i_1}, s_i), \mathbb{P}(x_{i_2}, s_i) > 0$ such that $\mathbb{P}(Y|x_{i_1}, s_i) = \mathbb{P}(Y|x_{i_2}, s_i)$.

Contextual relevances can be exactly and explicitly represented by special local dependence models in BNs, such as decision trees or decision graphs, which can express dependences at the value level (which is below the variable level) [6]. These models allow the explicit representation of the contexts defined by value configurations $C = c$ ($C \subset V$), in those cases in which X_i is conditionally independent of Y. Non-BN-based methods implicitly capable of addressing contextual relevance include multifactor dimensionality reduction (MDR) [39] and logic regression [47].

The identification of a contextually relevant factor X_i is particularly challenging, because it has an effect—by definition—in particular contexts $C = c$. If its effect strength is moderate in these contexts or if these contexts are very rare, then reliable identification of the relevance of X_i statistically requires prior knowledge and multivariate modeling. Furthermore, it is also possible, in the case of a purely epistatic factor, that a strongly relevant predictor has no main effect [16].

A practical consequence is that the statistical identification of a contextually relevant factor is not possible without observing other variables forming its context. In BNs lacking the ability to explicitly represent contextual independence, this phenomenon frequently leads to misinterpretations because of the incorrect representation of a contextually relevant factor X_i by an interactionist relevance relation $IR(X, Y)$ with target variable Y.[11] Using omics data, such as genome-wide association studies (GWAS) data or next-generation sequencing (NGS) data, the lack of observation of the other variables involved in the interaction is diminished. However, the problem of multiple hypothesis testing arises in the frequentist statistical framework, because of the high number of potential interactions: the reader is directed to Section 13.4 (page 344) for the discussion of multiple hypotheses in the Bayesian approach and to [19, 35, 65] for non-BN-based Bayesian methods to explore interactions.

13.2.4 Necessary Subsets and Sufficient Supersets in Strong Relevance

The MBM feature gives an overall characterization of strong relevance for each predictor separately but does not capture the joint relevance of predictors. At the other extreme, Markov boundary subsets characterize the joint strong relevance of predictors, but the number of possible MB sets is exponential, which is intractable computationally and statistically. The concept of k-ary Markov boundary subsets, focusing on k sized sets of variables, was introduced to support a constrained multivariate analysis of strong relevance, called *partial multivariate* analysis of relevance [4]. Here we complement this *sub-relevance* concept with an analogous concept of *sup-relevance*.

Definition 13.6
For a distribution $\mathbb{P}(V)$ with Markov boundary set mbs, a set of variables s is called sub-relevant if it is a k-ary Markov boundary subset (k-subMBS), i.e., $|s| = k$ and $s \subseteq$ mbs. A set of variables s is called sup-relevant if it is a k-ary Markov boundary superset (k-supMBS), i.e., $|s| = k$ and mbs $\subseteq s$.

The k-subMBS and k-supMBS concepts are more flexible than the MBS relation, because they can be used to capture the presence or the absence of relevant variables. A k-subMBS set s_{sub} contains those variables that are strongly relevant without specifying the status of the other variables in its complementary s_{sub}^c (i.e., a k-subMBS denotes a "necessary" set of variables). The complement of a k-supMBS set s_{sup}^c contains all variables that are not strongly relevant, but it does not state anything about the strong relevance of the variables in s_{sup} (i.e., a k-supMBS denotes a "sufficient" set of variables). Note that the probabilistically defined concept of sub-relevance can be seen as a multivariate extension of strong relevance for variables in Definition 13.1 (page 320). Note also that the concept of sup-relevance is equivalent to the concept of Markov blankets, but for symmetry and uniformity we use the "sub-" and "sup-" terminology. We use the term k-MBS to denote together the k-subMBS and k-supMBS concepts.

---

11 The interactionist relevance (IR) representation of a contextually relevant factor X_i for target Y using variable C as a common child is incorrect, because it implies that the effect of variable X_i on Y related to variable C is not present if we do not know C. This typically arises if the X_i and C variables in the interaction are dependent and X_i has negligible main effect on target Y. However, the IR representation is correct in the special case of a purely epistatic factor, which has no main effect [16].

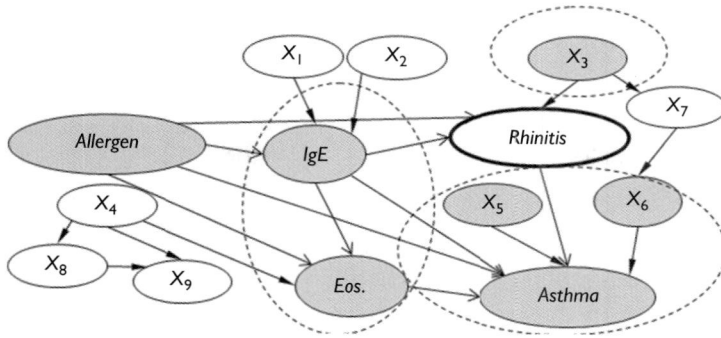

Fig 13.3 Sub-relevant sets for the target variable *Rhinitis* in stable distributions which are Markov compatible with the depicted Bayesian network structure G. Sub-relevant sets are denoted by dashed circles. The target variable is highlighted in bold. Sub-relevant sets: 1-subMBS: X_3; 2-subMBS: *IgE, Eosinophil (Eos.)*; 3-subMBS: X_5, X_6, *Asthma*. The Markov boundary of *Rhinitis* is denoted by shaded nodes, and any set including these nodes is a sup-relevant set (k-supMBS). Pairwise relevance relations are as follows: interactionist relevance (e.g., *Rhinitis* and X_6 have a common child), direct causal relevance (e.g., X_1 and *IgE* have a common edge), confounded relevance (e.g., *Rhinitis* and *Eosinophil* level have a common ancestor), and mixed conditional relevance e.g., X_7 and *IgE* are connected through an undirected path on which *Asthma* has two converging edges).

Furthermore, the k-subMBS and k-supMBS concepts form hierarchically related, overlapping hypotheses, e.g., a k-subMBS hypothesis can be extended to multiple, more complex $(k + 1)$-subMBS hypotheses. In fact, the scalable polynomial cardinality of the set of k-subMBSs and k-supMBSs bridges the linear cardinality of the features and the exponential complexity of the MBS cardinality $(\mathcal{O}(n) < \mathcal{O}(\binom{n}{k}) < \mathcal{O}(2^n))$, where n denotes the number of observed variables. Because the cardinality of the MBGs and DAGs is even higher [12], we can think of MBMs, k-subMBSs/k-supMBSs, MBSs, MBGs, and DAGs as more and more complex hypotheses about relevance. Indeed, they form a hierarchy of levels of abstraction to support a multilevel analysis, in which intermediate levels of k-MBS for varying k allow a scalable partial multivariate analysis focusing on k number of variables.

An application of these concepts to the problem domain of asthma is illustrated in Fig. 13.3.

13.2.5 Relevance for Multiple Targets

If there are multiple target variables Y which have to be examined together, and if the relations among them are irrelevant, one may ask for the variables relevant to the target set. The basic concepts of relevance for a single target variable can easily be extended to multiple targets [4].

Definition 13.7
A variable X_i is strongly (weakly) relevant to Y, if and only if it is strongly (weakly) relevant to any $Y_i \in Y$.

We introduce additional relations, such as exclusive or multiple relevance, in order to better characterize the relevance types of a predictor with respect to multiple target variables. The corresponding definitions are given in Table 13.2, using a Bayesian network representation. We refer the reader to Subsection 13.3.6 (page 340) for the application of these relations in asthma and allergy research.

Table 13.2 Subtypes of multitarget relevance. The other targets except Y' are denoted as Y^- ($Y^- = Y \setminus Y'$).

Direct relation(s) to	Notation	Graphical
One or more targets	$EdgeToAny(X, Y)$	There is an edge between X and Y.
Exactly one of the targets	$EdgeToExactlyOne(X, Y)$	There is exactly one edge between X and Y.
Two or more targets	$MultipleEdges(X, Y)$	There is more than one edge between X and Y.
One or more other targets	$EdgeToOthers(X, Y^-, Y')$	There is at least one edge between X and Y^-, but not between X and Y'.
MBM relation(s) to	**Notation**	**Graphical**
One or more targets	$MBMToAny(X, Y)$	There is a $Y_i \in Y$ with $MBM(X, Y_i)$.
Exactly one of the targets	$MBMToExactlyOne(X, Y)$	There is exactly one $Y_i \in Y$ with $MBM(X, Y_i)$.
Two or more targets	$MultipleMBMs(X, Y)$	There is at least two $Y_i \in Y$ with $MBM(X, Y_i)$.
One or more other targets	$MBMWithOthers(X, Y^-, Y')$	There is a $Y_i \in Y^-$ with $MBM(X, Y_i)$ and $\neg MBM(X, Y')$.

13.3 A Bayesian View of Relevance for Complex Phenotypes

The Bayesian network representation, along with the concept of Markov blanket sets and strong relevance, opened many research directions in feature learning, in the feature subset selection problem and in genetic association studies [48]. The "filter" approaches, later termed "local causal" approaches, emerged from Markov Blanket sets (MBSs) and strong relevance [1, 10, 28, 34, 63].[12] However, despite the rapid development of methods designed to identify an optimal MBS, the global significance of the optimal MBS in the frequentist framework, as well as the lack of dominant MBSs in the Bayesian framework, are still overlooked.

Bayesian methods are increasingly popular in genetic association studies for their ability to sufficiently characterize and explore weakly significant results and to cope with multiple hypothesis testing (for the general approach see [18, 24]; for the application to GAS see [52]; for methods see [19, 35, 65]).

The Bayesian approach aims at determining the posterior distribution of each of the relevance relations defined in Section 13.2 (page 320) given some data, which is used to update the

---

[12] In a similar vein, the local causal discovery methods apply well-selected local tests instead of using global scores or interventional data, e.g., see [11, 13].

prior probability distribution, i.e., our knowledge on these relations prior to obtaining those data. (Henceforth we will use the shorthand "posterior" both for the posterior probability of the specific value of a feature and for the posterior probability distribution of a feature in general if it is clear from the context.) These posterior distributions, such as the ones of Markov boundary graphs, Markov boundary sets, or Markov boundary memberships, are derived by the propagation of the DAG posterior to these "upper" levels by aggregation. Technically speaking, these posteriors are induced[13] by the posterior $\mathbb{P}(G|D_N)$ over the DAG space given the data set D_N, because $MBG(Y,G)$, $MBS(Y,G)$, and $MBM(Y,X_i,G)$ are functions (transformations) interpreted over DAGs (i.e., they map each DAG G to a unique value). Consequently, a (normalized) probability distribution over DAGs will induce a (normalized) distribution over the MBGs, the MBSs, and the MBM for each X_i as follows:

$$\mathbb{P}(mbg|D_N) = \mathbb{E}_{\mathbb{P}(G|D_N)}[1(MBG(Y,G) = mbg)] = \sum_{G:MBG(Y,G)=mbg} \mathbb{P}(G|D_N), \qquad (13.2)$$

$$\mathbb{P}(mbs|D_N) = \mathbb{E}_{\mathbb{P}(G|D_N)}[1(MBS(Y,G) = mbs)] = \sum_{G:MBS(Y,G)=mbs} \mathbb{P}(G|D_N), \qquad (13.3)$$

$$\mathbb{P}(MBM(Y,X_i,G)|D_N) = \mathbb{E}_{\mathbb{P}(G|D_N)}[1(MBM(Y,X_i,G))] = \sum_{G:MBM(Y,X_i,G)} \mathbb{P}(G|D_N) \qquad (13.4)$$

where $1(A)$ denotes the indicator function[14] of event A.

Note that $MBS(Y,G)$ and $MBM(Y,X_i,G)$ depend only on $MBG(Y,G)$; thus the functions $MBS(Y,G) : G \rightarrow mbs$, and $MBM(Y,X_i,G) : G \rightarrow \{0,1\}$ can be analogously defined over the MBGs $MBS(Y,MBG) : mbg \rightarrow mbs$ and $MBM(Y,X_i,MBG) : mbg \rightarrow \{0,1\}$. Thus, the DAG posterior $\mathbb{P}(G|D_N)$ is not necessary to define the MBS posterior ($\mathbb{P}(MBS(Y,G)|D_N)$) and MBM posteriors ($\mathbb{P}(MBM(Y,X_i,G)|D_N)$) because the MBG posterior ($\mathbb{P}(MBG(Y,G)|D_N)$) is sufficient. Analogously to equation (13.2), the MBG posterior induces the MBS posterior and MBM posterior, because of the functional dependence of MBSs and MBM for each X_i on MBGs.

$$\mathbb{P}(mbs|D_N) = \sum_{mbg:MBS(Y,mbg)=mbs} \mathbb{P}(mbg|D_N), \qquad (13.5)$$

$$\mathbb{P}(MBM(Y,X_i,G)|D_N) = \sum_{mbg:MBM(Y,X_i,mbg)} \mathbb{P}(mbg|D_N). \qquad (13.6)$$

Consequently, we can formulate the central task of the Bayesian analysis of relevance using BNs as the estimation of the posterior distribution of MBGs. Note that the posterior of k-subMBSs and k-supMBSs can also be induced by the MBG posterior. However, the MBG posterior is not sufficient to induce the posteriors of relevance types in Tables 13.1 (page 324) and 13.2 (page 328), and therefore these should be estimated individually. Thus, in addition to estimation of the MBG posterior, the general task also includes estimation of the expectation of arbitrary binary random

---

[13] For a brief definition of Bayesian model averaging, see Appendix 13.A (page 354).
[14] For the definition of the indicator function, see Appendix 13.A (page 354).

variables $1(F(V', G) = f)$ over the space of DAGs, such as the indicator functions corresponding to relevance types in Tables 13.1 and 13.2.[15] However, the exact computation of this expectation is generally not feasible due to the superexponential number of possible DAG structures; thus an approximation is applied using M randomly sampled DAG G_i from the posterior $\mathbb{P}(G|D_N)$:

$$\mathbb{E}_{\mathbb{P}(G|D_N)}[1(F(V', G) = f)] = \sum_{G:\, 1(F(V',G)=f)} \mathbb{P}(G|D_N) \approx \frac{1}{M} \sum_{i=1}^{M} 1(F(V', G_i) = f) \qquad (13.7)$$

There are several classes of methods designed to facilitate the estimation in equation (13.7). The Bayesian inference of structural properties of BNs was proposed in [7, 12]. In [37], Madignan et al. proposed a Markov chain Monte Carlo (MCMC) scheme, to approximate such Bayesian inference using an unnormalized posterior $\tilde{p}(G|D_N)$ and an auxiliary ordering of variables. In [22], Friedman et al. reported an MCMC scheme over the space of orderings. In [33], Koivisto et al. reported a method to perform exact full Bayesian inference over modular features.[16] Recent applications in GAS are reported in [31, 61]. Another approach using PGMs applied the bootstrap framework to provide uncertainty measures for arbitrary subgraphs [20, 43]. To support Bayesian inference over relevance relations, we reported specialized ordering MCMC methods designed to efficiently estimate posteriors of structural model properties, particularly of Markov Blanket Graphs [3, 4].

13.3.1 Estimating the Posteriors of Complex Features

As we discussed in the previous subsection, the posterior probability of any structural feature $F(V', G) = f$ can be estimated by MCMC methods if efficient computation exists for an unnormalized posterior $\tilde{p}(G|D_N)$ and feature value $F(V', G) = f$ (see equation (13.7)). However, the estimation of posterior distributions of complex features with high cardinality, such as the MBS, MBG, and k-MBS features, poses an additional challenge. Earlier we proposed a specialized ordering MCMC method to estimate MBG and MBS posteriors [3], but DAG MCMC methods are more suitable for general, potentially knowledge-intensive structural features; thus, we developed a two-step DAG-based process to estimate all the relations in the BN-BMLA methodology.

We assume a complete data set D_N, discrete variables, multinomial sampling, Dirichlet parameter priors ($\mathbb{P}(\Theta|G)$), and a uniform structure prior ($\mathbb{P}(G)$). We used two types of Bayesian Dirichlet parameter priors: the first is defined by the Cooper–Herskovits hyperparameters, and the second is defined by the Bayesian Dirichlet equivalent uniform (BDeu) hyperparameters with virtual sample size equal to one (for their definition, see [12, 30]; for the effect of various virtual sample sizes, see [50, 57]). These assumptions allow the derivation of an efficiently computable unnormalized posterior over the Bayesian network structures G (i.e., over directed acyclic graphs, (DAGs) [12, 30].

Using this unnormalized posterior, we perform a special random walk in the space of DAGs G by applying an MCMC sampling method, which inserts, deletes, and inverts edges [25]. The probability to apply different DAG operators in the proposal distribution is uniform, the length of the burn-in is 10^6, and the length of the sample collection (L) is 5×10^6.

---

[15] For a brief definition of indicator function, see Appendix 13.A (page 354).
[16] For a brief definition of modular features, see Appendix 13.A (page 354).

The MCMC process generates a dependent sequence of L DAGs D_L^G. Using the MCMC simulation, we estimate the posterior of the MBGs for the target variables according to equation (13.7) see Section 13.3 (page 328). In each MCMC step, we determine the boundary graph $bd(Y, G)$ corresponding to the DAG G in this step and update the relative frequency of this boundary graph. (We recall that the presence of the boundary graph $bd(Y, G)$ implies with probability one that the corresponding variables are the Markov boundary, see Subsection 13.2.2 (page 322).) This update is analogously done for all the pairwise relations in Table 13.1 (see Subsection 13.2.3, page 323) and in Table 13.2 (see Subsection 13.2.5, page 327); these relations are evaluated for every possible pair of variables at each MCMC step, and hence a counter is maintained for each possible instantiation of the given relation throughout the MCMC sampling. Note that there are no practical restrictions on the selection of target variables, i.e., within the same MCMC simulation we can evaluate multiple target sets simultaneously. The computational complexity of the evaluation of the structural features discussed in this chapter and the update of their relative frequency is $O(n^2)$ (n denotes the number of variables). In case of a complex phenotype with multiple descriptors, we can use the descriptors together as a joint target set, and each of the descriptors individually. An occasionally practical inversion is to select a predictor variable as the target, because we can explore all the phenotypes relevant for this predictor (for the application of this inversion in the frequentist framework, see [34]). Hence the set of evaluated features can be fully fitted to the needs, queries, or preconceptions of the expert performing the analysis.

In a second, "post-hoc"[17] phase, we compute various MBS related, marginal posteriors from the MBG posterior estimated in the first phase. The MBS and MBM posteriors are computed exactly from the estimated MBG posterior analogously to equation (13.5) (see Section 13.3, page 328). The posterior of a given k-subMBS set or k-supMBS set can be directly computed from the MBS posterior according to equations 13.9 and 13.10 (see Subsection 13.3.4, page 336). To find the highly probable k-subMBS and k-supMBS sets, we apply greedy algorithms, because the cardinalities of these sets grow polynomially ($\mathcal{O}(n^k)$). In the case of k-subMBSs, the starting state of the greedy search is the empty set, which can be regarded as a trivial 0-subMBS with probability one. The algorithm then expands the set into the $(k+1)$-subMBS with maximal posterior. In the case of k-supMBSs, the initial state of the search is the complete set U, from which the algorithm iteratively eliminates predictors to get the $(k-1)$-supMBS with maximal posterior.

In the MCMC simulation, we also calculate various quantitative measures of convergence and confidence of the posteriors of complex features. The following set of measures can be regarded as a standard set:

1 the Geweke Z-score, measuring convergence within one chain, i.e., the significance of the difference of the posteriors between the beginning and the end of the sampling [24]

2 the Gelman–Rubin R-score, measuring inter-chain convergence i.e., significance of the difference of independent sampling processes [24]

3 confidence intervals, based on the standard error of the MCMC[18]

---

[17] The goal of a post-hoc analysis is to search the primary, "raw" results (in our case the MBS posteriors) for patterns that are not expected a priori or that are too numerous. In our case, such patterns are the k-subMBS and the k-supMBS posteriors, that would be redundant and intractable if directly sampled during the MCMC run.

[18] Confidence intervals for MCMC estimation can be calculated e.g. according to the batch means method [9].

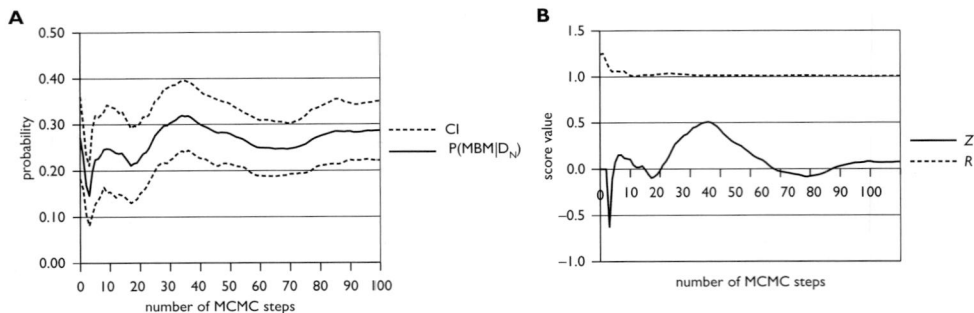

Fig 13.4 The trace plots of the convergence and confidence indicators for the MBM posterior of rs7928208 and *Asthma* using data set A. A) Estimated values of the MBM posterior ($P(MBM|D_N)$) and its 95% confidence interval (CI). B) Geweke's score (Z) and Gelman–Rubin score (R) for each 10000th step (x-axis) of the MCMC simulation.

Fig. 13.4 demonstrates the convergence of the applied MCMC sampling with respect to the burn-in phase. Note that these measures are different for each feature, i.e., the estimation of the MBM posteriors is typically faster than the estimation of the 3-subMBS posteriors or the MBG posterior. The values of the measures are calculated in each step l of the MCMC simulation using the MCMC samples of steps $1, \ldots, l$.

In the following, we demonstrate the application of this approach on a case study. The study involves 1201 unrelated individuals from the Hungarian (Caucasian) population. Four hundred and thirty-six asthmatic children, aged 3 to 18 were recruited for the study. The control group consisted of 765 subjects (mean age: 19 years, 405 males/360 females). We used three embedded data sets: (1) the asthma status was known in all cases (1201 subjects, data set A); (2) in 1100 cases, the status of *Rhinitis* was also known (data set RA) (only those subjects whose rhinitis status had been verified by specialists); and (3) in 200 cases, the status of rhinitis and the serum levels of *IgE* and *Eosinophil* were involved in this data set were also known (data set CLI).

13.3.2 Sufficiency of the Data for Full Multivariate Analysis

Regardless whether a Bayesian conditional method (such as Bayesian logistic regression) or the BN-based approach is used, the posterior probability distribution of sets of predictors indicating their joint relevance with respect to the selected model class (such as the MBS posterior in case of BNs) is typically flat for the settings of contemporary GASs, because of the relation of sample size, number of predictors, effect sizes, and model complexity including prior. Such flat MBS posteriors, ranked from the maximum a posteriori (MAP) MBS to the less probable ones, are illustrated in Fig. 13.5, which shows that there are several sets with only slightly lower probabilities than that of the MAP set. These also indicate that the MAP MBS is not dominant, as its posterior is negligible. Furthermore, the cumulative distribution functions in Fig. 13.5 also indicate the lack of dominant MBSs, i.e., the lack of a small number of MBSs with high posteriors such that their summed posterior is close to one. It needs to be emphasized that these results are the consequences of the power of the data and not of the priors; thus, they also indicate the lack of dominantly optimal model in the frequentist, maximum likelihood approach. In the case of the data set RA and the *Asthma* target variable, the MAP set has only a probability of 0.010688.

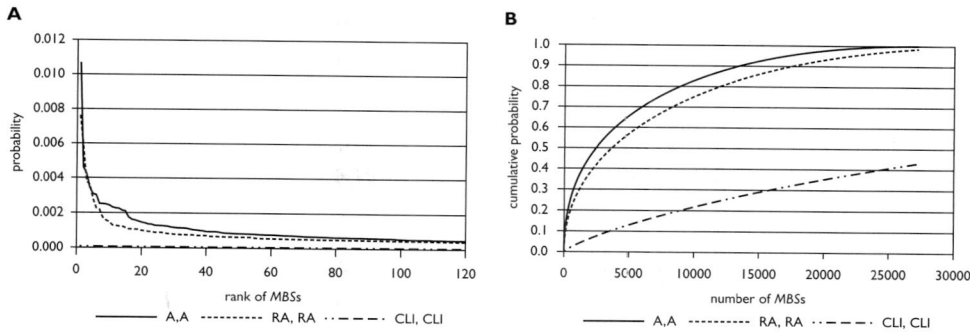

Fig 13.5 Posterior probability and cumulative posterior probability plotted against the rank of the MBSs and the number of MBSs, respectively, for three cases. The three cases are the following: (A,A): the target is *Asthma* and the data set is A; (RA,RA): the target set contains *Asthma* and *Rhinitis* and the data set is *RA*; (CLI,CLI): the target set contains *Eosinophil, IgE, Rhinitis*, and *Asthma*, the data set is *CLI*. Ranks go from the maximum a posteriori (MAP) MBS to the less probable ones.

Because of the smaller sample sizes, when *Asthma* and *Rhinitis* are the target variables (i.e., multi-target analysis), the probability of the MAP set is even lower, namely 0.007626. This phenomenon is even more pronounced in the case of data set *CLI*, where the corresponding probabilities of the MAP sets are 0.001496 (*Asthma* target), and 0.000073 (multitarget), respectively. These flat MBS posteriors are consistent with our earlier findings from the simulations in [4] suggesting that a 200-sized sample in general induces a very flat posterior distribution ("small sample size"), whereas a 1000-sized sample corresponds to a "medium sample size" with respect to our settings with 100 variables, which is typical in candidate GASs and partial genome screening studies.

The relation between the posterior value p of MBSs and the corresponding ranks r is well described by a power law (cf. Zipf's law [60]) meaning that the posterior p is proportional to $1/r^\alpha$, where $\alpha > 0$ (Fig. 13.5). The heavy-tailed distribution and the lack of a dominant set or a set of highly significant sets indicate the necessity of refined approaches to cope with weakly significant results, e.g., Fig. 13.5 shows that the cumulative posterior of the 10 000 most probable MBSs is below 0.22.

13.3.3 Rate of Learning: Effect of Feature and Model Complexity

The MBM, k-subMBS/k-supMBS, MBS, and MBG features form a hierarchy of increasing cardinality ($|MBM| < |k\text{-}MBS| < |MBS| < |MBG| < |BN|$), which is related to the learnability of these levels (for the sample complexity of learning BNs, see [23]). We investigated the effect of this increasing cardinality and computed various performance measures to characterize the learning rate of the simplest and other more complex structural features about strong relevance, namely MBM, MBS, and MBG features. To do this, we performed systematic studies varying the sample size and model complexity using an artificial reference model M_0 and an artificial data set generated from this model.

The artificial reference model M_0, used in our experiments throughout this chapter, is the maximum a posteriori Bayesian network learned from data set A. In the reference model M_0, the Markov blanket of the *Asthma* target variable consisted of 15 variables. Subsequently, we

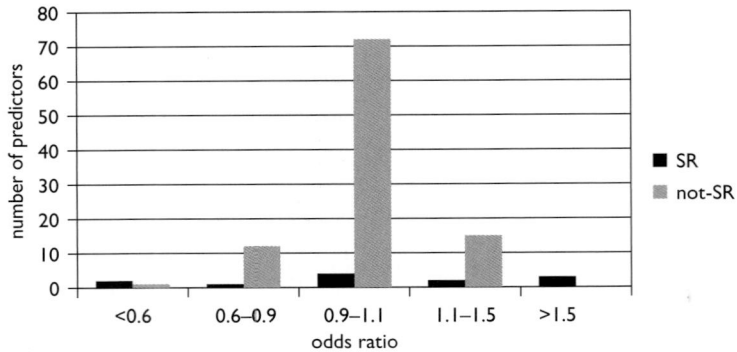

Fig 13.6 Histogram of the odds-ratios for the strongly relevant and the not-strongly relevant predictors in the reference model M_0. SR, strongly relevant (black); not-SR, not-strongly relevant (gray).

randomly generated an artificial data set from M_0, which contained 10 000 independent, identically distributed, random samples from the estimated distribution. From this data set, we created smaller ones by simply taking the first 50, 100, 250, 500, 1000, 2000, and 5000 samples, respectively. Additionally, we defined embedded sets of variables with sizes 20, 40, 60, 80, and 100 ($s_{20}, s_{40}, s_{60}, s_{80}, s_{100}$) to emulate different model complexities.[19] These sets always contained the 15 reference input variables and the target variable, which were supplemented with randomly drawn variables not in the reference MBS. Fig. 13.6 shows the effect strengths (odds-ratios[20]) of the predictors in the reference model M_0. Note that there are many weak associations in the reference model M_0, i.e., in the real-world data set.

The relative flatness of a posterior at a given level generally indicates uninformativeness, i.e., high level of uncertainty; thus the inapplicability of a given level, such as the inapplicability of the level of MBGs, MBSs, or k-subMBSs/k-supMBSs above a given k. The general uncertainty of a posterior distribution can be characterized by its entropy,[21] which will be high for a flat, nearly uniform, non-informative posterior distribution. Fig. 13.7 shows the entropy of the distribution of the MBM and MBS features for various model sizes and sample sizes.

---

[19] Model complexity also depends on the maximum number of parents and the local models used in the BN (A local model defines the conditional distribution for a variable given its parents).

[20] Odds-ratios are standard measures of the effect size of the binary predictor X on binary target Y. The odds-ratio is defined as the ratio of the odds of Y in the case when $X = 0$ and $X = 1$, respectively: $\big(\mathbb{P}(Y = 1|X = 1)/\mathbb{P}(Y = 0|X = 1)\big) / \big(\mathbb{P}(Y = 1|X = 0)/\mathbb{P}(Y = 0|X = 0)\big)$.

[21] The entropy $H(X)$ of a discrete random variable X with probability mass function $P(X)$ is the expected value of the negative logarithm of $P(X)$, that is: $H(X) = E[-\ln(P(X))]$. An important consequence of this definition is that a uniformly distributed random variable (i.e., whose distribution is completely "flat") has maximal entropy. Note that the entropy depends on the cardinality, e.g., in case of uniform distributions, entropy is higher for the MBG than for the MBS level. Also note that the MBM level consists of separate entropies for each $MBM(Y, X_i, G)$, which are summed together to provide an aggregate measure (this summation corresponds to the simplification that $MBM(Y, X_i, G)$ and $MBM(Y, X_j, G)$ are mutually independent).

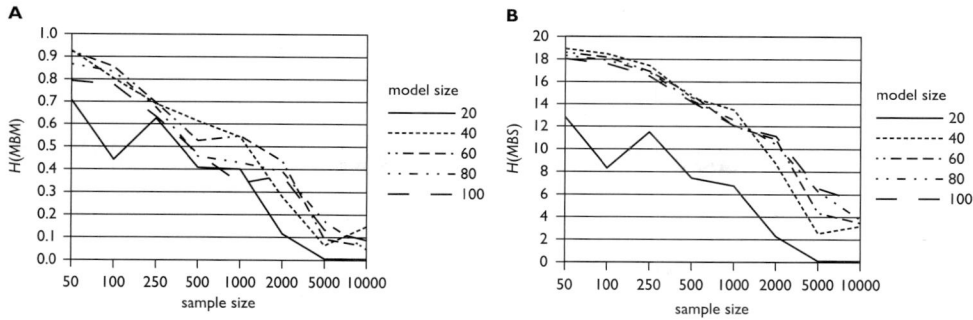

Fig 13.7 Entropy of the distribution of the MBM and MBS features as a function of sample size, for various model sizes. A) MBM. B) MBS. The various model sizes correspond to the different curves.

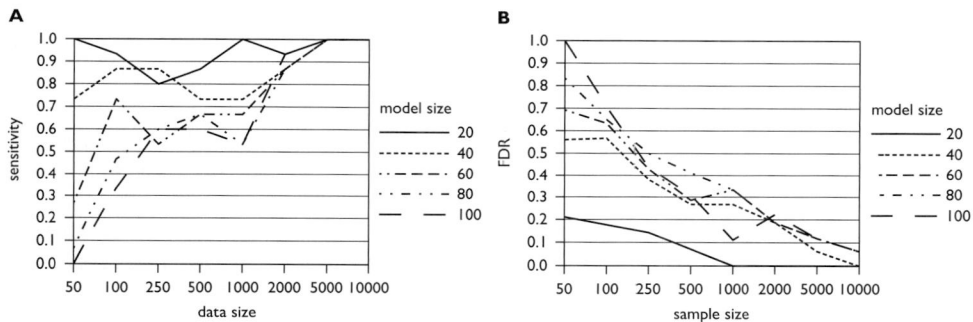

Fig 13.8 Sensitivity and FDR as a function of the sample size, for models with different sizes. A) Sensitivity. B) FDR. The various model sizes correspond to the different curves.

We now present results about learning rates using sensitivity (Fig. 13.8A), false discovery rate (FDR, Fig. 13.8B), misclassification rate (Fig. 13.9B), and area under the Receiver Operating Characteristic (ROC) curve (AUC; Fig. 13.9A).[22] Results are calculated at an acceptance threshold of $0.5 < \mathbb{P}(MBM(Y, X_i, G)|D_N)$; i.e., this is the probability above which we accept a factor X_i to be relevant.

The trends indicated in Figs. 13.8 and 13.9 (page 336) are summarized in Table 13.3 (page 336). For each performance measure one can select thresholds for acceptable and very good performances, which allow the definition of "small" and "large" sample sizes as the minimal number of samples to reach these thresholds. These thresholds can be selected in our case with a given number of variables and model complexity as follows: 0.6 and 0.9 for AUC, 0.1 and 0.5 for sensitivity, and 0.5 and 0.1 for false discovery rate. The inverse problem of selecting an optimal decision threshold for a given sample size is discussed in Section 13.4. The relatively modest performance and relatively high "small" and "large" sample sizes are the consequences of the abundance of weak associations in the reference model M_0.

---

[22] For the definition of AUC, see Appendix 13.A (page 354).

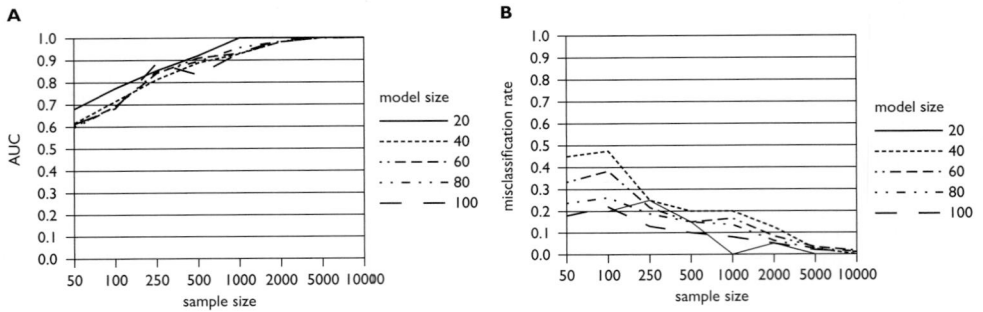

Fig 13.9 Area under the ROC curve score and misclassification rate as a function of sample size, for various model sizes. A) AUC. B) Misclassification rate. The various model sizes correspond to the different curves.

Table 13.3 The "small" and "large" sample sizes for various performance measures (the decision threshold for MBM posteriors is 0.5).

	Sensitivity		FDR		AUC	
	threshold	N	threshold	N	threshold	N
"small" sample size	0.1	100	0.5	250	0.6	250
"large" sample size	0.5	2000	0.1	5000	0.9	5000

Note: AUC, area under the receiver operating characteristic curve; FDR, false discovery rate.

These results are good examples of our experiences from other candidate GASs, which suggest that if the ratio of the number of samples and the number of variables is below ten then the MBS posterior is very flat but the MBM posteriors show peaks. These observations indicate the necessity of the intermediate level of k-subMBSs and k-supMBSs.

13.3.4 Bayesian network-based Bayesian Multilevel Analysis of Relevance

In Subsection 13.2.1 (page 320), we defined different relevance types that can be used to infer strongly relevant variables, either separately from each other (MBM) or jointly in a complete set (i.e., in an MBS). Moreover, in case of the MBG-based relevance type even the interactions among those strongly relevant variables can be investigated. In this section, we provide some characteristics of these relevance types and show how each of them can be used in a genetic association study to reason about the relevance of the predictors.

Two complementary approaches can be distinguished in relevance analysis depending on their focuses. The first approach provides an overall characterization of relevance relations using a limited number of features and feature values, such as pairwise edges and Markov blanket membership relations [20, 21]. The second approach is at the other extreme of feature learning as it provides a detailed characterization of dependence relations between the target variables and the

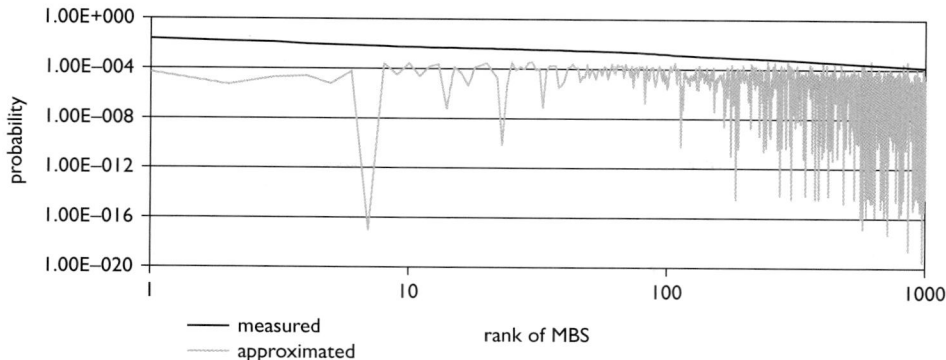

Fig 13.10 The ranked MBS posteriors and their MBM-based approximations. Ranked MBS posteriors: black line; MBM-based approximations: gray line.

predictors. Here we find the identification of MBSs and MBGs, which is also related to the learning of arbitrary subgraphs with statistical significance [43]. To understand this whole spectrum, first recall that the MBM, k-MBS, MBS, and MBG features are of increasing complexity. Consequently, the MBG and MBS posterior distribution is often too "flat" (i.e., there are hundreds of MBS or MBG features whose summed probability is relatively high but whose individual posteriors are low) even when the MBM posterior distributions $\mathbb{P}(MBM(Y, X_i, G)|D_N)$ show peaks. To illustrate the relation between the two approaches, we report the approximation of the (set-based) MBS posterior $\mathbb{P}(mbs(Y, G)|D_N)$ using the (pairwise) MBM posteriors as follows:

$$\mathbb{P}(mbs(Y, G)|D_N) \approx \prod_{X_i \in mbs(Y,G)} \mathbb{P}(MBM(Y, X_i, G)|D_N) \prod_{X_i \notin mbs(Y,G)} (1 - \mathbb{P}(MBM(Y, X_i, G)|D_N)). \quad (13.8)$$

Fig. 13.10 shows that the MBM-based approximation allows only a rough quantitative estimation, and the corresponding ranking differs significantly (the difference is particularly pronounced in most practical cases when the sample is relatively small).

The Bayesian multilevel analysis of relevance introduced scalable intermediate levels to provide a comprehensive view over multiple levels. It was motivated by the observation that typically—even when the MBG and MBS posterior distributions are flat—the most probable MBSs and MBGs share significant common patterns. We introduced the concept of sub-relevance, denoted in this chapter as k-subMBS (see Definition 13.6 (page 326) and [4]), to characterize the common elements. Typically, these common variables are present in MBSs with high posteriors, and they usually have larger effect sizes. The posterior probability of the sub-relevance of a subset s is:

$$\underline{p}(s|D_N) = \mathbb{P}(MBS(Y, G) = s|D_N) + \sum_{s':s \subset s'} \mathbb{P}(MBS(Y, G) = s'|D_N), \quad (13.9)$$

where the first term is the exact MBS posterior of s and subsequent terms, and behind the summation sign, are the MBS posteriors of each proper superset of s. The posterior $\underline{p}(s|D_N)$ expresses the probability that s contains only strongly relevant variables. If s is the common subset of almost all

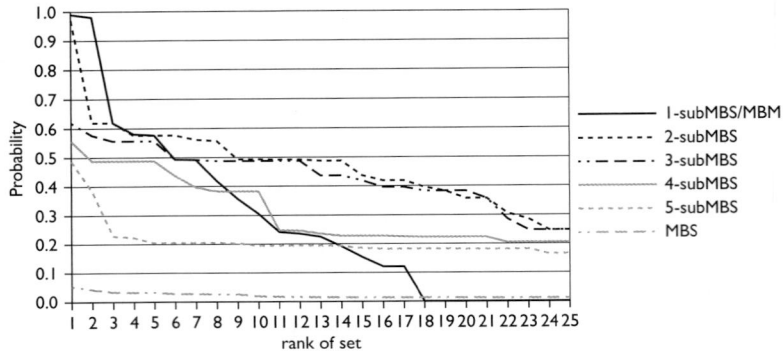

Fig 13.11 Posteriors for MBSs and k-subMBSs for $k = 1, \ldots, 5$ in the case of data set *CLI* and target *Asthma*.

highly probable MBSs, then its sub-relevance posterior will be high. On the contrary, if s is not the subset of the most probable MBSs, then the posterior will be low.

We demonstrate the use of the k-subMBS concept in the *Asthma* domain. In this domain, the MBS posterior distributions were very flat, whereas the MBM posterior distributions were very rough, which indicates that analysis at intermediate levels of k-subMBSs can unhinge significant results. Thus, we evaluated the partial multivariate results illustrated in Fig. 13.11. In the cases of $k = 1, 2, 3, 4$, the high maximum a posteriori probability (which corresponds to relatively rough posterior distribution) indicates that the sample size is sufficient to infer that these variables are strongly relevant jointly. In contrast, for $k > 4$, the maximum a posteriori multivariate features are weakly significant. These results are in line with the expectation that, with increasing feature cardinality, the posterior distribution progressively flattens out. The posteriors corresponding to the polynomially increasing cardinality of the k-subMBS bridge the gap between the flat posterior of the MBS and the MBM posteriors characterized by the presence of numerous peaks.

Analogously, we define the concept of k-supMBS to characterize the not strongly relevant variables typically not present in MBSs. The posterior probability of sup-relevance is computed analogously to that of sub-relevance (equation (13.9)):

$$\bar{p}(s|D_N) = \mathbb{P}(MBS(Y, G) = s|D_N) + \sum_{s':s \supset s'} \mathbb{P}(MBS(Y, G) = s'|D_N). \qquad (13.10)$$

The posterior $\bar{p}(s|D_N)$ expresses the joint probability that all strongly relevant variables are in set s ($X_i \in s$) (equation 13.11). Equivalently, none of the $X_i \notin s$ are strongly relevant (equation (13.12)):

$$\bar{p}(s|D_N) = \mathbb{P}(mbs : \forall X_i : X_i \in mbs \Rightarrow X_i \in s|D_N), \qquad (13.11)$$

$$\bar{p}(s|D_N) = \mathbb{P}(mbs : \forall X_i : X_i \notin s \Rightarrow X_i \notin mbs|D_N). \qquad (13.12)$$

In summary, a k-subMBS with high posterior contains variables that are strongly relevant with high confidence ("necessary" variables), and the complement of a highly probable k-supMBS

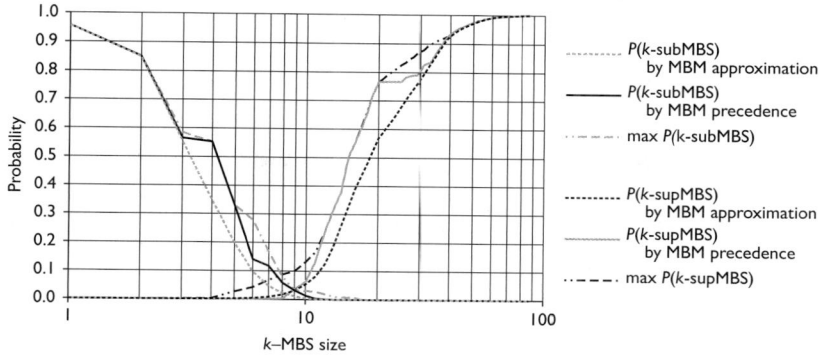

Fig 13.12 Illustration of sub- and sup-relevance posteriors using data set A and target *Asthma*. The sub-relevance p and the sup-relevance $\bar{p}$ of the k-sized sets containing the k variables with the k largest MBM posteriors for $k = 1, \ldots, 100$ are indicated by black and gray solid lines, respectively. The maximal sub-relevance p and the maximal sup-relevance $\bar{p}$ over the k sized sets are also indicated ("max $\mathbb{P}(k$-subMBS)", "max $\mathbb{P}(k$-supMBS)"). For these sets, the MBM-based approximations of their posteriors according to equation (13.8) are also shown, denoted by "$\mathbb{P}(k$-subMBS) by MBM approximation" and "$\mathbb{P}(k$-supMBS) by MBM approximation", respectively.

contains variables that are not strongly relevant with high confidence (i.e., k-supMBS contains a "sufficient" set of variables). The strong relevance of any individual $X_i \notin s$ can be excluded with this probability (although $1 - \mathbb{P}(MBM(Y, X_i, G))$ gives a sharper value as the exact posterior probability of being not strongly relevant).

Note that for a given precedence (ordering) of variables $\prec$, the sup-relevance $\bar{p}$ for $k = 0, 1, \ldots$ increases monotonously from 0 to 1, and p decreases monotonously from 1 to 0 as we progressively extend the empty set to the complete set following that precedence (see Fig. 13.12).

13.3.5 Posteriors for Multiple Target Variables

The definition of strong relevance to multiple targets is a straightforward extension from strong relevance to a single target variable (see Subsection 13.2.5, page 327, Definition 13.7). However, the posterior for a given target set Y cannot be calculated in general from the posteriors corresponding to the members of any partitioning of $Y = \bigcup_i Y_i$, although the posteriors corresponding to subsets of the target set can be used for an approximation. In the case of MBMs, the corresponding approximation is:

$$\mathbb{P}(MBM(X_j, Y)|D_N) \approx 1 - \prod_i (1 - \mathbb{P}(MBM(X_j, Y_i)|D_N)) \tag{13.13}$$

We computed and compared the posteriors of each directly relevant variable X_j to each target variable Y_i in order to decompose the relevance of genetic factors to various phenotypic target variables. These targets were *IgE* (level), *Eosinophil* (level), *Rhinitis*, and *Asthma*, which are believed to participate in a complex causal model with multiple paths (see Fig. 13.3, page 327). Since in this problem domain the target variables form a causal model with strong dependences, we can expect that the relevance of a factor X_j to a given target is not independent from that to

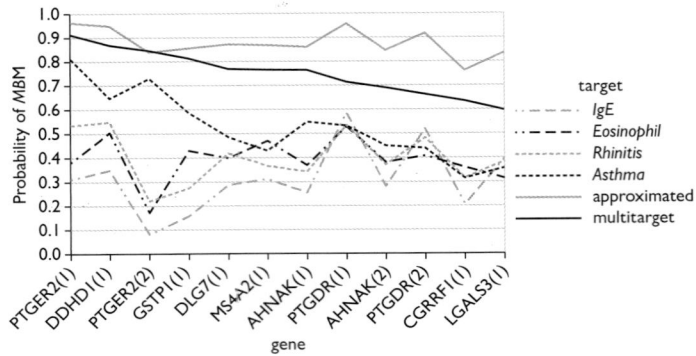

Fig 13.13 Posteriors of the MBM relevance for multiple target variables jointly (multitarget), separately (*IgE*, *Eosinophil*, *Rhinitis*, and *Asthma*) and with respect to the single-target approximation (approximation).

other targets. Indeed, the multitarget approach is motivated by the finding that equation (13.13) provides a rather poor approximation (Fig. 13.13). Note that the value of posteriors for multiple targets is higher than that for single targets due to the possibility of relevance to any of the targets, and so a quantitative comparison correction is necessary.

Using the rs17831682 SNP in the PTGDR gene as an example (see PTGDR(1) in Fig. 13.3), we demonstrate the main advantages of the multitarget approach, namely that it allows us to distinguish between subtypes of multitarget relevance, which were summarized previously in Table 13.2 (page 328). When multitarget relevance is ignored, the posterior of strong relevance (i.e., the MBM posterior) of rs17831682 to *IgE* (level), *Eosinophil* (level), *Rhinitis*, and *Asthma* is 0.58, 0.52, 0.53, and 0.53, respectively, which indicates a somewhat moderate relevance to each target. The posterior for being strongly relevant to at least one of them (*MBMToAny* relation in Table 13.2) is higher: 0.71 (the approximation according to equation (13.13) is 0.95). However, the posterior probability that rs17831682 is strongly relevant exclusively to *IgE*, *Eosinophil*, *Rhinitis*, or *Asthma* (*MBMToExactlyOne* relation) is only 0.06, 0.04, 0.05, and 0.05, respectively, which shows that this SNP is probably related to multiple targets. This hypothesis is also supported by the posterior that this SNP is strongly relevant to other targets but not to *IgE*, *Eosinophil* (level), *Rhinitis*, or *Asthma* (*MBMWithOthers* relation): 0.37, 0.42, 0.42, and 0.42, respectively. Finally, the posterior for rs17831682 being a relevant SNP for multiple phenotypic targets (*MultipleMBMs* relation) is high (0.51), suggesting that this SNP is strongly relevant to the set of targets and that the SNP plays roles in multiple mechanisms.

13.3.6 Subtypes of Strong and Weak Relevance

The distinction between different types of relevance is essential in revealing the possible causal and mechanistic path(s) that connect the relevant SNP to its target variable(s). Estimating the posterior probability of various relevance types enables us to decide whether a SNP is directly relevant or its association is mediated by other factors or both. We demonstrate the dissection of relevance types in a BN-BMLA analysis conducted on data set *RA*, which contains two phenotypic variables: *Asthma* and *Rhinitis*. Using *Asthma* as the sole target, the following posteriors were

Table 13.4 Posteriors for various types of relevance based on data set *RA* using *Asthma* as the only target. Each line contains the posteriors for a SNP identified by the first column, e.g., PTGDR(2) refers to the second SNP measured within the PTGDR gene. Subsequent columns correspond to relevance types in the following order: association (A), direct causal relevance (DCR), strong relevance (SR), interactionist relevance (IR), indirect causal relevance (ICR).

SNP	A	DCR	SR	IR	ICR
AHNAK(2)	0.643	0.029	0.736	0.708	0.535
TXNDC16(1)	0.309	0.008	0.722	0.713	0.189
PRPF19(1)	0.878	0.718	0.843	0.125	0.822
PTGDR(2)	0.923	0.326	0.362	0.035	0.747
PTGER2(2)	0.970	0.350	0.354	0.004	0.604
WDHD1(1)	0.964	0.115	0.242	0.127	0.202

estimated for each SNP: direct causal relevance (DCR), association (A), strong relevance (SR), interactionist relevance (IR), and indirect causal relevance (ICR). The corresponding posteriors for a number of SNPs are shown in Table 13.4.

In our current example, the SNPs can be clustered into four groups as shown in Fig. 13.14 (page 342). Note that association, direct relevance, transitive relevance, and interactionist relevance are complex, potentially overlapping events (see Fig. 13.2, page 325). The SNPs AHNAK(2) and TXNDC16(1) both have a moderately high posterior for strong relevance (0.736 and 0.722) but a very low direct causal relevance posterior (0.029 and 0.08). This means that the strong relevance of these SNPs to *Asthma* is not due to direct causal relationship but to a purely interactionist relevance with *Rhinitis*. Furthermore, the fact that the posterior for a transitive relationship with *Asthma* is relatively low (the posterior is 0.535 and 0.189 for AHNAK(2) and TXNDC16(1), respectively) indicates that interactionist relevance (posterior: 0.708 and 0.713) is the only relevance subtype underlying the association of these SNPs to *Asthma*. This means that these SNPs are relevant and associated only if the *Rhinitis* state is known.

In contrast, PRPF19(1) is associated with *Asthma* (0.822) not only transitively but also through a direct causal relationship (0.718) suggesting two distinct causal paths connecting PRPF19(1) to *Asthma*: one of the paths can be blocked by other factors but the other one cannot. Its posterior for interactionist relevance is low (0.125), suggesting the lack of a third causal path.

In the third group of SNPs, PTGDR(2) and PTGER2(2) have a very high probability of association with *Asthma* (0.923 and 0.970 respectively), which is induced by a transitive relation indicated by moderately high TCR posteriors (0.747 and 0.604). Note that all other posteriors are relatively low, indicating that TCR is the only significant relevance type in this case.

WDHD1(1) contrasts with all other SNPs in the previous groups because it has a high probability of association with *Asthma* (0.96) but none of its other posteriors are significant. This is possible in the case of a pure confounding relationship, in which a common cause influences both the SNP and the target (which are independent from each other otherwise). Note that if transitive

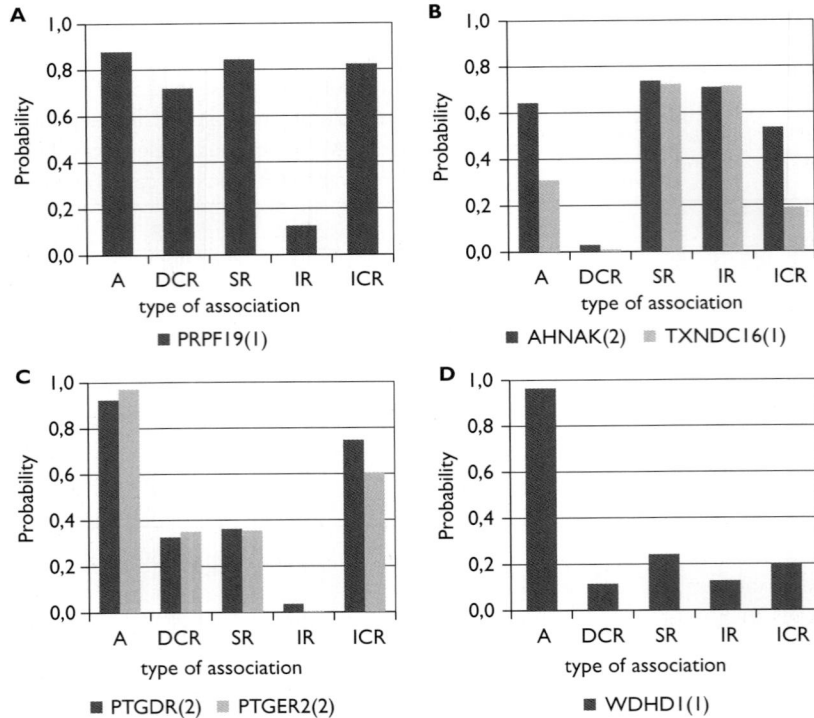

Fig 13.14 Comparison of posteriors for various relevance types of SNPs of table 13.4. The posteriors segregate the SNPs into four groups. A) Direct relevance: PRPF19(1). B) Interactionist relevance: AHNAK(2), TXNDC16(1). C) Transitive relevance: PTGDR(2), PTGER2(2). D) Confounded relevance: WDHD1(1). A, association; DCR, direct causal relevance; ICR, indirect causal relevance; IR, interactionist relevance; SR, strong relevance.

dependence and confounded dependence cannot be differentiated, such as in the case of linked SNPs, the *transitively relevant* and the *confounded groups* can be merged.

For a more detailed biomedical discussion of the application of this methodology in asthma and allergy, we refer the reader to [58].

13.3.7 Interaction-redundancy Scores Based on Posteriors of Strong Relevance

The MBM posterior probabilities $\mathbb{P}(MBM(Y, X_i, G))$ characterize the strong relevance of individual variables $X_i \in V$ (equation (13.2), page 329). However, $\mathbb{P}(MBM(Y, X_i, G))$ inherently reflects multivariate aspects as it is derived from a full Bayesian multivariate analysis through marginalization from the MBG posterior distribution (equation (13.6), page 329). It means that even a variable X_i which is independent of variable Y can have high MBM posterior. The reader is referred to Subsection 13.2.3 (page 323) for a discussion of Definition 13.5 (page 325) regarding the relationships connecting interactionist relevance, contextual relevance, and the relevance of a variable with no or negligible main effect.

However, the exploration of statistical interactions is an additional challenge, because the conceptualizations of statistical interactions are based on the joint, non-linear effect of the interacting variables (for overviews, see e.g., [14, 45]; for methods, see e.g., [35, 40, 49, 65]). In the BN-based Bayesian framework, the joint relevance of the set of variables is quantified by its MBS posterior. Thus, we can define an interaction-redundancy score based on the decomposability of the posterior of the strong relevance of a set of variables [4].

Definition 13.8

For a given data set D_N with sample size N, the features $X' = \{X_{i_1}, \ldots, X_{i_k}\}$ are structurally interacting (redundant), if the posterior $\mathbb{P}(k\text{-subMBS}(X', Y, G)|D_N)$ is larger (less) than the MBM posterior-based product $\prod_{j=1}^{k} \mathbb{P}(MBM(X_{i_j}, Y, G)|D_N)$.

The interaction-redundancy is quantified by the following score

$$IRS(X'; Y) = \log \frac{\mathbb{P}(k\text{-subMBS}(\mathbf{X}', \mathbf{Y}, G)|D_N)}{\prod_{j=1}^{k} \mathbb{P}(MBM(X_{i_j}, Y, G)|D_N)}. \tag{13.14}$$

The interaction-redundancy score is illustrated in Fig. 13.15. This model-level approach to interaction and redundancy formalizes the intuition that relevant input variables with decomposable roles at the parametric level appear independently in the model. If the k-subMBS posterior of set s is larger than its approximation based on MBM posteriors according to equation (13.8) and equation (13.9), it may indicate that the variables in set s have a joint parameterization expressing non-linear joint effects. In contrast, in the case of k-subMBS including redundant variables, the

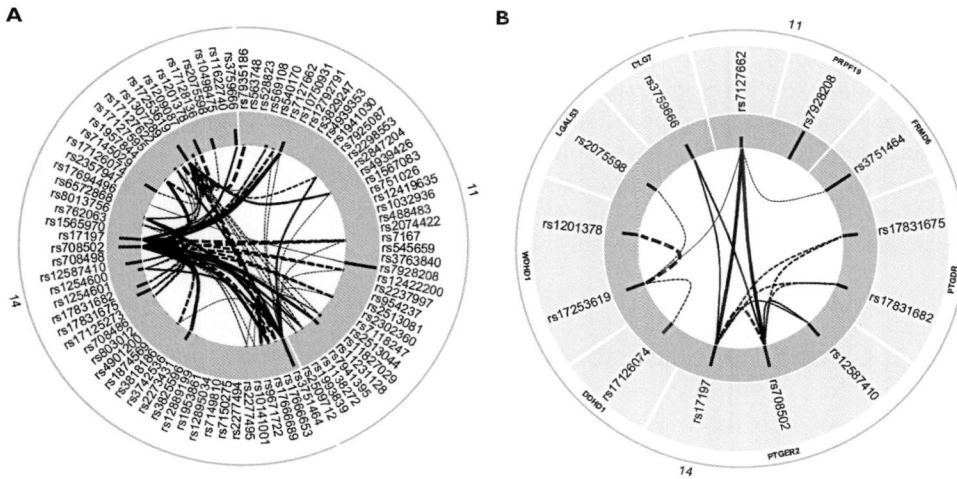

Fig 13.15 Interaction-redundancy scores based on the dependence of the posteriors being strongly relevant for the *Asthma* target using data set A. The thickness of the curved edges connecting two SNPs X_i, X_j shows the strength (absolute value) of the interaction-redundancy score $|IRS(X_i, X_j; Asthma)|$, and dashed and solid lines indicate interaction (positivity) and redundancy (negativity) of $IRS(X_i, X_j; Asthma)$ respectively. The bars on the internal circle show the MBM posteriors. The segments of the two external rings show the genes and chromosomes of the SNPs respectively. A) All SNPs are shown. B) Only a selected subset of SNPs with high MBM posteriors is shown.

posterior is smaller than its approximation based on MBM posteriors, because the joint presence of the redundant variables in the model is suppressed.

Note that the interaction-redundancy scores corresponding to a given target do not appear to be related to the genetic linkage between the SNPs. Fig. 13.15 clearly shows that there are several intragenic, intrachromosomal, and interchromosomal interactions in this domain exemplified by rs17197 and rs708502 in the PTGER2 gene (chromosome 14), rs12587410 in the PTGER2 gene and rs376966 in the DLG7 gene (both genes in chromosome 14), and rs11827029 in AHNAK (chromosome 11) and rs17831675 in the PTGDR gene (chromosome 14).

13.4 Bayes Optimal Decisions about Multivariate Relevance

The relatively high number of predictor variables in GASs poses a serious challenge, because of the multiple hypothesis testing problem: in the univariate approach, the number of hypotheses is linear in the number of variables. Additionally, the number of hypotheses can be exponential in the multivariate approach using a complex model class. Several approaches have emerged within the frequentist framework to handle the multiple hypothesis testing problem in both the univariate and the multivariate contexts. These approaches include correction methods, permutation test-based approaches, and involve concepts such as false discovery rate (FDR) and q-value[23] [54].

Because of its direct semantics, the Bayesian multivariate approach has a built-in automated correction for the multiple hypothesis testing problem: the posterior typically is more flat with increasing number of variables and with increasing model complexity, i.e., in a more complex hypothesis space.

Furthermore, the Bayesian decision theoretic framework permits optimal decisions about model properties, such as the optimal scientific reporting of results or the optimal continuation of the study (for Bayesian study design using BNs, e.g., see [2, 62]). First, we summarize the optimal decision problem of the relevance of variables based on their univariate posteriors and utilities. Second, we show the application of the Bayesian approach to construct a Bayesian FDR. Third, we consider the use of general informative-loss functions.

13.4.1 Optimal Decision about Univariate Relevance

As we discussed in Subsection 13.3.4 (page 336), the univariate view of relevance using the MBM posteriors allows an overview, but it provides only a raw approximation for the MBS posteriors (see Fig. 13.10, page 337). We can construct a 2×2 cost matrix for the errors of a positive/negative decision about univariate strong relevance, when the real status of relevance is positive/negative: $C_{1|0}$ is the cost of the false positive decision, $C_{0|1}$ is the cost of the false negative decision, and for

---

[23] For the definition of the q-value, see Appendix 13.A.

simplicity we assume that the other costs are $C_{0|0} = C_{1|1} = 0$. The positive decision about the strong relevance of a given variable X_i with respect to the target variable Y is optimal if

$$\tau = \frac{1}{1 + \frac{C_{1|0}}{C_{0|1}}} \leq \mathbb{P}(MBM(Y, X_i, G)|D_N), \tag{13.15}$$

where $\tau \in [0, 1]$ is the optimal decision threshold. For example, if the cost $C_{1|0}$ of doing further measurement on a non-relevant variable (a false positive decision) is equal to the cost $C_{0|1}$ of missing a discovery (a false negative decision), then the optimal decision threshold τ is 0.5.

Fig. 13.16 (page 346) shows the ROC curves of the MBM-based decision of relevance using the reference model M_0 from Subsection 13.3.3. The ROC curves show the sensitivity and specificity values for various sample sizes, considering reference variable sets $s_{20}, s_{40}, s_{60}, s_{80}$, and s_{100} with varying number of variables.

13.4.2 Optimal Bayesian Decision to Control FDR

The measures of classification performance, such as sensitivity, FDR, and AUC are valuable tools, but they need an external reference, a "gold standard," which is typically available in an evaluation context (for a recent comparison of measures, see e.g., [55]). The classical frequentist approach also assumes that there is an unknown reference set, the "true model" beneath our data. However, the Bayesian framework offers a natural solution for the lack of a reference model, based on Bayesian model averaging (BMA).

In the decision theoretic framework, sensitivity, specificity, FDR, negative predictive value (NPV) and other performance measures can be interpreted as special utility or cost functions based on the selected set $\hat{s}$ and the possible set s (for which posterior exists). As an example for a cost function, FDR is the proportion of members in the selected set $\hat{s}$ not present in the possible set s, and sensitivity (i.e., true positive rate, TPR) is the frequency of members in the selected set $\hat{s}$ present in the possible set s:

$$Sensitivity(\hat{s}|s) = |\hat{s} \cap s|/|s|, \tag{13.16}$$
$$Specificity(\hat{s}|s) = |\neg\hat{s} \cap \neg s|/|\neg s|, \tag{13.17}$$
$$FDR(\hat{s}|s) = |\hat{s} \cap \neg s|/|\hat{s}|, \tag{13.18}$$
$$NPV(\hat{s}|s) = |\neg\hat{s} \cap \neg s|/|\neg\hat{s}|. \tag{13.19}$$

In the Bayesian framework, the corresponding expected values (denoted with upper bar) are well defined and can be approximated using MCMC simulation (see Subsection 13.3.1, page 330). For example, the expected value of the FDR corresponding to the decision for selecting $\hat{s}$ is as follows:

$$\overline{FDR}(\hat{s}) = \mathbb{E}_{\mathbb{P}(MBS(Y,G)|D_N)}[FDR(\hat{s}|MBS(Y, G))]. \tag{13.20}$$

This calculation sums over possible scenarios with pairs $(\hat{s}, s)$, where the set $\hat{s}$ is fixed and the possible set s is changing. The probability of a scenario is defined by the posterior $\mathbb{P}(s|D_N)$ of the

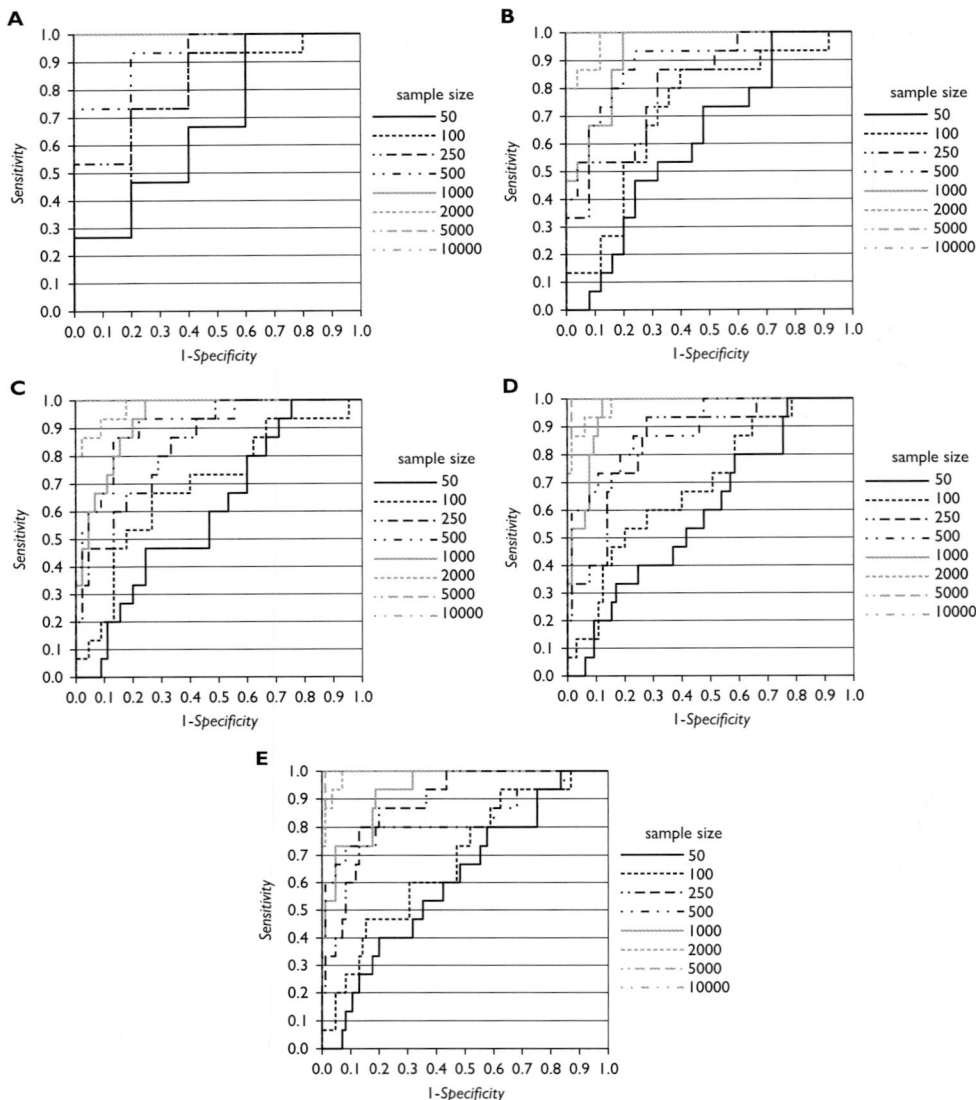

Fig 13.16 The ROC curves for various model sizes used in the evaluation framework from Subsection 13.3.3. The legend corresponds to the number of samples in the data set. Increasing the number of samples provides greatly increased sensitivity. A, B, C, D, and E correspond to models with 20, 40, 60, 80, and 100 variables, respectively.

corresponding possible set. Because the loss is defined by $FDR(\hat{s}|s)$, equation (13.20) defines the expected loss for a selected $\hat{s}$, allowing a search for an optimal set $\hat{s}*$ minimizing the expected FDR loss.

Fig. 13.17 shows the expected values of these measures using data set A and the *Asthma* target.

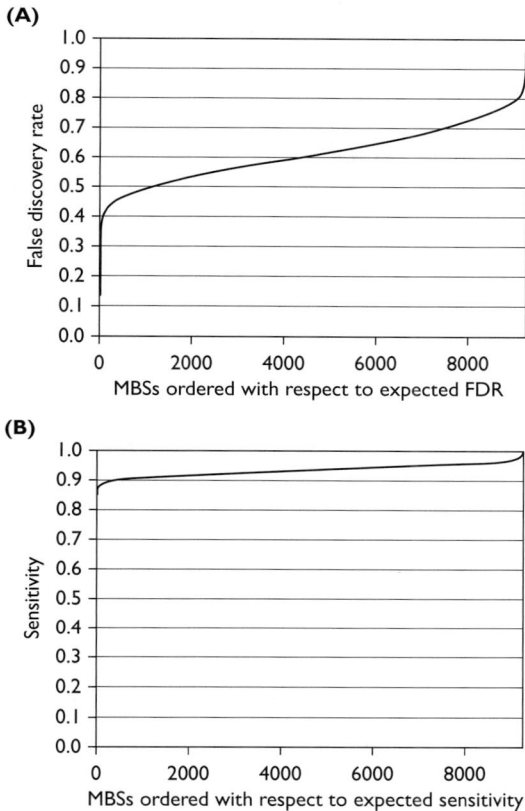

(A)

(B)

Fig 13.17 Expected value of FDR and *Sensitivity* for the 10 000 most probable MBS sets, using data set A and *Asthma* target. A) Expected value of FDR. B) Expected value of *Sensitivity*.

The availability of the FDR value for any set allows the definition of the Bayesian discovery threshold called the *d*-value, a Bayesian analogue of the *q*-value[24] [54] from the frequentist approach [54].

Definition 13.9
The d-value for a given variable X_i is the minimal expected FDR value corresponding to the multivariate selection of a relevant set s including X_i: d-value$(X_i) = \min_{s:X_i \in s} \overline{FDR}(s)$.

Note that the expected FDR value of a set $\hat{s}$ containing only one variable X_i is $1 - \mathbb{P}(MBM(Y, X_i))$, which is minimal for the variable with the maximal MBM posterior value ($X^{MBM} = \arg\max_{X_i} \mathbb{P}(MBM(Y, X_i)|D_N)$). However, the MBS posterior of this set $\hat{s} = \{X^{MBM}\}$ is typically very small, and sets with high MBS posteriors typically contain further variables with high MBM posteriors. The identification and construction of the set $\hat{s}$ with minimal FDR is supported by the observation that FDR can be decomposed as follows:

---

24 For the definition of *q*-value, see Appendix 13.A (page 354).

$$\overline{FDR}(\hat{s}) = \mathbb{E}_{\mathbb{P}(S|D_N)}[FDR(\hat{s}|S)] \tag{13.21}$$

$$= \sum_s \mathbb{P}(s|D_N)FDR(\hat{s}|s) \tag{13.22}$$

$$= \sum_s \mathbb{P}(s|D_N)\frac{1}{|\hat{s}|}\sum_{i=1}^{i=n} 1(\hat{s}_i \wedge \neg s_i) \tag{13.23}$$

$$= \frac{1}{|\hat{s}|}\sum_{i=1}^{i=n}\sum_s \mathbb{P}(s|D_N)1(\hat{s}_i \wedge \neg s_i) \tag{13.24}$$

$$= \frac{1}{|\hat{s}|}\sum_{i=1}^{i=n}\mathbb{E}_{\mathbb{P}(S|D_N)}[1(\hat{s}_i \wedge \neg S_i)], \tag{13.25}$$

where $\hat{s}_i$, s_i denote the presence of the predictor X_i in the set (as always, the capital S_i denotes the corresponding random variable).

In fact, the expected value of any linear performance measure $U(\hat{s}|s)$ can be computed efficiently, because the following proposition holds:

Proposition 13.1
Given the linear utility function $U(\hat{s}|s) = \sum_i \lambda_i U_i(\hat{s}_i|s_i)$, the expected value of a selected set $\hat{s}$ is $EU(\hat{s}) = \sum_i \lambda_i EU(\hat{s}_i)$, where $\hat{s}_i$ and s_i denote $X_i \in \hat{s}$ and $X_i \in s$ respectively, and $\lambda_i \in \mathbb{R}$.

Consequently, as the following theorem shows, the *d*-value(X_i) can be defined at the univariate level of the predictors using only their MBM posteriors $MBM(Y, X_i)$. Furthermore, the optimal set contains the variables with maximal univariate relevance.

Theorem 13.3
Because of the linear property of the FDR and proposition 13.1, the d-value(X_i) is $\frac{1}{|s|}\sum_{X_j:X_j \in s} 1 - \mathbb{P}(MBM(Y, X_j))$, where the optimal set s contains X_i and all X_j with higher posteriors (i.e., $s = \{X_i\}\bigcup\{X_j : \mathbb{P}(MBM(Y, X_i)) < \mathbb{P}(MBM(Y, X_j))\}$).

Fig. 13.18 shows the expected values of FDR for sets $s^*_{MBM,k}$ with increasing size containing k variables with highest MBM posteriors. Because of Theorem 13.3, these are the *d*-values. Note that these sets are typically different from the most probable MBSs (sets s with high MBS posterior, because of the difference between the MBS posterior ($\mathbb{P}(MBS(Y, G) = s|D_N)$) and its MBM posterior-based approximation in equation 13.8 (see Fig. 13.10, page 337). This is because of the potential dependences of the univariate relevances $MBM(X_i)$ and the estimation errors of the finite MCMC sampling process.

13.4.3 General Bayes Optimal Decision about Multivariate Relevance

The most important factor in a general utility/loss function $U(\hat{s}|s)$ designed to select the set of variables $\hat{s}$ is the penalty for the difference between sets $\hat{s}$ and s using various loss functions, e.g., the "0–1" (L_0), linear (L_1), or quadratic (L_2) losses. The effects of these loss functions can be illustrated in the continuous domain, where the optimal values corresponding to these losses are

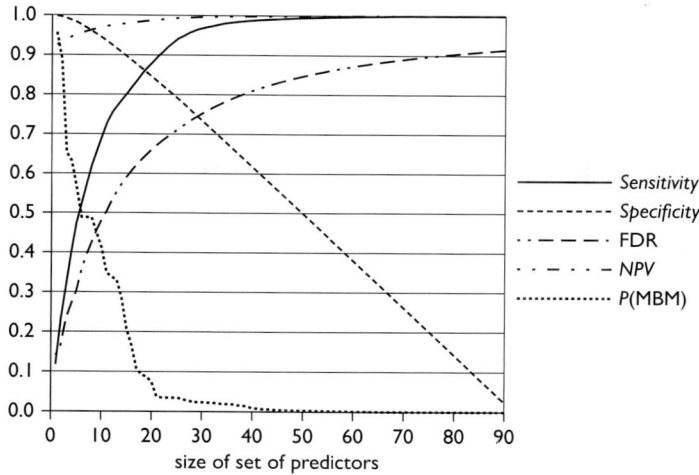

Fig 13.18 Expected values of *Sensitivity*, *Specificity*, FDR, and *NPV* corresponding to the sets $s^*_{MBM,\,k}$ with minimal FDR for various set sizes. The figure also shows the MBM posteriors ($\mathbb{P}(MBM)$) of the predictors. The FDR curve can also be interpreted as *d*-values for the predictors (see Theorem 13.3).

the maximum, median, and average values, respectively. For example a quadratic loss function with uniform weights and its corresponding expected value is defined as

$$U(\hat{s}|s) = |\hat{s} \cap s| + |\hat{s}^c \cap s^c| - (|s \setminus \hat{s}| + |\hat{s} \setminus s|)^2, \tag{13.26}$$

$$\bar{U}(\hat{s}|s) = \mathbb{E}_{\mathbb{P}(S|D_N)}[U(\hat{s}|S)], \tag{13.27}$$

where $()^c$ denotes the complement and the capital S denotes the corresponding random variable. Note that the terms in equation (13.26) are the true positive, true negative, false negative, and false positive counts, which directly correspond to the performance measures shown in Fig. 13.18 (see equation (13.6)). Fig. 13.19 (page 350) illustrates the non-monotonic relation of the MBS posteriors and the expected utilities corresponding to equation (13.26).

The MBS posteriors (y-axis) are plotted against the ranks (x-axis) of the sets, where ordering is defined by their expected $L2$ utilities (see equation (13.26)).

Other factors in a general utility/loss function $U(\hat{s}|s)$ can incorporate and score many aspects to select the set $\hat{s}$. Such aspects are as follows:

1 *Statistical measures:* To characterize the statistical aspects of the selected set, we can combine factors such as FDR and NPV from equation (13.16) and equation (13.20).

2 *Univariate importance:* Depending on the domain, the cost of false positive and false negative errors can be defined separately for each variable X_i, i.e., using a linear utility function from Proposition 13.1 (page 348).

3 *Structural aspects (interactions):* The utility of set $\hat{s}$ can also be defined based on the structural aspects of the reference using either the full network ($U(\hat{s}|G)$) or an mbg subgraph ($U(\hat{s}|mbg)$),

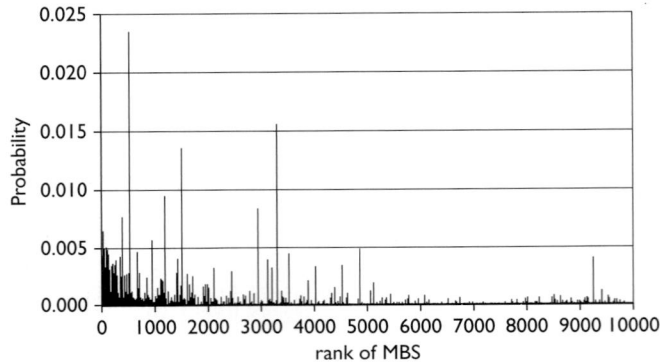

Fig 13.19 MBS posteriors plotted against the ranks of the sets, where the ordering is defined by the expected *L2* utilities. For the definition of the *L2* utility, see equation (13.26).

e.g., to express that set $\acute{s}$ as a set of variables includes factors of interactions represented by the graphical models.

4 *Informative consistency*[25] *and completeness*[26]: The internal consistency and completeness of set $\acute{s}$ can be quantified based on domain-specific knowledge, thus providing a factor $U(\acute{s})$ without referring to the possible set. It can, for example, express that predictors in set $\acute{s}$ correspond to and completely cover one or more biological entities or mechanisms.

13.5 Knowledge Fusion: Relevance of Genes and Annotations

In genetic association studies, a key challenge is the incorporation of a wide range of information to transform and interpret the data at higher abstraction levels, such as at the level of haplotypes, genes, and proteins through "positional" or "functional" integration; at the level of pathways and cascades through "functional" integration; and "temporal" integration of earlier results to approximate an ideal meta-analysis. Whereas the fusion of knowledge at the level of genes, proteins, and pathways is an open research question, the data analysis workflow suggests the following scenarios. Fusion can be done either in the preprocessing phase by creating new synthetic variables to be used in the data analysis, or in the interpretational (postprocessing) phase. The haplotype level analysis is a well-known example for the first approach, because it is regularly done off-line in a preprocessing phase or integrated into the data analysis itself (see e.g., HapScope [64], HAPLOT [27], GEVALT [17], and PHASE [53]). However, we can also construct synthetic variables $f_S(\underline{X})$ at the level of genes, proteins, and pathways, to represent various functional effects analogously to the preprocessing approach with haplotypes. If we use the synthetic variables

---

[25] Consistency expresses the lack of contradiction, e.g., that the set does not contain variables from competing, alternative theories.

[26] Completeness expresses the lack of omission, e.g., that the set contains all the variables corresponding to a given theory.

in the Bayesian relevance analysis, then we can estimate the posterior over their joint strong relevance, and the posterior for all the partial relevances, e.g., univariate relevances $\mathbb{P}(f_S|D_N)$ can be estimated.

In the postprocessing phase, a more common approach is to use synthetic variables $f_S(X)$ only to support interpretation. The mapping of the genetic variants to genes and then the use of Gene Ontology (GO) above the gene level is a practically important case. The most commonly used approach in the frequentist framework is to combine p-values for SNPs into an overall significance level to represent a gene and to combine p-values for the genes into an overall significance level to investigate the association of a pathway with the disease (such combination methods are the Fisher's, Sidak's, or Sime's combination tests [44]). Note that these combination methods are vulnerable to correlations between the entities, and they cannot cope with complex interactions.

The Bayesian framework and the Bayesian networks offer a principled solution for the multivariate propagation of uncertainties from the level of strong relevance of the predictors to the annotations of the predictors and to the taxonomies over the predictors, i.e., to induce a posterior over related terms from the MBS posterior allowing uncertain relations in the annotations and taxonomies as well.

Let us assume a taxonomic tree $\mathcal{T}$, where each leaf node corresponds to a given indicator variable representing the strong relevance of a given variable in the analysis, the internal nodes correspond to indicator variables of terms in the taxonomy, and the edges represent the taxonomical relations directed from specific to general forming a polytree, e.g., SNPs to genes. For each term $A_{j,\mathcal{T}}$ in the set of terms $\mathcal{A}_\mathcal{T}$ present in the taxonomic tree $\mathcal{T}$, the tree defines a corresponding indicator function over the predictors $X_i \in X$ $A_{j,\mathcal{T}}(X_i) : X \rightarrow \{0,1\}$ (if context allows, we omit the index $\mathcal{T}$ and use the term and its indicator function interchangeably):

$$A_{j,\mathcal{T}}(X_i) = 1(\text{there is a directed path from } X_i \text{ to concept } A_j \text{ in } \mathcal{T}). \qquad (13.28)$$

Based on these relations, a multivariate "semantic strong relevance" relation can also be defined representing the semantically related annotations $s^{\mathcal{A}}$ for a given set of predictors s, $SSR_\mathcal{T}(s)$: $2^X \rightarrow 2^{\mathcal{A}}$ as follows

$$(SSR_\mathcal{T}(s) = s^{\mathcal{A}}), \text{ if } ((X_i \in s) \wedge A_{j,\mathcal{T}}(X_i)) \Leftrightarrow (A_{j,\mathcal{T}} \in s^{\mathcal{A}}\mathcal{T}). \qquad (13.29)$$

We will generalize this logical taxonomy to a BN, but the current definition in equation (13.29) simply states that if the predictor X_i in set s is connected to the term A_j through a directed path, then A_j belongs to the set of "semantically related annotations" associated with s.

In the Bayesian statistical framework, the multivariate MBS posterior over sets of predictors $(\mathbb{P}(MBS(Y,G) = s|D_N))$ induces a multivariate posterior distribution over the sets of annotations $s^{\mathcal{A}}$ as follows:

$$\mathbb{P}(s^{\mathcal{A}}|D_N, \mathcal{T}) = \sum_{s:SSR_\mathcal{T}(s)=s^{\mathcal{A}}} \mathbb{P}(s|D_N), \qquad (13.30)$$

where the posterior of a given annotation/term set $s^{\mathcal{A}}$ is the sum of the posterior of the sets of predictors s for which $SSR_\mathcal{T}(s) = s^{\mathcal{A}}$.

The posterior probability of an annotation A_j is as follows:

$$\mathbb{P}(A_j|D_N, \mathcal{T}) = \sum_{G:(X_i \in bd(Y,G)) \wedge (A_j(X_i))} \mathbb{P}(G|D_N). \qquad (13.31)$$

The interpretation of equation (13.31) can be illustrated by the following example. Let us assume that $A_j()$ represents the correspondence of SNPs to a given gene g_j. Then equation (13.31) follows the intuition that the posterior relevance of the gene g_j is the posterior probability that at least one SNP in this gene is strongly relevant. The multivariate transformation defined in equation (13.31) is typically different from various ad hoc averaging and maximization-based aggregations of the MBM posteriors corresponding to the leaves of the annotation term. This is the consequence of the fact that the multivariate approach exactly takes into account the dependences between the strong relevances of the predictors corresponding to the leaves.

The structure of the taxonomy and prior domain knowledge can also be used to refine the semantic relevance relations. We can interpret the taxonomy $\mathcal{T}$ as a special BN, where the local parametric models are logical OR relations. In this model, the posterior for the multivariate semantic strong relevance relation can be interpreted as the result of an inference process with hard evidences at the leaves, which correspond to the indicator variables representing the strong relevance of analyzed variables. However, this BN representation, which mixes predictors (e.g., SNPs) and terms, allows the incorporation of more background knowledge, e.g., using Noisy-OR local parametric models, in which the effect of the true state of a given input is inhibited with a given "inhibition probability" [41]. The parameters in the Noisy-OR models can represent the in- and out-degrees in the taxonomy, e.g., if a given term is annotated by many genes, and therefore its in-degree is relatively high, then the parameters are set to smaller values to adequately model the generality of the term. Similarly, if a given gene is annotated by many terms, and therefore its out-degree is relatively high, then the parameters can be set to smaller values to model the higher frequency of the gene.

In Fig. 13.20, we demonstrate the results of the aggregation from the level of SNPs to the level of Gene Ontology biological process terms. The posterior probability of the MBSs calculated from data set A was aggregated to the level of genes by considering the physical loci and the functional roles of the SNPs. Then, we aggregated these results to the level of GO terms considering the annotations of the genes. The results can be visualized as a network whose nodes are the functional terms, and the connections between the nodes correspond to the hierarchy of the ontology. The sizes of the nodes are proportional to the posterior probability that the given functional term represented by the node has a functional role in the biological phenomena under investigation.

13.6 Conclusion

The Bayesian approach provides a unified framework for study design, integrated exploratory data analysis, optimal decision, and knowledge fusion in genetic association studies. Probabilistic graphical models, and particularly Bayesian networks, allow the decomposition and refinement of the overloaded concept of association. Bayesian networks in the Bayesian framework allow the inference of posteriors over multivariate strong relevance, interactions, global dependence, and causal relations, optionally with various specializations for multiple targets. Furthermore, the Bayesian network-based Bayesian MultiLevel Analysis (BN-BMLA) of relevance in GAS allows

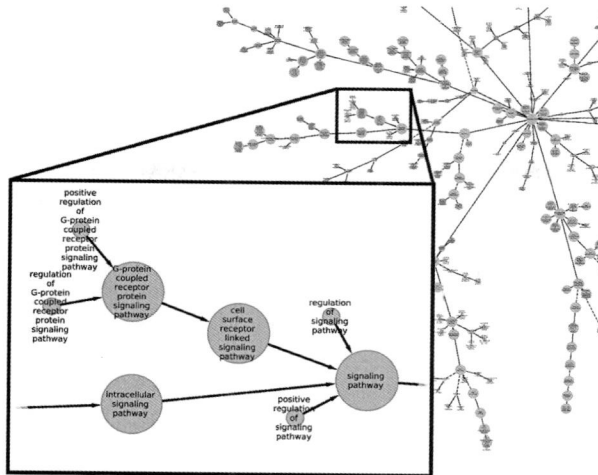

Fig 13.20 Relevant functional terms of the Gene Ontology biological process sub-ontology as calculated from data set A for the target variable *Asthma*. The posterior probability of the strong relevance of the GO terms was aggregated from the level of SNPs. The terms are visualized as a network. The nodes are the terms, and the connections indicate the hierarchy of the ontology. The sizes of the nodes are proportional to the posterior probability of the strong relevance of the term to the target variable *Asthma*.

scalable intermediate levels between univariate strong relevance and full multivariate relevance to interpret the results at partial multivariate levels; additionally, at each level, relevance can be analyzed from a dual point of view of necessity (*k*-subMBS) and sufficiency (*k*-supMBS).

The application of the Bayesian decision theoretic framework for the results of BN-BMLA from the data exploration phase opens up new possibilities to incorporate domain knowledge in supporting interpretation and potentially automating the discovery of interesting relations. The Bayesian framework also allows the principled and computationally efficient management of FDR and other performance measures.

The Bayesian statistical framework also offers a normative solution for the multiple hypothesis testing problem, which is caused by the high number of predictors and particularly by the number of interactions within the frequentist framework. This statement also remains true for the much richer hypothesis space of the new relevance relations defined in the language of BNs, such as the MBMs, *k*-subMBSs/*k*-supMBSs, MBSs, and MBGs. Within the Bayesian framework, the more-or-less flat posterior is the consequence of the high number of variables and high number of models, which is an analogue of the loss of power in the frequentist framework, because of the corrections for the high number of variables and high number of models. However, there is a fundamental difference between the two approaches, which is very valuable in the biomedical application: the Bayesian approach, specifically Bayesian model averaging,[27] provides a normative method for the derivation of posteriors for complex hypotheses, such as *k*-subMBSs/*k*-supMBSs, MBSs, MBGs or the semantic strong relevance. This is especially important in data and knowledge fusion, which is the main bottleneck in current biomedical/translational research.

---

[27] For a brief description of Bayesian model averaging, see Appendix 13.A.

Nonetheless, the posterior estimates for BN features, such as MBMs, k-subMBSs/k-supMBSs, MBSs, and MBGs, still suffer from the multiple hypothesis testing problem, because the MCMC process itself, i.e., their estimation, is done in the frequentist framework. But this problem is related mainly to the efficiency and length of the MCMC simulation, i.e., to the sampled DAGs in D_L^G, and not to the data set D_N. In other words, the Bayesian statistical framework transforms the statistically rooted multiple hypothesis testing problem into a computational task.

Fusion is an already well-recognized central challenge in genetic association studies. With the spread of next-generation sequencing technologies targeting rare variants, the importance of fusion will further increase. Genetic factors have a hierarchical taxonomy, starting from SNPs and moving up to genes then GO terms and pathways. We can expect a similar hierarchical taxonomy to emerge over the phenotypic descriptors as well, such as the Human Phenotype Ontology [46]. Because genetic factors are typically predictors and phenotypic descriptors are typically targets in the BN-BMLA methodology,[28] the methodology can be seen as a support to analyze relevance at multiple granularity and multiple abstraction levels.

The advantage of the direct probabilistic semantics of the Bayesian statistical approach allows a mathematically direct and biomedically interpretable way to combine the results of data analysis with logical prior knowledge (for the aggregation of BN-BMLA results at the SNP level to gene and pathway levels, see [36, 58, 59]). In addition to the propagation of the posteriors to upper levels by aggregation, it also allows the construction of Bayesian data analytic knowledge-bases to support the fusion of weakly significant results from multiple data analyses.

APPENDIX 13.A GLOSSARY

Area under the receiver operating characteristic curve (AUC) The AUC is a widely used measure of the ranking performance of a system with respect to a classification [29]. A plausible interpretation of the AUC can be explained as follows: if we select a random positive sample and a negative sample, the AUC is the probability that the method assigns a higher rank to the positive sample. The AUC value 0.5 corresponds to absolutely random selection; 1 corresponds to the perfect prediction.

Bayesian model averaging (BMA) The posterior distribution over Bayesian network structures given N observations $\mathbb{P}(G|D_N)$ allows averaging in prediction and in exploration of model properties. In the latter case, we can calculate the posterior probability of any structural feature $F(V', G) = f$ by summing the posterior probability of models with property (feature) f (see equation (13.7) in Section 13.3, page 328). In fact, if the structural feature $F(V', G)$ is interpreted as a discrete transformation (mapping) $t(x) : X \rightarrow Y$, then this corresponds to the inducement of a distribution over set Y from the distribution over X, i.e., transformation of the posterior distribution over the model space $\mathbb{P}(G|D_N)$ into a posterior distribution $\mathbb{P}(F|D_N)$ over simpler (less numerous) feature space (see equations (13.2) and (13.7) in Section 13.3, page 328).

---

[28] Note that the BN-BMLA methodology is neutral for the selection of target variables and that the MBM relation is symmetric. Consequently, a genetic variable or small group of genetic variables can also be the target to explore their phenotypic relevance in case of rich phenotypic descriptors.

Causal model A model of a given set of random variables is said to be causal when edges between two nodes are present, if and only if the parent is a direct cause of the child (for in depth treatment see [42, 51]).

Contextual independence Contextual independence is a specialized form of conditional independence, when the conditional independence of two variable sets A and B conditioned on a variable set Z is valid only for a certain value c (i.e., the context) of another disjoint set C.

d-separation d-separation is a graphical test of independence between variables in a Bayesian network structure. Two sets of variables A and B are d-separated by a third set Z if and only if any path between A and B is blocked by a variable in Z. A path is blocked by a variable z if it is part of the path with non-converging edges, or if neither z nor any of its descendants is part of the path with converging edges. In Bayesian networks, d-separation implies conditional independence.

Domain In the context of Bayesian networks, the domain is the segment of the world that a Bayesian network tries to model. A Bayesian network is said to represent a given domain if it models all the necessary knowledge of the domain. Represented/modeled entities of the domain appear as nodes in the Bayesian network.

Feature subset selection (**FSS**) FSS is the task of selecting an optimal subset of features/properties from a larger set according to some external optimality criterion. A good example is to select a subset of tests to perform so that one achieves optimal cost—information gain balance. FSS is often viewed as a search problem in a space of feature subsets. To carry out this search, one must specify a starting point, a strategy to traverse the space of subsets, an evaluation function (induction algorithm), and a stopping criterion.

Filter approach In the feature subset selection problem, the filter approach selects features using a preprocessing step that relies solely on properties of the data without taking into account the induction algorithm itself. This approach may lead to feature selections which in practice will perform poorly when evaluated by the induction algorithm. An example is to use the Pearson product–moment correlation coefficient to filter the predictors.

Independence map A directed acyclic graph G is an independence map of a distribution if the all Markov assumptions (independence relations) implied by G are satisfied by the distribution.

Indicator function An indicator function defined over a set X indicates membership of an element in a subset A of X, assigning the value 1 for all elements of A and the value 0 for all elements of X not in A.

Induction algorithm In the feature subset selection problem, the induction algorithm is used for evaluating a subset of the features.

Loss function In the context of decision theory, a loss function assigns a real number to the possible outcomes, reflecting one's preferences over these outcomes. Contrary to a utility function, greater value of a loss function represents dispreference of the corresponding outcome.

L1, L2 L1 and L2 are standard distances between two vectors. The L^f distance of two vectors x and y is computed as $(\sum_i |x_i - y_i|^p)^{\frac{1}{p}}$; L1 is the Manhattan distance, and L2 is the Euclidean distance.

Markov compatibility A joint probability distribution $\mathbb{P}$ and a DAG G (over the random variables in $\mathbb{P}$) are Markov compatible if each variable is conditionally independent of the set of all its non-descendants, given the set of all its parents.

Modular feature A mapping f from graph structures onto 0, 1 is modular if $f(G) = \prod_{i=1}^{n} f_i(G_i)$, where each f_i is a mapping from the subsets of the complementary set $V \setminus X_i$ onto 0, 1, i.e., f is modular if the graph structures with $f = 1$ can be defined as the free combinations of substructures with $f_i = 1$ for $i = 1, \ldots, n$. For example, the indicator of a directed edge between two nodes is modular, because the graphs with this edge can be defined as two indicators for these two nodes selecting the legal parental sets. Furthermore, the indicator of any subgraph is also modular, because the parental sets covering the subgraph can be defined independently.

Pairwise association Pairwise association is a form of statistical association between two random variables. In genetic association studies, the strength of association is measured between a single nucleotide polymorphism and trait status, based on samples that are informative for the trait of interest.

Observational equivalence class Two directed acyclic graphs (DAGs) are observationally equivalent if they imply the same set of independence relations. Observational equivalence is a transitive relation; the set of DAGs observationally equivalent to each other is called an observational equivalence class.

q-value The q-value is the false discovery rate (FDR) analogue of the p-value. The q-value of an individual hypothesis test is the minimum FDR at which the test may be called significant.

Sample complexity Sample complexity is the minimum number of samples a learning algorithm needs for choosing a hypothesis whose error is smaller than a given threshold (ε) with at least a given $(1 - \delta)$ probability.

Stable distribution A distribution is said to be stable if there exists a directed acyclic graph that exactly represents the dependences and independences entailed by the distribution.

Strong relevance A feature (attribute, variable) X_i is strongly relevant to Y if there exists some $X_i = x_i, Y = y$ and $s_i = x_1, \ldots, x_{i-1}, x_{i+1}, \ldots, x_n$ for which $\mathbb{P}(x_i, s_i) > 0$ such that $\mathbb{P}(y|x_i, s_i) \neq \mathbb{P}(y|s_i)$. Strong relevance of a variable with respect to a given target variable implies that the variable is indispensable, in the sense that sometimes it is essential for the prediction of the target.

Utility function In the context of decision theory, a utility function assigns a real number to the possible outcomes, reflecting one's preferences over these outcomes. In the context of reporting a given set of strongly relevant variables, a utility function declares how good the given set is.

Weak relevance A feature X_i is weakly relevant to Y if it is not strongly relevant, and there exists a subset of features S_i' of S_i for which there exists some x_i, y, and s_i' for which $\mathbb{P}(x_i, s_i') > 0$ such that $\mathbb{P}(y|x_i, s_i') \neq \mathbb{P}(y|s_i')$. Weak relevance of a variable with respect to a given target variable implies that the variable can sometimes improve the prediction accuracy of the target.

Wrapper approach In the feature subset selection problem, the wrapper approach means that the selection algorithm searches for a good subset using the induction algorithm itself (as a black box) as part of the function evaluating feature subsets. The feature subset with the highest evaluation is chosen as the final set on which to run the induction algorithm. Examples include standard likelihood-based forward selection and backward elimination methods in linear regression and logistic regression.

· ·

REFERENCES

[1] C.F. Aliferis, A. Statnikov, I. Tsamardinos, S. Mani, and X. Koutsoukos. Local causal and Markov blanket induction for causal discovery and feature selection for classification. *Journal of Machine Learning Research*, 11:171–284, 2010.

[2] P. Antal, G. Hajós, and P. Sárközy. Bayesian network based analysis in sequential partial genome screening studies. In *MODGRAPH 2009* Probabilistic Graphical Models for Integration of Complex Data and Discovery of Causal Models in Biology, *Satellite Meeting of JOBIM 2009*, 2009.

[3] P. Antal, G. Hullám, A. Gézsi, and A. Millinghoffer. Learning complex Bayesian network features for classification. In *Probabilistic Graphical Models*, pages 9–16, 2006.

[4] P. Antal, A. Millinghoffer, G. Hullám, C. Szalai, and A. Falus. A Bayesian view of challenges in feature selection: feature aggregation, multiple targets, redundancy and interaction. *Journal of Machine Learning Research – Proceedings Track*, 4:74–89, 2008.

[5] D.J. Balding, M. Bishop, and C. Cannings. *Handbook of Statistical Genetics*, 2007.

[6] C. Boutilier, N. Friedman, M. Goldszmidt, and D. Koller. Context-specific independence in Bayesian networks. In E. Horvitz and F. Verner Jensen, editors, *Proceedings of the Twelfth Conference on Uncertainty in Artificial Intelligence (UAI 96)*, pages 115–123. Morgan Kaufmann Publishers, 1996.

[7] W.L. Buntine. Theory refinement of Bayesian networks. In B. D'Ambrosio and P. Smets, editors, *Proceedings of the Seventh Conference on Uncertainty in Artificial Intelligence (UAI 91)*, pages 52–60. Morgan Kaufmann Publishers, 1991.

[8] M. Woodward. *Epidemiology: Study Design and Data Analysis*. Chapman and Hall, 1999.

[9] Shao Q. Ibrahim J.G. Chen, M. *Monte Carlo Methods in Bayesian Computation*. Springer-Verlag, 2000.

[10] G.F. Cooper, C.F. Aliferis, R. Ambrosino, J. Aronis, B. G. Buchanan, R. Caruana, M. J. Fine, C. Glymour, G. Gordon, B. H. Hanusa, J. E. Janosky, C. Meek, T. Mitchell, T. Richardson, and P. Spirtes. An evaluation of machine-learning methods for predicting pneumonia mortality. *Artificial Intelligence in Medicine*, 9:107–138, 1997.

[11] G. Cooper. A simple constraint-based algorithm for efficiently mining observational databases for causal relationships. *Data Mining and Knowledge Discovery*, 2:203–224, 1997.

[12] G.F. Cooper and E. Herskovits. A Bayesian method for the induction of probabilistic networks from data. *Machine Learning*, 9:309–347, 1992.

[13] G.F. Cooper and C. Yoo. Causal discovery from a mixture of experimental and observational data. In K. B. Laskey and H. Prade, editors, *Proceedings of the Fifteenth Conference on Uncertainty in Artificial Intelligence (UAI 99)*, pages 116–125. Morgan Kaufmann Publishers, 1999.

[14] H.J. Cordell. Detecting gene-gene interactions that underlie human diseases. *Nature Reviews Genetics*, 10:392–404, 2009.

[15] H.J. Cordell and D.G. Clayton. Genetic association studies. *Lancet*, 366:1121–1131, 2005.

[16] R. Culverhouse, B.K. Suarez, J. Lin, and T. Reich. A perspective on epistasis: limits of models displaying no main effect. *American Journal of Human Genetics*, 70:461–471, 2002.

[17] O. Davidovich, G. Kimmel, and R. Shamir. Gevalt: an integrated software tool for genotype analysis. *BMC Bioinformatics*, 8:2105–2112, 2007.

[18] Holmes C.C. Mallick B.K. Smith A.F.M. Denison, D.G.T. *Bayesian Methods for Nonlinear Classification and Regression.* Wiley & Sons, 2002.

[19] B.I. Fridley. Bayesian variable and model selection methods for genetic association studies. *Genetic Epidemiology*, 33:27–37, 2009.

[20] N. Friedman, M. Goldszmidt, and A. Wyner. On the application of the bootstrap for computing confidence measures on features of induced Bayesian networks. In *7th International Workshop on Artificial Intelligence and Statistics*, pages 197–202, 1999.

[21] N. Friedman and D. Koller. Being Bayesian about network structure. In C. Boutilier and M. Goldszmidt, editors, *Proceedings of the Sixteenth Conference on Uncertainty in Artificial Intelligence(UAI-2000)*, pages 201–211. Morgan Kaufmann Publishers, 2000.

[22] N. Friedman and D. Koller. Being Bayesian about network structure. *Machine Learning*, (50):95–125, 2003.

[23] N. Friedman and Z. Yakhini. On the sample complexity of learning Bayesian networks. In E. Horvitz and F. Verner Jensen, editors, *Proceedings of the Twelfth Conference on Uncertainty in Artificial Intelligence (UAI 96)*, pages 274–282. Morgan Kaufmann Publishers, 1996.

[24] A.Gelman, J.B. Carlin, H.S. Stern, and D.B. Rubin. *Bayesian Data Analysis.* Chapman & Hall, 1995.

[25] P. Giudici and R. Castelo. Improving Markov chain Monte Carlo model search for data mining. *Machine Learning*, (50):127–158, 2003.

[26] Cooper G.F. Glymour, C. *Computation, Causation, and Discovery.* AAAI Press, 1999.

[27] S. Gu, A.J. Pakstis, and K.K. Kidd. Haplot: a graphical comparison of haplotype blocks, tagSNP sets and SNP variation for multiple populations. *Bioinformatics*, 21(20):3938–3939, 2005.

[28] B. Han, M. Park, and X. Chen. A Markov blanket-based method for detecting causal SNPs in GWAS. *BMC Bioinformatics*, 11:5, 2010.

[29] D.J. Hand. *Construction and Assessment of Classification Rules.* Wiley & Sons, 1997.

[30] D. Heckerman, D. Geiger, and D. Chickering. Learning Bayesian networks: the combination of knowledge and statistical data. *Machine Learning*, (20):197–243, 1995.

[31] X. Jiang, M.M. Barmada, and S. Visweswaran. Identifying genetic interaction in genome-wide data using Bayesian networks. *Genetic Epidemiology*, (34):575–581, 2010.

[32] R. Kohavi and G.H. John. Wrappers for feature subset selection. *Artificial Intelligence*, 97:273–324, 1997.

[33] M. Koivisto and K. Sood. Exact Bayesian structure discovery in Bayesian networks. *Journal of Machine Learning Research*, (5):549–573, 2004.

[34] D. Koller and M. Sahami. Toward optimal feature selection. In *Thirteenth International Conference on Machine Learning*, pages 284–292. Morgan Kaufmann Publishers, 1996.

[35] C. Kooperberg and I. Ruczinski. Identifying interacting SNPs using Monte Carlo logic regression. *Genetic Epidemiology*, 28:157–170, 2005.

[36] O. Lautner-Csorba et al. Candidate gene association study in pediatric acute lymphoblastic leukemia evaluated by Bayesian network based Bayesian multilevel analysis of relevance. *BMC Medical Genomics*, 5:42, 2012.

[37] D. Madigan, S.A. Andersson, M. Perlman, and C.T. Volinsky. Bayesian model averaging and model selection for Markov equivalence classes of acyclic digraphs. *Communications in Statistics: Theory and Methods*, (25):2493–2520, 1996.

[38] C. Meek. Causal inference and causal explanation with background knowledge. In P. Besnard and S. Hanks, editors, *Proceedings of the Eleventh Conference on Uncertainty in Artificial Intelligence (UAI 95)*, pages 403–410. Morgan Kaufmann Publishers, 1995.

[39] J.H. Moore, J.C. Gilbert, C.T. Tsai, F.T. Chiang, T. Holden, N. Barney, and B.C. White. A flexible computational framework for detecting, characterizing, and interpreting statistical patterns of epistasis in genetic studies of human disease susceptibility. *Journal of Theoretical Biology*, 241:252–261, 2006.

[40] M.Y. Park and T. Hastie. Penalized logistic regression for detecting gene interactions. *Biostatistics*, (9):30–50, 2007.

[41] J. Pearl. *Probabilistic Reasoning in Intelligent Systems*. Morgan Kaufmann, 1988.

[42] J. Pearl. *Causality: Models, Reasoning, and Inference*. Cambridge University Press, 2000.

[43] D. Pe'er, A. Regev, G. Elidan, and N. Friedman. Inferring subnetworks from perturbed expression profiles. *Bioinformatics*, (17):215–224, 2001.

[44] G. Peng, L. Luo, H. Siu, Y. Zhu, P. Hu, S. Hong, J. Zhao, X. Zhou, J.D. Reveille, L. Jin, C.I. Amos, and M. Xiong. Gene and pathway-based second-wave analysis of genome-wide association studies. *European Journal of Human Genetics*, 18:111–117, 2010.

[45] P.C. Phillips. Epistasis - the essential role of gene interactions in the structure and evolution of genetic systems. *Nature Reviews Genetics*, 9:855–867, 2008.

[46] P.N. Robinson and S. Mundlos. The human phenotype ontology. *Clinical Genetics*, 77:525–534, 2010.

[47] I. Ruczinski, C. Kooperberg, and M. LeBlanc. Logic regression. *Journal of Computational and Graphical Statistics*, 12:475–511, 2003.

[48] Y. Saeys, I. Inza, and P. Larrañaga. A review of feature selection techniques in bioinformatics. *Bioinformatics*, 23:2507–2517, 2007.

[49] H. Schwender and K. Ickstadt. Identification of SNP interactions using logic regression. *Biostatistics*, (9):187–198, 2008.

[50] T. Silander, P. Kontkanen, and P. Myllymäki. On sensitivity of the MAP Bayesian network structure to the equivalent sample size parameter. In R. Parr and L. C. van der Gaag, editors, *Proceedings of the Twenty-third Conference on Uncertainty in Artificial Intelligence (UAI-07)*, pages 360–367. AUAI Press, 2007.

[51] P. Spirtes, C. Glymour, and R. Scheines. *Causation, Prediction, and Search*. MIT Press, 2001.

[52] M. Stephens and D.J. Balding. Bayesian statistical methods for genetic association studies. *Nature Review Genetics*, 10(10):681–690, 2009.

[53] M. Stephens and P. Donnelly. A comparison of Bayesian methods for haplotype reconstruction from population genotype data. *The American Journal of Human Genetics*, 73(5):1162–1169, 2003.

[54] J.D. Storey and R. Tibshirani. Statistical significance for genomewide studies. *Proceedings of the National Academy of Sciences of the United States of America*, 100(16):9440–9445, 2003.

[55] S.J. Swamidass, C. Azencott, K. Daily, and P. Baldi. A croc stronger than roc: measuring, visualizing and optimizing early retrieval. *Bioinformatics*, 26:1348–1356, 2010.

[56] I. Tsamardinos and C. Aliferis. Towards principled feature selection: relevancy, filters, and wrappers. In *Proceedings of the Ninth International Workshop on Artificial Intelligence and Statistics*. Morgan Kaufmann Publishers, 2003.

[57] M. Ueno. Learning networks determined by the ratio of prior and data. In P. Grünwald and P. Spirtes, editors, *Proceedings of the Twenty-Sixth Conference on Uncertainty in Artificial Intelligence (UAI 10)*, pages 598–605. AUAI Press, 2010.

[58] I. Ungvári, G. Hullám, P. Antal, P.S. Kiszel, A. Gézsi, É. Hadadi, V. Virág, G. Hajós, A. Millinghoffer, A. Nagy, A. Kiss, Á.F. Semsei, G. Temesi, B. Melegh, P. Kisfali, M. Széll, A. Bikov, G. Gálffy, L. Tamási, A. Falus, and C. Szalai. Evaluation of a partial genome screening of two asthma susceptibility regions using Bayesian network based Bayesian multilevel analysis of relevance. *PLOS ONE*, (7):e33573, 2012.

[59] G. Varga, A. Szekely, P. Antal, P. Sárközy, Z. Nemoda, Z. Demetrovics, and M. Sasvari-Szekely. Additive effects of serotonergic and dopaminergic polymorphisms on trait impulsivity. *American Journal of Medical Genetics Part B: Neuropsychiatric Genetics*, 159:281–288, 2012.

[60] R.E. Wyllys. Empirical and theoretical bases of Zipf's law. *Library Trends*, (30):53–64, 1981.

[61] H. Xing, P.D. McDonagh, J. Bienkowska, T. Cashorali, K. Runge, R.E. Miller, D. DeCaprio, B. Church, R. Roubenoff, I.G. Khalil, and J. Carulli. Causal modeling using network ensemble simulations of genetic and gene expression data predicts genes involved in rheumatoid arthritis. *PLOS Computational Biology*, 7(3):e1001105, 2011.

[62] C. Yoo and G. Cooper. An evaluation of a system that recommends microarray experiments to perform to discover gene-regulation pathways. *Artificial Intelligence in Medicine*, (31):169–182, 2004.

[63] L. Yu and H. Liu. Efficient feature selection via analysis of relevance and redundancy. *Journal of Machine Learning Research*, 5:1205–1224, 2004.

[64] J. Zhang, W.L. Rowe, J.P. Struewing, and K.H. Buetow. HapScope: a software system for automated and visual analysis of functionally annotated haplotypes. *Nucleic Acids Research*, 30(23):5213–5221, 2002.

[65] Y. Zhang and J.S. Liu. Bayesian inference of epistatic interactions in case-control studies. *Nature Genetics*, 39:1167–1173, 2007.

PART V

Epigenetics

CHAPTER 14

Bayesian Networks in the Study of Genome-wide DNA Methylation

MEROMIT SINGER AND LIOR PACHTER

This chapter explores the use of Bayesian networks in the study of genome-scale DNA methylation. It begins by describing different experimental methods for the genome-scale annotation of DNA methylation. The methyl-Seq method is detailed, and the biases induced by this technique, which constitute as many challenges for further analysis are depicted. These challenges are addressed introducing a Bayesian network framework for the analysis of methyl-Seq data. This previous model is extended to incorporate more information from the genomic sequence. Genomic structure is used as a prior on methylation status: unmethylated sites tend to cluster, and unmethylated sites are more conserved than other sites. A recurring theme is the interplay between the model used to glean information from the technology, and the view of methylation that drives the model specification. Finally, a study is described in which such models were used, leading to both interesting biological conclusions and to insights about the nature of methylation.

Probabilistic Graphical Models for Genetics, Genomics, and Postgenomics. First Edition. Christine Sinoquet & Raphaël Mourad (Eds). © Oxford University Press 2014. Published in 2014 by Oxford University Press.

14.1 Introduction to Epigenetics

Epigenetic mechanisms influence phenotype through heritable, but potentially reversible, regulation of gene expression. These mechanisms work at many levels including DNA methylation, histone modifications, nucleosome positioning, and replication timing [9]. All of these have been shown to be heritable across cell divisions and to differ across different tissues and cell types [14, 18, 19, 21, 43, 48]. There is a significant body of ongoing research on each of these different mechanisms, and in this section we summarize two of the most studied epigenetic features: DNA methylation and histone modifications.

DNA methylation DNA methylation relates to the covalent attachment of a methyl group to a cytosine (C) nucleotide, in place of the hydrogen atom on the 5 position on the pyrimidine ring, or to the addition of a methyl group to an adenine (A). Adenine methylation has been found only in bacterial genomes [18], and the term "DNA methylation" usually, and throughout this chapter, refers to cytosine methylation. Major events in the cell, such as cell differentiation, X-chromosome inactivation and retrotransposon silencing to name a few, are characterized by DNA methylation, and in mice it has been shown that inhibition of the methylation regulating enzymes results in embryonic lethality [33, 55]. Broadly speaking, DNA methylation is associated with a heterochromatin state and silencing of transcription. In vertebrates, DNA methylation is restricted mostly to CpG sites (Cs that are followed by guanines (Gs)). Most of the vertebrate genome is methylated, and the unmethylated sites tend to cluster together along the genome [4, 55]. Many such unmethylated clusters occur at promoter regions, and their methylation is associated with the silencing of the nearby gene [60].

Histone modification The N-terminal tails of histones are subject to various types of post-translational covalent modifications, including lysine and arginine methylation, lysine acetylation, ubiquitination, and serine phosphorylation [30]. A histone may harbor a number of different modifications at a given time, giving rise to many possible configurations of modifications, sometimes related to as a histone code [52, 57]. Histone modifications can influence the packing assembly of the DNA by moderating a histone's DNA-binding affinity and by recruiting further chromatin remodeling complexes [30]. By doing so, these modifications can affect gene expression; some modifications are associated with transcriptional repression, and others are associated with transcriptional activation. Both types of modification can be present at either promoter or intragenic regions [8].

Epigenetic phenomena can affect gene regulation and be inherited between cell divisions through a mechanism different than that by which they were initiated. This results in an ability to maintain a regulatory program across cell divisions, enabling a form of cell "memory." Epigenetics is therefore at the forefront of cell differentiation studies [10, 27, 39], and numerous epigenetic mutations have been associated with various diseases [46]. Specifically, many studies have reported association between altered methylation states and various cancers [3, 13, 28].

Another active field of research concerns deciphering the affects of environmental factors on epigenetic states, along with the extent to which changes in epigenetic states are stochastic. For example, it has been shown that monozygotic twins accumulate differences in DNA methylation throughout their lives, and that the accumulated differences affect their gene expression portrait [15]. Another interesting study has showed that prenatal tobacco smoke exposure affects DNA methylation in the fetus [7].

Transgenerational inheritance of DNA methylation has been observed in several loci in mice [6, 41, 44]. In plants, changes in DNA methylation that are inherited across generations have been shown to be rather frequent and seem to be a common way of adapting gene regulation to a changing environment [20, 23, 47]. The implications of these phenomena for theories of inheritance and evolution are currently being explored [51].

The genesis of high-throughput sequencing technologies (also referred to as "next-generation" sequencing) has enabled the study of epigenetic features on a genome-wide scale, giving rise to new fields of study. Genome-wide data sets allow for the characterization of epigenetic phenomena across all known genes and allow for broad comparison studies between genes, tissues, individuals, and species [26, 27, 35, 36, 62]. The presence of genome-scale epigenetic data enables association studies to look at both genomic and epigenomic features in the search for causal variants of disease, and the development of methods for epigenome-wide association studies is underway [45].

In this chapter, we explore the use of Bayesian networks in the study of genome-scale DNA methylation. We begin by describing different experimental methods for the genome-scale annotation of DNA methylation in Section 14.2 and focus on the challenges present in the analysis of a specific method, called methyl-Seq. In Section 14.3, we introduce a Bayesian network for the analysis of methyl-Seq data, and in Section 14.4, we extend this model to incorporate more information from the genomic sequence. A recurring theme is the interplay between the model used to glean information from the technology, and the view of methylation that drives the model specification. We describe a study in which such models were used in Section 14.5, and show in Section 14.6 how the models led to both interesting biological conclusions and to insights about the nature of methylation.

14.2 Next-generation Sequencing and DNA Methylation

The transformative impact of high-throughput sequencing on many fields of biology cannot be overstated. The ability to sequence many short DNA fragments at a low cost per base pair (bp), motivated initially by the goal of lowering the costs of generating the DNA sequence of an individual or species, has found unexpected applications in molecular biology. One of the fields in which sequencing technologies have initiated a revolution over the past few years is epigenetics, and specifically DNA methylation. In this section, we describe the different ways in which high-throughput sequencing is used to study DNA methylation, and explain why there is a relatively large number of methods in use today. Throughout this section we focus on the possibilities and challenges of studying DNA methylation through the use of high-throughput sequencing. We will first outline the different techniques that enable measurement of DNA methylation on a genome-wide scale, and then describe the main methods that make use of these techniques, explaining in detail the methyl-Seq protocol, which is the focus of this chapter.

There are several ways to detect DNA methylation, and a detailed description of the different techniques can be found in [31]. The major techniques currently available are illustrated in Fig. 14.1 (page 366) and include:

- *Enzyme digestion:* This technique uses restriction enzymes that in addition to being sequence-specific are methylation sensitive (cut only if their recognition site is unmethylated). A notable example is *HpaII*, which digests at CCGG sites at which the second cytosine is unmethylated

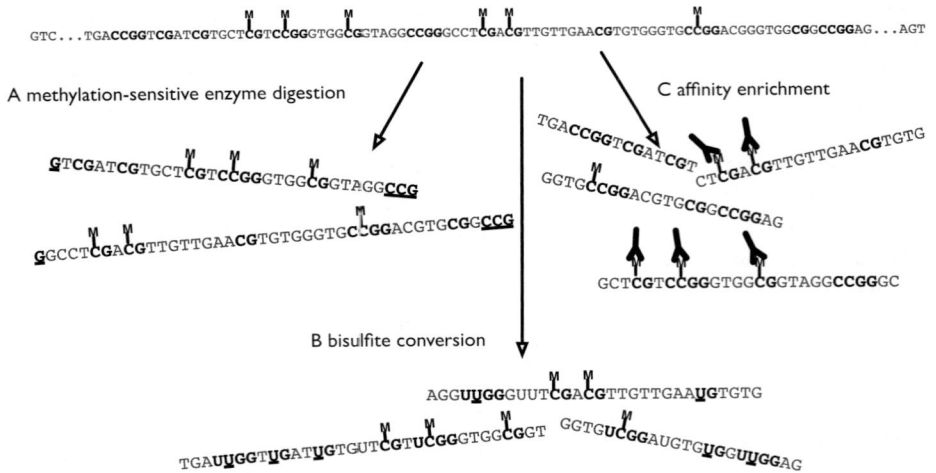

Fig 14.1 Three common techniques for genome-scale annotation of DNA methylation. (A) Enzyme digestion: the genomic DNA is digested with a methylation-sensitive restriction enzyme such as *Hpall*, which digests unmethylated CCGG sites. (B) Bisulfite conversion: converts cytosines that are not methylated to uracil. (C) Affinity enrichment: methylated cytosines in methylated regions are bound by antibodies or methyl-CpG binding proteins. M denotes methylation site. Reproduced in color in the color plate section.

(in this chapter we denote a C that is followed by a guanine (G) as CpG, and, to ease notation, denote a pair of Cs followed by a pair of Gs as CCGG, rather than CpCpGpG). The pattern of digestion by such enzymes can provide a read-out of the DNA methylation. A disadvantage of this technique is the possibility of incomplete digestion and for some enzymes, the lack of site-specific resolution.

- *Sodium bisulfite conversion:* Treatment of DNA molecules with sodium bisulfite converts un-methylated Cs to uracils [24, 59] Followed by a PCR step, the uracils are sequenced as thymines (Ts). This procedure enables the conversion of an epigenetic mark into a modification in the genomic sequence. Disadvantages of this technique include a reduction in sequence complexity (see details below), the need to chemically treat the DNA which can result in degradation, and the possibility of incomplete conversion.

- *Affinity enrichment:* In this process, the genome is digested, and the digest is enriched for methy-lated regions. This is achieved through the recognition of methylated regions by either antibodies [40] or methyl-CpG binding domain proteins [63] and the separation of DNA regions that were bound from those that were nct. Disadvantages of this technique include binding biases of the enzymes and the lack of site-specific resolution.

14.2.1 Assaying Genome-wide DNA Methylation

Methods for genome-wide measurement of DNA methylation generally use one of the afore-mentioned techniques coupled with either high-throughput sequencing or array hybridization. Potentially, methods may incorporate more than one of the techniques, but current approaches do not do so. These methods can be broadly classified as *methyltyping* versus *methylome sequencing*, in analogy with *genotyping* versus *genome sequencing* for DNA. In genotyping, only a small subset of an individual's nucleotides are assayed (SNP locations), while in genome sequencing the whole genome of an individual is sequenced. The advantage of genotyping is in its significantly lower cost compared to whole genome sequencing, allowing for the inclusion of many more individuals in an experiment of a fixed cost. Similarly, methyltyping technologies allow surveying of genome-scale methylation patterns by sampling a subset of CpG sites, and emphasize low cost at the expense of high resolution. In this section, we discuss different methods for methylome sequencing and methyltyping, along with the advantages and disadvantages associated with each method.

The different methods used for high-throughput genomics, including those for measuring DNA methylation, reduce to the ability of counting DNA fragments in a digest, obtained using either arrays or sequencing. Arrays are considered less accurate than sequencing for quantification purposes, mainly due to biases introduced in arrays by the variability in hybridization. Each of the techniques described can be followed by array hybridization, but affinity enrichment is by far the technique that is best suited for arrays [31]. The advantage of using arrays is the low cost, but disadvantages include the biases introduced in the hybridization step, as well as genome-scale array-based methods not being site-specific (there are array-based methods that use either enzyme digestion or bisulfite conversion and are site-specific, but since they sample a considerably smaller number of CpGs, we do not consider them as genome-scale methyltyping methods).

Advances in high-throughput sequencing in the past several years have brought about a large increase in the number of available methods for mapping DNA methylation on a genome-wide scale. Whole-genome bisulfite sequencing (WGBS) involves random digestion of the genome followed by bisulfite treatment, amplification, and sequencing, offering the ability to measure absolute levels of DNA methylation at a single-nucleotide resolution. While this procedure has been used in different studies [11, 34, 35, 62], its use is limited as it is the most expensive method for DNA methylation annotation, because it requires the sequencing of whole genomes. Moreover, sequencing a whole methylome is considerably more expensive than sequencing a whole genome because of the continuous nature of the methylation phenomenon (we would like to determine the proportion of cells in the digest in which a cytosine was methylated), requiring significantly higher coverage than genome sequencing. Therefore, this method cannot, in the foreseeable future, be used for large-scale comparison studies such as population studies and epigenetic association studies. Other disadvantages of the method that are not present in whole-genome sequencing include biases introduced in the polymerase chain reaction (PCR) amplification step and a reduction in sequence complexity. The PCR amplification step introduces biases that are due to unmethylated instances introducing AT-rich sequences. The sequence complexity is reduced because every T sequenced in a read could be mapped to either a T or a C in the reference genome, reducing the number of locations producing uniquely mapping reads. A second method for annotating methylation across the whole genome is that of affinity enrichment followed by high-throughput sequencing. This method has been used in [37], but requires extensive sequencing, is not site-specific, and is prone to binding biases.

The use of high-throughput sequencing has enabled several methods for methyltyping, two of which are discussed here: methyl-Seq and Reduced representation bisulfite sequencing (RRBS). In methyl-Seq the genome is digested with the methylation sensitive restriction enzyme *HpaII* that digests unmethylated CpGs that are within CCGG sites. This is followed by size selection of the fragments, a PCR amplification step, and sequencing the fragments' ends. Mapping sequenced reads back to a reference genome reveals unmethylated CpGs. methyl-Seq is a convenient methyltyping strategy because it is cost-effective due to the sequencing of only unmethylated sites (the minority of CpG sites in vertebrates [55]), requires only small amounts of material, and avoids bisulfite conversion. However, although the experiment is relatively simple, interpretation of the sequencing data is not straightforward due to the protocol resulting in a non-random segmentation of the genome which is followed by a size selection step. We describe the methyl-Seq protocol and the bias that is associated with it in greater detail in the next section.

A second methyltyping method, RRBS, is based on digestion with a methylation-insensitive enzyme followed by bisulfite sequencing [38]. The procedure is enzyme digestion followed by size selection of the fragments, bisulfite treatment, PCR amplification, and sequencing of the fragments' ends. The digestion is commonly performed with *MspI* [22, 39], which digests CCGG sites, to enrich for regions of the genome that are rich in CpG sites. RRBS is favorable due to being significantly less expensive to perform than whole-genome bisulfite sequencing, but suffers from the same disadvantages related to methods that use bisulfite conversion: bias introduced by the PCR amplification step and reduced sequence complexity.

When designing an experiment in which DNA methylation is measured on a genome-wide scale, one must consider the trade-off between the cost of the method and the type of coverage it produces. In addition to this, the application and analysis of the different methods are complicated by a number of other issues. The major complications of the different methods based on the technologies they use were discussed above. On top of this, the analysis requirements for different assays vary in difficulty, and the rapid development of the sequencing field calls for frequent reassessment of the comparison between methods. For these reasons, there has been a proliferation of methods whose pros and cons are constantly changing as sequencing technologies change. In [31], it was suggested that methyl-Seq is the method with the most favorable profile of pros and cons, with respect to the measures chosen for comparison (see Table 2 of [31]). We compare here the different methods presented, given the current stage of the field, but the reader should keep in mind that the comparison criteria will change as the field develops and as new techniques are introduced.

Table 14.1 summarizes the main features by which methods are compared. It is divided into methyltyping methods and whole-genome methods. We have omitted a column for a cost comparison due to the changing nature of the field (sequencing costs are rapidly decreasing), but it seems certain that in the foreseeable future, whole-genome methods will require significantly more sequencing than methyl-Seq and RRBS and that RRBS will require more sequencing than methyl-Seq. Methyl-seq retrieves information spanning more of the genome than RRBS, because of a more favorable profile of fragment sizes produced by *HpaII* relative to *MspI* [49].

In the next section, we explain in detail the methyl-Seq experimental procedure and the need for a statistical model dedicated to the analysis of the data produced, motivating the subsequent sections in the chapter.

14.2.2 The methyl-Seq Method

In this section, we describe the methyl-Seq experiment and the experimental bias it introduces. In the methyl-Seq experiment, the genomic DNA is digested with a methylation-sensitive restriction enzyme; in our case, *HpaII*. *HpaII* digests at CCGG sites in which the second C is unmethylated. This step obtains a digest of DNA fragments such that at all of the fragment ends there is an unmethylated *HpaII* site, and all *HpaII* sites within a fragment (not including the ends) are

Table 14.1 A summary of the characteristics of commonly used methods for methyltyping (top three) and whole methylome annotation (bottom two) in humans. For further information on the derivation of the values in the table see [49] and [31].

	Site-specific	Pre-selected Regions	Coverage of Human Genome	Coverage of CpG Islands	# CpGs Sampled	Analysis Challenges
methyl-Seq	yes	no	9.2%	92.9 %	~1.4 M	inference procedure needed
RRBS	yes	no	8.1%	69.8 %	~1.4 M	low complexity + PCR bias
Affinity-based Array	no	yes	pre-selected	pre-selected	-	binding biases
WGBS	yes	no	whole genome	whole genome	~28 M	low complexity + PCR bias
Affinity-based Seq	no	no	whole genome	whole genome	~28 M	binding biases + array biases

Note: RRBS, reduced representation bisulfite sequencing; WGBS, whole-gencme bisulfite sequencing.

methylated (assuming complete digestion). Then, the fragments are size-selected using a run on an agarose gel, where the common length accepted is between 50 and 300 bps. This size selection is important to achieve good sequencing throughput when using Illumina machines. A paired-end library is constructed from the fragments that pass the size-selection step and, after a PCR amplification step, the ends of the fragments are sequenced. The paired sequenced reads are then mapped to a reference genome using an aligner that supports paired-end reads, such as Bowtie [32]. Notice that by constructing a paired-end library, the sequencing experiment returns pairs of sequences, where each pair is the product of sequencing the ends of one fragment. Using this protocol one can conclude which *fragments* where present in the digest.

After the digestion step has completed, all unmethylated *HpaII* sites are present at the ends of digested fragments (see Fig. 14.1), but the size selection step required by the sequencing protocol limits sequencing to fragments of a narrow size range. This results in many cases in which one cannot determine the extent to which a site is methylated based on its read counts alone. Fig. 14.2 (page 370) shows three different types of epialleles and the fragments generated by them that pass the size selection step and are sequenced. When determining the methylation state of the highlighted site, cases 2 and 3 are indistinguishable if one considers only fragments originating at that site: in both cases, there are no fragments sequenced that have the highlighted site at either end. However, in case 2 we see that the sites on both sides of the highlighted site are unmethylated, and therefore if the highlighted site had been methylated, we would have sequenced a fragment 60 bps in length corresponding to the fragment in the middle (as sequenced in case 3). Since such a fragment was not sequenced, we can conclude that the site is unmethylated. In case 3, since the highlighted site was present in the interior of a fragment, we can conclude that it was methylated. While the examples in this figure assume binary methylation states, when relating to a population

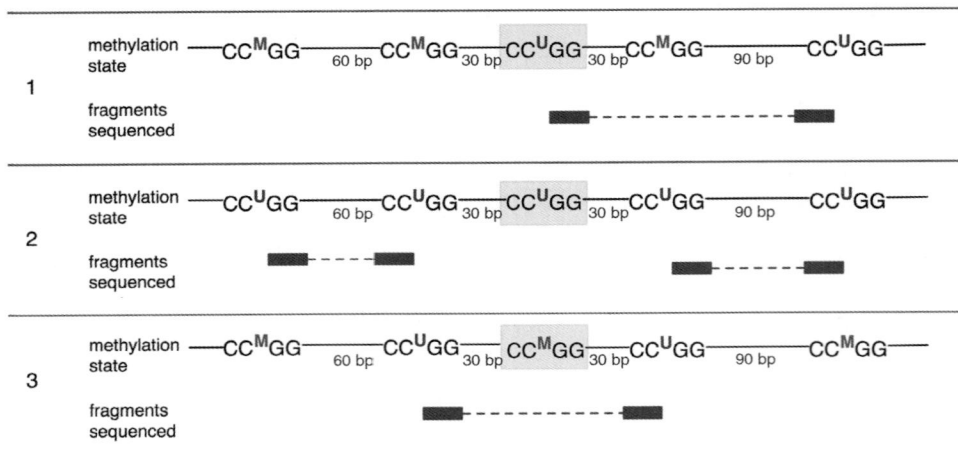

Fig 14.2 The methylation state of a site cannot always be determined from the number of fragments that originated at that site. In many cases, the methylation state of a site cannot be determined from the extent to which it was present at the end of sequenced fragments but can be determined by integrating sequencing data from its neighborhood. bp, base pair, M, methylated; U, unmethylated. Reproduced in color in the color plate section.

of cells, methylation states are assumed to be continuous variables, because the methylation states can be heterogeneous even within a population of a single cell type [1, 64]. It can easily be seen that the examples above can be generalized to the case of continuous variables, where one cannot determine the extent to which a site is methylated given the read counts at that site alone, but can do so when making use of the fragments generated from the site's neighborhood.

14.3 A Bayesian network for methyl-Seq Analysis

We have seen in the previous section that while methyl-Seq is a favorable method for methyl-typing, the bias present in the method needs to be corrected for. When performing a methyl-Seq experiment, we would like to gain knowledge of the extent to which each *HpaII* site is methylated. More precisely, we would like to know for each *HpaII* site the proportion of cells in the digest that were methylated at that site. What we observe, however, is the number of times each fragment was sequenced in the experiment. Using our knowledge of the experimental procedure, we take a generative approach to model the procedure by which the methylation states of the *HpaII* sites impact the number of times each fragment is sequenced. On the basis of this generative model, we can infer the expected methylation states of the *HpaII* sites, given the observed fragment counts.

In the next section, we discuss how the methyl-Seq experiment can be modeled as a Bayesian network, and how that Bayesian network can be used to infer the methylation extents for all *HpaII* sites given the fragment counts. We begin by introducing a generative model for our data that is based on a Bayesian network. We then describe how the model parameters can be learned using the expectation-maximization (EM) algorithm, and how we use this together with the observed fragment counts to infer the methylation states of the *HpaII* sites. In the following section, we discuss how expanding the model by adding an additional set of variables can achieve better site-specific inferences, as well as infer the location of unmethylated clusters.

14.3.1 Notation

In this chapter, we will denote random variables by uppercase letters, and a specific assignment to a random variable by the corresponding lowercase letter. For example, three coin tosses can be modeled by the random variables X_1, X_2, and X_3 for the first, second, and third toss, respectively, and an assignment to X_1 is denoted by x_1. We will denote sets of random variables by bold uppercase letters, and an assignment to the set of random variables by the corresponding bold lowercase letter. For example, we would denote by $\boldsymbol{X}$ the three random variables X_1, X_2, and X_3, and by $\boldsymbol{x}$ an assignment to all three variables.

14.3.2 A Generative Model

In this section, we describe a generative model for the methyl-Seq experiment. Given a reference genome, let F be the set of genomic sequences that have a *HpaII* site at each end. Notice that these regions may also have *HpaII* sites within them. We note that although theoretically F is the set of fragments that may be present in the digest of a methyl-Seq experiment, some fragments have probability of being sequenced which is essentially zero, like fragments spanning whole chromosomes, but we disregard that now to ease notation. For every *HpaII* site, we assign a random variable $\{Y_i\}_{i=1,\dots,N} \in [0,1]$, where N is the number of *HpaII* sites in the reference genome. Y_i represents the proportion of cells from the digest in which site i was methylated.

In our generative model, we assume that at each step a cell is drawn at random, and the methylation state of each *HpaII* site is determined as 1 (methylated) or 0 (unmethylated) using Bernoulli draws with probability y_i (in the case of diploid genomes, this sampling is done twice, once for each of the chromosomes). The methylation assignment to all *HpaII* sites of the chromosome determines what fragments will be generated from this cell. Let $\{Z_i\}_{i=1,\dots,|F|} \in [0,1]$ be random variables for the proportion of instances from which fragment f_i was generated. In other words, let $\{Z_i\}_{i=1,\dots,|F|}$ be random variables for the proportion of times that the methylation configuration of the *HpaII* sites of a chromosome generated fragment f_i. In this generative approach, $z_i \sim \mathcal{N}\left(\mu_i, \frac{\mu_i(1-\mu_i)}{wq}\right)$ where $\mu_i = (1 - y_s)(1 - y_e)\Pi y_m$ and y_s and y_e are the sites on the upstream and downstream boundaries of fragment f_i, y_m are the sites between y_s and y_e, q is the number of cells in the digest, and w is the number of chromosome copies in each cell (for human $w = 2$).[1]

Let $\{X_i\}_{i=1,\dots,|F|}$ be random variables for the number of times fragment f_i was sequenced (the number of paired-end reads that were sequenced from fragment f_i). In our generative model, X_i follows a Poisson process in which the expected number of reads to be sequenced is $\lambda_i = z_i\theta_{l_i}$, where θ_{l_i} is a factor determined by the length of the fragment, and depends on the specific experiment technicalities, such as the total amount of sequencing in the experiment. We condition X_i on the length of f_i since, as described in the previous section, the size selection step in the experiment is not precise, and moreover, fragments of different lengths have different probabilities of being

---

[1] The number of occurrences of fragment f_i in the solution follows a binomial distribution $B(wq, \mu_i)$, for which $\mathcal{N}(wq\mu_i, wq\mu_i(1-\mu_i))$ is a good approximation. Therefore, the proportion of samples from which f_i was generated can be approximated by the distribution $\mathcal{N}\left(\mu_i, \frac{\mu_i(1-\mu_i)}{wq}\right)$.

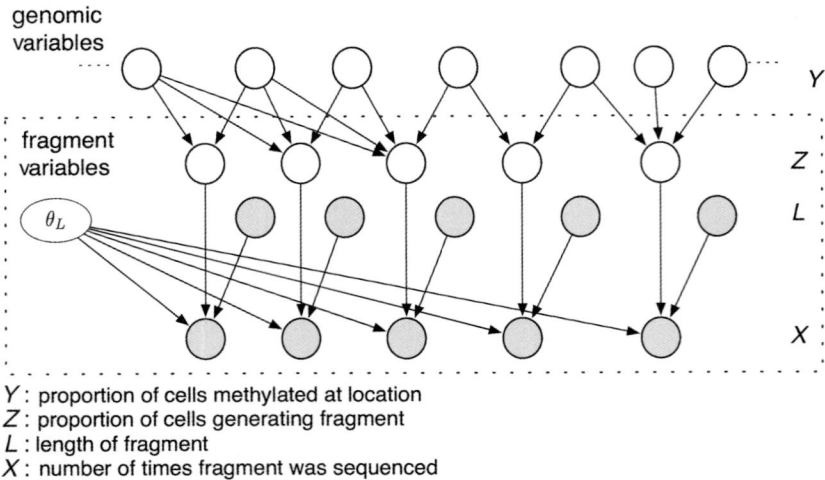

Y : proportion of cells methylated at location
Z : proportion of cells generating fragment
L : length of fragment
X : number of times fragment was sequenced

Fig 14.3 A Bayesian network representation for the generative model. Variables that are observed in the methyl-Seq experiment are filled in gray.

sequenced, due to technicalities of the sequencing machinery. Fig. 14.3 shows the dependences of our generative model in a graphical model representation.

We have made a few simplifying assumptions in this generative model because they greatly simplify the dependences among variables and result in a tractable model that is convenient to work with for inference purposes. The first simplifying assumption we make is that the methylation states of neighboring *HpaII* sites are independent. Second, in contrast to many generative models of RNA sequencing (RNA-seq) [42], in this model we assume X_i, the number of times fragment f_i is sequenced, depends on the proportion of cells from which f_i was generated (Z_i), but not on the relative proportion to which f_i is present in the digest $\left(\frac{z_i}{\sum_{j=1}^{|F|} z_j} \right)$. We allow ourselves to make this assumption in the methyl-Seq case because, in RNA-seq, the contribution of each fragment to the digest is bounded by the number of cells (each cell can contribute at most $w \cdot f_i$ fragments), making the differences between fragment frequencies much smaller than in RNA-seq experiments. This, together with the fact that in methyl-Seq the number of possible fragments (contributors) is much larger than the number of transcripts in RNA-seq experiments, brought us to relax the dependence on the relative proportion of f_i in the solution.

14.3.3 Parameter Learning and Inference of Posterior Probabilities

Now that we have a model in place, recall that our objective is to infer the posterior probability for each Y_i, given the observed X and L variables. We have described a generative model by which we assume the X_i's are generated (and are dependent on the Y_i's). In this section, we will describe how we can use learning and inference algorithms to attain estimates of the Y variables given the experimental results. For this purpose we will make a few simplifying changes to our model: we discretize the Y variables such that $\{Y_i\}_{i=1,\dots,N} \in \{0.1, 0.3, 0.5, 0.7, 0.9\}$, and we assume that the Z variables are determined deterministically by the Y variables, such that $z_i = \mu_i$.

To summarize, the list of variables, parameters and dependences in our model is as follows:

Variables:

$Y_1, \dots, Y_N \in \{0.1, 0.3, 0.5, 0.7, 0.9\}$: for each *HpaII* site, the proportion of cells in the digest that are methylated.

$Z_1, \dots, Z_{|F|} \in [0, 1]$: the proportion of cells from which fragment f_i was generated.

$L_1, \dots, L_{|F|} \in \{1, \dots, l_{max}\}$: the length in bps of fragment f_i.

$X_1, \dots, X_{|F|} \in \{1, \dots, x_{max}\}$: the number of times fragment f_i was sequenced.

Parameters:

$$\theta_L = (\theta_{1-20}, \theta_{21-40}, \dots, \theta_{381-400}, \theta_{\geq 400}).$$

Dependence functions:

$z_i = (1 - y_{s_i})(1 - y_{e_i}) \prod_{j=s_i+1}^{e_i-1} y_j$, where s_i and e_i are the indexes of the locations at which f_i begins and ends.

$\mathbb{P}(x_i | z_i, l_i) = \frac{\lambda^{x_i} \exp\{-\lambda\}}{x_i!}$, where $\lambda = z_i f(l_i)$, and $f : \{1, \dots, l_{max}\} \to \theta_L$ is a function that takes in l_i and returns θ_{s-e} such that $s \leq l_i \leq e$.

In practice, fragments that are very short or very long relative to the boundaries of the size selection step have probability zero of being observed, and we can omit the corresponding variables from the model. This significantly reduces the number of variables in the model, and the complexity of parameter learning and inference procedures.

Although we listed $\{Z_i\}_{i=1,\dots,|F|} \in [0, 1]$ for convenience, we note that when the Y variables are discrete, so are the Z variables. The number of variables a specific Z_i is directly dependent on is determined by the number of *HpaII* sites that are on the fragment Z_i represents. Due to the relatively small size-selection bounds (50–300 bps) and the *HpaII* restriction site being of length four (CCGG), the majority of Z variables are directly dependent on a small number of Y variables. This is important from practical reasons, because when conducting the inference step in the EM approach (see below), the Bayesian network is moralized,[2] and the running time of the inference procedure is dependent on the sizes of the cliques introduced (from a computational complexity perspective, the running time is exponential in the size of the largest clique). To avoid our moralized graph from encompassing large cliques we determine a priori a maximal number of parents we allow for any Z_i variable. Variables with more parents than allowed are not incorporated into the model. While this results in ignoring data from fragments that encompass too many *HpaII* sites, it has a crucial impact on lowering the computational complexity of the inference procedure.

The direct output of the experiment is a list of paired sequenced reads. Each pair of reads is mapped to a reference genome to determine which genomic fragment it originated from (see Subsection 14.2.2). In practice, some of the pairs cannot be mapped uniquely to one fragment of the genome. To resolve this, one may add another layer to the model, taking into account the uncertainty regarding which fragment the reads originated from. However a simpler approach is to disregard from the model fragments that can generate pairs of reads that do not uniquely map.

---

[2] In the moralization step, edges are added between all parents of each node, forming cliques, and all edges of the graph are changed to be non-directed. For further information on moralization and its importance for exact inference procedures, see [29].

In the same manner that we have a variable determining the length of each fragment, we can consider additional variables for characteristics that affect the extent of sequencing (such as GC content [2, 12]). However one must take into account that in the current setting, this will result in an exponential growth in the number of parameters.

THE EM APPROACH

We would like to compute for each Y_i its posterior distribution given the observed data. Namely, we are interested in $\mathbb{P}(Y_i|X, L)$ for all Y_i variables. Given some parameter assignment to θ_L, we can compute the posterior probabilities $\mathbb{P}(Y_i|X, L, \theta_L)$ in time that is exponential in the size of the largest clique of the moralized graph, using the junction tree algorithm [29]. This computation however requires a parameter assignment, which we do not know.

On the other hand, if we were to observe a full assignment of all variables in the Bayesian network ($Y, Z, L,$ and X), let such a data set be $\mathcal{D}$, we could learn a set of parameters for our model using the maximum likelihood approach by finding the set of parameters that maximizes the likelihood function relative to $\mathcal{D}$. We give an outline of these computations here, and leave the details for Appendix 14.A (page 381). From the Bayesian network structure we can construct the likelihood function:

$$L(\theta_L : \mathcal{D}) = \left[\Pi_{j=1}^{N} \mathbb{P}(y_j) \right] \cdot \left[\Pi_{i=1}^{|F|} \mathbb{P}(l_i)\mathbb{P}(z_i|y)\mathbb{P}(x_i|z_i, l_i) \right]$$

$$\propto \Pi_{i=1}^{|F|} \frac{(z_i f(l_i))^{x_i}}{x_i!} e^{-z_i f(l_i)},$$

assuming a uniform prior over Y and L and that the probability of $\mathcal{D}$ is not zero (that all z_i assignments are as expected given the y_i assignments). The function $f(l_i)$ is as defined in the "dependence function" annotation. The log-likelihood is therefore:

$$LL(\theta_L : \mathcal{D}) = \sum_{j=1}^{L} \left[\sum_{i=1}^{|F|} (x_i \log \theta_{L(j)} - z_i \theta_{L(j)}) \mathbf{1}\{f(l_i) = \theta_{L(j)}\} \right] + C,$$

where $\theta_{L(j)}$ is the jth element of θ_L and C is the part of the equation with elements that do not include instances from θ_L. We would like to find θ_L which maximizes the likelihood function. Each $\theta_{L(j)}$ can be optimized separately, and by differentiating the log-likelihood, setting to zero and solving for $\theta_{L(j)}$, we get the maximum likelihood parameters:

$$\hat{\theta}_{L(j)} = \frac{\sum x_i \mathbf{1}\{f(l_i) = \theta_{L(j)}\}}{\sum z_i \mathbf{1}\{f(l_i) = \theta_{L(j)}\}}.$$

To conclude, in the presence of a set of parameters we can compute the posterior distribution for the Ys, and in the presence of a fully observed data set, we can compute the maximum likelihood estimator for the parameters. This is a good setting to use the EM algorithm, which returns a parameter assignment for our model, given the observed data. Given an initial assignment of (arbitrary) parameters, the algorithm computes the expected values of the numerator and denominator used for the maximum likelihood parameter estimation. It then updates the parameters of the model using the expected values computed. This algorithm is guaranteed to converge to a stationary point of the log-likelihood function. We start with some (possibly random) parameter assignment, θ_L^1, and iterate between the following two steps:

Expectation step (E-step): In this step the algorithm uses the parameters from the previous maximization step (M-step), θ_L^t, to compute the expected values for the sufficient statistics used in the maximum likelihood parameter estimation. The expected values are computed for sufficient statistics that incorporate "hidden" (unobserved) variables of the model. The expected sufficient statistic that needs to be computed for $\theta_{L(j)}$ is

$$\mathbb{E}_{Z|X,L,\theta_L^t}\left[\sum_i Z_i \mathbf{1}\{f(l_i) = \theta_{L(j)}\}\right] = \sum_i \mathbb{E}_{Z|X,L,\theta_L^t}\left[Z_i \mathbf{1}\{f(l_i) = \theta_{L(j)}\}\right]$$

$$= \sum_i \sum_Z Z_i \mathbb{P}(z_i|X, L, \theta_L) \mathbf{1}\{f(l_i) = \theta_{L(j)}\},$$

and can be computed using the posterior distributions obtained by the junction-tree algorithm on the moralized graph.

Maximization step (M-step): In this step, we use the expected sufficient statistics from the E-step to perform the maximum likelihood estimation:

$$\theta_{L(j)}^{t+1} = \frac{\sum_i x_i \mathbf{1}\{f(l_i) = \theta_{L(j)}\}}{\mathbb{E}_{Z|X,\theta_L^t}\left[\sum_i z_i \mathbf{1}\{f(l_i) = \theta_{L(j)}\}\right]}.$$

The algorithm iterates between the E-step and the M-step until the increase in the likelihood function is smaller than a given threshold. After the algorithm has converged at θ_L^f, we can compute for each Y_i its posterior distribution using a final run of the junction tree algorithm.

14.4 Genomic Structure as a Prior on Methylation Status

In the generative model described in the previous section, we did not assume any prior knowledge about the methylation states of different *HpaII* sites. However, it is well known that one can make use of the genomic sequence to gain some information regarding the probability that a specific *HpaII* site is methylated. This is because in vertebrates, unmethylated sites tend to cluster together and CpG sites that are constitutively unmethylated are more conserved than other CpG sites. This results in the effect that regions that tend to be unmethylated are more CpG rich than regions that tend to be methylated [4]. There are several methods to annotate such CpG-rich regions, based on genomic sequences, in an effort to find regions that have interesting DNA methylation behavior. Such regions are called "CpG islands".

The precise definition of CpG island regions and their methylation states is a challenging task that has been the subject of much research throughout the years [4, 16, 17, 25, 50, 53, 54, 56, 61]. This is not surprising if we recall that DNA methylation states are more dynamic than the DNA sequence. For example, DNA methylation states may be different across the different tissues of

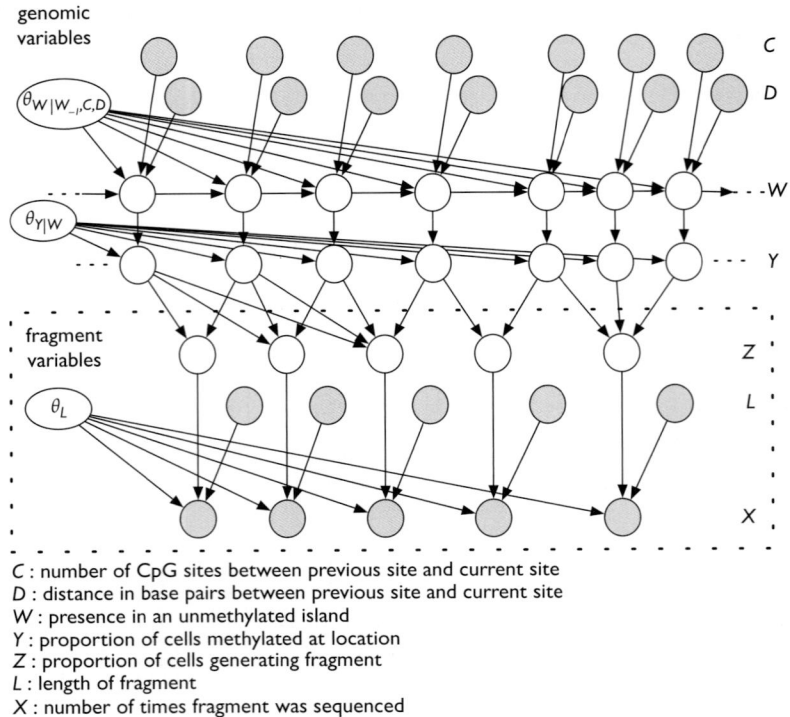

Fig 14.4 A Bayesian network representation for our generative model incorporating genomic structure. Variables that are observed in the methyl-Seq experiment are filled in gray.

C : number of CpG sites between previous site and current site
D : distance in base pairs between previous site and current site
W : presence in an unmethylated island
Y : proportion of cells methylated at location
Z : proportion of cells generating fragment
L : length of fragment
X : number of times fragment was sequenced

an individual [26, 53]. Thus, a purely sequence based definition of a "CpG island" is problematic. Nonetheless, in the case of experimental annotation of unmethylated sites, clusters of CpGs provide evidence against methylation that should ideally be incorporated into our model.

We therefore add a hidden random variable for each *HpaII* site, $\{W_i\}_{i=1,...,N} \in \{1, 0\}$, indicating whether the ith *HpaII* site is in an unmethylated cluster or not. We note that in our setting we assume that there are unmethylated clusters in the experiment, and the W variables denote the presence of a *HpaII* site in such a cluster. Different methyl-Seq experiments, on different tissues for example, may have a different set of unmethylated clusters. In addition, we add for each *HpaII* site two observed variables, denoted by **C** and **D**. The notation $\{C_i\}_{i=1,...,N} \in \{1, ..., c_{max}\}$ indicates the number of CpG sites between the $(i-1)^{\text{th}}$ *HpaII* site and the ith *HpaII* site (any CpGs that are not part of a CCGG site). The notation $\{D_i\}_{i=1,...,N} \in \{1, ..., d_{max}\}$ indicates the distance in bps between the $(i-1)^{\text{th}}$ *HpaII* site and the ith *HpaII* site. A graphical model representation of the model augmented with these additions can be seen in Fig. 14.4. The list of variables, parameters, and dependence functions added to our model is the following:

Variables:

$W_1, ..., W_N \in \{1, 0\}$: indicates for *HpaII* site i whether it is present in an unmethylated island (1) or not (0).

$C_1, \ldots, C_N \in \{1, \ldots c_{max}\}$: the number of CpG sites between the previous and current *Hpall* sites.
$D_1, \ldots, D_N \in \{1, \ldots, d_{max}\}$: the distance between the previous and current *Hpall* sites.

Parameters:

$\theta_{Y|W} = \{\theta_{y^0|w^0}, \ldots, \theta_{y^4|w^0}, \theta_{y^0|w^1}, \ldots, \theta_{y^4|w^1}\}$,
   where $0 \leq \theta_{y^i|w^j} \leq 1$, $\sum_i \theta_{y^i|w^0} = 1$ and $\sum_i \theta_{y^i|w^1} = 1$.
$\theta_{W|W_{-1},C,D} = \{\theta_{w^a|w^b_{-1},c_{(l_C,h_C)},d_{(l_D,h_D)}}\}$ where
   $a, b \in \{0, 1\}$,
   $(l_C, h_C) \in \{(0, 2), \ldots, (2^{k-1} + 1, 2^k), \ldots, (513, 1024), (\geq 1025)\}$,
   $(l_D, h_D) \in \{(0, 5), \ldots, (5 \cdot 2^{k-1} + 1, 5 \cdot 2^k), \ldots, (1280, 2560), (\geq 2561)\}$,
   and w_{-1} denotes the states of the W variable that is the parent of the W variable at hand.
   For all indexes, $0 \leq \theta_{w^a|w^b_{-1},c_{(l_C,h_C)},d_{(l_D,h_D)}} \leq 1$, and
$\sum_{t=\{0,1\}} \theta_{w^t|w^b_{-1},c_{(l_C,h_C)},d_{(l_D,h_D)}} = 1$.

Dependence functions:

All of the new dependence functions are fully described by the parameters added. The way in which
   every variable is dependent on its parents is fully captured by the parameters specified.

APPLYING THE EM ALGORITHM

In order to make inferences from this model, for the same reasons discussed in the previous
section, we make use of the EM algorithm. Assuming we have a fully observed data set, we can use
the likelihood function as in the previous section to derive the maximum likelihood estimators
for our model parameters:

$$\hat{\theta}_{L(i)} = \frac{\sum x_i \, \mathbf{1}\{f(l_i) = \theta_{L(j)}\}}{\sum z_i \, \mathbf{1}\{f(l_i) = \theta_{L(j)}\}},$$

$$\hat{\theta}_{y^a|w^b} = \frac{\sum_i \mathbf{1}\{y_i = y^a, w_i = w^b\}}{\sum_i \mathbf{1}\{w_i = w^b\}},$$

and

$$\hat{\theta}_{w^a|w^b_{-1},c_{l_C,h_C},d_{l_D,h_D}} = \frac{\sum_i \mathbf{1}\{w_i = w^a, w_{i-1} = w^b_{-1}, f_1(c_i) = c_{(l_C,h_C)}, f_2(d_i) = d_{(l_D,h_D)}\}}{\sum_i \mathbf{1}\{w_{i-1} = w^b_{-1}, f_1(c_i) = c_{(l_C,h_C)}, f_2(d_i) = d_{(l_D,h_D)}\}},$$

where $f_1 : \{1, \ldots c_{max}\} \to (l_C, h_C)$ takes in c_i and returns a pair, (l_C, h_C), such that $l_C \leq c_i \leq h_C$,
and $f_2 : \{1, \ldots, d_{max}\} \to (l_D, h_D)$ takes in d_i and returns a pair, (l_D, h_D), such that $l_D \leq d_i \leq h_D$.

As before, we start with some (possibly random) assignment of the parameters, and iterate
between the following two steps until we determine convergence. For convenience, we denote by
θ the complete parameter set for our model.

Expectation step (E-step): In this step, the algorithm uses the parameters from the last M-step,
θ^t, to compute the *expected* values for the sufficient statistics used in the maximum likelihood parameter estimation (that incorporate "hidden" variables). For $\theta_{L(i)}$, this is done as previously described.
Let $\mathbf{O} = (\mathbf{X}, \mathbf{L}, \mathbf{C}, \mathbf{D})$ be the set of observed variables, and $\mathbf{H} = (\mathbf{W}, \mathbf{Y}, \mathbf{Z})$ be the set of hidden

variables. Using the junction-tree algorithm on the moralized graph, we can compute the probability of any assignment to some variable, Q, and its parents, $\boldsymbol{U}$, given the observed data and θ^t. In other words, after a run of the junction-tree algorithm, we know $\mathbb{P}(\boldsymbol{q}, \boldsymbol{u}|\boldsymbol{O}, \theta^t)$ for every assignment of Q and $\boldsymbol{U}$, $(\boldsymbol{q}, \boldsymbol{u})$. We can then compute the needed expected sufficient statistics. For $\theta_{y^a|w^b}$ we compute

$$\mathbb{E}_{H|O,\theta^t}\left[\sum_i \mathbf{1}\{y_i = y^a, w_i = w^b\}\right] = \sum_i \mathbb{E}_{H|O,\theta^t}\left[\mathbf{1}\{y_i = y^a, w_i = w^b\}\right]$$

$$= \sum_i \mathbb{P}(y_i = y^a, w_i = w^b|O, \theta^t)$$

and

$$\mathbb{E}_{H|O,\theta^t}\left[\sum_i \mathbf{1}\{w_i = w^b\}\right] = \sum_i \mathbb{P}(w_i = w^b|O, \theta^t).$$

For $\theta_{w^a|w^b_{-1},c_{(l_C,h_C)},d_{(l_D,h_D)}}$ we compute

$$\mathbb{E}_{H|O,\theta^t}\left[\sum_i \mathbf{1}\{w_i = w^a, w_{i-1} = w^b_{-1}, f_1(c_i) = c_{(l_C,h_C)}, f_2(d_i) = d_{(l_D,h_D)}\}\right]$$

$$= \sum_i \mathbb{P}(w_i = w^a, w_{i-1} = w^b_{-1}, f_1(c_i) = c_{(l_C,h_C)}, f_2(d_i) = d_{(l_D,h_D)}|O, \theta^t)$$

and

$$\mathbb{E}_{H|O,\theta^t}\left[\sum_i \mathbf{1}\{w_{i-1} = w^b_{-1}, f_1(c_i) = c_{(l_C,h_C)}, f_2(d_i) = d_{(l_D,h_D)}\}\right]$$

$$= \sum_i \mathbb{P}(w_{i-1} = w^b_{-1}, f_1(c_i) = c_{(l_C,h_C)}, f_2(d_i) = d_{(l_D,h_D)}|O, \theta^t).$$

We recall that the running time complexity of the junction-tree algorithm is exponential in the size of the maximum sized clique of the moralized graph, and it is easy to see how the moralized network can be chorded such that, aside from the cliques generated in the previous model, only cliques of size three are introduced.

Maximization step (M-step): In this step we generate an updated set of parameters, θ^{t+1}, by using the expected sufficient statistics computed in the E-step. The computation of θ_L^{t+1} remains the same as before, and $\theta^{t+1}_{y^a|w^b}$ and $\theta^{t+1}_{w^a|w^b_{-1},d_{l_D,h_D},c_{l_C,h_C}}$ are updated as follows:

$$\theta^{t+1}_{y^a|w^b} = \frac{\sum_i \mathbb{P}(y_i = y^a, w_i = w^b|O, \theta^t)}{\sum_i \mathbb{P}(w_i = w^b|O, \theta^t)},$$

and

$$\theta^{t+1}_{w^a|w^b_{-1},c_{(l_C,h_C)},d_{(l_D,h_D)}}$$

$$= \frac{\sum_i \mathbb{P}(w_i = w^a, w_{i-1} = w^b_{-1}, f_1(c_i) = c_{(l_C,h_C)}, f_2(d_i) = d_{(l_D,h_D)}|O,\theta^t)}{\sum_i \mathbb{P}(w_{i-1} = w^b_{-1}, f_1(c_i) = c_{(l_C,h_C)}, f_2(d_i) = d_{(l_D,h_D)}|O,\theta^t)}.$$

The algorithm iterates between the E-step and the M-step until the increase in the likelihood function is smaller than a given threshold. After the algorithm has converged at θ^f, we can compute the posterior distribution for each Y_i by using a final run of the junction tree algorithm. We can now also compute for every W_i the probability that the *HpaII* site at index i is in an unmethylated cluster. We elaborate on this in the next section.

14.5 Application: Methyltyping the Human Neutrophil

In the previous sections, we have described a Bayesian network for the correction of bias introduced in high-throughput sequencing experiments in which the genome is digested in a non-random manner followed by a size-selection step, and have focused on the methyl-Seq experiment. In this chapter, we present in detail results from a study that has applied such a model for the purpose of characterizing the methyltypes of neutrophil cells from four human individuals [49].

In [49], the authors show that the methyltyping approach applied by methyl-Seq coupled with a Bayesian network correction procedure, called MetMap, is sufficient to survey methylation states across the genome and provides significant insight into the methylome, inside and outside of CpG islands, at site-specific resolution. In the study at hand, the methyltypes of four male human individuals were determined, using the standard methyl-Seq protocol. Site-specific methylation values could be assigned to 4.8% of the CpGs in the human reference genome, and of these, 20% were inside CpG islands.

The authors show that the precision achieved by using MetMap's correction was greatly increased over that of using the raw read counts. This was shown by determining the true methylation state of 46 *HpaII* sites using direct bisulfite sequencing and observing that the Pearson correlation of the MetMap inferred scores with the bisulfite validations, which was 0.90, was significantly higher than the Pearson correlation of the methylation values estimated by the raw fragment counts with the bisulfite validations, which was 0.67.

14.5.1 Unmethylated Clusters

As we have discussed throughout this chapter, unmethylated sites in vertebrate genomes tend to be clustered together. While this enables predicting locations of such clusters from the genomic sequence, if experimental methylation data are available, it is desirable to annotate unmethylated islands directly. This is because DNA methylation varies across tissues and conditions, and experimental data will reveal a more precise picture of the methylation state for the specific experiment.

In [49], the authors identified experiment-specific unmethylated islands, and called them SUMIs (strongly unmethylated islands). These unmethylated clusters are specific to the

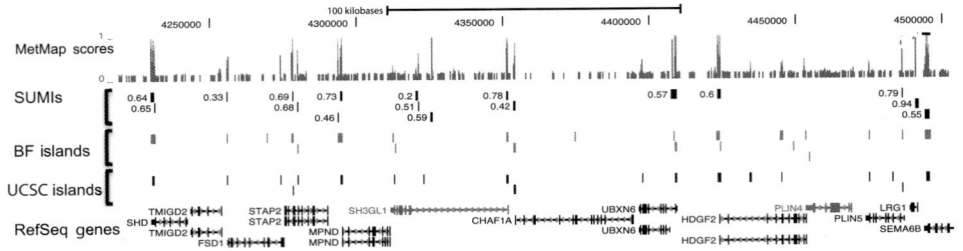

Fig 14.5 A section of the genome showing site-specific methylation scores (top panel) and unmethylated clusters (SUMIs, second panel) as inferred from one of human neutrophil samples. For the site-specific scores, a score of 0 determines a site as fully methylated. The third and fourth panels show BF islands as annotated by [5] and UCSC islands, respectively. While there is substantial overlap between SUMIs and the islands inferred by the sequence-based methods, a few novel SUMIs are seen in this figure, one of them at a transcription start site. RefSeq denotes genes annotated in the National Center for Biotechnology Information reference sequence database. Reproduced in color in the color plate section.

experiment at hand, and therefore to the human neutrophil. An interesting test was to investigate how many of the SUMIs found in the neutrophil experiment were also present in the regions inferred as unmethylated by sequence-based methods (see Fig. 14.5). The authors compared the 20 986 SUMIs they found with regions inferred to be unmethylated by two sequence-based methods: CpG islands found using the University of California, Santa Cruz (UCSC) browser (termed UCSC islands) and "bona fide" CpG islands (termed BF islands) [5]. The UCSC islands are the sequence-based set of islands that are most commonly used, while the BF islands were annotated using a more sophisticated, and potentially more exact, method. Of the 20 986 SUMIs present in at least one of the four individuals, 4 652 do not overlap the UCSC islands, and 7 055 do not overlap the BF islands.

The authors could validate the sensitivity and specificity of the SUMI set, compared to the sequence-based methods, by using direct bisulfite sequencing. Using this method, it was possible to annotate the methylation states of all the CpG sites at stretches of several hundred bps. As a validation of specificity, five regions that were annotated as part of both a UCSC island and a BF island, and did not overlap with SUMIs, were bisulfite sequenced. All regions were found to be methylated in the neutrophil sample. As a validation of sensitivity, four regions that overlapped SUMIs and UCSC islands but not BF islands were bisulfite sequenced. All regions were found to be unmethylated in the neutrophil sample. These results demonstrated that the true methylation state in a specific tissue may be different than the methylation state inferred by the genome sequence alone, and perhaps of the methylation state present in other cell types.

Interestingly, 3 797 SUMIs did not overlap with BF islands or UCSC islands, revealing new regions that are unmethylated in neutrophil cells. These regions showed strong association with regions associated with other functional features, such as open chromatin, high sequence conservation, gene regions, and the 5′ end of genes. Table 14.2 displays the association of the different sets of islands with these features. The authors point out that the proportion of SUMIs that map at a distance from transcription start sites is larger for novel SUMIs than for all SUMIs, but that novel SUMIs have a degree of association with open chromatin similar to that observed for all SUMIs; this suggests that novel SUMIs may often represent distal regulatory sequences.

Table 14.2 Percentages of neutrophil strongly unmethylated islands, CpG islands annotated by the University of California, Santa Cruz browser, and bona fide CpG islands that overlap regions associated with functionality. For details on how the functional regions were determined, see [49].

	SUMIs (%)	UCSC CpG Islands (%)	BF Islands (%)	Novel SUMIs (%)
Open chromatin	70.0%	52.9%	65.3%	61.0%
Conservation	71.1%	68.5%	76.2%	49.6%
Gene regions	76.9%	77.7%	79.7%	59.9%
TSS regions	59.8%	52.2%	61.4%	22.0%

Note: BF, bona fide; SUMI, strongly unmethylated islands; TSS, transcription start site; UCSC, University of California, Santa Cruz.

14.6 Conclusions

In this chapter, we have demonstrated how Bayesian networks can be utilized in the genome-scale study of DNA methylation. By designing a generative model that incorporates hidden variables for the biological states and observed variables for the generated experimental data we were able to predict the biological states, even when such inference was not straightforward.

The need for a Bayesian network comes from the experiment of interest (methyl-Seq) encompassing a non-random digestion of the genome followed by a size-selection step. Bayesian networks are suitable for such a case because they can be used to model the dependences between neighboring sites. The model presented can be adjusted to account for similar biases in several other methyltyping methods including the use of several different methylation-sensitive restriction enzymes. It is also expected to prove useful in other high-throughput sequencing experiments that are prone to such biases, for example some of the recent protocols developed for determining RNA secondary structure [58].

ACKNOWLEDGMENTS

We thank Dario Boffelli and David I.K. Martin for many extensive discussions on DNA methylation. M.S. and L.P. were funded in part by NIH grant R01 HG006129-01.

APPENDIX 14.A COMPUTATIONS USED FOR THE EM APPROACH

In this appendix, we give a detailed description of the computations involved in annotating the maximum likelihood parameters, given an assignment to all states of the Bayesian network shown in Fig. 14.3. Given some assignment $\{y, z, l, x\}$ to the variables Y, Z, L and X the Bayesian network implies the likelihood function:

$$L(\theta_L : \mathcal{D}) = \left[\Pi_{j=1}^{N} \mathbb{P}(y_j) \right] \cdot \left[\Pi_{i=1}^{|F|} \mathbb{P}(l_i) \mathbb{P}(z_i|\mathbf{y}) \mathbb{P}(x_i|z_i, l_i) \right]$$

$$= Q \cdot W \cdot \left[\Pi_{i=1}^{|F|} \mathbb{P}(x_i|z_i, l_i) \right]$$

$$\propto \left[\Pi_{i=1}^{|F|} \mathbb{P}(x_i|z_i, l_i) \right]$$

$$\propto \Pi_{i=1}^{|F|} \frac{(z_i f(l_i))^{x_i}}{x_i!} e^{-z_i f(l_i)},$$

where Q is the constant for $\left[\Pi_{j=1}^{N} \mathbb{P}(y_j) \right]$, W is the constant for $\left[\Pi_{i=1}^{|F|} \mathbb{P}(l_i) \right]$ and we assume that the probability of $\mathcal{D}$ is not zero ($\mathbb{P}(z_i|y) = 1$ for all Z_i of $\mathcal{D}$). The function $f(l_i)$ is as defined in the "dependence function" annotation. Notice that in this setting, we are assuming a full assignment to all random variables of the model, and therefore terms that do not consider parameters can be referred to as constants. Our goal is to find the assignment of parameters that maximize the above likelihood function. We can do this by finding the parameter assignment that maximizes the log-likelihood:

$$\ell(\theta_L : \mathcal{D}) = \log \left(Q \cdot W \cdot \Pi_{i=1}^{|F|} \mathbb{P}(x_i|z_i, l_i) \right)$$

$$= \log(\Pi_{i=1}^{|F|} \mathbb{P}(x_i|z_i, l_i)) + \log(Q \cdot W)$$

$$= \log \left(\Pi_{i=1}^{|F|} \frac{(z_i f(l_i))^{x_i}}{x_i!} e^{-z_i f(l_i)} \right) + \log(Q \cdot W)$$

$$= \sum_{i=1}^{|F|} \left[x_i \log f(l_i) - z_i f(l_i) \right] + \sum_{i=1}^{|F|} \left[x_i \log(z_i) - \log(x_i!) \right] + \log(Q \cdot W)$$

$$= \sum_{k=1}^{L} \left[\sum_{i=1}^{|F|} (x_i \log \theta_{L(k)} - z_i \theta_{L(k)}) \mathbf{1}\{f(l_i) = \theta_{L(k)}\} \right] + C,$$

where $\theta_{L(k)}$ is the kth element of θ_L and C is the part of the equation with elements that do not include instances from θ_L. We would like to find θ_L that maximizes the likelihood function, which we denote by $\hat{\theta}_L$. We denote the jth element of $\hat{\theta}_L$ by $\hat{\theta}_{L(j)}$. For each j we can find $\hat{\theta}_{L(j)}$ by differentiating the log-likelihood, setting to zero and solving for $\theta_{L(j)}$:

$$0 = \frac{\partial}{\partial \theta_{L(j)}} \left(\sum_{k=1}^{L} \left[\sum_{i=1}^{|F|} (x_i \log \theta_{L(k)} - z_i \theta_{L(j)}) \mathbf{1}\{f(l_i) = \theta_{L(k)}\} \right] + C \right)$$

$$= \sum_{i=1}^{|F|} \left[\left(\frac{x_i}{\theta_{L(j)}} - z_i \right) \mathbf{1}\{f(l_i) = \theta_{L(j)}\} \right].$$

Solving this for $\theta_{L(j)}$, we get:

$$\hat{\theta}_{L(j)} = \frac{\sum x_i \mathbf{1}\{f(l_i) = \theta_{L(j)}\}}{\sum z_i \mathbf{1}\{f(l_i) = \theta_{L(j)}\}}.$$

REFERENCES

[1] M.P. Ball, J.B. Li, Y. Gao, J. Lee, E.M. Leproust, I. Park, B. Xie, G. Q. Daley, and G.M. Church. Targeted and genome-scale strategies reveal gene-body methylation signatures in human cells. *Nature Biotechnology*, 27(4):361–368, 2009.

[2] Y. Benjamini and T.P. Speed. Summarizing and correcting the GC content bias in high-throughput sequencing. *Nucleic Acids Research*, 40(10): e72, 2012.

[3] B.P. Berman, D.J. Weisenberger, J.F. Aman, T. Hinoue, Z. Ramjan, Y. Liu, H. Noushmehr, C.P.E. Lange, C.M. van Dijk, R.A.E.M. Tollenaar, D. Van Den Berg, and P.W. Laird. Regions of focal DNA hypermethylation and long-range hypomethylation in colorectal cancer coincide with nuclear lamina associated domains. *Nature Genetics*, 44(1):40–46, 2011.

[4] A.P. Bird. CpG-rich islands and the function of DNA methylation. *Nature*, 321(6067):209–213, 1986.

[5] C. Bock, J. Walter, M. Paulsen, and T. Lengauer. CpG island mapping by epigenome prediction. *PLOS Computational Biology*, 3(6):e110, 2007.

[6] J. Borgel, S. Guibert, Y. Li, H. Chiba, D. Schübeler, H. Sasaki, T. Forné, and M. Weber. Targets and dynamics of promoter DNA methylation during early mouse development. *Nature Genetics*, 42(12):1093–1100, 2010.

[7] C.V. Breton, H.M. Byun, M. Wenten, F. Pan, A. Yang, and F.D. Gilliland. Prenatal tobacco smoke exposure affects global and gene-specific DNA methylation. *American Journal of Respiratory and Critical Care Medicine*, 180(5):462–467, 2009.

[8] E.I. Campos and D. Reinberg. Histones: annotating chromatin. *Annual Review of Genetics*, 43(1):559–599, 2009.

[9] H. Cedar and Y. Bergman. Epigenetics of haematopoietic cell development. *Nature Reviews Immunology*, 11(7):478–488, 2011.

[10] N.M. Cohen, V. Dighe, G. Landan, S. Reynisdóttir, A. Palsson, S. Mitalipov, and A. Tanay. DNA methylation programming and reprogramming in primate embryonic stem cells. *Genome Research*, 19(12):2193–2201, 2009.

[11] S.J. Cokus, S. Feng, X. Zhang, Z. Chen, B. Merriman, C.D. Haudenschild, S. Pradhan, S.F. Nelson, M. Pellegrini, and S.E. Jacobsen. Shotgun bisulphite sequencing of the Arabidopsis genome reveals DNA methylation patterning. *Nature*, 452:215–219, 2008.

[12] J.C. Dohm, C. Lottaz, T. Borodina, and H. Himmelbauer. Substantial biases in ultra-short read data sets from high-throughput DNA sequencing. *Nucleic Acids Research*, 36(16):e105, 2008.

[13] M. Ehrlich. DNA methylation in cancer: too much, but also too little. *Oncogene*, 21(35):5400–5413, 2002.

[14] J. Ernst, P. Kheradpour, T.S. Mikkelsen, N. Shoresh, L.D. Ward, C.B. Epstein, X. Zhang, L. Wang, R. Issner, M. Coyne, M. Ku, T. Durham, M. Kellis, and B.E. Bernstein. Mapping and analysis of chromatin state dynamics in nine human cell types. *Nature*, 473(7345):43–49, 2011.

[15] M.F. Fraga, E. Ballestar, M.F. Paz, S. Ropero, F. Setien, M.L. Ballestar, D. Heine-Suner, J.C. Cigudosa, M. Urioste, J. Benitez, M. Boix-Chornet, A. Sanchez-Aguilera, C. Ling, E. Carlsson, P. Poulsen, A. Vaag, Z. Stephan, T.D. Spector, Y.Z. Wu, C. Plass, and M. Esteller. Epigenetic differences arise during the lifetime of monozygotic twins. *Proceedings of the National Academy of Sciences of the United States of America*, 102(30):10604–10609, 2005.

[16] M. Gardiner-Garden and M. Frommer. CpG islands in vertebrate genomes. *Journal of Molecular Biology*, 196(2):261–282, 1987.

[17] J.L. Glass, R.F. Thompson, B. Khulan, M.E. Figueroa, E.N. Olivier, E.J. Oakley, G. Van Zant, E.E. Bouhassira, A. Melnick, A. Golden, M.J. Fazzari, and J.M. Greally. CG dinucleotide clustering is a species-specific property of the genome. *Nucleic Acids Research*, 35(20):6798–6807, 2007.

[18] M.G. Goll and T.H. Bestor. Eukaryotic cytosine methyltransferases. *Annual Review of Biochemistry*, 74(1):481–514, 2005.

[19] A. Goren and H. Cedar. Replicating by the clock. *Nature Reviews Molecular Cell Biology*, 4(1):25–32, 2003.

[20] R.T. Grant-Downton and H.G. Dickinson. Epigenetics and its implications for plant biology 2. The 'epigenetic epiphany': epigenetics, evolution and beyond. *Annals of Botany*, 97(1):11–27, 2006.

[21] S.I.S. Grewal and D. Moazed. Heterochromatin and epigenetic control of gene expression. *Science*, 301(5634):798–802, 2003.

[22] H. Gu, C. Bock, T.S. Mikkelsen, N. Jager, Z.D. Smith, E. Tomazou, A. Gnirke, E.S. Lander, and A. Meissner. Genome-scale DNA methylation mapping of clinical samples at single-nucleotide resolution. *Nature Methods*, 7(2):133–136, 2010.

[23] I.R. Henderson and S.E. Jacobsen. Epigenetic inheritance in plants. *Nature*, 447(7143):418–424, 2007.

[24] H. Hikoya. Discovery of bisulfite-mediated cytosine conversion to uracil, the key reaction for DNA methylation analysis—a personal account. *Proceedings of the Japan Academy, Series B*, 84(8):321–330, 2008.

[25] F. Hsieh, S.C. Chen, and K. Pollard. A nearly exhaustive search for CpG islands on whole chromosomes. *The International Journal of Biostatistics*, 5(1):14, 2009.

[26] R. Illingworth, A. Kerr, D. Desousa, H. Jørgensen, P. Ellis, J. Stalker, D. Jackson, C. Clee, R. Plumb, J. Rogers, S. Humphray, T. Cox, C. Langford, and A. Bird. A novel CpG island set identifies tissue-specific methylation at developmental gene loci. *PLOS Biology*, 6(1):e22, 2008.

[27] H. Ji, L.I.R. Ehrlich, J. Seita, P. Murakami, A. Doi, P. Lindau, H. Lee, M.J. Aryee, R.A. Irizarry, K. Kim, D.J. Rossi, M.A. Inlay, T. Serwold, H. Karsunky, L. Ho, G.Q. Daley, I.L. Weissman, and A.P. Feinberg. Comprehensive methylome map of lineage commitment from haematopoietic progenitors. *Nature*, 467(7313):338–342, 2010.

[28] P.A. Jones and S.B. Baylin. The epigenomics of cancer. *Cell*, 128(4):683–692, 2007.

[29] D. Koller and N. Friedman. *Probabilistic Graphical Models: Principles and Techniques*. The MIT Press, 2009.

[30] T. Kouzarides. Chromatin modifications and their function. *Cell*, 128(4):693–705, 2007.

[31] P. W. Laird. Principles and challenges of genome-wide DNA methylation analysis. *Nature Reviews Genetics*, 11(3):191–203, 2010.

[32] B. Langmead, C. Trapnell, M. Pop, and S. Salzberg. Ultrafast and memory-efficient alignment of short DNA sequences to the human genome. *Genome Biology*, 10(3):R25, 2009.

[33] E. Li, T. H. Bestor, and R. Jaenisch. Targeted mutation of the DNA methyltransferase gene results in embryonic lethality. *Cell*, 69(6):915–26, 1992.

[34] R. Lister, R. C. O'Malley, J. Tonti-Filippini, B. D. Gregory, C. C. Berry, H. A. Millar, and J. R. Ecker. Highly integrated single-base resolution maps of the epigenome in Arabidopsis. *Cell*, 133(3):523–536, 2008.

[35] R. Lister, M. Pelizzola, R.H. Dowen, R.D. Hawkins, G. Hon, J. Tonti-Filippini, J.R. Nery, L. Lee, Z. Ye, Q.M. Ngo, L. Edsall, J. Antosiewicz-Bourget, R. Stewart, V. Ruotti, A.H. Millar, J.A. Thomson, B. Ren, and J.R. Ecker. Human DNA methylomes at base resolution show widespread epigenomic differences. *Nature*, 462(7271):315–322, 2009.

[36] D.I. Martin, M. Singer, J. Dhahbi, G. Mao, L. Zhang, G. P. Schroth, L. Pachter, and D. Boffelli. Phyloepigenomic comparison of great apes reveals a correlation between somatic and germline methylation states. *Genome Research*, 21(12):2049–2057, 2011.

[37] A.K. Maunakea, R.P. Nagarajan, M. Bilenky, T.J. Ballinger, C. D'Souza, S.D. Fouse, B.E. Johnson, C. Hong, C. Nielsen, Y. Zhao, G. Turecki, A. Delaney, R. Varhol, N. Thiessen, K. Shchors, V.M. Heine, D.H. Rowitch, X. Xing, C. Fiore, M. Schillebeeckx, S.J.M. Jones, D. Haussler, M.A. Marra, M. Hirst, T. Wang, and J.F. Costello. Conserved role of intragenic DNA methylation in regulating alternative promoters. *Nature*, 466(7303):253–257, 2010.

[38] A. Meissner, A. Gnirke, G.W. Bell, B. Ramsahoye, E.S. Lander, and R. Jaenisch. Reduced representation bisulfite sequencing for comparative high-resolution DNA methylation analysis. *Nucleic Acids Research*, 33(18):5868–5877, 2005.

[39] A. Meissner, T. S. Mikkelsen, H. Gu, M. Wernig, J. Hanna, A. Sivachenko, X. Zhang, B. E. Bernstein, C. Nusbaum, D. B. Jaffe, A. Gnirke, R. Jaenisch, and E. S. Lander. Genome-scale DNA methylation maps of pluripotent and differentiated cells. *Nature*, 454:766–770, 2008.

[40] F. Mohn, M. Weber, D. Schübeler, and T.C. Roloff. Methylated DNA Immunoprecipitation (MeDIP). In J. M. Walker and J. Tost, editors, *DNA Methylation*, volume 507 of *Methods in Molecular Biology*, pages 55–64. Humana Press, 2009.

[41] H.D. Morgan, H.G. Sutherland, D.I. Martin, and E. Whitelaw. Epigenetic inheritance at the agouti locus in the mouse. *Nature Genetics*, 23(3):314–318, 1999.

[42] L. Pachter. Models for transcript quantification from RNA-Seq. arXiv:1104.3889v2, 2011.

[43] C. A. Perry, C. D. Allis, and A. T. Annunziato. Parental nucleosomes segregated to newly replicated chromatin are underacetylated relative to those assembled de novo. *Biochemistry*, 32(49):13615–13623, 1993.

[44] V.K. Rakyan, S. Chong, M.E. Champ, P.C. Cuthbert, H.D. Morgan, K.V.K. Luu, and E. Whitelaw. Transgenerational inheritance of epigenetic states at the murine Axin(Fu) allele occurs after maternal and paternal transmission. *Proceedings of the National Academy of Sciences of the United States of America*, 100(5):2538–2543, 2003.

[45] V.K. Rakyan, T.A. Down, D.J. Balding, and S. Beck. Epigenome-wide association studies for common human diseases. *Nature Reviews Genetics*, 12(8):529–541, 2011.

[46] K.D. Robertson. DNA methylation and human disease. *Nature Reviews Genetics*, 6(8):597–610, 2005.

[47] R.J. Schmitz, M.D. Schultz, M.G. Lewsey, R.C. O'Malley, M.A. Urich, O. Libiger, N.J. Schork, and J.R. Ecker. Transgenerational epigenetic instability is a source of novel methylation variants. *Science*, 334(6054):369–373, 2011.

[48] Y. Shufaro, O. Lacham-Kaplan, B.Z. Tzuberi, J. McLaughlin, A. Trounson, H. Cedar, and B.E. Reubinoff. Reprogramming of DNA replication timing. *Stem Cells*, 28(3):443–449, 2010.

[49] M. Singer, D. Boffelli, J. Dhabhi, A. Schönhuth, G.P. Schroth, D.I. Martin, and L. Pachter. Met-Map enables genome-scale methyltyping for determining methylation states in populations. *PLOS Computational Biology*, 6(8):e1000888, 2010.

[50] M. Singer, A. Engström, A. Schönhuth, and L. Pachter. Determining coding CpG islands by identifying regions significant for pattern statistics on Markov chains. *Statistical Applications in Genetics and Molecular Biology*, 10(1):43, 2011.

[51] M. Slatkin. Epigenetic inheritance and the missing heritability problem. *Genetics*, 182(3):845–850, 2009.

[52] B. D. Strahl and C. D. Allis. The language of covalent histone modifications. *Nature*, 403(6765):41–45, 2000.

[53] R. Straussman, D. Nejman, D. Roberts, I. Steinfeld, B. Blum, N. Benvenisty, I. Simon, Z. Yakhini, and H. Cedar. Developmental programming of CpG island methylation profiles in the human genome. *Nature Structural and Molecular Biology*, 16(5):564–571, 2009.

[54] Y. Sujuan, A. Asaithambi, and Y. Liu. CpGIF: an algorithm for the identification of CpG islands. *Bioinformation*, 2(8):335–8, 2008.

[55] M.M.M. Suzuki and A. Bird. DNA methylation landscapes: provocative insights from epigenomics. *Nature Reviews Genetics*, 9:465–476, 2008.

[56] D. Takai and P.A. Jones. Comprehensive analysis of CpG islands in human chromosomes 21 and 22. *Proceedings of the National Academy of Sciences of the United States of America*, 99(6):3740–3745, 2002.

[57] B.M. Turner. Defining an epigenetic code. *Nature Cell Biology*, 9(1):2–6, 2007.

[58] J.G. Underwood, A.V. Uzilov, S. Katzman, C.S. Onodera, J.E. Mainzer, D.H. Mathews, T.M. Lowe, S.R. Salama, and D. Haussler. FragSeq: transcriptome-wide RNA structure probing using high-throughput sequencing. *Nature Methods*, 7(12):995–1001, 2010.

[59] R.Y. Wang, C.W. Gehrke, and M. Ehrlich. Comparison of bisulfite modification of 5-methyldeoxycytidine and deoxycytidine residues. *Nucleic Acids Research*, 8(20):4777–4790, 1980.

[60] M. Weber, I. Hellmann, M.B. Stadler, L. Ramos, S. Pääbo, M. Rebhan, and D. Schübeler. Distribution, silencing potential and evolutionary impact of promoter DNA methylation in the human genome. *Nature Genetics*, 39(4):457–466, 2007.

[61] H. Wu, B. Caffo, H.A. Jaffee, R.A. Irizarry, and A.P. Feinberg. Redefining CpG islands using Hidden Markov Models. *Biostatistics*, 11(3):499–514, 2010.

[62] A. Zemach, I.E. McDaniel, P. Silva, and D. Zilberman. Genome-wide evolutionary analysis of eukaryotic DNA methylation. *Science*, 328(5980):916–919, 2010.

[63] X. Zhang, J. Yazaki, A. Sundaresan, S. Cokus, S.W.L. Chan, H. Chen, I.R. Henderson, P. Shinn, M. Pellegrini, S.E. Jacobsen, and J.R. Ecker. Genome-wide high-resolution mapping and functional analysis of DNA methylation in Arabidopsis. *Cell*, 126(6):1189–1201, 2006.

[64] Y. Zhang, C. Rohde, S. Tierling, T.P. Jurkowski, C. Bock, D. Santacruz, S. Ragozin, R. Reinhardt, M. Groth, J. Walter, and A. Jeltsch. DNA methylation analysis of chromosome 21 gene promoters at single base pair and single allele resolution. *PLOS Genetics*, 5(3):e1000438, 2009.

CHAPTER 15

Latent Variable Models for Analyzing DNA Methylation

E. ANDRÉS HOUSEMAN

DNA methylation is tightly linked with cellular differentiation. For instance, it has been observed that DNA methylation in tumor cells encodes phenotypic information about the tumor. Thus, the understanding of tumor biology is fruitfully enhanced by the study of the multivariate structure of DNA methylation data. To the extent that such data possess discrete latent structure, it can be viewed as encoding different tumor subtypes (in cancer studies) or tissue types (more generally). However, in some cases, there may be more evidence of continuous latent structure reflecting a continuous range of variation. This chapter discusses several specific latent variable models that have been used in the last decade to analyze DNA methylation data. First are discussed approaches for modeling DNA methylation data in low-dimensional settings such as in candidate gene studies. Discrete and continuous latent variables are distinguished, the former consisting of finite mixture models or their non-parametric analogs (often described as clustering) and the latter consisting of latent trait models that typically involve numerical integration. Then approaches for modeling DNA methylation in high-dimensional settings, such as in data obtained from expression microarrays, are discussed. In this setting, continuous latent variables become difficult to operationalize and interpret, so the focus is set only on methods that are computationally efficient for clustering. The Recursively Partitioned Mixture Model (RPMM) presents classes in a hierarchical format similar to that of hierarchical clustering, thus allowing for tree pruning in a manner not supported by conventional algorithms. However, RPMM cannot handle data obtained from nowadays standard methylation arrays. RPMM was therefore equipped with a preprocessing step to produce semi-supervised RPMM (SS-RPMM), a scalable procedure.

Probabilistic Graphical Models for Genetics, Genomics, and Postgenomics. First Edition. Christine Sinoquet & Raphaël Mourad (Eds). © Oxford University Press 2014. Published in 2014 by Oxford University Press.

15.1 Introduction

Epigenetics is the study of heritable changes in gene function that cannot be explained by changes in DNA sequence. The development of any multisystem life form is fundamentally grounded in systematic cellular differentiation, defined by lineage commitment of cells whose origin can be traced to a pluripotent progenitor.[1] It is now established that this lineage commitment is marked by mitotically heritable epigenetic changes that reflect complex transcriptional programming. One such epigenetic mark is the addition of methyl groups to the cytosine (C) of DNA basepairs C and guanine (G) (CpG dinucleotides); this methylation mark is tightly associated with alterations in the chromatin and nucleosome DNA scaffold, which are, in turn, responsible for the coordination of gene expression within the individual cell [20, 23, 32]. Thus, DNA methylation is tightly linked with cellular differentiation, and the DNA methylation profile over all of its CpG dinucleotides in essence encodes information about its cellular function. For example, Fig. 15.1, from [8], depicts a clustering heat map showing the degree to which 11 different tissue types cluster based on their DNA methylation profiles at 500 CpG dinucleotides. Fig. 15.1 demonstrates that measures of DNA methylation, viewed from a multivariate perspective, retain much information about cell type.

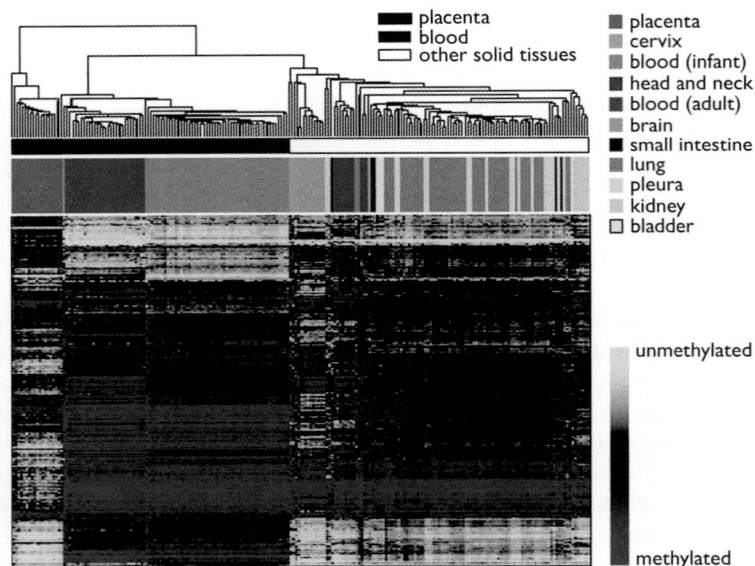

Fig 15.1 Clustering heat map showing DNA methylation patterns for 11 normal tissues [8]. Each cell represents an average beta value from the GoldenGate assay (Illumina). Rows represent one of 500 CpG dinucleotides, columns represent one of 211 individual samples. Reproduced in color in the color plate section.

---

[1] A stem-like cell that has the capacity to differentiate into one of many cell types.

This fact has been exploited in cancer biology to distinguish tumors that appear histopathologically similar on the basis of their molecular profile. In particular, the CpG Island Methylator Phenotype (CIMP) has been a popular construct for understanding cancer [16, 35, 42, 43]. Essentially, it has been observed that DNA methylation in tumor cells, measured in promoter regions of a small number of genes, encodes phenotypic information about the tumor. Thus, understanding of tumor biology is fruitfully enhanced by the study of the multivariate structure of a vector Y of DNA methylation measurements. To the extent that Y possesses discrete latent structure, it can be viewed as encoding different tumor subtypes (in cancer studies) or tissue types (more generally). However, in some cases there may be more evidence of continuous latent structure reflecting a continuous range of variation [29, 30].

For a given CpG, a clonal population of cells can be expected to have a categorical methylation state, 0, 1/2, or 1, depending on whether the CpG is methylated at neither chromosome, one chromosome, or both chromosomes, respectively. Since hemi-methylation (DNA methylation on only one chromosome) is relatively rare, analysis of DNA methylation on a clonal cell population can be viewed as a latent variable problem on an observed vector of dichotomous variables. While tumor populations are typically clonal, it is often in practice difficult to separate tumor cells from surrounding normal cells or stromal tissue without labor-intensive microdissection techniques. Additionally, there has been an interest in assaying DNA methylation in whole blood [6, 31, 39], whose DNA methylation states are defined by mixtures of distinct circulating blood cells [34, 44]. Thus, DNA methylation measurements may, in practice, be effectively continuous variables in the unit interval, either bimodal (in the case of relatively clonal populations of cells) or unimodal (in the case of extremely heterogeneous mixtures such as blood).

The distribution assumed for the components of Y will depend on technology. There are a wide variety of assays and microarray platforms available for measuring DNA methylation; a comprehensive list is provided in a recent review by Peter Laird [25]. Popular examples include methylation-specific polymerase chain reaction (PCR) [15], COBRA [45], PCR followed by pyrosequencing [41], MethyLight [12], Sequenom [21], Illumina methylation arrays [3–5], and MeDIP [36]. The first five of these are examples of labor-intensive approaches that have been appropriate for candidate gene studies or validation of microarray results, while the latter two admit high-throughput analysis. Methylation-specific PCR (MSP) produces a dichotomous value for each interrogated locus (gene), unmethylated (coded as 0) and methylated (coded as 1). COBRA, pyrosequencing, MethyLight, Sequenom, and Illumina methylation arrays produce interval-scaled values between 0 and 1 for each interrogated locus. Some platforms, such as MethyLight, can in principle produce values that are arbitrarily large, lying beyond 1, although such events, which are difficult to interpret, are typically rare occurrences. MeDIP, on the other hand, produces fluorescence intensity values measured at regularly spaced intervals with very dense coverage of the genome, and is typically reduced to peak intensities associated with p-values; interpretation of peaks is meant to be dichotomous.

Fig. 15.2 (page 390) illustrates the types of models considered in this chapter. DNA methylation measurements Y_{ij} from a potentially large number, J, of individual loci are assumed to have correlation induced by their dependence on a common unobserved variable, Z_i. The variable Z_i, which may be either a categorical or a univariate continuous variable, represents either a molecular subtype or an overall methylation propensity and may itself be influenced by covariates denoted $\mathbf{x}_i$. From a statistical modeling perspective, this is a deceptively simple formulation, but the potentially large number of loci, J, can complicate estimation and interpretation.

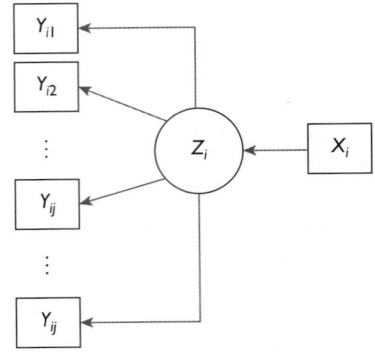

Fig 15.2 Schematic diagram depicting latent variable models for DNA methylation data. Here, Y_{ij} represents a DNA methylation measurement for subject i at locus j, Z_i represents a common latent variable influencing all measurements, and x_i represents covariates that are assumed to affect Z_i through biological processes of interest.

In general, for DNA methylation assay data Y_{ij} from subject i, $i \in \{1, 2, \ldots, n\}$ at CpG dinucleotide j, $j \in \{1, 2, \ldots, J\}$, we assume distribution $f(Y_{ij} = y | Z_i = z; \theta_j)$ for some latent variable Z_i; here θ_j is a parameter vector that governs the dependence of Y_{ij} on Z_i and CpG j, the precise nature of the dependence and corresponding dimension of θ_j being determined by the specific model employed. It follows that the marginal distribution of Y_i is $\int f(y|z; \theta_1, \theta_2, \ldots, \theta_J) \, d(z)$, and that the latent variable Z_i can be extracted as $\mathbb{E}(Z_i | Y_i; \theta_1, \theta_2, \ldots, \theta_J)$. A common additional assumption is conditional independence of loci, i.e., that Y_{ij} and $Y_{ij'}$ are independent conditional on Z_i if $j \neq j'$. Relaxation of this independence assumption is a topic of current research, especially in high-dimensional settings where unstructured correlation models are difficult to implement and incorporation of known biological information is crucial.

In the remainder of the chapter, we discuss several specific such latent variable models that have been used in the last decade to analyze DNA methylation data. In Section 15.2, we discuss approaches for modeling DNA methylation data in low-dimensional settings (small J), e.g., candidate gene studies. Here we distinguish *discrete* from *continuous* latent variables, the former consisting of *finite mixture models* or their non-parametric analogs (often described as *clustering*), and the latter consisting of *latent trait models* that typically involve numerical integration. In Section 15.3, we discuss approaches for modeling DNA methylation in high-dimensional settings (large J), e.g., data obtained from expression microarrays. In this setting, continuous latent variables become difficult to operationalize and interpret, so we focus only on computationally efficient methods for clustering.

In the mathematical exposition that follows, matrices are always indicated as bold, upper-case letters. Upper-case is also used to designate a random variable, while lower-case is reserved for fixed quantities. Boldface letters also designate vectors (random or fixed), while normal type designates a scalar quantity. All vectors are assumed to be column vectors unless otherwise noted, with a row vector written as a transposed column vector (e.g., $\boldsymbol{\theta}^{\mathrm{T}}$).

15.2 Latent Variable Methods for DNA Methylation in Low-dimensional Settings

It is possible to construct rich graphical models when the number of loci J has relatively small value. In this section, we describe several approaches that have been used for small candidate gene studies, mostly appearing in the middle of the first decade of the 21st century.

15.2.1 Discrete Latent Variables

When $\mathbf{Z}_i$ is a discrete variable, the corresponding latent variable analysis leads to a clustering problem. Although many popular non-parametric clustering techniques, such as K-means or hierarchical clustering, have been used in the analysis of DNA methylation data [35], there is some evidence that likelihood-based approaches may provide superior results [37]. Consequently, we focus in this chapter mostly on such likelihood-based methods, which are often referred to as *model-based clustering* [13, 14]. These methods rest ultimately on the finite mixture model, which we now describe.

We assume that each subject i belongs to one of K classes, where K is, for the moment, pre-specified. The latent variable model is then, more explicitly, defined in terms of conditional distributions $f(Y_{ij} = y | C_i = k; \boldsymbol{\theta}_{jk}) = g(y; \boldsymbol{\theta}_{jk})$, where $C_i \in \{1, 2, \ldots, K\}$ denotes the (unknown) class to which subject i belongs, and $\mathbb{P}(C_i = k) = \eta_k$, $\sum_{k=1}^{K} \eta_k = 1$. Note that there are only $K - 1$ independent η quantities, and that $\boldsymbol{\theta}_{jk}$ may be multivariate (e.g., representing mean and standard deviation parameters for a normal distribution). As shown in Appendix 15.A (page 401), with an additional assumption that the Y_{ij}'s are independent conditional on C_i, the resulting data log-likelihood $LL(\boldsymbol{\theta}; \mathbf{y}_1, \ldots, \mathbf{y}_n) \equiv \log(\mathbb{P}(\mathbf{Y}_1 = \mathbf{y}_1, \ldots, \mathbf{Y}_n = \mathbf{y}_n; \boldsymbol{\theta}))$ is

$$LL(\boldsymbol{\theta}; \mathbf{y}_1, \ldots, \mathbf{y}_n) = \sum_{i=1}^{n} \log \left\{ \sum_{k=1}^{K} \eta_k \prod_{j=1}^{J} g(y_{ij}; \boldsymbol{\theta}_{jk}) \right\}, \tag{15.1}$$

where $\boldsymbol{\theta} = (\eta_1, \ldots, \eta_{K-1}, \boldsymbol{\theta}_{11}^{\mathrm{T}}, \ldots, \boldsymbol{\theta}_{JK}^{\mathrm{T}})$ is the vector of all unknown parameters. The expectation-maximization (EM) algorithm [11] is typically used to fit (15.1). Briefly, maximization of (15.1) is equivalent to maximization of

$$LL(\boldsymbol{\theta}; \mathbf{y}_1, \ldots, \mathbf{y}_n) = \sum_{k=1}^{K} \sum_{j=1}^{J} \left\{ \sum_{i=1}^{n} w_{ik}(\boldsymbol{\theta}) \log g(y_{ij}; \boldsymbol{\theta}_{jk}) \right\}, \tag{15.2}$$

where

$$w_{ik}(\boldsymbol{\theta}) = \frac{\eta_k \prod_{j=1}^{J} g(y_{ij}; \boldsymbol{\theta}_{jk})}{\sum_{k=1}^{K} \eta_k \prod_{j=1}^{J} f(Y_{ij} = y_{ij} | C_i = k; \boldsymbol{\theta}_{jk})} \tag{15.3}$$

is the empirical Bayes prediction of $\mathbb{P}(C_i = k | \mathbf{Y}_i = \mathbf{y}_i)$. Equation (15.2) shows that maximization of (15.1) can be decomposed into JK separate, manageable weighted-likelihood problems, provided the weights $w_{ik}(\boldsymbol{\theta})$ are known. The EM algorithm proceeds from initial estimates of w_{ik} by calculating

$$\eta_k = \frac{1}{n} \sum_{i=1}^{n} w_{ik} \tag{15.4}$$

and maximizing each inner term in (15.2) with respect to $\boldsymbol{\theta}_{jk}$. Then the weights are recomputed via (15.3), the η_k's are recomputed via (15.4), and (15.2) is again maximized, iterating until some convergence criterion is met. Common choices for families of distributions g include Bernoulli for discrete dichotomous measures of methylation and beta or normal for continuous measures, with an appropriate data transform in the normal case: logit or arcsine-square-root for interval-scaled measures and log for measures that can exceed one. Siegmund et al. (2004) demonstrate via simulation that finite mixture models tend to outperform non-parametric methods such as hierarchical clustering [37], due to full use of statistical information; they propose several choices of g for MethyLight, including Bernoulli (for dichotomized values), log normal, and Bernoulli log normal mixtures.

Typically the number of classes K is unknown; consequently, model fit statistics such as Akaike information criterion (AIC) and Bayesian information criterion (BIC) are often employed to select the value of K [13, 14]. AIC and BIC are measures of prediction error for a given model, i.e., the extent to which the model accurately predicts a newly observed outcome, and are in general used extensively for model comparison and to guard against over-fitting. In an application of such methods, Marsit et al. (2006) present a latent variable analysis of methylation-specific PCR (i.e., dichotomous) data collected on $J = 15$ loci from 344 bladder cancers, 346 head and neck squamous cell carcinomas (HNSCC), 146 non-small-cell lung cancer (NSCLC) tumors and 71 malignant pleural mesotheliomas (MPM), using (15.1) with Bernoulli g and values of K from 2 to 6. AIC values were computed as $2d_K - 2LL$, where d_K is the number of independent unknown parameters corresponding to a model with K classes, specifically $d_k = |\boldsymbol{\theta}| = K(J + 1) - 1$ (i.e., JK mean values plus $K - 1$ independent η parameters). In 2006, Marsit et al. [28] found that AIC was smaller for larger values of K, indicating that larger values of K led to better fitting models. With large values of K, model (15.1) can be viewed as an approximation to a continuous latent variable [27], thus motivating a continuous latent variable model for these data.

15.2.2 Continuous Latent Variables

Instead of using a discrete latent variable $\mathbf{Z}_i$, it is possible instead to assume $\mathbf{Z}_i$ is continuous. In this case, $\mathbf{Z}_i$ represents one or more continuous scales of variation, akin to principal components or factors in the more familiar domains of principal components analysis or factor analysis; greater the dependence of individual loci on $\mathbf{Z}_i$ reflect a greater degree of correlation among these loci.

In addition to finite mixture models, Marsit et al. [28] presented a model with normal, univariate Z_i, resulting in a *latent trait model* with a log-likelihood LL of the form

$$LL(\boldsymbol{\theta}; \mathbf{y}_1, \ldots, \mathbf{y}_n) = \sum_{i=1}^{n} \log \left\{ \int_{-\infty}^{\infty} \prod_{j=1}^{J} g_j(y_{ij}; \boldsymbol{\theta}, z) \phi(z) dz \right\}, \tag{15.5}$$

where $\phi(z)$ is the standard normal density, $\boldsymbol{\theta} = (\boldsymbol{\beta}_0^{\mathrm{T}}, \boldsymbol{\beta}_1^{\mathrm{T}}, \boldsymbol{\gamma}^{\mathrm{T}})$ is the vector of unknown parameters; $\boldsymbol{\beta}_0$ and $\boldsymbol{\beta}_1$ are vectors of intercepts and slopes, respectively, each of dimension $J \times 1$; and $\boldsymbol{\gamma}$ is a vector of dimension $q \times 1$, with q chosen to match the dimension of $\mathbf{x}_i$, a covariate vector having dimension $q \times 1$. Here $\mathbf{x}_i^{\mathrm{T}} \boldsymbol{\gamma}$ is a "centering" term that adjusts the mean latent variable, which in turn adjust the success probabilities of Y_{ij}. Specifically, $g_j(y_{ij}; \boldsymbol{\theta}, z) = p_{ij}(\boldsymbol{\theta}, z)^{y_{ij}} [1 - p_{ij}(\boldsymbol{\theta}, z)]^{1-y_{ij}}$, where $p_{ij}(\boldsymbol{\theta}, z) = \mathrm{logit}^{-1}[\beta_{0j} + \beta_{1j}(\mathbf{x}_i^{\mathrm{T}} \boldsymbol{\gamma} + z)]$ is a Bernoulli probability. In this application, $q = 3$ and $\mathbf{x}_i$ consisted of indicator variables for tumor type, with bladder cancer as reference. Equation (15.5) is a form of *Item Response Theory* model, used extensively in the education literature to capture the common effect of a latent competency (Z_i) on the responses to individual test items (Y_{ij}), which vary in difficulty (β_{1j}); here, β_{1j} reflects the degree of correlation between locus j and latent methylation propensity $\mathbf{x}_i^{\mathrm{T}}\boldsymbol{\gamma} + z$, with the central tendency of the latent trait potentially modified by covariates $\mathbf{x}$. In other words, large values of $|\beta_{1j}|$ indicate sensitivity of methylation at locus j to an overall methylation characteristic of the specimen, which is itself potentially related to external covariates ($\mathbf{x}_i$), disease type in this instance.

The integral in (15.5) was approximated by Gauss–Hermite quadrature [22], i.e., discretizing the problem using the approximation $\int_{-\infty}^{\infty} \prod_{j=1}^{J} g_j(y_{ij}; \boldsymbol{\theta}, z) \phi(z) \, dz \approx \sum_{r=1}^{R} w_r \prod_{j=1}^{J} g_j(y_{ij}; \boldsymbol{\theta}, z_r)$ for a certain choice of weights $w_1, \ldots, w_R$ and evaluation points $z_1, \ldots, z_R$ depending upon an integer R (chosen for desired precision). In addition to the AIC values computed for the finite mixture models (15.1) with different values of K, Marsit et al. [28] also reported the AIC value for (15.5), calculated as $2(2J + q) - 2LL$. Latent trait model (15.5) produced the lowest (best) AIC, indicating that it fit the data best. Marsit et al. reported estimates, standard errors, and p-values for the $\boldsymbol{\theta}$ parameters, some of which are reproduced in Table 15.1. From the estimates of the $\boldsymbol{\gamma}$ parameters, it was concluded that DNA methylation at the $J = 15$ loci does not distinguish tumor type among bladder cancers, HNSCC and MPM, but that NSCLC tumors are distinguished by significantly higher methylation overall. In addition, from the differences in estimates of the slope parameters $\boldsymbol{\beta}_1$, it was concluded that there is a moderate level of variation in the degree to which an individual gene informs the latent variable representative of overall methylation for the entire set of genes. In other words, some genes were more correlated than others. Finally, Marsit and colleagues compared the model-fitting results with hierarchical clustering and a naive Bayes approach, demonstrating that all models lead to similar overall conclusions, but that the latent variable approach provides more opportunities for modeling interesting features of the data, most notably the correlated response in DNA methylation. For example, the naive Bayes approach obscures the extent to which methylation is correlated among genes and the extent to which the strength of correlation varies from gene to gene (a phenomenon that is related to the CpG Island Methylator Phenotype concept); on the other hand, hierarchical clustering provides limited insight into the nature of clustering, and little potential for model extension of the form discussed below. Following [28], Marsit et al. used a slight variation of (15.5) for modeling DNA methylation in bladder tumors [30]. In a low-dimensional candidate-gene study in which the genes had been carefully selected to correspond to a known biological process, the values of the slope parameters β_{1j} in (15.5) could be expected to be somewhat similar, each lying close to a mean slope parameter $\bar{\beta}_1$. Consequently, Marsit et al. [30] used penalized likelihood to impose constraints on the value of $\boldsymbol{\beta}_1 = (\beta_{11}, \ldots, \beta_{1J})^{\mathrm{T}}$. Specifically, the quadratic penalty $\lambda \boldsymbol{\beta}_1^{\mathrm{T}} \mathbf{Q} \boldsymbol{\beta}_1$ was subtracted from (15.5), where $\lambda > 0$ is a tuning parameter and $\mathbf{Q}$ is a positive, semi-definite, symmetric matrix whose non-zero eigenvalues are equal to one and whose nullspace is spanned by $\mathbf{1} = (1, \ldots, 1)^{\mathrm{T}}$ (see Appendix 15.A for details). While the objective is to estimate the β and γ parameters, a choice of tuning parameter must be made apart from the likelihood maximization (as discussed in Appendix 15.A); thus, the tuning parameter λ is chosen by minimizing AIC computed with degrees-of-freedom equal to $\mathrm{Tr}([\mathbf{H} + 2\lambda \mathbf{Q}]^{-1} \sum_{i=1}^{n} \boldsymbol{\Psi}_i \boldsymbol{\Psi}_i^{\mathrm{T}})$, where the $\boldsymbol{\Psi}_i$'s are the first derivatives of the individual log-likelihood contributions in (15.5) evaluated at the parameter estimates, $\mathbf{H}$ is the corresponding second derivative matrix of the entire log-likelihood (i.e., the sum of the derivatives of $\boldsymbol{\Psi}_i$), and $\mathrm{Tr}(\mathbf{A})$ is the *trace* of matrix $\mathbf{A}$, i.e., the sum of its diagonal entries. Note that $\mathbf{H} = \mathbb{E}(\sum_{i=1}^{n} \boldsymbol{\Psi}_i \boldsymbol{\Psi}_i^{\mathrm{T}})$, as shown using standard probability theory, and that $\mathrm{Tr}(\mathbf{H}^{-1}\mathbf{H})$ is equal to the number of model parameters d, so that the formula for degrees of freedom agrees approximately with the unpenalized likelihood case corresponding to $\lambda = 0$. The justification for this

Table 15.1 Results of fitted latent trait models. In [29], methylation over four different solid tumor types was compared, while [30] focused on bladder cancer and factors that covaried with methylation: age, sex, TP53 staining in the tumor, and smoking status.

	Marsit et al. (2006) [29]			Marsit et al. (2007) [30]		
	Estimate	Std. Err.	Z-score	Estimate	Std. Err.	Z-score
Slope for gene (β_1)						
APC	0.03	0.08	0.4	1.29	0.23	5.6
CDH1	0.94	0.12	7.8	0.71	0.44	1.6
CDH13	1.64	0.19	8.6	2.36	0.50	4.7
CDKN2A	0.32	0.08	4.0	0.30	0.13	2.3
DAPK	0.72	0.11	6.5	0.52	0.23	2.3
GSTP1	0.60	0.20	3.0			
LAMC2	0.99	0.13	7.6	0.77	0.22	3.5
MGMT				0.90	0.29	3.1
MLH1	0.31	0.15	2.1	0.13	0.26	0.5
PRSS3				0.95	0.18	5.3
PYCARD	0.57	0.17	3.4	1.07	0.38	2.8
RAR	1.21	0.13	9.3	1.38	0.27	5.1
RASSF1	0.42	0.08	5.3	0.94	0.19	4.9
SFRP1	1.71	0.19	9.0	1.91	0.36	5.3
SFRP2	1.92	0.25	7.7	2.47	2.88	0.9
SFRP4	0.70	0.11	6.4	0.86	0.29	3.0
SFRP5	1.61	0.17	9.5	1.92	0.33	5.8
Trait centering covariates (γ)						
Bladder		Referent				
HNSCC	0.07	0.14	0.5			
NSCLC	1.76	0.16	11.0			
MPM	0.27	0.16	1.7			

Continued

Table 15.1 (Continued).

	Marsit et al. (2006) [29]			Marsit et al. (2007) [30]		
	Estimate	Std. Err.	Z-score	Estimate	Std. Err.	Z-score
Age (years)						
<58					Referent	
58–65				0.19	0.21	0.9
66–70				0.14	0.20	0.7
>70				0.61	0.21	2.9
Sex						
Female					Referent	
Male				0.41	0.18	2.3
TP53 staining						
No/low					Referent	
High				0.37	0.14	2.6
Smoking status						
Never/former					Referent	
Current				0.43	0.15	2.9

HNSCC, head and neck squamous cell carcinomas; MPM, malignant pleural mesotheliomas; NSCLC, non-small-cell lung cancer; Std. Err., standard error.

form of degrees-of-freedom results from a Taylor approximation to the penalized log-likelihood. The overall approach is justified by Houseman et al. [19], who propose the penalized likelihood method as an approximation to a hierarchical random effects model where $\beta_{1j} \sim \mathcal{N}(\bar{\beta}_1, 1/2\lambda)$ and demonstrate that the penalized likelihood approximation returns estimates close to the posterior mean parameters obtained using Markov chain Monte Carlo (MCMC) techniques implemented in WinBUGS [38], but with greater computational efficiency. While MCMC is often used to obtain faithful representations of posterior distributions, which are often informally interpreted in a manner similar to frequentist sampling distributions, [19] suggests that reasonable frequentist results are obtained via penalized likelihood, which was found in [19] to produce estimates for a single value of λ approximately 60 times faster than those obtained from only 200 MCMC draws for the similar hierarchical random effects model. Marsit et al. [30] reported that there was heterogeneity in the effect β_{j1} of latent trait Z_i on DNA methylation at various loci and that the loci differed from each other in the level of information they carried about latent disease status; these results are similar to the moderate heterogeneity found in [29]. In addition, Marsit and colleagues reported significant recentering of the latent trait (via γ) by age, gender, smoking status, and TP53 immunohistochemistry staining; their model suggests that older subjects, males, smokers, and subjects with high staining values have greater DNA methylation overall. Table 15.1 presents some of the parameter estimates from this study.

15.3 Latent Variable Methods for DNA Methylation in High-dimensional Settings

Modern DNA methylation research makes use of high-density microarrays or even sequencing-based technologies, leading to high-dimensional data sets with extremely large values of J. Table 15.2 presents the dimensions of commonly available arrays in comparison to dimensions encountered in the applications described above in Section 15.2. In this setting, a one-dimensional continuous Z_i is implausible, and multivariate $\mathbf{Z}_i$ of dimension adequate to model $\mathbf{Y}_i$ leads to computational infeasibility using strictly likelihood-based approaches. Even the finite-mixture model (15.1) is time consuming to fit using standard software such as Mplus [13]. Consequently, tractable algorithms for analyzing high-dimensional genomic data require simplifying assumptions. In this section, we describe some approaches that have been used to analyze DNA methylation arrays.

15.3.1 Model-based Clustering: Recursively Partitioned Mixture Models

When the number of classes K of a finite mixture model (15.1) is not known in advance, K is typically selected by optimizing AIC or BIC, in which case the time complexity of the EM algorithm is

Table 15.2 Dimensions of commonly available DNA methylation arrays, along with subject-matter papers employing a latent variable model for analyzing data from the given platform. Both the total number of loci and the number of autosomal loci are given, since sex chromosome loci are often excluded in applications involving both male and female subjects, due to anticipated differences in methylation of sex chromosomes and consequent potential for confounding. Numbers of loci on products by Illumina are obtained from product specifications made available by the company.

Platform	Examples of Applications	Array dimension	
		Total	Autosomal
Custom MSP arrays	[29, 30]	~ 15 gene promoter regions	~ 15
Illumina GoldenGate	[7, 9, 10, 28]	1 505 CpG loci	1421
Illumina Infinium Methyl27K	[2, 31]	27 578 CpG loci	26 486
Illumina Infinium Methyl450K	(none to date)	485 577 CpG loci	473 929

MSP: Methylation-specific PCR.

approximately $O(nJK_{max}^2)$, where values $K \in \{1, 2, \ldots, K_{max}\}$, the set of all candidate values to be fit sequentially. Other approaches have involved Dirichlet-process mixtures employing MCMC techniques, but these are even less computationally efficient. Houseman et al. present an alternative, recursive approach [18] that retains some of the non-parametric features of Dirichlet process mixtures but has, on average, approximate time complexity $O(nJK_0)$, where K_0 is the true number of classes, and performs demonstrably faster than conventional approaches with sequential trials of $K \in \{1, 2, \ldots, K_{max}\}$. Furthermore, the *Recursively Partitioned Mixture Model* (RPMM) presents classes in a hierarchical format similar to that of hierarchical clustering, thus allowing for tree pruning in a manner not supported by conventional algorithms.

The algorithm, illustrated in Fig. 15.3, fits a recursive sequence of models enumerated $l = 1, 2, \ldots$, described as follows. It begins by fitting model $l = 1$, a finite mixture model (15.1) with $K = 2$ to obtain $\widehat{\boldsymbol{\theta}}^{(1)}$, and comparing the result with that obtained from fitting (15.1) with $K = 1$. In each case, the BIC statistic $(K - 1 + JKr) \log n - 2LL$ is computed, where r is the number of parameters required to characterize the assumed family of distributions g; typically $r = 2$ for continuous data, e.g., mean and standard deviation or the two parameters for a beta distribution. Here, $K - 1 + JKr$ is the number of parameters in the model, i.e., $K - 1$ independent η parameters plus JKr independent parameters for characterizing the individual $\boldsymbol{\theta}_{jk}$. If the BIC for $K = 1$ is better than that for $K = 2$, the algorithm stops. On the other hand, if $K = 2$ produces a better BIC, the algorithm continues by further "splitting" each of the two child classes, with an implicit interpretation on class membership that is characterized as follows: the final weights $\omega_i^{(k+1)} = w_{ik}(\widehat{\boldsymbol{\theta}}^{(1)})$ computed by (15.3) are assumed to represent fuzzy cluster assignments of each

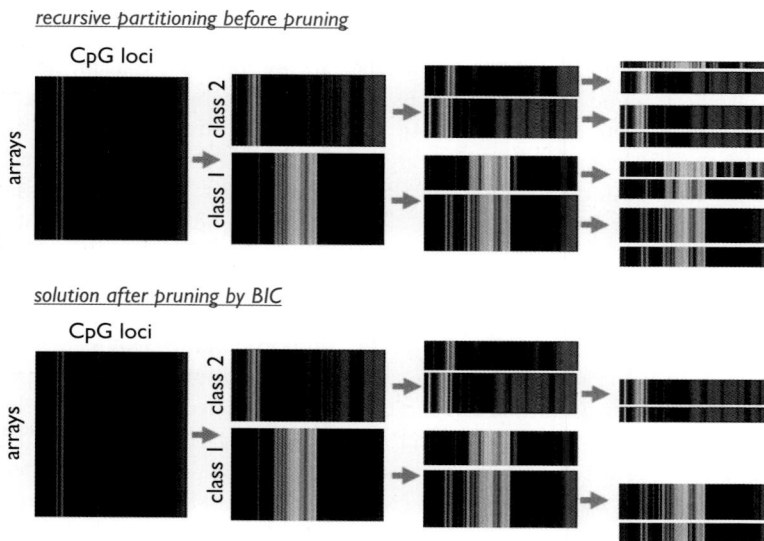

Fig 15.3 Schematic representation of the RPMM. Rows represent individual specimens or arrays, columns represent individual CpG loci. Initially, each array is assumed to be drawn from the same multivariate distribution consisting of a distinct distribution for each CpG (indicated by color). The data set is partitioned recursively into component data subsets using a two-part mixture model. Along the way, BIC is used to prune the tree, so that partitions that are likely to be unstable are never attempted. Reproduced in color in the color plate section.

subject i to the two child classes obtained from the initial model fit. In other words, $\omega_i^{(2)}$ represents the proportion of subject i assigned to the "left" child (class 1) of model $l = 1$, and $\omega_i^{(3)}$ represents the complementary proportion assigned to the "right" child (class 2) of model $l = 1$. Subsequently, each of these children spawns new models; two additional finite mixture models are fit ($l = 2$ and $l = 3$), one for each child cluster. In each such model, (15.2) is substituted with a similar weighted-likelihood formulation:

$$LL(\boldsymbol{\theta}; \boldsymbol{\omega}, \mathbf{Y}_1, \ldots, \mathbf{Y}_n) = \sum_{k=1}^{K} \sum_{j=1}^{J} \left\{ \sum_{i=1}^{n} \omega_i w_{ik}(\boldsymbol{\theta}) \log g(Y_{ij}; \boldsymbol{\theta}_{jk}) \right\},$$

where $\omega_i = \omega_i^{(2)}$ in model $l = 2$ and $\omega_i = \omega_i^{(3)}$ in model $l = 3$. The process continues recursively, with $\omega_i^{(2l+h)} = \omega_i^{(l)} w_{i(h+1)}(\widehat{\boldsymbol{\theta}}^{(l)})$, $h \in \{0, 1\}$ and $l \in \{1, 2, \ldots [(n-1)/2]\}$. At each new stage $2l + h$, BIC comparisons are made ($K = 1$ versus $K = 2$), with recursion terminating whenever $K = 1$ produces a better BIC. In practice, model fitting becomes unstable for values $l < [(n-1)/2]$, so that recursion also terminates when the $\omega_i^{(l)}$'s decrease beyond some preset threshold (e.g., $n_l \equiv \sum_{i=1}^{n} \omega_i^{(l)} < 10$). Note also that for each model-fitting stage, the sample size used to calculate BIC is assumed to be the "local" sample size n_l. While the original paper assumed $g(y; \boldsymbol{\theta})$ corresponding to a beta distribution, it is easily adapted to normal distributions [24] as well as others. Because simple analytic solutions exist for problems involving normal distributions (e.g., weighted sample means and variances) but not for problems involving beta distributions, which require numerical optimization, models involving normal distributions are much faster. In addition, when data are appropriately transformed, normal assumptions have been found to produce results that are not meaningfully different from those obtained by assuming beta distributions [24]. In [17], a motivation is provided for the beta distribution, though an equally compelling argument could be made for assuming a normal distribution for log-transformed expression data (see Appendix 15.A). A discussion of distribution assumptions for DNA methylation data and corresponding transformations appears in [25].

RPMM has been employed in a number of settings. In its debut [18], it was used to cluster the DNA methylation profiles of the 211 tissues depicted in Fig. 15.1, resulting in latent classes having a high degree of correspondence with tissue type. In a study examining DNA methylation profiles from pleural mesothelioma and normal pleura, RPMM classes displayed a high degree of correspondence with tumor versus normal classification, were correlated with lung asbestos burden, and predicted patient survival [7]. In another study examining HNSCC tumors [28], RPMM classes distinguished normal and tumor samples, as shown in Fig. 15.4, and correlated well with other etiologic factors, as shown in Fig. 15.5 (page 400). The RPMM class assignments corresponding to a subset of these HNSCC tumors also correlated with classes of copy number alterations [33], as shown in Fig. 15.6 (page 400). In a breast cancer study, RPMM classes of methylation correlated with tumor size, alcohol, and dietary folate intake [9], and in a study of glioma, RPMM classes were found to correlate with survival and with the mutation of the IDH1 gene [10]. Similar associations have been found in colon cancer between RPMM methylation classes and mutations [16].

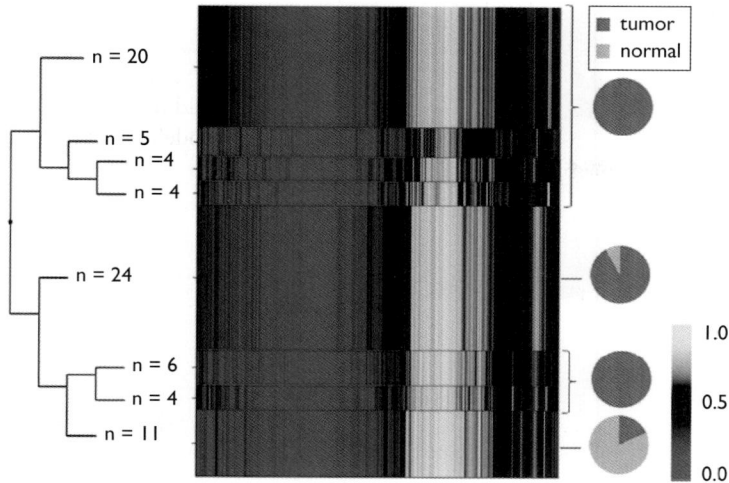

Fig 15.4 Recursive partitioning mixture model classification of normal and tumor head and neck tissues. The model was based on methylation values of 1413 autosomal loci measured using the GoldenGate assay produced by Illumina, and resulted in eight classes whose average methylation values are represented in the heat map. Distribution of normal and tumor samples within each class is depicted in pie charts on the right. Reproduced from [28], Figure 1B. Reproduced in color in the color plate section.

15.3.2 Semi-Supervised Recursively Partitioned Mixture Models

The RPMM algorithm was designed to address the requirements of analyzing the GoldenGate methylation array (Illumina), which interrogates $J = 1505$ CpG sites. However, the standard methylation arrays from Illumina now interrogate anywhere from 27 578 to 485 577 CpG sites, far exceeding the computational capacity of even RPMM. In addition, for any given clinical or epidemiological problem, it may be difficult even to formulate a simple latent variable model that can be meaningfully related to the outcome of interest.

Motivated by the "semi-supervised" methodology proposed by Bair and Tibshirani [1], RPMM was equipped with a preprocessing step to produce *Semi-Supervised* RPMM (SS-RPMM) [24]. In this approach, each CpG is ranked according to the degree to which the DNA methylation assay values are informative with respect to a clinical outcome. The M most informative CpGs are selected for inclusion in an RPMM algorithm, where the value of M is selected via ten-fold cross-validation to optimize predictive performance. The results are validated in an independent data set (possibly obtained by splitting a larger sample into training and test sets prior to ranking the CpGs): RPMM class assignments are predicted via empirical Bayes for the subjects in the independent data set, and association between predicted class membership and outcome is assessed. The independent or split-sample validation is used to circumvent the potential for overfitting that might result from preselecting the loci using data that will be used to conduct final predictive inference.

SS-RPMM has been used in a bladder cancer case control study to predict case status in DNA methylation assayed from whole blood using the Infinium 27K microarray (Illumina) [31]. In addition, it has been used on data from the same array technology to examine associations between DNA methylation in human placentas and infant growth restriction [2].

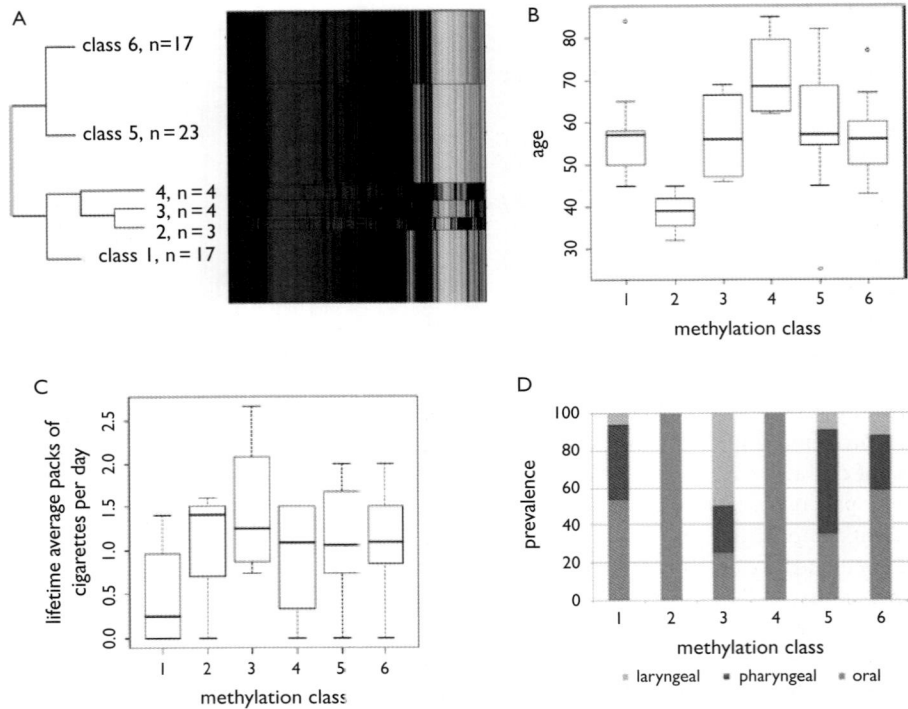

Fig 15.5 Recursive partitioning mixture model classification of HNSCCs. A) Six classes with average methylation values across loci depicted in the heat map. Associations of class membership with age, lifetime average packs of cigarettes smoked per day, and tumor location are shown in B, C, and D, respectively. Reproduced from [28], Figure 2A–D. Reproduced in color in the color plate section.

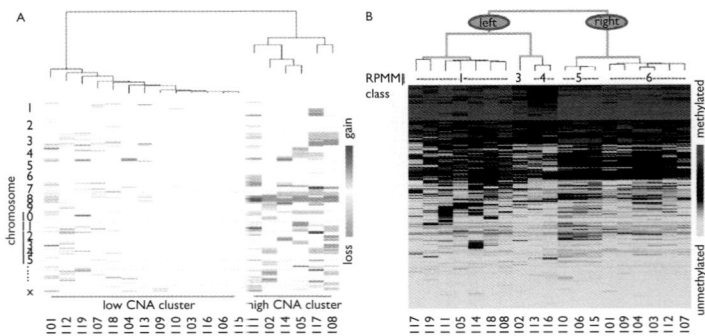

Fig 15.6 A) DNA copy number states are arranged by chromosome for 500 000 SNP loci. Copy number is red for amplified regions with three or more copies, white for two normal copies, and green for allele loss (no copies). Tumors are ordered by unsupervised hierarchical clustering and are dichotomized into low/high clusters of copy number alterations (CNAs). B) Methylation loci (more methylated = blue, less methylated = yellow) are grouped by Euclidean distance, and tumors samples are ordered first by RPMM class structure (green branches) then by simple hierarchical clustering (black branches). Tumor IDs are provided below each plot and "high CNA" samples are colored orange for reference. Reproduced from [33]. Reproduced in color in the color plate section.

15.4 Conclusion

We have presented an overview of the latent variable models that have been employed in the analysis of DNA methylation data. The two primary types of models used have been the *finite mixture model* (15.1), an essentially discrete latent variable modeling approach, and variations on the *latent trait* model (15.5). Over time, the former approach has emerged as more useful than the latter, in that its results have proven easier to interpret and are consistent with the ideas of the CpG Island Methylator Phenotype (CIMP) which posit "molecular subtypes" inherent in DNA methylation profiles, or similar ideas in the context of analyzing mRNA expression data [1]. Additionally, finite mixture models more readily extend to high-dimensional settings where adequate characterization of correlation via continuous latent traits would require multivariate latent continuous dimensions of moderate or high dimension, thus complicating numerical quadrature and leading to MCMC approaches that have been demonstrated to be cumbersome even with moderate dimension [19]. However, recent work has applied singular value decompositions [26] or independent components analysis [40] to mRNA expression and DNA methylation data, with a goal of characterizing technical noise and accounting for it. The high dimensions considered in these papers require locus-selection techniques similar to those proposed in [24], and thus represent compromises that are similar to those discussed in Section 15.3 above.

In low-dimensional settings, model (15.1) could accommodate covariate adjustments $\mathbf{x}^T\boldsymbol{\gamma}$ in a manner similar to that appearing in (15.5). However, in high dimensions, the requirement that $\boldsymbol{\gamma}$ remain constant for all classes would destroy the computational gains achieved by partitioning the data set recursively. Thus, a preferred approach has been to estimate covariate-dependence of class membership in a second stage analysis; simple bivariate approaches appear in [7, 28], while a multinomial logistic regression approach used to model more complex multivariate associations appears in [9]. Current research involves the use of more complicated multivariate distributions g to account for correlations among CpG loci associated with the same gene, or the same CpG locus measured by two different arrays (e.g., in different tissues or in the same tissue at different points in time); also under investigation is more biologically driven selection of CpG loci in approaches that are similar to SS-RPMM.

APPENDIX 15.A MATHEMATICAL DETAILS

In this appendix, we present mathematical details for several topics mentioned in this chapter.

APPENDIX 15.A.1 EXPECTATION-MAXIMIZATION OF FINITE MIXTURE MODEL LIKELIHOODS

Here we demonstrate the equivalence of (15.1) and (15.2) in the context of maximum likelihood. The basic principle of the EM algorithm is to maximize the expected likelihood for all variables, conditional on observed variables, i.e., integrating the *full data likelihood* over the latent variables conditional on the observed variables. In the present case, this requires computation of

$$\bar{LL}(\boldsymbol{\theta}) \equiv \mathbb{E}_{\boldsymbol{\theta}}\left[LL(\boldsymbol{\theta}; C_1, \ldots, C_n, \mathbf{y}_1, \ldots, \mathbf{y}_n) | \mathbf{Y}_1 = \mathbf{y}_1, \ldots, \mathbf{Y}_n = \mathbf{y}_n\right],$$

where the expectation is taken over the class membership variables C_i. Note that $\mathbb{E}_{\boldsymbol{\theta}}(C_i = k | \mathbf{Y}_i = \mathbf{y}_i) = \mathbb{P}_{\boldsymbol{\theta}}(C_i = k | \mathbf{Y}_i = \mathbf{y}_i) = w_{ik}(\boldsymbol{\theta})$, so that the required expectation is

$$
\bar{LL}(\boldsymbol{\theta}) = \mathbb{E}_{\boldsymbol{\theta}}\left(\sum_{i=1}^{n} \sum_{k=1}^{K} \mathbb{1}(C_i = k) \log \left\{ \prod_{j=1}^{J} g(y_{ij}; \boldsymbol{\theta}_{jk}) \right\} \middle| \mathbf{Y}_1 = \mathbf{y}_1, \ldots, \mathbf{Y}_n = \mathbf{y}_n \right)
$$

$$
= \sum_{i=1}^{n} \sum_{k=1}^{K} w_{ik}(\boldsymbol{\theta}) \log \left\{ \prod_{j=1}^{J} g(y_{ij}; \boldsymbol{\theta}_{jk}) \right\}
$$

$$
= \sum_{i=1}^{n} \sum_{k=1}^{K} w_{ik}(\boldsymbol{\theta}) \sum_{j=1}^{J} \log \left\{ g(y_{ij}; \boldsymbol{\theta}_{jk}) \right\},
$$

which, after rearranging summations, is equal to (15.2). Alternatively, one recalls that maximization of (15.1) entails setting its derivative equal to zero, or

$$
\mathbf{0} = \frac{\partial}{\partial \boldsymbol{\theta}_{jl}} LL(\boldsymbol{\theta}; \mathbf{y}_1, \ldots, \mathbf{y}_n)
$$

$$
= \sum_{i=1}^{n} \frac{\partial}{\partial \boldsymbol{\theta}_{jl}} \log \left\{ \sum_{k=1}^{K} \eta_k \prod_{j=1}^{J} g(y_{ij}; \boldsymbol{\theta}_{jk}) \right\}
$$

$$
= \sum_{i=1}^{n} \sum_{k=1}^{K} \eta_k \frac{(\partial/\partial \boldsymbol{\theta}_{jl}) \prod_{j=1}^{J} g(y_{ij}; \boldsymbol{\theta}_{jk})}{\sum_{k=1}^{K} \eta_k \prod_{j=1}^{J} g(y_{ij}; \boldsymbol{\theta}_{jk})}
$$

$$
= \sum_{i=1}^{n} \sum_{k=1}^{K} \eta_k \frac{[\partial g(y_{ij}; \boldsymbol{\theta}_{jl})/\partial \boldsymbol{\theta}_{jl}] \left[\prod_{j=1}^{J} g(y_{ij}; \boldsymbol{\theta}_{jk}) \right] / g(y_{ij}; \boldsymbol{\theta}_{jl})}{\sum_{k=1}^{K} \eta_k \prod_{j=1}^{J} g(y_{ij}; \boldsymbol{\theta}_{jk})}
$$

$$
= \sum_{k=1}^{K} \sum_{j=1}^{J} \left\{ \sum_{i=1}^{n} w_{ik}(\boldsymbol{\theta}) \frac{\partial}{\partial \boldsymbol{\theta}_{jl}} \log g(y_{ij}; \boldsymbol{\theta}_{jk}) \right\},
$$

demonstrating that (15.1) and (15.2) are equivalent in terms of maximization, provided that the parameter $\boldsymbol{\theta}_{jk}$ governs only a single component of $\mathbf{Y}_i$ and $w_{ik}(\boldsymbol{\theta})$ is considered fixed. The second derivation shows explicitly that the "separation of variables" approach that makes the EM algorithm tractable will fail if $\boldsymbol{\theta}_{jk}$ governs a large number of the components of $\mathbf{Y}_i$, which would be required for most modern approaches to penalized likelihood, e.g., [19]. Thus, in high-dimensional settings, penalized-likelihood approaches become infeasible.

Note that when g is the probability density function of a normal variable, the estimated mean and variance for locus j and class k are, respectively, $\widehat{\mu}_{jk} = \sum_{i=1}^{n} w_{ik}(\boldsymbol{\theta}) y_{ij} / \sum_{i=1}^{n} w_{ik}(\boldsymbol{\theta})$ and $\widehat{\sigma}_{jk}^2 = \sum_{i=1}^{n} w_{ik}(\boldsymbol{\theta}) y_{ij}^2 / \sum_{i=1}^{n} w_{ik}(\boldsymbol{\theta}) - \widehat{\mu}_{jk}^2$, which are trivial to compute.

APPENDIX 15.A.2 PENALIZED LIKELIHOOD

In many settings, it is desirable to impose data-driven constraints on parameters. Penalized likelihood is a popular method for achieving such constrained maximization. Suppose maximization of log-likelihood $LL(\theta)$ subject to a constraint $Q(\theta) - Q_0 = 0$ is desired. The method of Lagrange multipliers motivates maximization of the "penalized" log-likelihood $LL(\theta) - \lambda Q(\theta)$ for some fixed $\lambda \geq 0$, where the term $\lambda Q(\theta)$ (with $Q(\theta) \geq 0$) is understood to be a penalty for values of θ which result in large values of $Q(\theta)$. Typically $Q(\theta)$ is taken to be a quadratic form on a regression coefficient subvector β of θ, i.e., $Q(\theta) = \beta^T \mathbf{Q} \beta$. When $\mathbf{Q}$ is positive semidefinite (i.e., has non-negative eigenvalues), values of β that are large in vector magnitude (in at least some directions of the parameter space) result in large penalties to the objective function, excluding their admissibility as solutions. In the specific case that $\mathbf{Q1} = \mathbf{0}$, where $\mathbf{1}$ is a vector of ones (i.e., $\mathbf{1}$ is in the kernel of the linear transformation represented by $\mathbf{Q}$), large intercept terms have no impact on the penalized log-likelihood. When $\mathbf{1}$ is the only eigenvector of $\mathbf{Q}$ with zero eigenvalue, then only large deviations from an intercept are penalized, so that the model effectively constrains the components of β to be similar.

Note that if $LL(\theta) - \lambda Q(\theta)$ is maximized with respect to θ and λ simultaneously, the result will be $\widehat{\lambda} = 0$ and $\widehat{\theta}$ equal to the maximum likelihood estimator. Thus, the value of λ must be determined independently of the maximization of the penalized log-likelihood. This is typically achieved using the conventions of model selection, e.g., minimizing a model-fit statistic obtained after $\widehat{\theta}$ has been found. It can be shown via Taylor series approximation that

$$-2 \sum_{i=1}^{n} LL_i(\widehat{\theta}_{-i}) = 2\mathrm{Tr}\left([\mathbf{H} + 2\lambda \mathbf{Q}]^{-1} \sum_{i=1}^{n} \Psi_i \Psi_i^T \right) - 2LL(\widehat{\theta}),$$

where $\mathbf{H}$ is the second derivative matrix of the log-likelihood $LL(\theta)$ and the left-hand side expresses a jackknife estimator of model fit, with $LL_i(\theta)$ representing the log-likelihood contribution from a single subject i (thus $\partial LL_i / \partial \theta = \Psi_i$) and $\widehat{\theta}_{-i}$ denoting the estimator obtained by deleting subject i. The right-hand side resembles the familiar form of the AIC statistic, with degrees of freedom equal to $\mathrm{Tr}\left([\mathbf{H} + 2\lambda \mathbf{Q}]^{-1} \sum_{i=1}^{n} \Psi_i \Psi_i^T\right)$, and yields results consistent with the conventional formulation, since $\lambda = 0$ corresponds to degrees of freedom equal to $\mathrm{Tr}(\mathbf{I}_{|\theta|}) = |\theta|$.

APPENDIX 15.A.3 DISTRIBUTION OF ASSAY MEASUREMENTS OF DNA METHYLATION

As described in [17], bead-array measures of DNA methylation are obtained by comparing the fluorescent intensities from two different probes. One probe corresponds to a methylated CpG, and the other corresponds to an unmethylated locus; the measurement of the former will be denoted M_{ij}, while the latter will be denoted U_{ij}. Illumina products report a so-called "average beta" value, which is computed as $B_{ij} = M_{ij}/(M_{ij} + U_{ij} + \varepsilon)$, where $\varepsilon = 100$ is a constant that is negligible relative to intensity values from a successful experiment (typically 10 000 to 65 535) but included to protect against divide-by-zero errors occurring when $M_{ij} = U_{ij} = 0$. Taking $\varepsilon \approx 0$, $B_{ij} \approx M_{ij}/(M_{ij} + U_{ij})$. If $M_{ij} \sim Gamma(\alpha_j, \xi_j)$ and $U_{ij} \sim Gamma(\beta_j, \xi_j)$, then $B_{ij} \sim Beta(\alpha_j, \beta_j)$;

additionally, beta-distributed B_{ij} is consistent with a biological view that the observed measure of DNA methylation is a binomial variable with a large number of trials (DNA molecules), with a potentially variable, beta-distributed DNA methylation proportion. On the other hand, if $\log(M_{ij}) \sim \mathcal{N}(\mu_{Mj}, \sigma_{Mj}^2)$ and $\log(U_{ij}) \sim \mathcal{N}(\mu_{Uj}, \sigma_{Uj}^2)$, then $\text{logit}(B_{ij}) \sim \mathcal{N}(\mu_{Mj}-\mu_{Uj}, \sigma_{Mj}^2+\sigma_{Uj}^2)$, so that normality of $\text{logit} B_{ij}$ corresponds to fairly general lognormality assumptions on the probe signals (even when M_{ij} and U_{ij} are correlated, though the variance has an addition term contributed from the cross-covariance).

• •

REFERENCES

[1] E. Bair, T. Hastie, D. Paul, and R. Tibshirani. Prediction by supervised principle components. *Journal of the American Statistical Association*, 101:119–137, 2006.

[2] C.E. Banister, D.C. Koestler, M.A. Maccani, J.F. Padbury, E.A. Houseman, and C.J. Marsit. Infant growth restriction is associated with distinct patterns of DNA methylation in human placentas. *Epigenetics*, 6(7):920–927, 2011.

[3] M. Bibikova and J.B. Fan. Goldengate assay for DNA methylation profiling. *Methods in Molecular Biology*, 507:149–63, 2009.

[4] M. Bibikova, J. Le, B. Barnes, S. Saedinia-Melnyk, L. Zhou, R. Shen, and K.L. Gunderson. Genome-wide DNA methylation profiling using infinium assay. *Epigenomics*, 1:177–200, 2009.

[5] M. Bibikova, Z. Lin, L. Zhou, E. Chudin, E.W. Garcia, B. Wu, D. Doucet, N.J. Thomas, Y. Wang, E. Vollmer, T. Goldmann, C. Seifart, W. Jiang, D.L. Barker, M.S. Chee, J. Floros, and J.B. Fan. High-throughput DNA methylation profiling using universal bead arrays. *Genome Research*, 16(3):383–93, 2006.

[6] N. Borghol, M. Suderman, W. McArdle, A. Racine, M. Hallett, M. Pembrey, C. Hertzman, C. Powerand, and M. Szyf. Associations with early-life socio-economic position in adult DNA methylation. *International Journal of Epidemiology*, 41:62–74, 2012.

[7] B.C. Christensen, E.A. Houseman, J.J. Godleski, C.J. Marsit, J.L. Longacker, C.R. Roelofs, M.R. Karagas, M.R. Wrensch, R.F. Yeh, H.H. Nelson, J.L. Wiemels, S. Zheng, J.K. Wiencke, R. Bueno, D.J. Sugarbaker, and K.T. Kelsey. Epigenetic profiles distinguish pleural mesothelioma from normal pleura and predict lung asbestos burden and clinical outcome. *Cancer Research*, 69(1):227–234, 2009.

[8] B.C. Christensen, E.A. Houseman, C.J. Marsit, S. Zheng, M.R. Wrensch, J.L. Wiemels, H.H. Nelson, M.R. Karagas, J.F. Padbury, R. Bueno, D.J. Sugarbaker, R.F. Yeh, J.K. Wiencke, and K.T. Kelsey. Aging and environmental exposures alter tissue-specific DNA methylation dependent upon CpG island context. *PLOS Genetics*, 5:e1000602, 2009.

[9] B.C. Christensen, K.T. Kelsey, S. Zheng, E.A. Houseman, C.J. Marsit, M.R. Wrensch, J.L. Wiemels, H.H. Nelson, M.R. Karagas, L.H. Kushi, M.L. Kwan, and J.K. Wiencke. Breast cancer DNA methylation profiles are associated with tumor size and alcohol and folate intake. *PLOS Genetics*, 6(7):e1001043, 2010.

[10] B.C. Christensen, A.A. Smith, S. Zheng, D.C. Koestler, E.A. Houseman, C.J. Marsit, J.L. Wiemels, H.H. Nelson, M.R. Karagas, M.R. Wrensch, K.T. Kelsey, and J.K. Wiencke. DNA methylation, isocitrate dehydrogenase mutation, and survival in glioma. *Journal of the Nationall Cancer Institute*, 103(2):143–153, 2011.

[11] A. Dempster, N. Laird, and D. Rubin. Maximum likelihood from incomplete data via the EM algorithm (with discussion). *Journal of the Royal Statistical Society, Series B*, 39:1–38, 1977.

[12] C.A. Eads, K.D. Danenberg, K. Kawakami, L.B. Saltz, C. Blake, D. Shibata, P.V. Danenberg, and P.W. Laird. Methylight: a high-throughput assay to measure DNA methylation. *Nucleic Acids Research*, 28(8):E32, 2000.

[13] C. Fraley and A.E. Raftery. Model-based clustering, discriminant analysis, and density estimation. *Journal of the American Statistical Association*, 97:611–631, 2002.

[14] C. Fraley and A.E. Raftery. Bayesian regularization for normal mixture estimation and model-based clustering. Technical report, Department of Statistics, University of Washington, 2005.

[15] J.G. Herman, J.R. Graff, S. Myohanen, B. Nelkin, and S.B. Baylin. Methylation-specific PCR: a novel PCR assay for methylation status of CpG islands. *Proceedings of the National Academy of Sciences of the United States of America*, 93:9821–9826, 1996.

[16] T. Hinoue, D.J. Weisenberger, C.P. Lange, H. Shen, H.M. Byun, D. Van Den Berg, S. Malik, F. Pan, H. Noushmehr, C.M. van Dijk, R.A. Tollenaar, and P.W. Laird. Genome-scale analysis of aberrant DNA methylation in colorectal cancer. *Genome Research*, 22(2):271–282, 2012.

[17] E.A. Houseman, B.C. Christensen, M.R. Karagas, M.R. Wrensch, H.H. Nelson, J.L. Wiemels, S. Zheng, J.K. Wiencke, K.T. Kelsey, and C.J. Marsit. Copy number variation has little impact on bead-array-based measures of DNA methylation. *Bioinformatics*, 25:1999–2005, 2009.

[18] E.A. Houseman, B.C. Christensen, R.F. Yeh, C.J. Marsit, M.R. Karagas, M. Wrensch, H.H. Nelson, J. Wiemels, S. Zheng, J.K. Wiencke, and K.T. Kelsey. Model-based clustering of DNA methylation array data: a recursive-partitioning algorithm for high-dimensional data arising as a mixture of beta distributions. *BMC Bioinformatics*, 9:365, 2008.

[19] E.A. Houseman, C. Marsit, M. Karagas, and L.M. Ryan. Penalized item response theory models: application to epigenetic alterations in bladder cancer. *Biometrics*, 63(4):1269–1277, 2007.

[20] H. Ji, L.I. Ehrlich, J. Seita, P. Murakami, A. Doi, P. Lindau, H. Lee, M.J. Aryee, R.A. Irizarry, K. Kim, D.J. Rossi, M.A. Inlay, T. Serwold, H. Karsunky, L. Ho, G.Q. Daley, I.L. Weissman, and A.P. Feinberg. Comprehensive methylome map of lineage commitment from haematopoietic progenitors. *Nature*, 467(7313):338–342, 2011.

[21] C. Jurinke, M.F. Denissenko, P. Oeth, M. Ehrich, D. van den Boom, and C.R. Cantor. A single nucleotide polymorphism based approach for the identification and characterization of gene expression modulation using MassARRAY. *Mutation Research*, 573:83–95, 2005.

[22] W.J. Kennedy and J.E. Gentle. *Statistical Computing*. Marcel Dekker, Inc., New York, 2007.

[23] D.A. Khavari, G.L. Sen, and J.L. Rinn. DNA methylation and epigenetic control of cellular differentiation. *Cell Cycle*, 9(19):3880–3883, 2011.

[24] D.C. Koestler, C.J. Marsit, B.C. Christensen, M.R. Karagas, R. Bueno, D.J. Sugarbaker, K.T. Kelsey, and E.A. Houseman. Semi-supervised recursively partitioned mixture models for identifying cancer subtypes. *Bioinformatics*, 26(30):2578–2585, 2010.

[25] P. Laird. Principles and challenges of genome-wide DNA methylation analysis. *Nature Reviews Genetics*, 11:191–203, 2010.

[26] J.T. Leek and J.D. Storey. Capturing heterogeneity in gene expression studies by surrogate variable analysis. *PLOS Genetics*, 3:1724–1735, 2007.

[27] B. Lindsay, C.C. Clogg, and J. Grego. Semiparametric estimation in the rasch model and related exponential response models, including a simple latent class model for item analysis. *Journal of the American Statistical Association*, 86:96–107, 1991.

[28] C.J. Marsit, B.C. Christensen, E.A. Houseman, M.R. Karagas, M.R. Wrensch, R.F. Yeh, H.H. Nelson, J.L. Wiemels, S. Zheng, M.R. Posner, M.D. McClean, J.K. Wiencke, K.T. Kelsey, J.J. Godleski, J.L. Longacker, C.R. Roelofs, R. Bueno, and D.J. Sugarbaker. Epigenetic profiling reveals etiologically distinct patterns of DNA methylation in head and neck squamous cell carcinoma. *Carcinogenesis*, 30(3):416–422, 2009.

[29] C.J. Marsit, E.A. Houseman, B.C. Christensen, K. Eddy, R. Bueno, D.J. Sugarbaker, H.H. Nelson, M.R. Karagas, and K.T. Kelsey. Examination of a CpG island methylator phenotype and implications of methylation profiles in solid tumors. *Cancer Research*, 66(21):10621–10629, 2006.

[30] C.J. Marsit, E.A. Houseman, A.R. Schned, M.R. Karagas, and K.T. Kelsey. Promoter hypermethylation is associated with current smoking, age, gender and survival in bladder cancer. *Carcinogenesis*, 28(8):1745–1751, 2007.

[31] C.J. Marsit, D.C. Koestler, B.C. Christensen, M.R. Karagas, E.A. Houseman, and K.T. Kelsey. DNA methylation array analysis identifies profiles of blood-derived DNA methylation associated with bladder cancer. *Journal of Clinical Oncology*, 29(9):1133–1139, 2011.

[32] G. Natoli. Maintaining cell identity through global control of genomic organization. *Immunity*, 33(1):12–24, 2011.

[33] G.M. Poage, B.C. Christensen, E.A. Houseman, M.D. McClean, J.K. Wiencke, M.R. Posner, J.R. Clark, H.H. Nelson, C.J. Marsit, and K.T. Kelsey. Genetic and epigenetic somatic alterations in head and neck squamous cell carcinomas are globally coordinated but not locally targeted. *PLOS ONE*, 5(3):e9651, 2010.

[34] J. Sehouli, C. Loddenkemper, T. Cornu, T. Schwachula, U. Hoffmuller, A. Grutzkau, P. Lohneis, T. Dickhaus, J. Grone, M. Kruschewski, A. Mustea, I. Turbachova, U. Baron, and S. Olek. Epigenetic quantification of tumor-infiltrating T-lymphocytes. *Epigenetics*, 6(2):236–246, 2011.

[35] L. Shen, M. Toyota, Y. Kondo, E. Lin, L. Zhang, Y. Guo, N.S. Hernandez, X. Chen, S. Ahmed, K. Konishi, S.R. Hamilton, and J.P. Issa. Integrated genetic and epigenetic analysis identifies three different subclasses of colon cancer. *Proceedings of the National Academy of Sciences of the United States of America*, 104(47):18654–18659, 2007.

[36] Y. Shen, S.D. Fouse, and G. Fan. Genome-wide DNA methylation profiling: the mDIP-chip technology. *Methods in Molecular Biology*, 568:203–216, 2009.

[37] K.D. Siegmund, P.W. Laird, and I.A. Laird-Offringa. A comparison of cluster analysis methods using DNA methylation data. *Bioinformatics*, 20:1896–1904, 2004.

[38] D.J. Spiegelhalter, A. Thomas, N.G. Best, and D. Lunn. *WinBUGS Version 1.4 User Manual*. Cambridge: Medical Research Council Biostatistics Unit, 2003.

[39] A.E. Teschendorff, U. Menon, A. Gentry-Maharaj, S.J. Ramus, S.A. Gayther, S. Apostolidou, A. Jones, M. Lechner, S. Beck, I.J. Jacobs, and M. Widschwendter. An epigenetic signature in peripheral blood predicts active ovarian cancer. *PLOS ONE*, 4(12):e8274, 2009.

[40] A.E. Teschendorff, J. Zhuang, and M. Widschwendter. Independent surrogate variable analysis to deconvolve confounding factors in large-scale microarray profiling studies. *Bioinformatics*, 27(11):1496–1505, 2011.

[41] J. Tost, H. El abdalaoui, and I.G. Gut. Serial pyrosequencing for quantitative DNA methylation analysis. *Biotechniques*, 40(6):721–726, 2006.

[42] M. Toyota, N. Ahuja, M. Ohe-Toyota, J.G. Herman, S.B. Baylin, and J.P. Issa. CpG island methylator phenotype in colorectal cancer. *Proceedings of the National Academy of Sciences of the United States of America*, 96(15):8681–8686, 1999.

[43] M. Toyota and J.P. Issa. CpG island methylator phenotypes in aging and cancer. *Seminars in Cancer Biology*, 9(5):349–357, 1999.

[44] G. Wieczorek, A. Asemissen, F. Model, I. Turbachova, S. Floess, V. Liebenberg, U. Baron, D. Stauch, K. Kotsch, J. Pratschke, A. Hamann, C. Loddenkemper, H. Stein, H.D. Volk, U. Hoffmuller, A. Grutzkau, A. Mustea, J. Huehn, C. Scheibenbogen, and S. Olek. Quantitative DNA methylation analysis of FOXP3 as a new method for counting regulatory T cells in peripheral blood and solid tissue. *Cancer Research*, 69(2):599–608, 2009.

[45] Z. Xiong and P.W. Laird. COBRA: a sensitive and quantitative DNA methylation assay. *Nucleic Acids Research*, 25(12):2532–2534, 1997.

PART VI

Detection of Copy Number Variations

Detection of Copy Number Variations from Array Comparative Genomic Hybridization Data Using Linear-chain Conditional Random Field Models

XIAOLIN YIN AND JING LI

To gain more understanding of the role of inheritable copy number polymorphisms (CNPs) in determining disease phenotypes, systematic mapping and cataloging of CNPs are being carried out. Array comparative genomic hybridization (aCGH) allows identification of copy number alterations across genomes. The key computational challenge in analyzing copy number variations (CNVs) using aCGH data is the detection of segment boundaries of copy number changes and inference of the copy number state for each segment. This chapter proposes a novel undirected graphical model based on conditional random fields for CNV detections in aCGH data. This model effectively combines data preprocessing, segmentation, and copy number state decoding into one unified framework. The approach depicted, termed CRF-CNV, provides great flexibility in defining meaningful feature functions. Therefore, it can effectively integrate local spatial information of arbitrary sizes into the model. For model parameter estimation, the conjugate gradient (CG) method has been adopted to optimize the likelihood, and

Probabilistic Graphical Models for Genetics, Genomics, and Postgenomics. First Edition. Christine Sinoquet & Raphaël Mourad (Eds). © Oxford University Press 2014. Published in 2014 by Oxford University Press.

efficient forward/backward algorithms have been developed within the CG framework. The final step of copy number decoding is based on the Viterbi algorithm. The method was evaluated using real data as well as data simulated with realistic assumptions. Finally, the method is compared with two popular, publicly available programs.

16.1 Introduction

Structure variations in DNA sequences such as inheritable copy number alterations have been reported to be associated with numerous diseases. It has also been observed that somatic chromosomal aberrations (i.e., amplifications and deletions) in tumor samples have shown different clinical or pathological features in different cancer types or subtypes [3, 6, 14]. To gain more understanding of the role of inheritable copy number polymorphisms (CNPs) in determining disease phenotypes, systematic mapping and cataloging of CNPs are needed and are being carried out. Identification of somatic copy number aberrations in cancer samples may lead to the discovery of important oncogenes or tumor suppress genes. With remarkable capacity from existing technologies in assessing copy number variants (CNVs), there is a great wave of interest recently from the research community to investigate inheritable as well as somatic CNVs [1, 3, 4, 6, 14, 16, 23].

Broadly speaking, there are essentially three technological platforms for CNV detection: array-based technology (including array comparative genomic hybridization (aCGH) [13, 20], as well as many other variants), single nucleotide polymorphism (SNP) genotyping technology [1, 14] and next-generation sequencing technology [2]. Array-based technology measures DNA copy number alterations from a disease/test sample relative to a normal/reference sample. In principle, for each gene (or a genomic segment), each individual inherits one copy from its father and one copy from its mother. The total number of copies is two. However, when there are copy number mutations in an individual, the total number of copies might be one (i.e., deletions), or three or more (i.e., amplifications/insertions). For each DNA fragment or "clone," aCGH can indirectly measure the number of copies through fluorescence intensities, which represent the abundance of DNAs hybridized to the segment. The intensity ratio of the test and the reference sample, measured in log scale, is expected to be proportional to the copy number change of the test sample relative to the reference sample (assumed to have two copies), though significant noise can be introduced from various sources during the process. Array-based technologies are primarily for large segments of amplifications and deletions, though different experimental platforms and designs using clones with different sizes may give very different resolutions and genome coverage [14]. The other two platforms, SNP genotyping and next-generation sequencing, use different techniques to measure copy number changes, and have gained popularity in recent years [2, 17]. We primarily focus on aCGH data analysis in this chapter.

Different platforms have different challenges. Naturally, one should use different approaches for these platforms by taking advantage of special properties from different data sets. Not surprisingly, various algorithms have been proposed for different data in recent years. On the other hand, the primary goal of all such studies is to identify contiguous sets of segments that share the same copy numbers. Therefore, the essential computational task for processing data from different platforms is the same: segment genome into discrete regions of the same copy number. Once the above segmentation step is done, each segment will be assigned a copy number. The latter task is called "classification". One important commonality in performing segmentation for data from different platforms is the spatial correlation among clones. Many existing approaches have taken advantage of such a property by utilizing the same methodology, Hidden Markov Models (HMMs), which can conveniently model spatial dependence using a chain structure. Results have shown initial success of HMMs [1, 4, 16, 23]. However, there is an inherited limitation for all these HMMs, i.e., they are all first-order HMMs and cannot take into consideration long-range dependence.

It has been shown that, in a variety of domains, conditional random fields (CRFs) consistently outperform HMMs, mainly because CRFs can potentially integrate all information from data [21]. This property makes CRFs particularly appealing to model CNV data since one can define feature functions using data from a region rather than a single or two data points for emissions and transitions in HMMs. We have proposed a novel undirected graphical model based on CRFs for CNV detections for aCGH data [25]. Our model effectively combines data preprocessing, segmentation, and copy number state decoding into one unified framework. Our approach (termed CRF-CNV) provides great flexibilities in defining meaningful feature functions; therefore it can effectively integrate local spatial information of arbitrary sizes into the model. For model parameter estimations, we have adopted the conjugate gradient (CG) method for likelihood optimization and developed efficient forward/backward algorithms within the CG framework. The final step of copy number decoding is based on the Viterbi algorithm. The method is evaluated using real data with known copy numbers as well as simulated data with realistic assumptions, and compared with two popular publicly available programs. Experimental results have demonstrated that CRF-CNV outperforms a Bayesian Hidden Markov Model-based approach on both data sets in terms of copy number assignments. When compared to a nonparametric approach, CRF-CNV has achieved much greater precision while maintaining the same level of recall on the real data, and the performances of the two methods are comparable when applied to simulated data.

The remainder of this chapter is organized as follows. In Section 16.2, we give a brief overview of aCGH data and existing approaches for detecting CNVs from aCGH data. Details about CRF model developments for CNVs and implementations are provided in Section 16.3. Our experimental results on two data sets and comparisons with other programs are presented in Section 16.4. We conclude this chapter with a few discussions in Section 16.5.

16.2 aCGH Data and Analysis

16.2.1 aCGH Data

Though, theoretically, our approach can be applied to data from different experimental platforms, we focus primarily on aCGH data in this analysis. Mathematically, aCGH data usually consist of an array of log_2 intensity ratios for a set of clones, as well as the physical position of each clone along a genome. Here, each clone is one data point in the array and all the clones are ordered according to their physical positions on the genome. Fig. 16.1(page 412) plots the normalized log_2 ratio of a human cancer cell line (i.e., sample) and normal reference DNA analyzed by Snijders et al. [18]. Each data point represents one genome segment/clone and the y-axis represents normalized log_2 intensity ratio. The primary goal in CNV detection based on aCGH is to segment a genome into discrete regions that share the same mean log_2 ratio pattern (i.e., have the same copy numbers). In a normal sample from humans, the copy number is usually two for all the autosomes. Ideally, the log_2 ratio of a clone should be 0 if a test sample (e.g., a cancer cell line) also has two copies of DNA, and the value should be 0.585 (or −1), if it has one single-copy gain

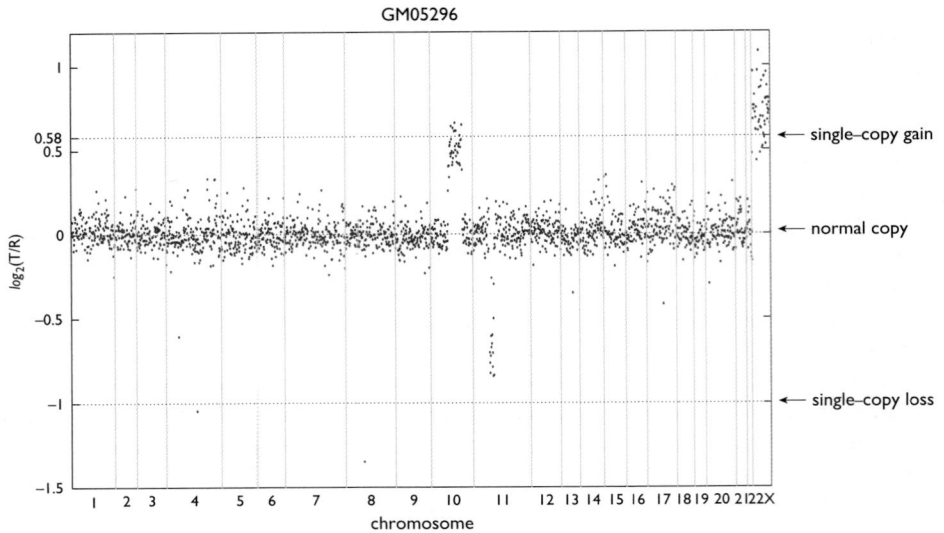

Fig 16.1 An aCGH profile of a Corriel cell line (GM05296). The borders between chromosomes are indicated by gray vertical lines. Dotted lines indicate the expected log_2 ratio $\frac{T}{R}$ (where T represents the gene expression level in the testing sample and R represents the gene expression level in the reference sample) for single-copy loss, same number of copies, and single-copy gain of this cell line relative to reference DNA.

(or single-copy loss), etc. The dotted lines in Fig. 16.1 from bottom to up show the three values for one, two, and three copies, respectively. A straightforward approach is to use some global thresholds to assign the number of copies for each clone based on its log_2 ratio. However, as shown in Fig. 16.1, aCGH data can be quite noisy with vague boundaries between different segments. It may also have complex local spatial dependence structure. These properties make the segmentation problem intrinsically hard. Approaches using global thresholds generally do not work well in practice.

16.2.2 Existing Algorithms

In general, a number of steps are needed to detect copy number changes from aCGH data. First, raw log_2 ratio data usually needs some preprocessing, including normalization and smoothing. Normalization is an absolutely necessary step to alleviate systemic errors due to experimental factors. Usually the input data are normalized by taking advantage of some control data sets where there are no differences in test and reference DNA. The purpose of normalization is to make the median or mean log_2 ratios in control data sets to be 0. Smoothing is used to reduce noise that is due to random errors or abrupt changes. Smoothing methods generally filter the data using a sliding window, attempting to fit a curve to the data while handling abrupt changes and reducing random errors. In addition to methods based on sliding windows, several other techniques have been proposed, which include quantile smoothing, wavelet smoothing, etc. (see [5, 8]).

The second step in analyzing aCGH data is referred to as *segmentation* and it aims to identify segments that share the same mean log_2 ratio. Broadly speaking, there are two related estimation problems. One is to infer the number and statistical significance of the alterations; the other is to locate their boundaries accurately. A few different algorithms have been proposed to solve these two estimation problems. Olshen et al. [12] have proposed a non-parametric approach based on the recursive circular binary segmentation (CBS) algorithm. Hupe et al. [9] have proposed an approach called GLAD, which is based on a median absolute deviation model, to separate outliers (observations that are numerically distant from the rest of the data) from their surrounding segments. Willenbrock and Fridlyand [24] have compared the performance of CBS (implemented in DNACopy) and GLAD using a realistic simulation model, and it was concluded that CBS in general is better than GLAD. We have adopted the simulation model from [24] in our experiment study. After obtaining the segmentation outcomes, a post-processing step is needed to combine segmentations with similar mean levels and to classify them as single-copy gain, single-copy loss, normal, multiple gains, etc. Methods such as GLADMerge [9] and MergeLevels [24] can post-process the segmentation results and label them accordingly.

As noted by Willenbrock and Fridlyand [24], it is more desirable to perform segmentation and classification simultaneously. An easy way to merge these two steps is to use a linear chain HMM. The underlying hidden states are the real copy numbers. Given a state, the log_2 ratio can be modeled using a Gaussian distribution. The transition from one state to another state reveals the likelihood of copy number changes between adjacent clones. Given observed data, standard algorithms (forward/backward, Baum–Welch and Viterbi) can be used to estimate parameters and to decode hidden states. A few variants of HMMs have been proposed for aCGH data in recent years [7, 16]. Lai et al. [11] have shown that HMMs performed the best for small aberrations given a sufficient signal/noise ratio. Guha et al. [7] have proposed a Bayesian HMM which can impose biological meaningful priors on the parameters. Shah et al. [16] have extended this Bayesian HMM by incorporating outliers and location-specific priors, which can be used to model inheritable copy number polymorphisms.

Notice that all these models are first-order HMMs which cannot capture long-range dependence. Intuitively, it makes sense to consider high-order HMMs to capture informative local correlation, which is an important property observed from aCGH data. However, considering higher orders will make HMMs more complex and computationally intensive.

16.3 Linear-chain CRF Model for aCGH Data

To overcome the limitations of HMMs, we propose a new model based on the theory of linear-chain conditional random fields (CRFs) [10, 21]. CRFs are undirected graphical models designed for calculating the conditional distribution of output random variables Y given input variables X. The term "random field," which is equivalent to "Markov network" in graphical theory, has been widely used in statistical physics and computer vision. CRFs are also known as conditional Markov networks [22]. Because CRFs are conditional models, dependences among variables X do not need to be explicitly specified. Therefore, one can define meaningful feature functions that can effectively capture local spatial dependence by using input variables X. CRFs have been widely applied to language processing, computer vision, and bioinformatics, with remarkable performance when compared with directed graphical models including HMMs.

In general, a linear-chain CRF (see Fig. 16.2, page 414) is defined as the conditional distribution

$$\mathbb{P}(Y|X) = \frac{1}{Z_\theta(X)} \exp \left\{ \sum_{i=1}^{n-1} \sum_{j=1}^{S(i)} \theta_{ij} f_{ij}(Y_i, Y_{i+1}, \tilde{X}_i) \right\},$$

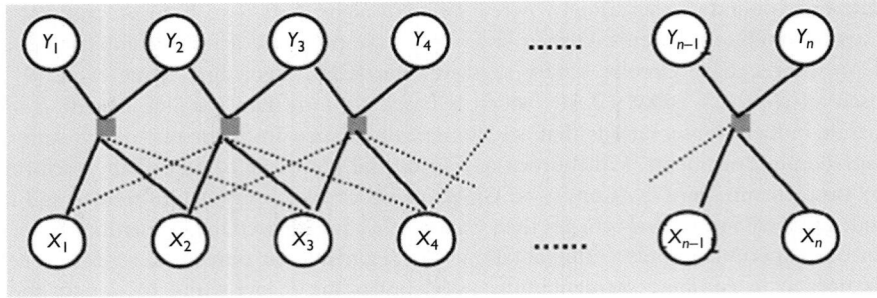

Fig 16.2 A linear-chain conditional random field model, where Y_i's are output variables and X_i's are input variables. They are connected through gray square nodes (which are not variables, but factors in factor graphs), indicating feature functions will be defined on those variables connected by square nodes. The dotted lines mean that for two adjacent output variables Y_i and Y_{i+1}, an arbitrary set of consecutive input variables consisting of the two corresponding input variables $(\{\ldots, X_i, X_{i+1}, \ldots\})$ can be linked to them.

where the partition function

$$Z_\theta(X) = \sum_Y \exp \left\{ \sum_{i=1}^{n-1} \sum_{j=1}^{S(i)} \theta_{ij} f_{ij}(Y_i, Y_{i+1}, \tilde{X}_i) \right\}.$$

Here $\theta(\theta = \{\theta_{ij}\})$ are parameters. Functions $\{f_{ij}\}$ are feature functions. $\tilde{X}_i$ is a neighbor set of X_i that is needed for computing features relating clones i and $i+1$. $S(i)$ is the total number of feature functions related to Y_i and Y_{i+1}. The variables connected by each square node in Fig. 16.2 are used to define potential feature functions. Each feature function will operate on pairs of adjacent label variables Y_i and Y_{i+1} and a neighbor set of X_i.

We define the linear-chain CRF model for aCGH data for one sample (actually one chromosome in one sample because different chromosomes can be treated independently) as follows. The general model for a set of observations can be easily obtained assuming independence between observations. Let $X = (X_1, \ldots, X_n)$ denote the normalized log_2 ratio intensities along one chromosome for a sample/cell line, where X_i is the log_2 ratio for clone i. One can assume that these n clones are sequentially positioned on a chromosome. Let $Y = (Y_1, \ldots, Y_n)$ denote the corresponding hidden copy number states, where Y_i belongs to $\{1, .., s\}$ and s is the total number of copy number states. These states usually indicate deletion, single-copy loss, neutral, single-copy gain, two-copy gain, or multiple-copy gain. The exact number of states and their meaning need to be specified based on specific input data. The conditional probability of Y given the observed log_2 ratio X based on our linear-chain CRF structure is defined as

$$\mathbb{P}(Y|X) = \frac{1}{Z_\theta(X)} \exp \left\{ \sum_{i=1}^{n} \sum_{j=1}^{s} [\lambda_j f_j(Y_i, \tilde{X}_i(u)) + \mu_j g_j(Y_i, \tilde{X}_i(u))] \right.$$

$$\left. + \sum_{j=1}^{s} \omega_j l_j(Y_1, \tilde{X}_1(u)) + \sum_{i=1}^{n-1} \sum_{j=1}^{s} \sum_{k=1}^{s} v_{jk} h_{jk}(Y_i, Y_{i+1}, \tilde{X}_{i,i+1}(u)) \right\} \qquad (16.1)$$

where the partition function is:

$$Z_\theta(X) = \sum_Y \exp\left\{ \sum_{i=1}^{n} \sum_{j=1}^{s} [\lambda_j f_j(Y_i, \tilde{X}_i(u)) + \mu_j g_j(Y_i, \tilde{X}_i(u))] \right.$$

$$\left. + \sum_{j=1}^{s} \omega_j l_j(Y_1, \tilde{X}_1(u)) + \sum_{i=1}^{n-1} \sum_{j=1}^{s} \sum_{k=1}^{s} v_{jk} h_{jk}(Y_i, Y_{i+1}, \tilde{X}_{i,i+1}(u)) \right\}.$$

Here $\theta = \{\lambda_j, \mu_j, \omega_j, v_{jk}\}$ are parameters that need to be estimated. Functions f_j, g_j, l_j, and h_{jk} are feature functions that need to be defined. $\tilde{X}_i(u)$ is defined as a neighbor set of X_i around clone i, i.e., $\tilde{X}_i(u) = \{X_{i-u}, \ldots, X_{i-1}, X_i, X_{i+1}, \ldots, X_{i+u}\}$, where u is a hyperparameter meant to define the dependence length. In principle, u can be defined based on physical distances between clones instead of the number of clones, which may better capture local dependence when clones are unevenly distributed on a chromosome. For simplicity, we use the number of clones here. For $i \leq u$ or $i \geq n - u$, $\tilde{X}_i(u)$ is redefined by proper truncation. For example $\tilde{X}_1(3) = \{X_1, \ldots, X_4\}$. Similarly, we define $\tilde{X}_{i,i+1}(u) = \{X_{i-u}, \ldots, X_{i-1}, X_i, X_{i+1}, X_{i+2}, \ldots, X_{i+1+u}\}, \tilde{X}_{i,i+1}^-(u) = \{X_{i-u}, \ldots, X_{i-1}, X_i\}$ and $\tilde{X}_{i,i+1}^+(u) = X_{i+1}, X_{i+2}, \ldots, X_{i+1+u}\}$. The two latter terms will be further used.

As illustrated in Fig. 16.2, although we also use a chain structure in our CRF model dedicated to CNV detection, the feature functions to be defined rely on observed data from a region. We define our feature functions by integrating all information from these neighborhood sets. Therefore, these feature functions can capture abundant local spatial dependence. The dependence length u plays a similar role as the width of the sliding window in smoothing methods. In addition, by using a CRF, we can effectively combine smoothing, segmentation, and classification into one unified framework. For notational simplification, we drop the parameter u in our subsequent discussions and write $\tilde{X}_i(u)$ as $\tilde{X}_i$, $\tilde{X}_{i,i+1}(u)$ as $\tilde{X}_{i,i+1}$, and so on.

16.3.1 Feature Functions

One important step for building our model is to define meaningful feature functions that can capture critical information from input data. Essentially, we define two types of feature functions, analogous to the emission and transition probabilities in HMMs. However, our feature functions can be of any form. Therefore, our model can provide much more flexibility and be able to capture long-range dependence. The emission feature functions $f_j(Y_i, \tilde{X}_i)$ and $g_j(Y_i, \tilde{X}_i)$ are defined as follows:

$$f_j(Y_i, \tilde{X}_i) = \begin{cases} med\ \tilde{X}_i & \text{if } Y_i = j \\ 0 & \text{otherwise,} \end{cases}$$

$$g_j(Y_i, \tilde{X}_i) = \begin{cases} (med\ \tilde{X}_i)^2 & \text{if } Y_i = j \\ 0 & \text{otherwise,} \end{cases}$$

where $med\ \tilde{X}_i$ is defined as the median value of set $\tilde{X}_i$. Our emission features serve two purposes. First they are used as median filters that will automatically smooth the input data. More importantly, the feature functions based on the first-order and second-order median statistics are robust sufficient statistics one can derive from a normal distribution.

The transition feature function $h_{jk}(Y_i, Y_{i+1}, \tilde{X}_{i,i+1})$ and the initial feature function $l_j(Y_1, \tilde{X}_1)$ are defined as follows:

$$h_{jk}(Y_i, Y_{i+1}, \tilde{X}_{i,i+1}) =$$

$$
\begin{cases}
\dfrac{(a_{j+1} - a_j)/2 + (a_{k+1} - a_k)/2}{(a_{j+1} - a_j)/2 + (a_{k+1} - a_k)/2 + med\,\tilde{X}^-_{i,i+1} - a_j + med\,\tilde{X}^+_{i,i+1} - a_k} & \text{if } Y_i = j, med\,\tilde{X}^-_{i,i+1} \geq a_j \text{ and} \\
& Y_{i+1} = k, med\,\tilde{X}^+_{i,i+1} \geq a_k, \\[2ex]
\dfrac{(a_{j+1} - a_j)/2 + (a_k - a_{k-1})/2}{(a_{j+1} - a_j)/2 + (a_k - a_{k-1})/2 + med\,\tilde{X}^-_{i,i+1} - a_j + a_k - med\,\tilde{X}^+_{i,i+1}} & \text{if } Y_i = j, med\,\tilde{X}^-_{i,i+1} \geq a_j \text{ and} \\
& Y_{i+1} = k, med\,\tilde{X}^+_{i,i+1} < a_k, \\[2ex]
\dfrac{(a_j - a_{j-1})/2 + (a_{k+1} - a_k)/2}{(a_j - a_{j-1})/2 + (a_{k+1} - a_k)/2 + a_j - med\,\tilde{X}^-_{i,i+1} + med\,\tilde{X}^+_{i,i+1} - a_k} & \text{if } Y_i = j, med\,\tilde{X}^-_{i,i+1} < a_j \text{ and} \\
& Y_{i+1} = k, med\,\tilde{X}^+_{i,i+1} \geq a_k, \\[2ex]
\dfrac{(a_j - a_{j-1})/2 + (a_k - a_{k-1})/2}{(a_j - a_{j-1})/2 + (a_k - a_{k-1})/2 + a_j - med\,\tilde{X}^-_{i,i+1} + a_k - med\,\tilde{X}^+_{i,i+1}} & \text{if } Y_i = j, med\,\tilde{X}^-_{i,i+1} < a_j \text{ and} \\
& Y_{i+1} = k, med\,\tilde{X}^+_{i,i+1} < a_k, \\[2ex]
0 & \text{otherwise,}
\end{cases}
$$

$$
l_j(Y_1, \tilde{X}_1) =
\begin{cases}
\dfrac{(a_{j+1} - a_j)/2}{(a_{j+1} - a_j)/2 + med\,\tilde{X}_1 - a_j} & \text{if } Y_1 = j, med\,\tilde{X}_1 \geq a_j, \\[2ex]
\dfrac{(a_j - a_{j-1})/2}{(a_j - a_{j-1})/2 + a_j - med\,\tilde{X}_1} & \text{if } Y_1 = j, med\,\tilde{X}_1 < a_j, \\[2ex]
0 & \text{otherwise.}
\end{cases}
$$

Here a_j denotes the mean log_2 ratio for clones with copy number state $j (j = 1, \ldots, s)$, while a_0 and a_{s+1} denote the greatest lower bound of log_2 ratio for clones with copy number state 1 and the least upper bound of log_2 ratio for clones with copy number state s, respectively. Without loss of generality, we assume $a_0 < a_1 < \cdots < a_{s+1}$. We define the initial feature function $l_j(Y_1, \tilde{X}_1)$ in a way such that data from the clone set $\tilde{X}_1$ will only provide information to its own state. Furthermore, when Y_1 is equal to j, the closer the value of $med\,\tilde{X}_1$ to a_j, the higher the value for $l_j(Y_1, \tilde{X}_1)$, the more information the data will provide, resulting in a higher contribution to parameter ω_j. The function $l_j(Y_1, \tilde{X}_1)$ will achieve its highest value of 1 when $med\,\tilde{X}_1$ is equal to a_j. The other terms (i.e., $(a_{j+1} - a_j)/2$ and $(a_j - a_{j-1})/2$) in $l_j(Y_1, \tilde{X}_1)$ are scale factors, which make $l_j(Y_1, \tilde{X}_1)$ to be 1/2 when the value of $med\,\tilde{X}_1$ is half way between a_{j-1} and a_j or half way between a_{j+1} and a_j (i.e., $med\,\tilde{X}_1 = (a_{j-1} + a_j)/2$, or $med\,\tilde{X}_1 = (a_{j+1} + a_j)/2$). The transition feature function $h_{jk}(Y_i, Y_{i+1}, \tilde{X}_{i,i+1})$ is similarly defined using the clone set $\tilde{X}_{i,i+1}$. When Y_i equals j and Y_{i+1} equals k, the closer the value of $med\,\tilde{X}^-_{i,i+1}$ to a_j and the value of $med\,\tilde{X}^+_{i,i+1}$ to a_k, the higher the value of $h_{jk}(Y_i, Y_{i+1}, \tilde{X}_{i,i+1})$ and the more information the data will contribute to v_{jk}. Those scale factors serve the same purpose as those in the emission feature functions. Clearly, both types of our feature functions can capture the local spatial dependence over a set of adjacent clones and thus potentially provide more robust inference about hidden copy number states.

16.3.2 Parameter Estimation

Unlike the standard algorithms for HMM training, there are significant computational challenges for efficiently and accurately estimating parameters for CRFs. For example, the partition function $Z_\theta(X)$ in model 16.1 needs to be explicitly calculated in order to estimate parameters. In contrast, there is no such partition function to be considered for HMMs. The implementation of the training algorithms for our proposed linear-chain CRF model requires sophisticated statistical and numerical algorithms. To our best knowledge, no existing implementations can be used to solve this problem. We propose the following algorithm for the parameter estimation.

In general, given a set of training data $\mathcal{D}$ consisting of D observations, i.e., individuals $\mathcal{D} = \{(X^{(d)}, Y^{(d)}), d = 1, \ldots, D\}$, to estimate parameter θ ($\theta = \{\lambda_j, \mu_j, \omega_j, \nu_{jk}\}$) in model 16.1, one needs to maximize a penalized conditional log-likelihood which is defined as follows:

$$
\begin{aligned}
L_\theta &= \sum_{d=1}^{D} \log(\mathbb{P}(Y^{(d)}|X^{(d)})) - \frac{\|\theta\|^2}{2\sigma^2} \\
&= \sum_{j=1}^{s} \lambda_j \sum_{d=1}^{D} \sum_{i=1}^{n} f_j\left(Y_i^{(d)}, \tilde{X}_i^{(d)}\right) + \sum_{j=1}^{s} \mu_j \sum_{d=1}^{D} \sum_{i=1}^{n} g_j\left(Y_i^{(d)}, \tilde{X}_i^{(d)}\right) \\
&\quad + \sum_{j=1}^{s} \omega_j \sum_{d=1}^{D} l_j\left(Y_1^{(d)}, \tilde{X}_1^{(d)}\right) + \sum_{j=1}^{s} \sum_{k=1}^{s} \nu_{jk} \sum_{d=1}^{D} \sum_{i=1}^{n-1} h_{jk}\left(Y_i^{(d)}, Y_{i+1}^{(d)}, \tilde{X}_{i,i+1}^{(d)}\right) \\
&\quad - \sum_{d=1}^{D} \log\left(Z_\theta(X^{(d)})\right) - \frac{\|\theta\|^2}{2\sigma^2}.
\end{aligned}
\tag{16.2}
$$

Here $\|\theta\|$ is the L2-norm of θ.[1] The penalization term $\|\theta\|^2/2\sigma^2$ is added for regularization, which is used to prevent overfitting. Before one can solve the optimization problem, one has to first specify an additional set of hyperparameters that include the dependence length u, the mean log_2 ratios $\{a_j, j = 0, \ldots, s + 1\}$ and the penalization coefficient σ^2. For each j belonging to $\{1, \ldots, s\}$, the set of $\{a_j\}$ can be directly estimated given the training data set $\mathcal{D}$, i.e., the maximum likelihood estimate of a_j is just the mean value of the log_2 ratios for all clones with copy number state j in $\mathcal{D}$ for $j = 1, \ldots, s$. In addition, a_0 and a_{s+1} can be imputed using the minimum log_2 ratio of all clones with copy number state 1, and the maximum value from all clones with copy number state s, respectively. For the dependent length u and the penalization coefficient σ^2, we rely on a grid-search approach through cross-validation, which performs an exhaustive search by partitioning the search space into small grids. More specifically, the original training set $\mathcal{D}$ will first be partitioned into two sets $\mathcal{D}_1$ and $\mathcal{D}_2$. We call $\mathcal{D}_1$ the new training set and $\mathcal{D}_2$ the validation set. For a given range of (discrete) parameter values of u and σ^2, we train the model on $\mathcal{D}_1$ and get estimates of θ for each fixed pair of (u_0, σ_0^2). The exact procedure to estimate θ given

---

[1] Recall that for a vector $b = (b_1, b_2, \ldots, b_n)$ of length n, L2-norm of b is computed as $\sqrt{b_1^2 + b_2^2 + \cdots + b_n^2}$.

(u_0, σ_0^2) will be further discussed shortly. We then apply the trained model with estimated parameters on the validation set $\mathcal{D}_2$ and record the prediction error under the current model. The model with the smallest prediction error as well as the associated parameters (u, σ^2, θ) will be kept. The prediction error is defined as the mean absolute error for all samples in the validation set $\mathcal{D}_2$. The absolute error for a clone i is defined as $| Y_i - \hat{Y}_i |$, where Y_i is the known copy number and $\hat{Y}_i$ is the predicted copy number. This measure not only captures whether a prediction is exactly the same as the real copy number but also reflects how close these two numbers are.

For a given set of hyperparameters $\{a_j\}$, u, and σ^2, the optimization of L_θ in equation 16.2 can be solved using gradient-based numerical optimization methods [19]. We choose the non-linear conjugate gradient (CG) method in our implementation, which generalizes the CG method to non-linear optimization. To find a local optimum of a non-linear function, the non-linear CG method only needs to calculate the gradient, i.e., the first derivatives of L_θ in our problem. The outline of our non-linear CG method is as follows. Given the objective function L_θ, we search in the gradient direction

$$d^{(0)} = \frac{\partial L_\theta}{\partial \theta} \Big|_{\theta=\theta^{(0)}},$$

where $\frac{\partial L_\theta}{\partial \theta}$ is the first derivative function of L_θ and $\theta^{(0)}$ is the initial value of θ. Then, we perform a line search in this direction until L_θ reaches the maximum and update the parameter θ accordingly:

$$\alpha^{(0)} = \arg\max_\alpha L\left(\theta^{(0)} + \alpha d^{(0)}\right),$$

$$\theta^{(1)} = \theta^{(0)} + \alpha^{(0)} d^{(0)}.$$

After this first iteration in the gradient direction $d^{(0)}$, the following steps constitute one iteration of moving along a subsequent conjugate direction $d^{(i)}$ and update θ accordingly:

1 Calculate the gradient direction $\xi^{(i)} = \frac{\partial L_\theta}{\partial \theta} \Big|_{\theta=\theta^{(i)}}$,

2 compute $\beta^{(i)} = \frac{(\xi^{(i)})' \xi^{(i)}}{(\xi^{(i-1)})' \xi^{(i-1)}}$,

3 update the conjugate direction $d^{(i)} = \xi^{(i)} + \beta^{(i)} d^{(i-1)}$,

4 perform a line search: optimize $\alpha^{(i)} = \arg\max_\alpha L\left(\theta^{(i)} + \alpha d^{(i)}\right)$,

5 update $\theta^{(i+1)} = \theta^{(i)} + \alpha^{(i)} d^{(i)}$.

The CG procedure runs until convergence. The final θ is regarded as the optimal parameter. The first-order derivatives of L_θ with respect to $\{\lambda_j\}$, $\{\mu_j\}$, $\{\omega_j\}$, and $\{v_{jk}\}$ are given by

$$\frac{\partial L_\theta}{\partial \lambda_j} = \sum_{d=1}^{D} \sum_{i=1}^{n} f_j\left(Y_i^{(d)}, \tilde{X}_i^{(d)}\right) - \sum_{d=1}^{D} \sum_{i=1}^{n} \sum_y \mathbb{P}\left(Y_i = y | X^{(d)}\right) f_j\left(y, \tilde{X}_i^{(d)}\right) - \frac{\lambda_j}{\sigma^2},$$
$$j = 1, \ldots, s,$$

$$\frac{\partial L_\theta}{\partial \mu_j} = \sum_{d=1}^{D} \sum_{i=1}^{n} g_j\left(Y_i^{(d)}, \tilde{X}_i^{(d)}\right) - \sum_{d=1}^{D} \sum_{i=1}^{n} \sum_y \mathbb{P}\left(Y_i = y | X^{(d)}\right) g_j\left(y, \tilde{X}_i^{(d)}\right) - \frac{\mu_j}{\sigma^2},$$
$$j = 1, \ldots, s,$$

$$\frac{\partial L_\theta}{\partial \omega_j} = \sum_{d=1}^{D} l_j\left(Y_1^{(d)}, \tilde{X}_1^{(d)}\right) - \sum_{d=1}^{D} \sum_{y} \mathbb{P}\left(Y_1 = y | X^{(d)}\right) l_j\left(y, \tilde{X}_1^{(d)}\right) - \frac{\omega_j}{\sigma^2},$$
$$j = 1, \ldots, s,$$

$$\frac{\partial L_\theta}{\partial v_{jk}} = \sum_{d=1}^{D} \sum_{i=1}^{n} h_{jk}\left(Y_i^{(d)}, Y_{i+1}^{(d)}, \tilde{X}_{i,i+1}^{(d)}\right) - \sum_{d=1}^{D} \sum_{i=1}^{n-1} \sum_{y,y^*} \mathbb{P}\left(Y_i = y, Y_{i+1} = y^* | X^{(d)}\right)$$
$$h_{jk}\left(y, y^*, \tilde{X}_{i,i+1}^{(d)}\right) - \frac{v_{jk}}{\sigma^2}, j, k = 1, \ldots, s.$$

The marginal distributions $\mathbb{P}(Y_i = y | X^{(d)}), (i = 1, \ldots, n)$ and $\mathbb{P}(Y_i = y, Y_{i+1} = y^* | X^{(d)})$, $(i = 1, \ldots, n-1)$ in the above derivatives can be efficiently computed using the forward–backward algorithms. First, define the following forward variables γ_i and backward variables η_i for $i = 1, \ldots, n$:

$$\gamma_1(y, d) = F\left(Y_1 = y, \tilde{X}_1^{(d)}\right) G\left(Y_1 = y, \tilde{X}_1^{(d)}\right) L\left(Y_1 = y, \tilde{X}_1^{(d)}\right), y = 1, \ldots, s,$$

$$\gamma_i(y, d) = F\left(Y_i = y, \tilde{X}_i^{(d)}\right) G\left(Y_i = y, \tilde{X}_i^{(d)}\right) \sum_{y^*=1}^{s} \Big\{ \gamma_{i-1}(y^*, d)$$
$$H\left(Y_{i-1} = y^*, Y_i = y, \tilde{X}_{i-1,i}^{(d)}\right) \Big\}, i = 2, \ldots, n; \ y = 1, \ldots, s,$$

$$\eta_n(y, d) = 1, \ y = 1, \ldots, s,$$

$$\eta_i(y, d) = \sum_{y^*=1}^{s} \Big\{ \eta_{i+1}(y^*) F\left(Y_{i+1} = y^*, \tilde{X}_{i+1}^{(d)}\right) G\left(Y_{i+1} = y^*, \tilde{X}_{i+1}^{(d)}\right)$$
$$H\left(Y_i = y, Y_{i+1} = y^*, \tilde{X}_{i,i+1}^{(d)}\right) \Big\}, i = n-1, \ldots, 1; y = 1, \ldots, s,$$

where the functions F, G, L and H are defined based on the feature functions:

$$F\left(Y_i, \tilde{X}_i\right) = \exp\left\{ \sum_{j=1}^{s} \lambda_j f_j\left(Y_i, \tilde{X}_i\right) \right\},$$

$$G\left(Y_i, \tilde{X}_i\right) = \exp\left\{ \sum_{j=1}^{s} \mu_j g_j\left(Y_i, \tilde{X}_i\right) \right\},$$

$$L\left(Y_1, \tilde{X}_1\right) = \exp\left\{ \sum_{j=1}^{s} \omega_j f_j\left(Y_1, \tilde{X}_1\right) \right\},$$

$$H\left(Y_{i-1}, Y_i, \tilde{X}_{i-1,i}\right) = \exp\left\{ \sum_{j=1}^{s} \sum_{k=1}^{s} v_{jk} h_{jk}\left(Y_{i-1}, Y_i, \tilde{X}_{i-1,i}\right) \right\}.$$

The marginal distributions can be obtained by combining forward and backward variables:

$$\mathbb{P}\left(Y_i = y | X^{(d)}\right) = \gamma_i(y,d)\,\eta_i(y,d)/Z_\theta\left(X^{(d)}\right), y = 1,\ldots,s; i = 1,\ldots,n, \tag{16.3}$$

$$\mathbb{P}\left(Y_i = y, Y_{i+1} = y^* | X^{(d)}\right) = \gamma_i(y,d)\,F\left(Y_{i+1} = y^*, \tilde{X}_{i+1}^{(d)}\right) G\left(Y_{i+1} = y^*, \tilde{X}_{i+1}^{(d)}\right)$$
$$H\left(Y_i = y, Y_{i+1} = y^*, \tilde{X}_{i,i+1}^{(d)}\right) \eta_{i+1}(y^*,d)/Z_\theta\left(X^{(d)}\right),$$
$$y, y^* = 1,\ldots,s; i = 1,\ldots,n-1 \tag{16.4}$$

and $Z_\theta(X^{(d)})$ can also be efficiently computed using forward variables $Z_\theta(X^{(d)}) = \sum_{y=1}^{s} \gamma_n(y,d)$.

We noticed that the calculation of the forward variables γ_i and backward variables η_i is not stable. It may suffer from overflows due to numerous products of exponentials. To achieve numerical stability, we take a *log* transformation in calculating the recursions. Similar techniques have been used in HMMs [15]. To calculate $\gamma_i(y,d)$ according to the formula defined earlier, first, let $y_0 = \arg\max_y \log \gamma_{i-1}(y,d)$. We will then divide each component in the recursive formula of $\gamma_i(y,d)$ before exponentiation by $\gamma_{i-1}(y_0,d)$. After taking *log* transformation, the new recursive relationship becomes:

$$\log \gamma_i(y,d) = \log \gamma_{i-1}(y_0,d) + \log\Big\{F\left(Y_i = y, \tilde{X}_i^{(d)}\right) G\left(Y_i = y, \tilde{X}_i^{(d)}\right)$$
$$H\left(Y_{i-1} = y_0, Y_i = y, \tilde{X}_{i-1,i}^{(d)}\right)\Big\} + \log\Big\{1 + \sum_{y^* \neq y_0} \exp\big[\log \gamma_{i-1}(y^*,d)$$
$$+ \log\left(H(Y_{i-1} = y^*, Y_i = y, \tilde{X}_{i-1,i}^{(d)})\right) - \log \gamma_{i-1}(y_0,d)$$
$$- \log\left(H(Y_{i-1} = y_0, Y_i = y, \tilde{X}_{i-1,i}^{(d)})\right)\big]\Big\}, i = 2,\ldots,n; y = 1,\ldots,s.$$

The backward variables $\eta_i(y,d)$ can be transformed similarly:

$$\log \eta_i(y,d) = \log \eta_{i+1}(z_0,d) + \log\Big(F\left(Y_{i+1} = z_0, \tilde{X}_{i+1}^{(d)}\right) G\left(Y_{i+1} = z_0, \tilde{X}_{i+1}^{(d)}\right)$$
$$H\left(Y_i = y, Y_{i+1} = z_0, \tilde{X}_{i,i+1}^{(d)}\right) + \log\Big\{1 + \sum_{y^* \neq z_0} \exp\big[\log \eta_{i+1}(y^*,d)$$
$$+ \log(F(Y_{i+1} = y^*, \tilde{X}_{i+1}^{(d)}) G(Y_{i+1} = y^*, \tilde{X}_{i+1}^{(d)}) H(Y_i = y, Y_{i+1} = y^*, \tilde{X}_{i,i+1}^{(d)}))$$
$$- \log \eta_{i+1}(z_0,d) - \log(F(Y_{i+1} = z_0, \tilde{X}_{i+1}^{(d)}) G(Y_{i+1} = z_0, \tilde{X}_{i+1}^{(d)})$$
$$H(Y_i = y, Y_{i+1} = z_0, \tilde{X}_{i,i+1}^{(d)}))\big]\Big\}, i = 1,\ldots,n-1; y = 1,\ldots,s,$$

where $z_0 = \arg\max_y \log \eta_{i+1}(y,d)$.

The marginal distributions in formulae 16.3 and 16.4 can be calculated using stable $\log \gamma_i$ and $\log \eta_i$. For example, $\mathbb{P}(Y_i = y | X^{(d)}) = \exp\{\log \gamma_i(y,d) + \log \eta_i(y,d) - \log Z_\theta(X^{(d)})\}, y = 1,\ldots,s; i = 1,\ldots,n$, where $\log Z_\theta(X^{(d)}) = \log \gamma_n(y_0,d) + \log(1 + \sum_{y \neq y_0} \exp[\log \gamma_n(y,d) - \log \gamma_n(y_0,d)])$, $y_0 = \arg\max_y \log \gamma_n(y,d)$.

It is important to discuss the computational cost of training. The partition function $Z_\theta(X)$ in the likelihood and the marginal distributions $\mathbb{P}(Y_i = y | X^{(d)})$, $\mathbb{P}(Y_i = y, Y_{i+1} = y^* | X^{(d)})$ in the gradient can be efficiently computed using the standard forward–backward algorithms. The forward–backward algorithm required for the three computations has an $O(ns^2)$ complexity, where n is the number of clones and s is the number of hidden states. However, each training data set will have its own partition function and marginal distributions, so we need to run a forward–backward procedure for each observation in the training data set and for each gradient computation. Therefore, the total cost for training is $O(ns^2dg)$, where d is the number of observations in the training data set, and g is the number of gradient computations required by the optimization procedure.

For graphical-model-based approaches such as HMMs, many researchers pool data across all chromosomes from all samples, which can dramatically reduce the number of parameters needed without sacrificing much on inference accuracy. We also take a similar approach. This is reflected by our homogeneous linear-chain CRF structure.

16.3.3 Evaluation Methods

We implemented the above proposed approach as a Matlab package termed CRF-CNV and evaluated its performance using a publicly available real data set with known copy numbers [18] and a synthetic data set generated in [24]. Notice that many clones have normal (two) copies of DNA; therefore, the number of correctly predicted state labels is not a good measure of performance of an algorithm. Instead, we compare the performance of CRF-CNV with two popular programs in terms of the number of predicted segments and the accuracy of segment boundaries, referred to as breakpoints. To summarize the performance of an algorithm over multiple chromosomes and individuals, we use a single value called F-measure, which is a combination of precision and recall. Recall that given the true copy number state labels and predicted labels, precision (P) is defined as $\frac{ntp}{np}$ and recall (R) is defined as $\frac{ntp}{nt}$, where ntp is the number of true positive (i.e., correctly predicted) breakpoints, np is the number of predicted breakpoints, and nt is the number of true breakpoints. F-measure is defined as $F = 2PR/(P + R)$, which aims to find a balance between precision and recall. The two programs we chose are CBS [12] and CNA-HMMer [16], both of which have been implemented as Matlab tools. As mentioned earlier, CBS is one of the most popular segmentation algorithms, and different groups have shown it in general performs better than many other algorithms. CNA-HMMer was chosen because we wanted to compare the performance of our CRF model with HMMs, and CNA-HMMer is an implementation of a Bayesian HMM model with high accuracy [16].

16.4 Experimental Results

16.4.1 A Real Example

The Coriell data are regarded as a well-known "gold standard" data set which was originally analyzed in [18]. The data can be downloaded from http://www.nature.com/ng/journal/v29/n3/suppinfo/ng754_S1.html and have been widely used in testing new algorithms and in comparing different algorithms. The CBS algorithm was applied to this data set in the original paper

[12]. The Coriell data consist of 15 cell lines, named GM03563, GM00143, ..., GM01524. We simply use the numbers $1, 2, \ldots, 15$ to represent these cell lines. For this particular data set, there are only three states ($s = 3$), i.e., single-copy loss, normal copy, and single-copy gain. Notice that unlike CBS, CRF-CNV requires training data to estimate parameters. It is unfair to directly compare the prediction results of CRF-CNV on training data with results from CBS. We took a simple approach which divides the 15 samples into three groups. Each group consists of 5 samples. In the first run, we used group 1 as training data and group 2 as validation data to obtain model parameters (as discussed in Subsection 16.3.2). We then used the model to predict data in group 3 (testing data), and recorded the prediction results. In the second and third run, we alternated the roles of groups 1–3 and obtained prediction results for samples in group 1 and group 2, respectively. Finally we summarized our results over all 15 samples. For example, for the first run, we first obtained $\{a_j, j = 0, \ldots, 4\}$ directly based on samples in group 1. The estimates for $\{a_j\}$ were $\{-1.348, -0.682, -0.001, 0.497, 0.810\}$. To estimate the penalization coefficient σ^2 and the dependent length u, we defined the search space as $A \times B = \{0, 1, 2, \ldots, 30\} \times \{0, 1, \ldots, 5\}$. For each value m in A, we let $\sigma^2 = 400 \times 0.8^m$ and $u = u_0$. Essentially, to search σ^2 in a broad range, we used a geometric decay. The upper bound on u was set as 5 because for aCGH data such as the Coriell data set, each clone can cover quite a long range of DNA. The optimal σ^2 and u were chosen by minimizing the prediction errors on samples in group 2 (the validation set). Our results indicated that the model where u and m are respectively equal to 1 and 21 achieves the lowest prediction error. The values of θs corresponding to the best model with the lowest prediction errors were recorded. We then applied Viterbi's algorithm to find the most possible hidden copy number states for samples in group 3, as well as the number and boundaries of segments. Run 2 and run 3 obtained results on group 1 and group 2. For the CNA-HMMer, one could either use its default priors, or use training data to obtain informative priors. We tested the performance of CNA-HMMer both with and without informative priors.

Table 16.1 shows the number of segments identified along the genome, for each individual from the gold standard. Table 16.1 similarly describes the predicted outcomes of the three algorithms CRF-CNV, CBS, and CNA-HMMer. The number of segments detected by CRF-CNV is exactly the same as the gold standard for almost all samples (except for samples 9 and 10). Further examination of samples 9 and 10 (see Fig. 16.3) reveals that the segment that we missed in sample 9 only has one clone, which has been smoothed out by our algorithm. The segment

Table 16.1 Comparison of the number of segments identified along the genome, by three algorithms, for each sample of the Coriell data set.

Method/Sample	1	2	3	4	5	6	7	8	9	10	11	12	13	14	15	Sum
Gold standard	5	3	5	3	5	3	5	5	5	5	3	5	2	3	3	60
CRF-CNV	5	3	5	3	5	3	5	5	3	3	3	5	2	3	3	56
CBS	17	42	7	6	9	5	5	6	5	5	13	7	17	3	9	156
CNA-HMMer(default)	9	83	9	7	11	3	7	5	11	11	21	5	16	10	16	224
CNA-HMMer (trained)	3	3	5	3	5	3	5	5	3	3	3	5	2	3	3	54

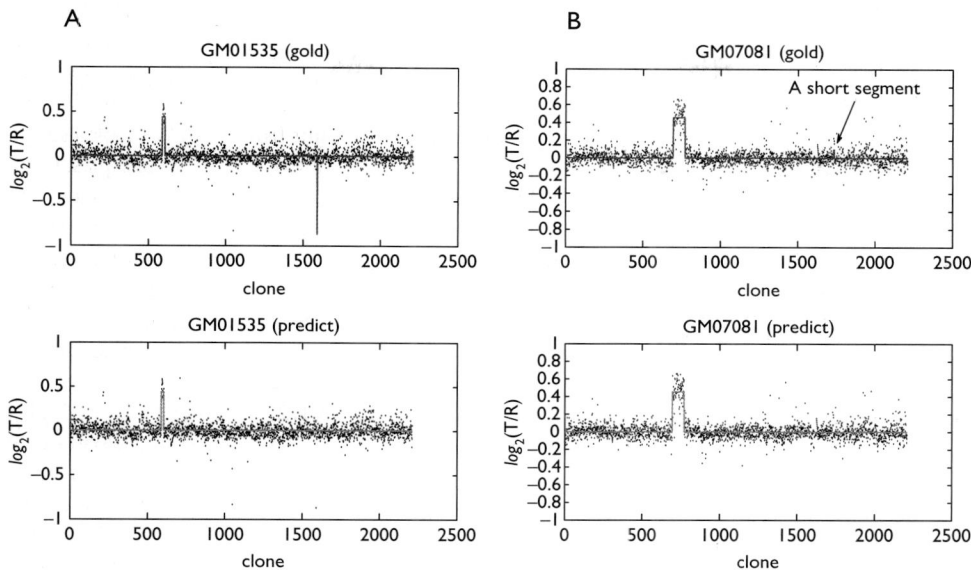

Fig 16.3 Predicted breakpoints by CRF-CNV (bottom) versus true breakpoints (top) on the two cell lines A) GM01535 and B) GM07081. R, gene expression level in the reference sample; T, gene expression level in the testing sample. Reproduced in color in the color plate section.

missed in sample 10 is also very short, and the signal is very weak. Our results clearly show that CBS generated many more segments comparing to the ground truth, which is consistent with the results in the original paper [18]. The overall number of segments reported by CNA-HMMer with default priors was even greater than the total number from CBS. On the other hand, once we used training data to properly assign informative priors for CNA-HMMer, it returned almost the same number of segments as CRF-CNV. The only exception was that CNA-HMMer missed one breakpoint in sample 1. This illustrates that by using correctly labeled training data, both CRF-CNV and CNA-HMMer can effectively eliminate all false positives in this data set. For the subsequent experiments, in the comparisons, we only report the results of CNA-HMMer with proper training.

As a comparison measure, the number of segments is a very rough index because it does not contain information about breakpoints. To further examine with which accuracy the breakpoints are predicted by each approach, we pooled all the breakpoints from all the samples and used the F-measure to compare the performances of the three algorithms. Notice that even though exact matches were possible, shifting by a few clones around boundaries was also likely given noisy input data. Therefore we used a match extent index D to allow some flexibility in defining matches of predicted breakpoints to those given by the gold standard. Table 16.2 shows F-measures given different match extent values for CRF-CNV, CNA-HMMer, and CBS. Clearly, CBS has the worst performance regardless of the match extent values. This partially reflects that it yields many false positives. The result for CNA-HMMer is quite accurate when no match extent is allowed. Then, the F-measure shows a modest increase as the value of D increases from 0 to 1. The result of CRF-CNV lies in between when the match index D is equal to 0. However, the performance of CRF-CNV is greatly enhanced when D is equal to 1, and finally it outperforms CNA-HMMer when D

Table 16.2 Comparison of *F*-measures with different match extent values for three algorithms.

Method/Match Extent	0	1	2	3	4
CRF-CNV	0.638	0.914	0.948	0.967	0.967
CNA-HMMer	0.877	0.947	0.947	0.947	0.947
CBS	0.333	0.500	0.519	0.519	0.519

is greater than or equal to 2. The primary reason that CRF-CNV shifted one or a few positions for many breakpoints is due to the automatic median smoothing step. In contrast, CNA-HMMer directly models outliers using prior distributions.

16.4.2 Simulated Data

Though results on the real data show that CRF-CNV has a higher performance than CBS and is comparable with CNA-HMMer when the match extent index is relaxed to 2, the experiment is limited because the data set size is very small. To further evaluate the performance of CRF-CNV, we tested the three algorithms using a simulated data set obtained from [24]. The data set consists of 500 samples each with 20 chromosomes. Each chromosome contains 100 clones. Each clone can be affected one of six possible copy number states. The authors generated these samples by sampling segments from a data set of primary breast tumor tissues and used several mechanisms (e.g., the fraction of cancerous cells in each sample, the variation of intensity values given a copy number state) to control the noise level. By using simulated data from the literature, we further evaluated the performance of CRF-CNV.

To train CRF-CNV, we divided the 500 samples into three groups as usual. This time, the training set group 1 contained samples 1–50, the validation set group 2 contained samples 51–100, and the test set group 3 contained samples 101–500. We used the same grid search approach as discussed earlier to obtain hyperparameters $\{a_j\}$, u, and σ^2. For each fixed set of hyperparameters, we used the CG method to obtain parameter θ. Finally, we used Viterbi's algorithm to decode the most possible hidden copy number state labels for samples in group 3 and compared the results with CNA-HMMer and CBS. In addition, we also compared the predictions by CRF-CNV for group 2 and group 3 to see with new testing data how much deterioration our model might incur based on sub-optimal parameters inferred from small number of samples. For easy comparison,

Table 16.3 Comparison of the numbers of segments predicted by three different approaches run on simulated data.

Method/Data	Group 2	Group 3
Gold	997	8299
CRF-CNV	966	8868
CNA-HMMer	784	6692
CBS	867	7430

Table 16.4 Comparison of *F*-measures with different match extent values for three algorithms and different test sets.

Data	Method/Match Extent	0	1	2	3	4
Group 2	CRF-CNV	0.590	0.792	0.875	0.900	0.906
	CNA-HMMer	0.702	0.801	0.832	0.852	0.855
	CBS	0.436	0.850	0.885	0.900	0.909
Group 3	CRF-CNV	0.568	0.786	0.864	0.889	0.896
	CNA-HMMer	0.697	0.805	0.840	0.858	0.869
	CBS	0.436	0.847	0.893	0.911	0.918

the results from CBS and CNA-HMMer are presented separately for these two groups. We also used group 1 as training data to assign informative priors for CNA-HMMer.

Table 16.3 shows the total numbers of segments in group 2 and group 3 predicted by CRF-CNV, CBS, and CNA-HMMer, respectively, and compares them with the known total number of segments. Interestingly, for this simulated data, both CBS and CNA-HMMer predicted a smaller number of segments. CRF-CNV predicted a smaller number of segments on group 2 and a greater number of segments in group 3. However, the number of segments does not provide a whole picture. We therefore examined the accuracy of boundary prediction by each method using the *F*-measure for both group 2 and group 3. Table 16.4 shows the *F*-measures for different methods, different groups, and different match extents. As expected, the *F*-measure increases as *D* increases from 0 to 4 for all methods and for both data groups. It is also not surprising to see that the results of CBS on group 2 and group 3 are consistent, like those for CNA-HMMer with group 2 and group 3. Interestingly, the performance of CRF-CNV with group 3 is also very close to its own performance with group 2. This property is desirable because it illustrates the robustness of CRF-CNV. The performance with new testing data was almost the same as the performance with validation data, which were used to select optimal model parameters. Notice that the sizes of training data and validation data are also very small. One can expect that with a training data set of small size, our approach can be used to reliably predict other data generated under the same experimental conditions. In terms of the performance of the three approaches, CNA-HMMer is more accurate than CRF-CNV, and CBS is the worst for the case of exact match. However, when we relax the matching criterion by increasing the value of *D*, both CBS and CRF-CNV achieve better performance than CNA-HMMer. The results of CNA-HMMer and CRF-CNV are consistent with those from the real data. CBS shows significantly higher performances compared to those observed for real data. However this might be attributed to the simulation process, because CBS was used to segment the 145 samples from the primary breast tumor data set [24].

16.5 Conclusion

The problem of detecting copy number variations has drawn much attention in recent years, and many computational approaches have been proposed to solve this issue. Among these computational developments, CBS has gained much popularity, and it has been shown that it generally performs better than other algorithms on simulated data [24]. However, as shown in the original paper (as well as rediscovered by our experiments), CBS was reported to yield many more false positives on copy number changes in the standard Coriell data set identified by spectral karyotyping [18]. Another commonly used technique for segmentation is the HMM. HMM approaches have the advantage of performing parameter (i.e., means and variances) estimation and copy number decoding within one framework, and their performance is expected to improve with more observations. However, almost all HMMs dedicated to aCGH are first-order Markov models and thus cannot incorporate long-range spatial correlations within data.

We have presented a novel computational model based on the theory of conditional random fields. Our model provides great flexibility in defining meaningful feature functions over an arbitrary neighborhood. Therefore, more local information can be incorporated, and more robust results are expected. In this chapter, we have defined feature functions using robust sufficient statistics that can effectively incorporate smoothing into our unified framework. We have also developed effective forward–backward algorithms within the conjugate gradient method for efficient computation of model parameters. We evaluated our approach using real data as well as simulated data, and the results show that our approach performed better than a Bayesian HMM on both data sets when a small shift is allowed while mapping breakpoints. When compared with CBS, our approach has fewer false positives with the real data set. With the simulated data set, the performance of our approach is comparable to that of CBS, which has been shown to be the best among the three popular segmentation approaches.

Like any other CRFs, in order to train our model, one has to rely on some training data. To be practically useful, Bayesian HMMs such as CNA-HMMer also need training data for proper assignment of informative priors. We argue that the problem is not as serious as it initially appears to be, primarily for two reasons. First, as illustrated in our experiments, our algorithm is indeed very robust and performs consistently, even if one cannot find optimal estimates for model parameters. For example, we used a simplified procedure in the analysis of the simulated data set by randomly picking one subset for training. Theoretically, parameters estimated from such a procedure might heavily depend on this particular subset and might not be necessarily globally optimal. However, the results shown in Table 16.4 demonstrate that the performance with new testing data is almost the same as with the validation data, which were used to tune the parameters. A 10-fold cross-validation is currently being performed to further verify this observation. Furthermore, the training size required by our algorithm is very small, as illustrated by both the real and the simulated data. The primary reason that our algorithm reported no false positives for segment numbers with the Coriell data is mainly that we learned the structure within the data by training. The drastic differences in performance by CNA-HMMer with and without training data also support this observation. Secondly, there indeed exist some experimentally validated data (such as the Coriell data set). One would expect more and more such data because eventually, for any prediction approaches, one has to experimentally verify some of their predictions. In addition, the robustness of our approach suggests that we most likely only need to train our algorithm for each specific platform. The parameters can then be used for future data to be generated on the same platform.

In terms of computation costs, CRF-CNV has two separate portions: time for training and time for prediction. The training requires intensive computations to optimize the log-likelihood and to determine the hyperparameters. In addition, one can also perform k-fold cross-validations, which will require much more computational time. In contrast, once the parameters have been estimated, the prediction phase is rather efficient. Fortunately, the training phase of our algorithm only requires a small data set, which makes the algorithm still practically useful. Possible extensions and applications of our algorithm can be applied to other high-throughput technologies for detecting copy number alterations.

REFERENCES

[1] N.P. Carter. Methods and strategies for analyzing copy number variation using DNA microarrays. *Nature Genetics*, 39(7 Suppl):S16–21, 2007.

[2] D.Y. Chiang, G. Getz, D.B. Jaffe, M.J. O'Kelly, X. Zhao, S.L. Carter, C. Russ, C. Nusbaum, M. Meyerson, and E.S. Lander. High-resolution mapping of copy-number alterations with massively parallel sequencing. *Nature Methods*, 6(1):99–103, 2009.

[3] E.K. Cho, J. Tchinda, J.L. Freeman, Y.J. Chung, W.W. Cai, and C. Lee. Array-based comparative genomic hybridization and copy number variation in cancer research. *Cytogenetic and Genome Research*, 115(3–4):262–272, 2006.

[4] S. Colella, C. Yau, J.M. Taylor, G. Mirza, H. Butler, P. Clouston, A.S. Bassett, A. Seller, C.C. Holmes, and J. Ragoussis. QuantiSNP: an objective Bayes hidden-Markov model to detect and accurately map copy number variation using SNP genotyping data. *Nucleic Acids Research*, 35(6):2013–2025, 2007.

[5] P.H. Eilers and R.X. de Menezes. Quantile smoothing of array CGH data. *Bioinformatics*, 21(7):1146–1153, 2005.

[6] J.L. Freeman, G.H. Perry, L. Feuk, R. Redon, S.A. McCarroll, D.M. Altshuler, H. Aburatani, K.W. Jones, C. Tyler-Smith, M.E. Hurles, N.P. Carter, S.W. Scherer, and C. Lee. Copy number variation: new insights in genome diversity. *Genome Research*, 16(8):949–961, 2006.

[7] S. Guha, Y. Li, and D. Neuberg. Bayesian hidden Markov modeling of array CGH data. *Journal of the American Statistical Association*, 103(482):485–497, 2008.

[8] L. Hsu, S.G. Self, D. Grove, T. Randolph, K. Wang, J.J. Delrow, L. Loo, and P. Porter. Denoising array-based comparative genomic hybridization data using wavelets. *Biostatistics*, 6(2):211–226, 2005.

[9] P. Hupe, N. Stransky, J.P. Thiery, F. Radvanyi, and E. Barillot. Analysis of array CGH data: from signal ratio to gain and loss of DNA regions. *Bioinformatics*, 20(18):3413–3422, 2004.

[10] J. Lafferty, A. McCallum, and F. Pereira. Conditional random fields: probabilistic models for segmenting and labeling sequence data. In *Eighteenth International Conference on Machine Learning*, pages 282–289. Morgan Kaufmann Publishers, 2001.

[11] W.R. Lai, M.D. Johnson, R. Kucherlapati, and P.J. Park. Comparative analysis of algorithms for identifying amplifications and deletions in array CGH data. *Bioinformatics*, 21(19):3763–3770, 2005.

[12] A.B. Olshen, E.S. Venkatraman, R. Lucito, and M. Wigler. Circular binary segmentation for the analysis of array-based DNA copy number data. *Biostatistics*, 5(4):557–572, 2004.

[13] D. Pinkel, R. Seagraves, D. Sudar, S. Clark, I. Poole, D. Kowbel, C. Collins, W.L. Kuo, C. Chen, Y. Zhai, S.H. Dairkee, B.M. Ljung, J.W. Gray, and D.G. Albertson. High resolution analysis of DNA copy number variation using comparative genomic hybridization to microarrays. *Nature Genetics*, 20(2):207–211, 1998.

[14] R. Redon, S. Ishikawa, K.R. Fitch, L. Feuk, G.H. Perry, T.D. Andrews, H. Fiegler, M.H. Shapero, A.R. Carson, W. Chen, E.K. Cho, S. Dallaire, J.L. Freeman, J.R. Gonzalez, M. Gratacos, J. Huang, D. Kalaitzopoulos, D. Komura, J.R. MacDonald, C.R. Marshall, R. Mei, L. Montgomery, K. Nishimura, K. Okamura, F. Shen, M.J. Somerville, J. Tchinda, A. Valsesia, C. Woodwark, F. Yang, J. Zhang, T. Zerjal, L. Armengol, D.F. Conrad, X. Estivill, C. Tyler-Smith, N.P. Carter, H. Aburatani, C. Lee, K.W. Jones, S.W. Scherer, and M.E. Hurles. Global variation in copy number in the human genome. *Nature*, 444(7118):444–454, 2006.

[15] S.L. Scott. Bayesian methods for Hidden Markov Models: recursive computing in the 21st century. *Journal of the American Statistical Association*, 97(457):337–351, 2002.

[16] S.P. Shah, X. Xuan, R.J. DeLeeuw, M. Khojasteh, W.L. Lam, R. Ng, and K.P. Murphy. Integrating copy number polymorphisms into array CGH analysis using a robust HMM. *Bioinformatics*, 22(14):e431–e439, 2006.

[17] F. Shen, J. Huang, K.R. Fitch, V.B. Truong, A. Kirby, W. Chen, J. Zhang, G. Liu, S.A. McCarroll, K.W. Jones, and M.H. Shapero. Improved detection of global copy number variation using high density, non-polymorphic oligonucleotide probes. *BMC Genetics*, 9(27):1471–2156, 2008.

[18] A.M. Snijders, N. Nowak, R. Segraves, S. Blackwood, N. Brown, J. Conroy, G. Hamilton, A.K. Hindle, B. Huey, K. Kimura, S. Law, K. Myambo, J. Palmer, B. Ylstra, J.P. Yue, J.W. Gray, A.N. Jain, D. Pinkel, and D.G. Albertson. Assembly of microarrays for genome-wide measurement of DNA copy number. *Nature Genetics*, 29(3):263–264, 2001.

[19] J.A. Snyman. *Practical Mathematical Optimization: An Introduction to Basic Optimization Theory and Classical and New Gradient-based Algorithms*. Springer, New York, 2005.

[20] S. Solinas-Toldo, S. Lampel, S. Stilgenbauer, J. Nickolenko, A. Benner, H. Dohner, T. Cremer, and P. Lichter. Matrix-based comparative genomic hybridization: biochips to screen for genomic imbalances. *Genes, Chromosomes and Cancer*, 20(4):399–407, 1997.

[21] C. Sutton and A. McCallum. An introduction to conditional random fields for relational learning. *Introduction to Statistical Relational Learning*, chapter 93:142–146, 2007.

[22] B. Taskar, P. Abbeel, and D. Koller. Discriminative probabilistic models for relational data. In A. Darwiche and N. Friedman, editors, *Proceedings of the Eighteenth Conference in Uncertainty in Artificial Intelligence (UA02)*, pages 485–492. Morgan Kaufmann Publishers, 2002.

[23] K. Wang, M. Li, D. Hadley, R. Liu, J. Glessner, S.F. Grant, H. Hakonarson, and M. Bucan. PennCNV: an integrated hidden Markov model designed for high-resolution copy number variation detection in whole-genome SNP genotyping data. *Genome Research*, 17(11):1665–1674, 2007.

[24] H. Willenbrock and J. Fridlyand. A comparison study: applying segmentation to array CGH data for downstream analyses. *Bioinformatics*, 21(22):4084–4091, 2005.

[25] X.L. Yin and J. Li. Detecting copy number variations from array CGH data based on a conditional random field model. *Journal of Bioinformatics and Computational Biology*, 8(2):295–314, 2010.

PART VII

Prediction of Outcomes from High-dimensional Genomic Data

CHAPTER 17

Prediction of Clinical Outcomes from Genome-wide Data

SHYAM VISWESWARAN

Prediction is a key component of clinical care, including individual risk assessment, diagnosis, prognosis, and selection of therapy. Probabilistic models of various types have been developed for making predictions from clinical data. Predictions using genome-wide data have the potential to improve clinical care. This chapter describes a probabilistic inference algorithm for prediction of clinical outcomes from genome-wide data by efficiently averaging over a large number of models. Bayesian model averaging (BMA) is the standard Bayesian approach wherein the prediction is obtained from a weighted average of the predictions of a set of models, with better models influencing the prediction more than others. The model-averaged naive Bayes (MANB) algorithm described here predicts clinical outcomes from genome-wide association study (GWAS) data by averaging over the predictions of all 2^n naive Bayes models, weighted by the posterior probability of each model. The MANB approach is then evaluated on a genome-wide Alzheimer's disease data set and compared with the naive Bayes method.

17.1 Introduction

Prediction is a key component of clinical care, including individual risk assessment, diagnosis, prognosis, and selection of therapy. Improvements in predictive performance have the potential to significantly improve patient outcomes and reduce healthcare costs. Probabilistic models of various types have been developed for making predictions from clinical data [6].

Probabilistic Graphical Models for Genetics, Genomics, and Postgenomics. First Edition. Christine Sinoquet & Raphaël Mourad (Eds). © Oxford University Press 2014. Published in 2014 by Oxford University Press.

Genome-wide patient-specific data are likely to become available to inform clinical care in the foreseeable future and provide significant opportunities for development of statistical models to improve prediction over what is currently possible from clinical data alone. The sheer magnitude of the number of features in genome-wide data (in the hundreds of thousands) presents formidable computational and modeling challenges. This chapter describes a probabilistic algorithm for prediction of clinical outcomes from genome-wide data by efficiently averaging over a large number of models.

The chapter has the following sections: Section 17.2 identifies some of the challenges in learning prediction models from genome-wide data; Section 17.3 provides related background information; Section 17.4 describes an efficient model-averaged naive Bayes (MANB) algorithm for prediction of clinical outcomes from genome-wide data; Section 17.5 provides details about the evaluation protocol; Section 17.6 provides the results from an evaluation of the MANB algorithm on a genome-wide Alzheimer's disease data set; and Section 17.7 concludes this chapter.

17.2 Challenges with Genome-wide Data

The commonest genetic variation in the human genome is the single nucleotide polymorphism (SNP), when at a specific location on the genome, only two variants (alleles) are observed across individuals in a population. The human genome is estimated to have about 20–30 million SNPs that constitute approximately 0.1% of the genome. With the development of gene chips that can measure a half-million SNPs or more (e.g., the Affymetrix 500K GeneChip), it is now possible to obtain a snapshot of the variation across the entire human genome.

The availability of gene-chip technology has led to a flurry of GWASs. The typical goal of GWASs has been twofold: (1) to identify SNPs (and corresponding genes) that are associated with a trait or disease and (2) to identify molecular pathways that are involved in disease. Thus, GWASs hope to elucidate the genetic causes and mechanisms that underlie traits or diseases. Fewer studies have investigated how well SNPs of an individual predict clinical outcomes about that individual, such as his or her risk of developing a disease. The focus of this chapter is on learning prediction models from GWAS data; this is in contrast to identifying explanatory models of disease that is the typical focus in GWASs.

Developing prediction models from GWAS data presents several challenges. One challenge is the high dimensionality of the data, which makes feature selection difficult. In addition, such data may have only a few strongly predictive features but many weak ones, and identifying all predictive features makes the problem even more challenging. When a subset of features is strongly predictive of the target outcome, then identifying those features will likely result in excellent prediction. However, when there are few or no strong features and many weak features, the effects of all features may have to be combined to achieve good prediction. One approach to feature selection that can adapt to both of these scenarios is model averaging. In model averaging, the final prediction is obtained by averaging the predictions of a number of models that contain different sets of features. In the scenario when there are several strong features, model averaging will behave like standard feature selection; in the other scenario, when there are many weak features, model averaging will aggregate the predictive effects of these features. A second challenge is that prediction models used in healthcare should have both good discrimination and good calibration. Discrimination is the ability to correctly separate the target outcome classes, while calibration measures how closely the predicted probabilities agree with the actual outcomes. It is obvious that prediction models should have good discrimination to provide accurate predictions; however, it is less obvious, though equally important, that well-calibrated predictions are needed for making rational decisions that lead to optimal clinical outcomes. A third challenge is computational tractability. Computationally efficient methods are needed to learn highly predictive and well-calibrated models from high-dimensional GWAS data.

This chapter describes a model-averaged naive Bayes (MANB) algorithm that predicts clinical outcomes from GWAS data by performing Bayesian model averaging over an exponential number of naive Bayes (NB) models. MANB averages over the predictions of every possible NB model with a distinct set of features, weighted by the posterior probability of each model. Compared to NB, MANB addresses all three challenges of GWAS data described above. It detects both strong and weak features, has better calibration, and is computationally as efficient as NB.

17.3 Background

This section provides background information about the NB model, Bayesian model averaging and Alzheimer's disease, as the MANB algorithm will be applied to predict Alzheimer's disease from GWAS data later on in the chapter.

17.3.1 The Naive Bayes Model

The naive Bayes is a probabilistic model that makes the simplifying assumption that any feature $X_1, X_2, \ldots, X_n$ in the set X is conditionally independent of any other given the value of the target variable T. Thus, for all values of $X_1, X_2, \ldots, X_n$ and T:

$$\mathbb{P}(X|T) \equiv \mathbb{P}(X_1, X_2, \ldots, X_n|T) = \prod_{i=1}^{n} \mathbb{P}(X_i|T). \qquad (17.1)$$

Given the prior probability distribution $\mathbb{P}(T)$ and the conditional probability distributions $\mathbb{P}(X_i|T)$, the posterior probability distribution $\mathbb{P}(T|x)$, where x is an instantiation of X, is obtained by applying Bayes' theorem:

$$\mathbb{P}(T|x) = \frac{\mathbb{P}(x|T)\mathbb{P}(T)}{\sum_t \mathbb{P}(x|t)\mathbb{P}(t)}. \qquad (17.2)$$

In equation (17.2), t is an instantiation of T, and the summation in the denominator is done over all possible values that T takes. An example of a small NB model with two features is shown in Fig. 17.1A.

The NB model has been used widely for prediction and classification in many domains because (1) it is learned efficiently from data; (2) it is compact, requiring modest amounts of memory; (3) it performs rapid predictions; and (4) it often has good discrimination in practice. The main disadvantage of the NB model is that its predictions are often miscalibrated, and the miscalibration is often accentuated when there are large numbers of features. Due to this problem, an NB model that is learned from high-dimensional GWAS data is likely to make predictions with posterior probabilities that are very close to 0 or 1.

17.3.2 Bayesian Model Averaging

Typically, methods that learn prediction models from data perform model selection in which a single good model is selected that summarizes the data well. This model is then used to make future predictions. However, given finite data, there is uncertainty in choosing one model to the exclusion of all others, and this can be especially problematic when the selected model is one of several distinct models that all summarize the data more or less equally well. A coherent approach to dealing with the uncertainty in model selection is Bayesian model averaging (BMA). BMA is the standard Bayesian approach wherein the prediction is obtained from a weighted average of the predictions of a set of models, with better models influencing the prediction more than others. There is also a biological rationale for the model-averaging approach. In GWASs, a fundamental problem is genetic heterogeneity due to some mutations affecting only some patients. Model averaging will likely include the effects of such variants that are found only in a subpopulation.

Theoretically, BMA is expected to have better predictive performance than any single model as described by [11]. This result is supported empirically by a wide variety of case studies [12]. For example, [17] applied BMA to select genes from DNA microarray data to predict prognosis in breast cancer, differentiate between two leukemia subtypes, and distinguish among three types of hereditary breast cancer and showed that BMA identified smaller numbers of relevant genes, with comparable prediction accuracy to other methods that identified larger numbers of genes. In addition, [8] provides several clinical case studies in which BMA performed better than various types of model selection. A good overview of BMA in general is provided in [8] and of BMA in the context of Bayesian networks is provided in [10].

17.3.3 Alzheimer's Disease

Alzheimer's disease is the commonest neurodegenerative disease and the commonest cause of dementia associated with aging [7]. Alzheimer's disease is characterized by adult onset of progressive dementia that typically begins with subtle memory failure and progresses to a variety of cognitive deficits like confusion, language disturbance, and poor judgment.

Alzheimer's disease is divided into early-onset familial Alzheimer's disease, in which the disease begins before 65 years of age, and late-onset Alzheimer's disease (LOAD), in which it begins at 65 years of age or later [1]. Early-onset Alzheimer's disease is a rare disease, and its genetic basis is well established. Most cases of early-onset familial Alzheimer's disease are caused by mutations in one of three genes: amyloid precursor protein gene, presenelin 1, or presenelin 2.

LOAD is the more common form of Alzheimer's disease and is widespread, affecting almost half of all people over the age of 85 years. LOAD has both genetic and environmental factors, and its genetic basis is more complex than that of early-onset Alzheimer's disease. Elucidating the role of genetic variants in the pathogenesis and development of LOAD has been a major focus of LOAD GWASs. The APOE SNP rs429358 is the most well-known SNP associated with increased risk of developing LOAD. In addition, in the past several years, GWASs have identified several other genetic loci associated with LOAD. The AlzGene website is a comprehensive and updated resource of Alzheimer's disease genetic studies and provides an updated list of LOAD-associated SNPs identified by GWASs [2].

17.4 The Model-Averaged Naive Bayes (MANB) Algorithm

The model averaged naive Bayes (MANB) algorithm averages over the predictions of every possible NB model with a distinct set of features, weighted by the posterior probability of each model. The MANB algorithm was initially described by [5] and was later applied to genomic data by [15]. An overview of the algorithm is provided first, followed by a more detailed description.

17.4.1 Overview of the MANB Algorithm

Inference with a NB model with features X and target T consists of deriving $\mathbb{P}(T|x)$ for a test instance with feature values x and is given by equation (17.2). The MANB algorithm derives $\mathbb{P}(T|x)$ by model averaging over all 2^n NB models, where n is the cardinality of X (i.e., the total number of features in the data set). For example, in Fig. 17.1, n equals 2, and there are 2^2, that is, four models over which MANB averages.

The simple example shown in Fig. 17.1 is now used to illustrate Bayesian model averaging. BMA is based on the notion of averaging over a set of possible models and weighting the prediction (inference) of each model according to its probability given training data set D where $D = X \cup T$. The model-averaged prediction for a test instance with feature values x is given by the following equation:

$$\mathbb{P}(T|x) = \sum_M \mathbb{P}(T|x, M)\mathbb{P}(M|D). \tag{17.3}$$

Consider the four NB models on two features X_1 and X_2 in Fig. 17.1, and suppose that, given D, the models a, b, c, and d are assigned probabilities of 0.5, 0.1, 0.3, and 0.1, respectively. Suppose further that for a test instance with feature values $x = \langle true, false \rangle$, the models a, b, c, and d predict $T = true$ as 0.9, 0.5, 0.8, and 0.7, respectively. Then, according to equation (17.3), the model-averaged estimate of $\mathbb{P}(true | \langle true, false \rangle)$ is $(0.5 \times 0.9) + (0.1 \times 0.5) + (0.3 \times 0.8) + (0.1 \times 0.7) = 0.81$.

As the number of features increases, the number of NB models increases exponentially. For example, for n equal to 100, 2^{100} is close to 10^{30}, which is far too many models to average over in an exhaustive way. This implies that it is not feasible to perform model averaging by explicitly performing inference with each NB model and averaging them to obtain $\mathbb{P}(T|x)$ as illustrated in the above example.

Fortunately, the independence relationships inherent in the NB models allow considerably more efficient model averaging. In the case of Bayesian networks, Buntine describes how to use a single conditional probability distribution to compactly represent the model-averaged relationship between a child node and its parent nodes [3]. In the NB model, each feature is a child node, with a single parent node, which is the target. Using Buntine's compact representation, [5] explains how it can be used to efficiently perform model averaging over all NB models on a set of features.

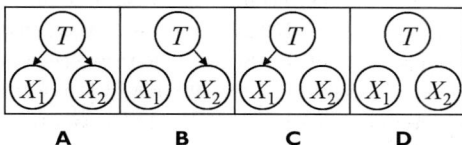

Fig 17.1 Four possible naive Bayes models for a data set with two features. T is the target node, and X_1 and X_2 are feature nodes.

By using this method, MANB inference becomes linear in the number of features and is as efficient as NB inference. Thus, rather than requiring time on the order of $O(2^n)$ to perform inference using model averaging, it only requires time of the order of $O(n)$.

The efficiency obtained by MANB inference is based on several assumptions including multinomial features, no missing values in the training data set, parameter independence, and structure modularity. Further details about these assumptions are provided in Appendix 17.A (page 440).

17.4.2 Details of the MANB Algorithm

This subsection provides a set of equations that describe the MANB algorithm. It first describes the main inference task and then successively decomposes it into its constituent parts.

Let X denote a set of n discrete-valued features, namely $\{X_1, X_2, \ldots, X_n\}$, and let x denote the values (an instantiation) of the features in X in a test instance. Suppose feature X_i has r_i possible values that are coded by the integers from 1 to r_i. Then r_i is the dimensionality of X_i. For example, X_i could represent a SNP that has three genotype values, and those values could be encoded as 1, 2, and 3. Let x_i be the value of feature X_i in an instance. Let T denote the discrete-valued target variable to be predicted and let r_T denote the dimensionality of T. For example, T could represent a disease such as LOAD that exhibits the value *absent* or *present* that could be encoded as 1 and 2. The inference task is to compute a model-averaged posterior probability for each value t of T conditioned on x.

The following equation is obtained by applying Bayes theorem:

$$\mathbb{P}_a(t|x) = \frac{\mathbb{P}_a(x|t)\mathbb{P}(t)}{\sum_{t'=1}^{r_T} \mathbb{P}_a(x|t')\mathbb{P}(t')}. \tag{17.4}$$

The subscript a in equation (17.4) denotes a model-averaged probability. Assuming that the features are conditionally independent of each other, given the value of the target T, the term $\mathbb{P}_a(x|t)$ is factored as:

$$\mathbb{P}_a(x|t) = \prod_{i=1}^{n} \mathbb{P}_a(x_i|t). \tag{17.5}$$

Each of the terms in equation (17.5) is estimated using a training data set and prior probabilities that are described below. Let the training data set D where $D = X \cup T$ contain m instances, and let D_i denote the part of the data set D that contains the values for just feature X_i and target variable T. In [5], it is proven that BMA over all 2^n NB models is equivalent to using the following value for each term in equation (17.5):

$$\mathbb{P}_a(x_i|t) = \mathbb{P}(T \rightarrow X_i|D_i)\,\mathbb{P}(x_i|t, D_i) + \mathbb{P}(T \ldots X_i|D_i)\,\mathbb{P}(x_i|D_i), \tag{17.6}$$

where $T \rightarrow X_i$ denotes that T and X_i are probabilistically dependent and $T \ldots X_i$ denotes that they are independent. When they are dependent, the conditional probability $\mathbb{P}(x_i|t, D_i)$ is used to estimate $\mathbb{P}_a(x_i|t)$. When they are independent, $\mathbb{P}(x_i|D_i)$ is used. Equation (17.6) can be viewed as using model averaging (regarding whether a relationship between T and X_i is present or not) to provide smoothing of the probability $\mathbb{P}_a(x_i|t)$ that is being estimated by equation (17.6). This smoothing is in addition to the smoothing that will be done in estimating $\mathbb{P}(x_i|t, D_i)$ and $\mathbb{P}(x_i|D_i)$ (see below), which also appears in equation (17.6).

Once $\mathbb{P}_a(x_i|t)$ has been derived for each value x_i (of feature X_i) and each value t (of target T), those probabilities can be used in equations (17.5) and (17.4) to calculate the posterior probability distribution of T conditioned on any instantiation x of X.

The derivation of each of the terms in equation (17.6) is described after introducing some notation. Let N_{ijk} denote the number of times in D_i that feature X_i has the value k when target T has the value j. To keep the notation simple, assume that x_i equals the value k and t equals the value j. Let $N_{ij} = \sum_{k=1}^{r_i} N_{ijk}$ and let $N_i = \sum_{j=1}^{r_T} N_{ij}$. Note that for all i, N_i equals N, where N is the total number of instances in D. Finally, let $N_{i*k} = \sum_{j=1}^{r_T} N_{ijk}$.

The term $\mathbb{P}(x_i|t, D_i)$ is computed using the Bayesian approach. A uniform prior distribution is specified for $\mathbb{P}(x_i|t)$ which is updated (using the training data) to obtain a posterior distribution for $\mathbb{P}(x_i|t, D_i)$. The mean of the posterior distribution is the Bayesian estimate for the term $\mathbb{P}(X_i = k|T = j, D_i)$ and is given by the following equation:

$$\mathbb{P}(X_i = k|T = j, D_i) = \frac{N_{ijk} + 1}{N_{ij} + r_i}. \tag{17.7}$$

The prior distribution specified for $\mathbb{P}(x_i|t)$ to derive equation (17.7) is known as the *parameter prior* [4]. Equation (17.7) provides an alternative way to estimate the parameters of a NB model with features X and target T instead of the commonly used maximum likelihood estimates.

In a manner similar to equation (17.7), the Bayesian estimate for $\mathbb{P}(X_i = k|D_i)$ is given by the following equation:

$$\mathbb{P}(X_i = k|D_i) = \frac{N_{i*k} + 1}{N_i + r_i}. \tag{17.8}$$

Next, the terms $\mathbb{P}(T \rightarrow X|D_i)$ and $\mathbb{P}(T \ldots X|D_i)$ in equation (17.6) are derived.

$$\mathbb{P}(T \rightarrow X_i|D_i) = \frac{\mathbb{P}(D_i|T \rightarrow X_i)\,\mathbb{P}(T \rightarrow X_i)}{\mathbb{P}(D_i|T \rightarrow X_i)\,\mathbb{P}(T \rightarrow X_i) + \mathbb{P}(D_i|T \ldots X_i)\,\mathbb{P}(T \ldots X_i)}, \tag{17.9}$$

$$\mathbb{P}(T \ldots X_i|D_i) = \frac{\mathbb{P}(D_i|T \ldots X_i)\,\mathbb{P}(T \ldots X_i)}{\mathbb{P}(D_i|T \rightarrow X_i)\,\mathbb{P}(T \rightarrow X_i) + \mathbb{P}(D_i|T \ldots X_i)\,\mathbb{P}(T \ldots X_i)}. \tag{17.10}$$

The terms on the right side of equation (17.9) are calculated as follows. The term $\mathbb{P}(T \rightarrow X_i)$ is the prior probability that T and X_i are probabilistically dependent and $\mathbb{P}(T \ldots X_i) = 1 - \mathbb{P}(T \rightarrow X_i)$. The prior probability is also known as the *structure prior*. A uniform structure prior is specified by q/n where n is the total number of features and q is the expected number of features that are predictive of T.

The term $\mathbb{P}(D_i|T \rightarrow X_i)$ in equation (17.9) is given by the following equation based on assumptions described in Appendix 17.A (page 440):

$$\mathbb{P}(D_i|T \rightarrow X_i) = \prod_{j=1}^{r_T} \left(\frac{(r_i - 1)!}{(N_{ij} + r_i - 1)!} \prod_{k=1}^{r_i} N_{ijk}! \right). \tag{17.11}$$

The details of the derivation of equation (17.11) are provided in Appendix 17.A. In a manner similar to equation (17.11), $\mathbb{P}(D_i|T \ldots X_i)$ in equation (17.9) is derived as follows:

$$\mathbb{P}(D_i|T\ldots X_i) = \frac{(r_i - 1)!}{(N_i + r_i - 1)!} \prod_{k=1}^{r_i} N_{i*k}!. \tag{17.12}$$

Now, all the terms in equation (17.6) have been derived, and the model-averaged posterior probability for each value t of T conditioned on x is estimated using equation (17.4).

17.5 Evaluation Protocol

This subsection briefly describes the Alzheimer's disease GWAS data set and the evaluation protocol used to assess the performance of the MANB and NB algorithms.

17.5.1 Data set

Several GWASs for LOAD have been conducted, and the data set used for evaluating the MANB algorithm is from one such study. The LOAD GWAS data were collected and analyzed originally by [13]. The genotype data were collected on 1411 individuals, of which 861 had LOAD and 550 did not. The target variable that is predicted is the LOAD status of the individual; it is binary and has the values *absent* and *present*. For each individual, the genotype data consist of 502 627 SNPs that were measured on an Affymetrix chip, reduced to 312 316 SNPs after applying quality controls by the original investigators. In addition, two APOE-related SNPs, namely rs429358 and rs7412, were separately genotyped. In total, 312 318 SNPs were used as features.

17.5.2 Protocol

The performances of the algorithms were evaluated using fivefold cross-validation. The data set was randomly partitioned into five approximately equal subsets, such that each subset had a similar proportion of individuals who had LOAD. For each algorithm, a model was trained on four subsets and was applied to obtain a LOAD prediction for each instance (individual) in the remaining subset; this process was done once for each of the five subsets. The predictions on the five subsets were then pooled to obtain a LOAD prediction for each of the 1411 individuals in the data set. The performance measures reported are based on those 1411 predictions.

Two performance measures are used: one measures discrimination and the other measures calibration. The area under the Receiver Operating Characteristic (ROC) curve (AUC) was used for measuring discrimination and the Hosmer-Lemeshow goodness-of-fit statistic was used for measuring calibration. Details of these two measures are provided in Appendices 17.B and 17.C (pages 443–444).

In addition, each algorithm's computation times are reported. The fivefold cross-validation process generated five models for each algorithm. For a given algorithm, the training time results reported are the average computation times for the five models learned by the algorithm.

17.6 Results

The performance of MANB was compared to that of NB on AUC, calibration, and computation time. For MANB, the parameters were estimated from the data with equations (17.7) and (17.8), and for NB, the parameters were estimated from the data with equation (17.7). For MANB, the structure prior m/n was set to 20/312318. The value 20 was chosen subjectively, informed by the number of strongest SNP predictors of LOAD that have been reported in the literature.

Fig. 17.2 shows that MANB has an AUC of about 0.72, while NB has an AUC of about 0.59, which is significantly different from that of the AUC of MANB ($p < 0.00001$). NB predicted almost all the test instances as having a posterior probability for LOAD very close to 0 or 1; such extreme predictions tend to occur with NB when there are a large number of features in the model. Fig. 17.3 (page 440) shows that NB (triangles) is very poorly calibrated, while MANB is better calibrated (circles) than NB.

In terms of computation time, both MANB and NB required only about 16 seconds to train a model (not including about 27 seconds to load the data into the main memory on a computer with 2.33GHz Intel processor and 2GB RAM). For both algorithms, the time required to predict each test instance was less than 0.1 second.

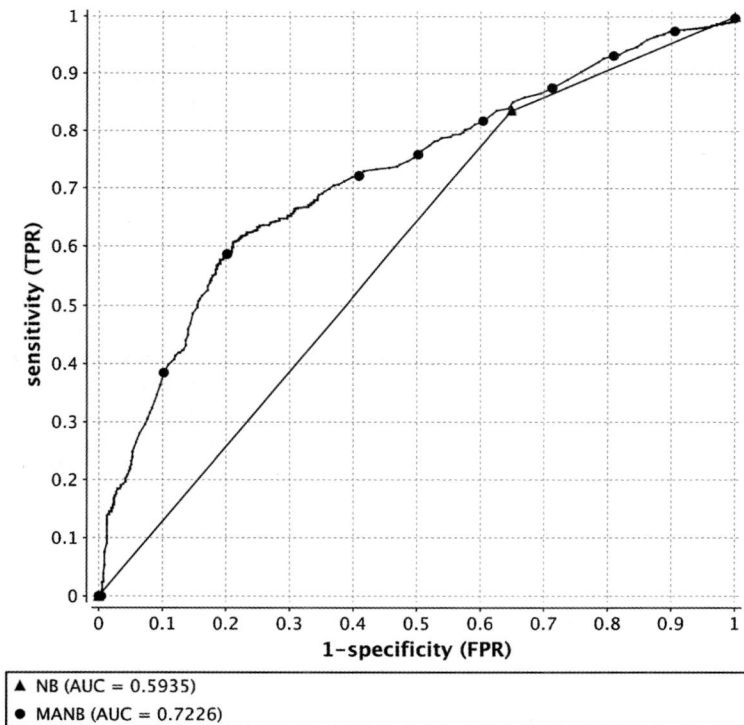

▲ NB (AUC = 0.5935)
● MANB (AUC = 0.7226)

Fig 17.2 The ROC curves and AUCs for MANB and NB algorithms. NB: naive Bayes; MANB: model-averaged naive Bayes; TPR: true positive rate; FPR: false positive rate.

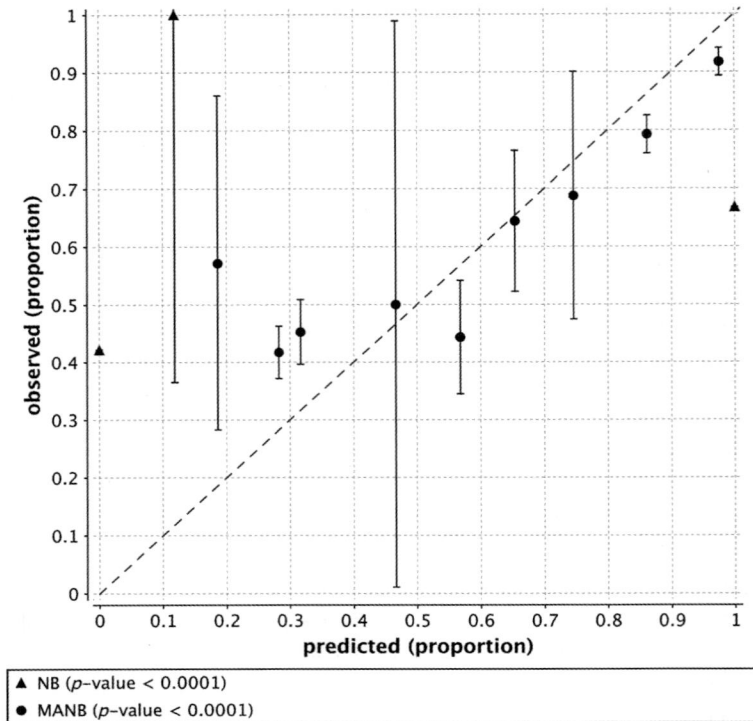

Fig 17.3 The mean calibration with 95% confidence intervals for MANB and NB algorithms. The p-values are for the Hosmer–Lemeshow goodness-of-fit statistic. NB: naive Bayes; MANB: model-averaged naive Bayes.

17.7 Conclusion

Developing prediction models from GWAS data present several challenges including feature selection in high-dimensional data, learning models with good discrimination, and calibration and computational efficiency. This chapter described the MANB algorithm that is designed to predict outcomes from GWAS data by averaging over a large number of NB models. The results show that, when evaluated on a LOAD GWAS data set, MANB performed significantly better than NB, in terms of both AUC and calibration, and that MANB was computationally as efficient as NB.

APPENDIX 17.A DERIVATION OF THE BAYESIAN SCORE

This section describes the derivation of the Bayesian score expressed in equation (17.12) in Subsection 17.4.2 (page 436) and the assumptions made in deriving it.

Let X be a set of n discrete-valued features, namely $\{X_1, X_2, \ldots, X_n\}$. Let feature X_i have r_i possible values, and let the target T have r_T possible values. Let the training data set D where $D = X \cup T$ contain m instances, and let D_i denote the part of the data set D that contains the values for just X_i and T. Let N_{ijk} denote the number of times in D_i that feature X_i has the value k when target T has the value j. Let $N_{ij} = \sum_{k=1}^{r_i} N_{ijk}$ and $N_i = \sum_{j=1}^{r_T} N_{ij}$.

Equation (17.11) in Subsection 17.4.2 (page 436) is rewritten as:

$$\mathbb{P}(D_i|S) = \prod_{j=1}^{r_T} \left(\frac{(r_i - 1)!}{(N_{ij} + r_i - 1)!} \prod_{k=1}^{r_i} N_{ijk}! \right), \tag{17.13}$$

where S denotes the Bayesian network structure $T \rightarrow X_i$; the arc indicates that T and X_i are probabilistically dependent.

The following assumptions are made in deriving equation (17.11) [4, 5]:

1 Multinomial variables: The features in X and the target T are multinomial and hence discrete valued.

2 Independent and identically distributed data: Instances in the training data set D are independent and identically distributed.

3 Complete data. The training data set D is complete; that is, each feature and the target has a value in each instance in D.

4 Parameter independence. The conditional probability distribution $\mathbb{P}(X_i|t, D)$ for the feature X_i given a target value t is independent of the conditional probability distribution $\mathbb{P}(X_i|t', D)$ for the same feature X_i given any other target value t' and the conditional probability distribution $\mathbb{P}(X_j|t, D)$ for any other feature X_j.

5 Dirichlet priors. The prior distributions for the conditional probabilities $\mathbb{P}(x_i|t, D)$ are specified by Dirichlet distributions. Moreover, the Dirichlet prior distributions are assumed to be uniform before observing D.

6 Structure modularity. The prior probability of an arc being present from target T to the feature X_i is independent of the prior probability of an arc from T to any other feature X_j.

The following equation is obtained by applying assumption 1:

$$\mathbb{P}(S, D_i) = \int_V \mathbb{P}(D_i|S, V) f(V|S) \mathbb{P}(S) \, dV, \tag{17.14}$$

where V is a vector whose values denote the conditional probability values for the BN structure S, and f is the conditional probability density function over V given S. The integral is over all possible values that V takes.

Since $\mathbb{P}(S)$ is a constant, it can be moved outside the integral to obtain

$$\mathbb{P}(S, D_i) = \mathbb{P}(S) \int_V \mathbb{P}(D_i|S, V) f(V|S) \, dV. \tag{17.15}$$

From equation (17.15), $\mathbb{P}(D_i|S)$ is obtained as follows:

$$\mathbb{P}(D_i|S) = \frac{\mathbb{P}(S, D_i)}{\mathbb{P}(S)} = \int_V \mathbb{P}(D_i|S, V) f(V|S) \, dV. \tag{17.16}$$

From the independence of instances expressed in assumption 2, it follows that

$$\mathbb{P}(D_i|S) = \int_V \left(\prod_{h=1}^m \mathbb{P}(C_h|S, V) \right) f(V|S) \, dV. \tag{17.17}$$

where C_h is the hth instance in D_i and m is the number of instances in D_i.

Let x_{ih} denote the value of X_i, and t_h denote the value of T in instance h. Applying assumption 3, which states that the instances in D have no missing values, equation (17.17) is rewritten as

$$\mathbb{P}(D_i|S) = \int_V \left(\prod_{j=1}^{r_T} \prod_{k=1}^{r_i} \mathbb{P}(X_i = k|T = j, V)^{N_{ijk}} \right) f(V|S) \, dV. \tag{17.18}$$

Let θ_{ijk} denote the conditional probability $\mathbb{P}(X_i = k|T = j, V)$. An assignment of numerical values to θ_{ijk} for $k = 1$ to r_i is a probability distribution that will be represented by the list $(\theta_{ij1}, \ldots, \theta_{ijr_i})$. In addition, for a given j let $f(\theta_{ij1}, \ldots, \theta_{ijr_i})$ denote the probability density function over $(\theta_{ij1}, \ldots, \theta_{ijr_i})$. The density function $f(\theta_{ij1}, \ldots, \theta_{ijr_i})$ is called a second-order probability distribution because it is a probability distribution over a probability distribution. Assumption 4 can be expressed as

$$f(V|S) = \prod_{j=1}^{r_T} f(\theta_{ij1}, \ldots, \theta_{ijr_i}), \tag{17.19}$$

which states that the values of a second-order probability distribution $f(\theta_{ij1}, \ldots, \theta_{ijr_i})$ are independent of the values of any other second-order probability distribution $f(\theta'_{ij1}, \ldots, \theta'_{ijr_i})$.

Substituting θ_{ijk} for $\mathbb{P}(X_i = k|T = j, V)$ in equation (17.18), and substituting equation (17.19) into equation (17.18) gives the following:

$$\mathbb{P}(D_i|S) = \int_{\theta_{ijk}} \cdots \int \left(\prod_{j=1}^{r_T} \prod_{k=1}^{r_i} \theta_{ijk}^{N_{ijk}} \right) \left(\prod_{j=1}^{r_T} f(\theta_{ij1}, \ldots, \theta_{ijr_i}) \right) d\theta_{i11}, \ldots, d\theta_{ijk} \ldots, d\theta_{ir_T r_i}, \tag{17.20}$$

where the integral is taken over all θ_{ijk} for $j = 1$ to r_T and $k = 1$ to r_i, and for every i and j the following condition holds: $\sum_k \theta_{ijk} = 1$. Since the terms in equation (17.20) are independent, the integral of products is converted to a product of integrals to obtain the following equation:

$$\mathbb{P}(D_i|S) = \prod_{j=1}^{r_T} \int_{\theta_{ijk}} \cdots \int \left(\prod_{k=1}^{r_i} \theta_{ijk}^{N_{ijk}} \right) f(\theta_{ij1}, \ldots, \theta_{ijr_i}) d\theta_{ij1}, \ldots, d\theta_{ijr_i}. \tag{17.21}$$

Assumption 5 states that before observing D, all possible values for the conditional probabilities $\theta_{ij1}, \ldots, \theta_{ijr_i}$ are equally likely. It therefore follows that $f(\theta_{ij1}, \ldots, \theta_{ijr_i}) = C_{ij}$ for some constant C_{ij}. Since $f(\theta_{ij1}, \ldots, \theta_{ijr_i})$ is a probability density function, it follows that

$$\int \cdots \int_{\theta_{ijk}} C_{ij} \, d\theta_{ij1}, \ldots, d\theta_{ijr_i} = 1. \tag{17.22}$$

Solving equation (17.22) for C_{ij} yields $C_{ij} = (r_i-1)!$, which implies that $f(\theta_{ij1}, \ldots, \theta_{ijr_i}) = (r_i-1)!$. Substituting this result in equation (17.20) gives

$$\mathbb{P}(D_i|S) = \prod_{j=1}^{r_T} \int \cdots \int_{\theta_{ijk}} \left(\prod_{k=1}^{r_i} \theta_{ijk}^{N_{ijk}} \right) (r_i - 1)! \, d\theta_{ij1}, \ldots, d\theta_{ijr_i}. \tag{17.23}$$

Since $(r_i - 1)!$ is a constant in equation (17.23), it can be moved outside the integral to obtain

$$\mathbb{P}(D_i|S) = \prod_{j=1}^{r_T} (r_i - 1)! \int \cdots \int_{\theta_{ijk}} \left(\prod_{k=1}^{r_i} \theta_{ijk}^{N_{ijk}} \right) d\theta_{ij1}, \ldots, d\theta_{ijr_i}. \tag{17.24}$$

The multiple integral in equation (17.24) has the following solution [16]:

$$\int \cdots \int_{\theta_{ijk}} \left(\prod_{k=1}^{r_i} \theta_{ijk}^{N_{ijk}} \right) d\theta_{ij1}, \ldots, d\theta_{ijr_i} = \frac{\prod_{k=1}^{r_i} N_{ijk}!}{(N_{ij} + r_i - 1)!}. \tag{17.25}$$

Substituting equation (17.25) into equation (17.24) gives the following result and completes the derivation:

$$\mathbb{P}(D_i|S) = \prod_{j=1}^{r_T} \left(\frac{(r_i - 1)!}{(N_{ij} + r_i - 1)!} \prod_{k=1}^{r_i} N_{ijk}! \right). \tag{17.26}$$

APPENDIX 17.B ROC CURVE AND AUC

The Receiver Operating Characteristic (ROC) curve measures how well a model's predictions (e.g., posterior probabilities) discriminate between the two values of a binary target variable. The ROC curve is a plot of sensitivity (also known as the *true positive rate*) versus 1-specificity (also known as the *false positive rate*) for different cut-points. A cut-point specifies a value; if the predicted value for an instance is above the cut-point, then it is assigned to one target value, and if it is below the cut-point, then it is assigned to the other target value. Each point on the ROC curve represents a pair of sensitivity and specificity values that corresponds to a particular cut-point. Given two values, namely, 0 and 1, that the target variable can take, sensitivity is defined as the probability of predicting correctly an instance that has target value 1, and specificity is defined as the probability of predicting correctly an instance that has target value 0. The area under the ROC curve (AUC) is typically used as a summary statistic of discrimination. The AUC is equivalent to the probability that a randomly chosen instance from instances with target value 0 will have a smaller predicted probability for value 1 than a randomly chosen instance from instances with target value 1. Higher values of AUC indicate better discrimination. A detailed description of the ROC curve and the AUC is provided in [14].

APPENDIX 17.C CALIBRATION

Calibration measures how good a model's predictions (e.g., posterior probabilities) are over a wide range of predictions for a binary target variable. Given two values, namely, 0 and 1, that the target variable can take, calibration assesses the goodness of predicted probabilities for the target value 0 (or 1). A model is well calibrated if the predicted probability for target value 0 for an instance corresponds closely to the observed proportion of instances with target value 0 in a set of similar instances. If, for example, a calibrated prediction model predicts that $\mathbb{P}(0|x) = 0.7$, then for a set of instances with values x for the features in X, the observed proportion of instances with target value 0 is approximately 0.7 (i.e., the target takes the value 0 about 70% of the time).

There are several measures available for assessing calibration, with the Hosmer–Lemeshow goodness-of-fit statistic being a popular one [9]. The Hosmer–Lemeshow goodness-of-fit statistic compares the average predicted probabilities to the observed proportion in ten defined intervals of probabilities. In order to calculate the value of the statistic, the instances are sorted in ascending order of their corresponding predicted probabilities. The instances are then categorized into ten groups using equal probability intervals, namely, 0–$0.1, 0.1$–$0.2, \ldots, 0.9$–1.0. For each group, the observed proportion is the number of instances with target value 0 (or 1) divided by the total number of instances in the group, and the average predicted probability is the average of the predicted probabilities for target value 0 (or 1) for the same instances. The ten pairs of values are then compared using the chi-square test. For well-calibrated predictions, the observed proportions will be close to the average predicted probabilities. This will result in a small chi-square value and a non-significant p-value ($p > 0.05$). In a calibration plot, each point represents a pair of observed proportion and average predicted probability values. For well-calibrated predictions, the plotted points are close to the 45 degree line (shown as a dotted line in Fig. 17.3). A detailed description of the calibration plot is provided in [14].

REFERENCES

[1] L. Bertram, C.M. Lill, and R.E. Tanzi. The genetics of Alzheimer disease: Back to the future. *Neuron*, 68(2):270–281, 2010.

[2] L. Bertram, M.B. McQueen, K. Mullin, D. Blacker, and R.E. Tanzi. Systematic meta-analyses of Alzheimer disease genetic association studies: The AlzGene database. *Nature Genetics*, 39(1):17–23, 2007.

[3] W. Buntine. Theory refinement on Bayesian networks. In B. D'Ambrosio and P. Smets, editors, *Proceedings of the Seventh Conference on Uncertainty in Artificial Intelligence (UAI 91)*, pages 52–60. Morgan Kaufmann Publishers, 1991.

[4] G.F. Cooper and E. Herskovits. A Bayesian method for the induction of probabilistic networks from data. *Machine Learning*, 9(4):309–347, 1992.

[5] D. Dash and G.F. Cooper. Exact model averaging with naive Bayesian classifiers. In *Nineteenth International Conference on Machine Learning*, pages 91–98. Morgan Kaufmann Publishers, 2002.

[6] S. Dreiseitl and L. Ohno-Machado. Logistic regression and artificial neural network classification models: a methodology review. *Journal of Biomedical Informatics*, 35:352–359, 2002.

[7] M. Goedert and M. G. Spillantini. A century of Alzheimer's disease. *Science*, 314(5800):777–781, 2006.

[8] J.A. Hoeting, D. Madigan, A.E. Raftery, and C.T. Volinsky. Bayesian model averaging: a tutorial. *Statistical Science*, 14(4):382–417, 1999.

[9] D.W. Hosmer and S. Lemeshow. *Applied Logistic Regression*. Wiley, 2nd edition, 2000.

[10] D. Koller and N. Friedman. *Bayesian Model Averaging. Probabilistic Graphical Models*. MIT Press, 2009.

[11] D. Madigan and A.E. Raftery. Model selection and accounting for model uncertainty in graphical models using Occam's window. *Journal of the American Statistical Association*, 89:1535–1546, 1994.

[12] A.E. Raftery, D. Madigan, and C.T. Volinsky. Accounting for model uncertainty in survival analysis improves predictive performance. In J.M. Bernardo, J.O. Berger, A.P. Dawid, and A.F.M. Smith, editors, *Bayesian Statistics 5*, pages 323–349. Oxford University Press, 1995.

[13] E.M. Reiman, J.A. Webster, A.J. Myers, J Hardy, T Dunckley, V.L. Zismann, K.D. Joshipura, J.V. Pearson, D. Hu-Lince, M.J. Huentelman, D.W. Craig, K.D. Coon, W.S. Liang, R.H. Herbert, T. Beach, K.C. Rohrer, A.S. Zhao, D. Leung, L. Bryden, L. Marlowe, M. Kaleem, D. Mastroeni, A. Grover, C.B. Heward, R. Ravid, J. Rogers, M.L. Hutton, S. Melquist, R.C. Petersen, G.E. Alexander, R.J. Caselli, W. Kukull, A. Papassotiropoulos, and D.A. Stephan. GAB2 alleles modify Alzheimer's risk in APOE epsilon4 carriers. *Neuron*, 54(5):713–720, 2007.

[14] M. Vuk and T. Curk. ROC curve, lift chart and calibration plot. *Metodološki zvezki*, 3(1):89–108, 2006.

[15] W. Wei, S. Visweswaran, and G.F. Cooper. The application of naive Bayes model averaging to predict Alzheimer's disease from genome-wide data. *Journal of the American Medical Informatics Association*, 18(4):370–375, 2011.

[16] S.S. Wilks. *Mathematical Statistics*. Wiley, 1962.

[17] K.Y. Yeung, R.E. Bumgarner, and A.E. Raftery. Bayesian model averaging: Development of an improved multi-class, gene selection and classification tool for microarray data. *Bioinformatics*, 21:2394–2402, 2005.

INDEX